W9-CUK-425

CRYSTAL STRUCTURES

Ralph W. G. Wyckoff, *University of Arizona, Tucson, Arizona*

Second Edition, VOLUME 1

INTERSCIENCE PUBLISHERS

a division of John Wiley & Sons, New York • London • Sydney

Library of Congress Catalog Card Number 48-9169
Printed in the United States of America by Mack Printing Co., Easton, Pa.

Contents

Chapter I

PREFACE AND INTRODUCTION

This is the first volume of a revision of the compilation *Crystal Structures* (1) which, including several Supplements, has been appearing in loose-leaf form over the last fifteen years. As such it is a continuation of the abbreviated statement and grouping of known atomic arrangements in crystalline solids that was begun forty years ago in the author's *Structure of Crystals* (2). A measure of the way our knowledge of crystal structures has been increasing is given by the fact that whereas as recently as 1934 they could all be described in about 300 pages, *Crystal Structures* has run to five large volumes and these do not include the hundreds of compounds that have been investigated since 1955.

When it was decided to continue this work, a decision had to be taken as to whether or not the loose-leaf format should be continued. It was chosen in the first place as an experiment in publishing designed to keep up-to-date and in one place the information pertaining to a rapidly expanding field of knowledge. In practice it proved to be a feasible way to assemble and present the many results that had accumulated over a period of twenty years and to integrate them with what had previously been learned. But as the bulk increased the loose-leaf approach became less manageable. There are a number of reasons for this, but the most compelling probably were the mounting difficulties of introducing new material and properly indexing it for ready reference. In the preparation of *Crystal Structures* it turned out that individual chapters were revised about every five years. Since the work is in five volumes this means that by making maximal use of the material already in loose-leaf form and publishing about one volume per year in book form, the same kind of coverage could be achieved without an undue increase in the work of preparation. The present book is the first of such a contemplated series of volumes.

This new program involves one curtailment in scope. *Crystal Structures* dealt both with determinations of atomic arrangements and with those numerous studies by x rays which led only to unit cells and possible space groups. The latter have now become very numerous and they are being covered in the *Structure Reports* (3) which are serving as a continuation of the *Strukturbericht* (4). The field of the present and of future volumes will therefore be restricted to determinations which define the positions of most, if not all, the atoms in a crystal. There is a regrettable tendency on the part of some workers to consider that the summary of a paper read at a scientific meeting, not followed by the usual account of the data upon which it has been based, is a sufficient description of a structure. Since no checks of any sort can be made of work thus described, it is the author's policy not to include structures which have been presented only at meetings.

The previous compilation did not seek to treat all intermetallic structures, and this exclusion has been continued. A complete description of the many intermetallic systems that have been investigated could scarcely be included in a work having the present objectives. The growth of our knowledge of semiconductors tends to destroy any sharp line of demarcation between intermetallic and non-intermetallic compounds; and data on many compounds between metallic elements have been introduced either because of this uncertainty or because they have the same atomic arrangements as well-known salts. There are no plans to expand this coverage to include nonstoichiometric alloy phases but it is proposed at a later date to add a chapter dealing with the structures of all definite intermetallic compounds.

What is wanted by different users of a survey of structural data is so varied that one can scarcely hope to satisfy all. An adequate bibliography is, however, required for all purposes, as is a statement of atomic positions. These are provided, and in a uniform fashion. The visualization of structures is greatly aided for most people by illustrations which show the way the atoms "pack" together in forming crystals and there is further help in making this mode of representation as uniform as possible. The drawings originally prepared for the *Structure of Crystals* provided the nucleus of such a uniform set. They were added to in the loose-leaf edition of *Crystal Structures* and they are being continued and extended here.

The description of a crystal structure in terms of its underlying space group is now general practice. The *International Tables* (5) incorporated the original space group descriptions of Niggli (6) and of the writer (7) and supplemented them with information helpful in determinations of structure. The set of *International Tables* (8) now replacing the earlier ones do not introduce great changes in terminology and space group description. In the present work the space group symbols of Schoenflies (9) will continue to be used together with those preferred in the *International Tables;* they fill the real need that often arises for a space group designation that does not depend on axial sequence.

The description of possible atomic positions derived from the various space groups are, except for occasionally different choices of origin, substantially the same in all these Tables. The designations of special positions used here are in general accord with the *International Tables.* Certain abbreviations have been used in stating the atomic positions. Thus in order to save space, only half the coordinates of body-centered arrangements are given. The designation B.C. (body-centered) after a group of coordinates indicates that they are to be repeated about $^1/_2\,^1/_2\,^1/_2$; and F.C. (face-centered) means that the stated coordinates are to be repeated about the remaining three lattice points $0\,^1/_2\,^1/_2$; $^1/_2\,0\,^1/_2$; $^1/_2\,^1/_2\,0$. Where positions of a rhombohedral arrangement are described in terms of a hexagonal pseudo unit, the symbol *rh* means that the coordinates with respect to hexagonal axes are to be repeated about $^1/_3\,^2/_3\,^2/_3$ and $^2/_3\,^1/_3\,^1/_3$. Further to conserve space, only half the coordinates are commonly stated if the origin falls in a center of symmetry: then coordinates are enclosed in parentheses, preceded by a plus-minus sign. In cubic crystals, and also when rhombohedral axes are used to describe the atomic positions of a crystal belonging to the rhombohedral division of the hexagonal system, the operation of a three-fold axis upon an atom at *xyz* gives others at *yzx* and *zxy*: it frequently saves space to designate this tripling of atomic positions by the symbol *tr* (meaning transposition).

After some experimentation with other imaginable ways of classifying structures for description, the general arrangement based on chemical composition first employed in the *Structure of Crystals* has been continued. There would be many advantages in a grouping that placed together all crystals having the same type of structure (for example, all the NaCl-like crystals) irrespective of chemical composition, but so many structures fall outside this pattern that it has not seemed a practical mode of classification. Within each chapter an effort has been made to group together those crystals with similar structures. These groups continue to be experimental and vary from chapter to chapter—ultimately it may be worthwhile to try to standardize them. The paragraph designations used in each chapter are dependent on this classification and are equally tentative. They are needed to facilitate cross reference between text, tables, and index, but no deeper significance is to be attached to them.

As in the loose-leaf edition, each chapter is provided with master tables which list the crystals considered in it and give the paragraph references involved in their description; these are the bibliography tables at the end of the text. Most compounds fit naturally into one or another of the chapters, but a few might reasonably be expected in more than one. They may be referred to in the tables of their several chapters but a structure has been described in only one.

The bibliography is based on that already published. It has, however, been abbreviated to omit those references which do not contribute to a knowledge of atomic arrangement. There are now in the literature additional papers stating the cell dimensions and unanalyzed x-ray diffraction patterns of more thoroughly studied crystals; these have for the most part been omitted. As before, all articles have been identified by the year and the authors' initials. In the loose-leaf version these literature references were made part of the initial master table of each chapter. It has now seemed better to create for them a table placed in each chapter just before the bibliography itself.

In the earlier days of crystal analysis, when methods of experimentation were under development, some sort of critical evaluation seemed requisite to a statement of atomic positions. Even if this were still desirable, the complicated crystals now being investigated and the enormous volume of data used in such studies would preclude any overall criticism of results in a general survey; and none has been attempted here. For the same reason, no estimate has been offered of the accuracy of spacing measurements. The cell dimensions that are given are those which seem to the author the best currently available and those structures have been included which he considers to be correct or to have a reasonable chance of being correct. Some years ago it was common practice among many students of crystal structure to attribute all sorts of failures to explain the observed data to "rotation" of atomic groups in a crystal and it has taken much serious work to ascertain when this was legitimate. Now "defects in structure" have become a similar whipping boy, and we must have genuinely thorough studies before really knowing what we mean by such statements as "an R_2X_3 crystal with the NiAs structure." In the meantime, our picture of the true state of affairs in many of these defect structures and the solid solutions they seem to form must remain far from complete; no attempt has been made to evaluate them in this compilation.

In a few instances efforts have been made to define in great detail the sort of disorder encountered in a crystal. Usually the results do not lend themselves to concise description and it has seemed better not to include them in the present work. Another omission which has been made in the interest of keeping the compilation within a manageable scope concerns the corrections for thermal motion to which the x-ray data point. They are becoming more and more an integral part of an adequate description of structure and it may be desirable later to include them in even as summary a description of atomic arrangements as that aimed at here.

Among the most valuable pieces of information supplied by a knowledge of where the atoms are in a crystal are the atomic separations that prevail. It was very early recognized that simple relationships of an approximately additive character exist between the atomic separations in many crystals. An evident possible interpretation of this additivity was that the different kinds of atoms had constant dimensions and in a crystal were in contact with one another. This led to the development of various systems of atomic "radii" based on early crystal structure results and much attention was given to trying to decide which of these systems corresponded most closely to reality and how constant these atomic "radii" actually were. It was soon recognized that the "radius" of an atom depended strongly on the type of chemical bonding that prevailed in a crystal and was also influenced, less strongly, by such factors as the number and nature of its neighbors. With the increase in data about structures and with attempts to deduce from theory the atomic dimensions and separations to be expected in different types of crystals, there gradually came to be accepted systems of "radii" applicable to different types of chemical bonding. Thus there are metallic radii, ionic radii, and "neutral" radii applicable to homopolar bonds of one sort or another. These have proved to be valid to a first approximation and are highly useful as an indication of the kind of bonding that prevails in a particular crystal, and often of the probable correctness of a disputed structure. A detailed discussion of how these radii are derived, of their limits of accuracy, and of how they can be corrected to take account of the number of atomic neighbors (the coordination) does not belong in this book.

For atoms in which the bonding is predominantly homopolar through the "sharing" of electrons, the number of surrounding atoms is determined by the chemical valence. This is the case outstandingly for the compounds of carbon. Where the bonding is largely heteropolar and more or less completely ionic, relative atomic sizes are influential in determining how these ions pack together to form a crystal. Especially with the chemically simpler compounds treated in the first chapters, atomic arrangement is fixed by the ratio of the ionic sizes—the radius ratio—and it is convenient to make use of this fact in grouping together compounds for description.

Where there are large differences in size among the atoms of a compound and the bonding is not homopolar, the structure may depend on the most effective way the large atoms can pack together; and this is true even though such atoms may be relatively strongly ionized and carry the same sign. Such considerations of "ionic" close-packing will also be used in the chapters that follow to select related structures for description.

Which of these factors will bring together the largest number of related compounds depends on the chemical complexity and, as a consequence, sometimes one and sometimes another will be employed in arranging the compounds to be discussed in different chapters. No attempt has been made to follow the same scheme of classification throughout the entire compendium.

Packing drawings of most of the structures are now included. A strong impression of depth is produced when they are examined with one eye through a reducing lens, or more simply through a short tube or even a hand curled to restrict the field of vision. Packings show this three-dimensional effect in very different degrees but all are improved when viewed in this way.

BIBLIOGRAPHY

1. R. W. G. Wyckoff, *Crystal Structures*, Interscience, New York, 1948, 1951, 1953, 1957, 1958, 1959, 1960.
2. R. W. G. Wyckoff, *The Structure of Crystals*, Chemical Catalog Co., New York, 1924; Reinhold, New York, 1931, 1935.
3. A. J. C. Wilson et al., *Structure Reports*, Oosthoek, Utrecht, 1951–. Volumes appearing at irregular intervals.
4. P. P. Ewald et al., *Strukturbericht*, Akademische Verlagsgesellschaft m.b.H., Leipzig, 1931–1939. Seven volumes. Photo-lithoprint reproduction by Edwards Bros., Inc., Ann Arbor, Michigan.
5. *Internationale Tabellen zur Bestimmung von Kristallstrukturen*, Gebrüder Borntraeger, Berlin, 1935. Two volumes. Photo-lithoprint reproduction by Edwards Bros., Inc., Ann Arbor, Michigan.
6. P. Niggli, *Geometrische Kristallographie des Diskontinuums*, Leipzig, 1919.
7. R. W. G. Wyckoff, *An Analytical Expression of the Results of the Theory of Space Groups*, Carnegie Institution of Washington, 1922, 1930.
8. *International Tables for X-ray Crystallography*, Kynoch Press, Birmingham, 1952–.
9. A. Schoenflies, *Krystallsysteme und Krystallstructur*, Leipzig, 1891; *Theorie der Kristallstruktur*, Berlin, 1923.

Chapter II

STRUCTURES OF THE ELEMENTS

Crystal structures have been determined for nearly all the elements. Most of these fall into a few simple types and are therefore conveniently treated together. As might be expected, structural complexity, and the tendency towards molecular formation which is partly responsible for it, depend on the position of an element in the periodic table. Nearly all the strongly metallic elements have simple structures, and the majority have their atoms in an arrangement that is substantially a close-packing of spheres. Weakly metallic elements tend to have their atoms more or less closely packed but with distortions that bring some neighbors much closer than others. The most electronegative elements, like the halogens and sulfur, have crystals that contain easily recognized molecular associations.

Because so many elements have simple structures, it is convenient to describe them first. For the rest, complexity in atomic arrangement is on the whole greater for elements farther to the right in the periodic table. They will be described in this order.

In this chapter, as in succeeding chapters, the bibliography will be preceded by a bibliography table which gives for each substance first the paragraph describing its structure and then a list of the papers (identified by year and author initials) from which the data have been drawn.

There are two closely related and equally dense close-packings of spheres which together represent the structures of about half the elements. One of these is cubic in symmetry, the other hexagonal.

II,a. In the *cubic close-packing*, atoms are in a face-centered array, the four atoms in the unit cube having the coordinates:

$$(4a)\quad 000;\ \tfrac{1}{2}\,0\,\tfrac{1}{2};\ \tfrac{1}{2}\,\tfrac{1}{2}\,0;\ 0\,\tfrac{1}{2}\,\tfrac{1}{2},\quad \text{or}\quad (4a)\quad 000;\ \text{F.C.}$$

As can be seen by reference to Figure II,1, each atom has 12 equidistant neighbors: four in the same cube-face plane at a distance of half a face diagonal, and eight in two planes, one above, the other below the chosen cube-face plane. This atomic arrangement, looked at along a threefold instead of a fourfold axis, is shown in Figure II,2. In any arbitrarily chosen plane normal to such an axis, atoms will be distributed as indicated by the full circles of this figure. The next parallel atomic layer (dotted circles) is just like this but is displaced horizontally so that its atoms fit into some of

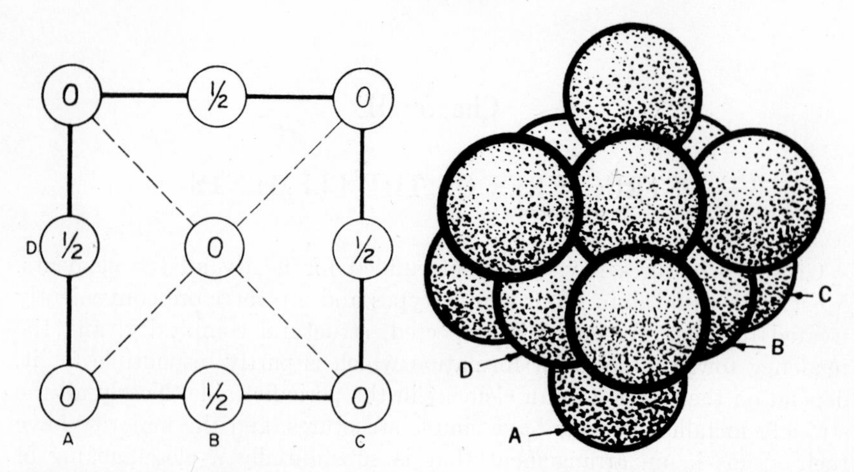

Fig. II,1a (left). The closest packed cubic arrangement projected on a cube face. In this and all subsequent projection drawings, numbers within the circles give the fractional distances in terms of cell axes of atoms above the plane of projection. The letters *A*, *B*, *C*, and *D* refer to correspondingly labeled atoms in the following packing drawing.

Fig. II,1b (right). A perspective packing drawing of the closest packed cubic arrangement.

the holes between atoms of the first layer. The third consecutive layer, shown as dashed circles, is like the others but is displaced so that its atoms fall into holes of the second layer and over those holes of the first layer that were not covered by the second layer. In a crystal built in this fashion the fourth layer is over the first and the sequence is repeated indefinitely. As can be seen from Figure II,2, six of the 12 equidistant neighbors of an atom are in the same (111) plane, three are in the plane below, and the remaining three are in the plane above. It is often convenient to represent this packing by designating the layer shown by full circles as *1*, the dotted layer as *2*, and the dashed layer as *3;* if this is done, the sequence of atomic planes in the cubic close-packing along a threefold axis can be written as:

$$1,2,3,1,2,3,1,2,3,1 \ldots$$

The lengths of the unit cubes of the elements known to have this structure are collected in Table II,1.

II,b1. In the very closely related *hexagonal close-packing* two atoms are contained in a unit prism. As can be seen from Table II,2, which gives the cell dimensions of the elements with this structure, the axial ratio is usually

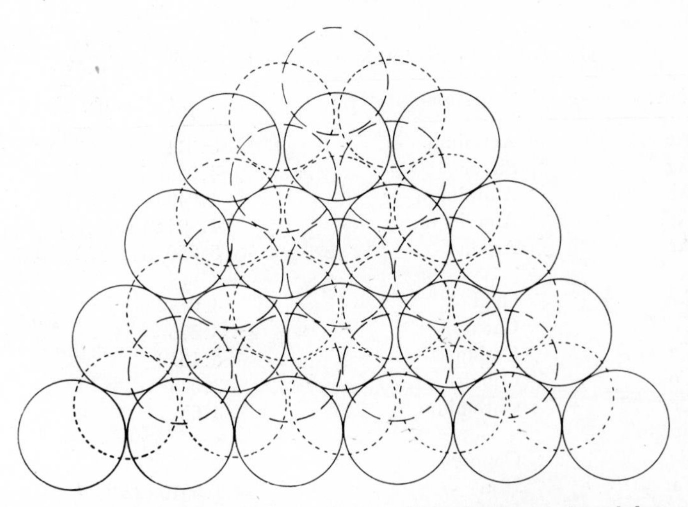

Fig. II,2. A drawing to show the cubic close-packing of spheres viewed down a three-fold axis. Successive layers normal to this axis are shown by full, dotted, and dashed circles.

near to $c/a = 1.633$. The atomic coordinates, developed from the space group D_{6h}^4 ($P6_3/mmc$) are

$$(2c) \quad {}^1/_3\,{}^2/_3\,{}^1/_4;\; {}^2/_3\,{}^1/_3\,{}^3/_4$$

or more simply, through a change in origin,

$$(2c) \quad 000;\; {}^1/_3\,{}^2/_3\,{}^1/_2$$

In this arrangement (Fig. II,3), as in the preceding one, an atom has 12 neighbors, six in the same plane (normal to the principal axis) and three each in adjacent parallel planes. When the axial ratio departs from 1.633, packing is not perfect for atoms considered as spherical; and then any atom's six neighbors in its own plane are at a different distance from it than are the two sets of three neighbors lying in adjacent planes.

Layers normal to the principal axis have the same atomic distribution as in the cubic close-packing of Figure II,2 but the sequence of these planes is different. If one starts with the layer outlined by full lines, the next layer can be the dotted layer *2*. In the hexagonal close-packing this is not followed by the dashed layer *3* but by another layer *1*, which then alternates with *2* to form the crystalline array. The planar sequence in this arrangement, represented in Figure II,4, can thus be designated as

1,2,1,2,1,2,1 . . .

TABLE II,1
Elements with the Cubic Close-Packed Structure (II,a)

Element	Name	Cell edge, a_0, A.
Ac	Actinium	5.311
Ag	Silver	4.0862
Al	Aluminum	4.04958 (25°C.)
Am	Americium	4.894 (see II,b2)
Ar	Argon	5.256 (4.2°K.)
Au	Gold	4.07825 (25°C.)
Ca	Calcium	5.576
Ce	Cerium	5.1612
Co	Cobalt	3.548
Cr	Chromium	3.68
Cu	Copper	3.61496 (18°C.)
Fe	Iron	3.5910 (22°C.)
Ir	Iridium	3.8394 (26°C.)
Kr	Krypton	5.721 (58°K.)
La	Lanthanum	5.296
Li	Lithium	4.404 (4.379) (−195°C.)
Mo	Molybdenum	4.16
Ne	Neon	4.429 (4.2°K.)
		4.471 for Ne^{20}
		4.455 for Ne^{22}
Ni	Nickel	3.52387 (25°C.)
Pb	Lead	4.9505
Pd	Palladium	3.8898 (25°C.)
Pr	Praseodymium	5.161
Pt	Platinum	3.9231 (25°C.)
Pu	Plutonium	4.6370 (320°C.)
Rh	Rhodium	3.8031 (25°C.)
Sc	Scandium	4.541
Sr	Strontium	6.0847 (25°C.)
Th	Thorium	5.0843 (25°C.)
Xe	Xenon	6.197 (58°K.)
Yb	Ytterbium	5.4862

In almost all hexagonal close-packed metals the axial ratio is slightly under 1.633, lying between 1.63 and ca. 1.57. That is, atoms in the same basal plane are usually a little farther apart than those immediately above and below. With beryllium as the only apparent exception, the axial ratios of

TABLE II,2
Elements with the Hexagonal Close-Packed Structure (**II,b1**)

Element	Name	a_0, A.	c_0, A.
Be	Beryllium	2.2866	3.5833 (ca. 22°C.)
β-Ca	Calcium	3.98	6.52 (450°C.)
Cd	Cadmium	2.97887	5.61765 (26°C.)
Ce	Cerium	3.65	5.96
α-Co	Cobalt	2.5071	4.0686 (20°C.)
Cr	Chromium	2.722	4.427
Dy	Dysprosium	3.584	5.668 (49°K.)
		3.5903	5.6475 (ca. 20°C.)
Er	Erbium	3.550	5.590 (43°K.)
		3.5588	5.5874 (ca. 20°C.)
Gd	Gadolinium	3.629	5.796 (106°K.)
		3.6360	5.7826 (ca. 20°C.)
He	Helium	3.57	5.83 (−271°C.)
Hf	Hafnium	3.1967	5.0578 (26°C.)
Ho	Holmium	3.5773	5.6158 (ca. 20°C.)
La	Lanthanum	3.75	6.07 (see II,b2)
Li	Lithium	3.111	5.093 (78°K.)
Lu	Lutetium	3.5031	5.5509 (ca. 20°C.)
Mg	Magnesium	3.20927	5.21033 (25°C.)
Na	Sodium	3.767	6.154 (5°K.)
Nd	Neodymium	3.657	5.902 (see II,b2)
Ni	Nickel	2.65	4.33
Os	Osmium	2.7352	4.3190 (20°C.)
Pr	Praseodymium	3.669	5.920 (see II,b2)
Re	Rhenium	2.7608	4.4582 (25°C.)
Ru	Ruthenium	2.70389	4.28168 (20°C.)
Sc	Scandium	3.3090	5.2733 (ca. 20°C.)
β-Sr	Strontium	4.32	7.06 (248°C.)
Tb	Terbium	3.6010	5.6936 (ca. 20°C.)
Ti	Titanium	2.950	4.686 (25°C.)
Tl	Thallium	3.456	5.525
		3.438	5.478 (5°K.)
Tm	Thulium	3.5375	5.5546 (ca. 20°C.)
Y	Yttrium	3.6474	5.7306 (ca. 20°C.)
Zn	Zinc	2.6648	4.9467 (25°C.)
α-Zr	Zirconium	3.232	5.147 (25°C.)

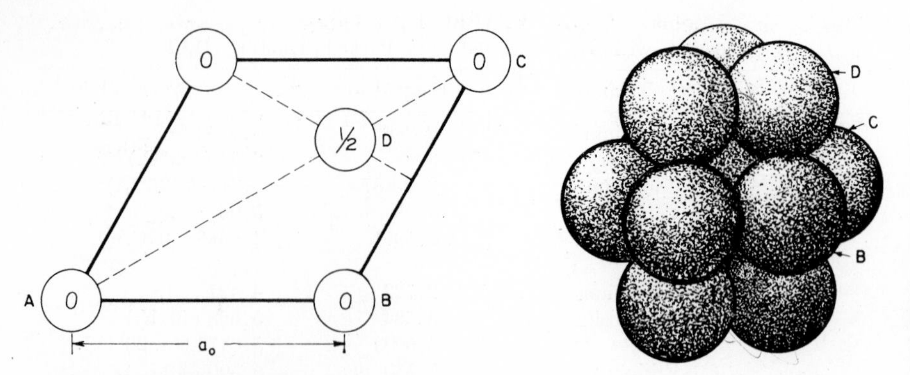

Fig. II,3a (left). A basal projection of the atomic arrangement in a unit prism of the hexagonal closest packing.

Fig. II,3b (right). A perspective drawing that shows several spheres in a hexagonal closest packed grouping. Spheres designated *B*, *C*, and *D* correspond to atoms labeled in this way in Figure II,3a.

these elements increase with temperature, heat thus tending to equalize the interatomic distances. Axial ratios for zinc and cadmium are exceptions to the foregoing statements. Their ratios are far greater than 1.63, being about 1.86, and this makes atomic separations in the basal planes more than 10% less than between planes. Their axial ratios also increase with temperature, but the effect of this is to exaggerate the departure from spherical close-packing.

Some metals occur in both types of close-packing. The energy associated with the two is substantially the same, and there is little tendency to change from one to the other. The hexagonal form of calcium is obtained above 450°C., and then only if the metal is especially pure. For cobalt and nickel the hexagonal form seems the more stable at room temperature, although the cubic form occurs more commonly. The cubic modification of nickel can be made hexagonal by annealing for several days at 170°C. Cobalt prepared by reduction above 450°C. is cubic and remains so on cooling, but grinding will bring about a change to the hexagonal form.

Some of the x-ray reflections in many preparations of cobalt are broad, while others are sharp. All the sharp lines have indices with $l = 0$ and $(h - k)/3$ as an integer; reflections with l even are more diffuse than those with l odd. This will be the case if the structure producing them is a mixture, on the atomic scale, of the two close-packed arrangements. The repetitions of atomic planes along the three- (or six) fold axes will be, for a time, according to the hexagonal close-packed sequence, and then, for a time, they will be according to the cubic close-packed sequence outlined above.

This is the simplest example of a mixed packing encountered more and more often among crystals.

The data on solid helium remain rather conflicting, and the situation is given added complexities by the apparent differences between He^3 and He^4. Four structures have been described for He^3 under different conditions of temperature and pressure. They are:

α-He^3: $a_0 = 3.045$ A., $c_0 = 4.986$ A. at 16°K., 1340 atm. The structure is presumably hexagonal close-packed.

α'-He^3: $a_0 = 4.01$ A. at 1.9°K., 100 kg./cm.^{2} The structure is described as body-centered cubic (**II,c**).

β-He^3: $a_0 = 3.46$ A., $c_0 = 5.60$ A. at 3.3°K., 183 kg./cm.^{2} The symmetry is given as hexagonal.

γ-He^3: $a_0 = 4.242$ A. at 19°K., 1693 atm. The structure is stated to be cubic close-packed.

The following two structures have been described for He^4:

α-He^4: $a_0 = 3.531$ A., $c_0 = 5.693$ A. or 5.76 A. at 1.5°K., 66 atm. This seems analogous to β-He^3.

β-He^4: $a_0 = 4.240$ A. at 16°K., 1250 atm. The structure is said to be analogous to γ-He^3, and cubic close-packed (**II,a**).

Obviously, much more work is needed on the difficult system presented by solid helium.

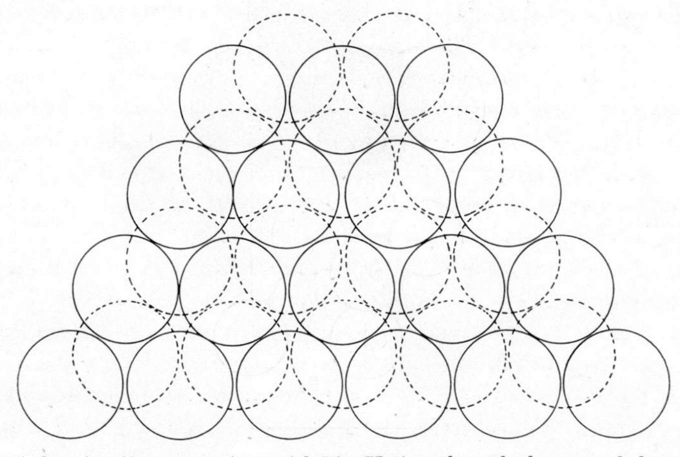

Fig. II,4. A drawing (for comparison with Fig. II,2) to show the hexagonal close-packing of spheres viewed down a threefold axis.

II,b2. When *americium* metal, Am, is prepared by reduction of AmF_3 with barium at 1000°C. followed by 10–16 hours of vacuum-cooling, it is hexagonal with a unit containing four atoms and having the edges:

$$a_0 = 3.474 \text{ A.}, \qquad c_0 = 11.25 \text{ A.}$$

The structure is described as a double hexagonal close-packing with atoms in the following positions of D_{6h}^4 ($P6_3/mmc$):

$$(2a) \quad 000;\ 0\,0\,{}^1/_2 \quad \text{and} \quad (2d) \quad {}^1/_3\,{}^2/_3\,{}^3/_4;\ {}^2/_3\,{}^1/_3\,{}^1/_4$$

The Am–Am distance is 3.47 A., corresponding to an americium metallic radius of 1.73 A. Figure II,5 shows the difference between this packing and the simple hexagonal one of the preceding paragraph.

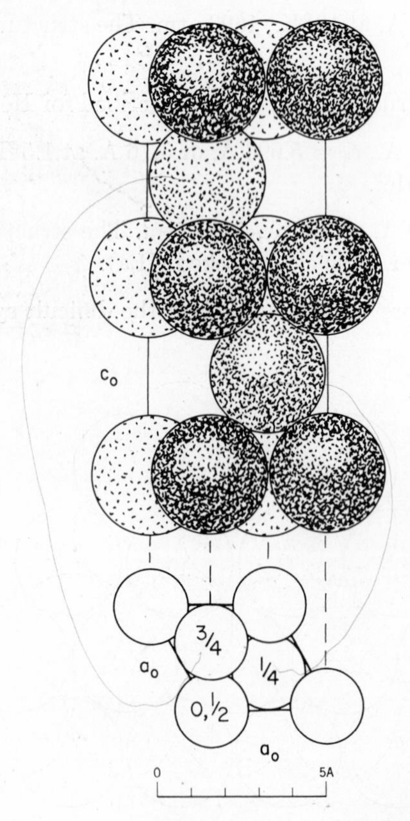

Fig. II,5. Two projections of the double hexagonal close-packing typified by metallic americium.

Cell data indicative of this type of double hexagonal packing have also been obtained from preparations of the following three rare-earth metals:

La: $a_0 = 3.770$ A., $c_0 = 12.159$ A.

Nd: $a_0 = 3.6579$ A., $c_0 = 11.7992$ A.

Pr: $a_0 = 3.6725$ A., $c_0 = 11.8354$ A.

Samples obtained by the vaporization of americium metal have not had the foregoing structure. Instead, they gave the patterns of intimate mixtures of cubic and hexagonal packings with the cubic predominating.

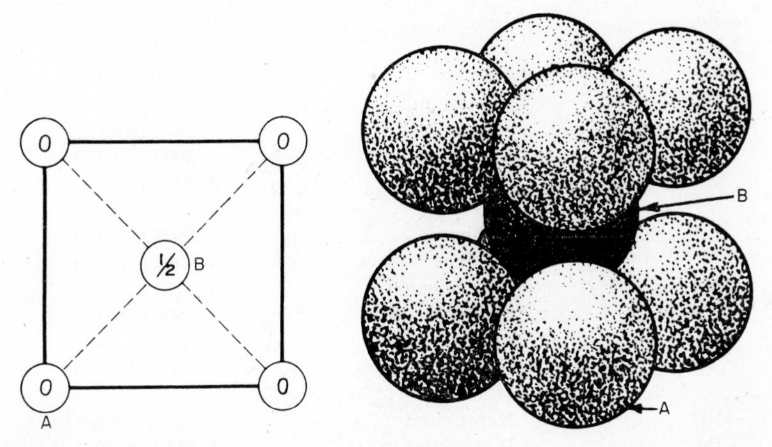

Fig. II,6a (left). The unit cell of the cubic body-centered arrangement projected on a cube face.

Fig. II,6b (right). A perspective drawing of the atoms associated with a unit cube of the body-centered structure. Atoms *A* and *B* refer to similarly labeled atoms in the projection of Figure II,6a.

II,c. The third common metallic structure is *body-centered cubic* with atoms at the corners and the center of its unit (Fig. II,6). The coordinates of these atoms then are:

$$(2a) \quad 000;\ {}^1/_2\,{}^1/_2\,{}^1/_2$$

Elements with this grouping are collected in Table II,3.

In this structure each atom has eight equidistant neighbors at the centers of surrounding cubes and six additional atoms at the corners of adjacent cubes which are only about 15% more distant. This split 14-fold coordination results in an atomic environment not very different from the environment of the 12-fold coordinations of the two close-packed structures.

TABLE II,3
Elements with the Cubic Body-Centered Structure (**II,c**)

Element	Name	a_0, A.
Ba	Barium	5.025 (26°C.)
		5.000 (5°K.)
γ-Ca	Calcium	4.38 (ca. 500°C.)
Cr	Chromium	2.8839 (25°C.)
Cs	Cesium	6.045 (5°K.)
		6.067 (78°K.)
Eu	Europium	4.606
		4.551 (5°K.)
α-Fe	Iron	2.8665 (25°C.)
β-Fe	Iron	2.91 (800°C.)
δ-Fe	Iron	2.94 (1425°C.)
K	Potassium	5.225 (5°K.)
		5.247 (78°K.)
Li	Lithium	3.5093 (20°C.)
		3.491 (78°K.)
Li^6	Lithium	3.5107 (20°C.)
Li^7	Lithium	3.5092 (20°C.)
Mo	Molybdenum	3.1473 (25°C.)
Na	Sodium	4.2906 (20°C.)
		4.225 (5°K.)
Nb	Niobium	3.3004 (18°C.)
γ-Np	Neptunium	3.52 (ca. 600°C.)
ε-Pu	Plutonium	3.638 (500°C.)
Rb	Rubidium	5.585 (5°K.)
		5.605 (78°K.)
γ-Sr	Strontium	4.85 (614°C.)
Ta	Tantalum	3.3058 (25°C.)
Th	Thorium	4.11 (1450°C.)
β-Ti	Titanium	3.3065 (900°C.)
Tl	Thallium	3.882
γ-U	Uranium	3.474
V	Vanadium	3.0240 (25°C.)
W	Tungsten	3.16469 (25°C.)
Zr	Zirconium	3.62 (850°C.)

II,d. *Hydrogen* (and its isotopes) is the only element belonging to the first three vertical columns of the periodic table whose crystalline structure is completely in doubt. Diffraction data exist, however, and according to these the structures of hydrogen, deuterium, and tritium are not necessarily alike.

Hydrogen is reported to be dimorphous with the two modifications having the cell dimensions:

$$\text{Tetragonal } H_2: \quad a_0 = 4.42 \text{ A.}; c_0 = 3.75 \text{ A.}$$

$$\text{Hexagonal } H_2: \quad a_0 = 3.75 \text{ A.}; c_0 = 6.49 \text{ A.}$$

The atomic positions have yet to be established.

Deuterium and tritium have been described as tetragonal with similar cells which, however, are unlike those of the tetragonal form of hydrogen:

$$\text{Tetragonal } D_2: \quad a_0 = 3.35 \text{ A.}; \quad c_0 = 5.86 \text{ A.}$$

$$\text{Tetragonal } T_2: \quad a_0 = 3.3 \text{ A.}; \quad c_0 = 5.78 \text{ A.}$$

Obviously much more work is required.

II,e1. *Mercury*, Hg, remains the only other element belonging to columns I and II whose atomic arrangement is not described by one of the simple structures already listed. The usual, α, form of solidified mercury is hexagonal, like that of its analogs, zinc and cadmium, but its atomic grouping is different. Its unit containing but one atom (at the origin) is a rhombohedron with

$$a_0 = 2.9863 \text{ A.}, \alpha = 70°44.6' \text{ at } 5°\text{K.}$$

and

$$a_0 = 2.9925 \text{ A.}, \alpha = 70°44.6' \text{ at } 78°\text{K.}$$

The corresponding triatomic hexagonal cells are

$$a_0' = 3.457 \text{ A.}, \quad c_0' = 6.663 \text{ A.} \quad \text{at } 5°\text{K.}$$

and

$$a_0' = 3.464 \text{ A.}, \quad c_0' = 6.677 \text{ A.} \quad \text{at } 78°\text{K.}$$

A cell of this shape expresses a cubic close-packing deformed by compression along a body diagonal. The larger four-atomic pseudocell diagonal to it that shows this relation to structure **II,a** has $a_0' = 4.581$ A. and an axial angle of 98°13′.

The zinc and cadmium distortions of the hexagonal close-packing make atoms in adjacent planes farther apart than those in the same plane; the compressive distortion in mercury, on the other hand, brings atoms in neighboring layers especially close together. The mercury–mercury separation within a rhombohedral (111) plane is 3.47 A.; between the nearer atoms of adjacent planes it is only 3.00 A. An atom has six neighbors at each of these distances.

II,e2. A beta form of mercury produced at high pressures is stable at ordinary pressures below 79°K. It is tetragonal with a diatomic cell having the edges:

$$a_0 = 3.995 \text{ A.}, \quad c_0 = 2.825 \text{ A.} \quad (77°\text{K.})$$

The atoms are in the body-centered positions 000; $\frac{1}{2}\frac{1}{2}\frac{1}{2}$.

This is a structure which, like the alpha form, can be considered as a distorted cubic close-packing. In this case, however, the distortion is a compression along a cubic rather than a trigonal axis. The shortest Hg–Hg = 2.825 A. creates strings along the c_0 axis; the next longer distance, 3.16 A., applies to eight neighbors.

II,e3. Several other metals have atomic arrangements that are distortions of one of the simple structures already described.

Thus *indium*, In, has the tetragonal unit:

$$a_0 = 3.244 \text{ A.}, \quad c_0 = 4.938 \text{ A.}$$

with two atoms in the body-centered positions 000; $\frac{1}{2}\frac{1}{2}\frac{1}{2}$. The shape of this cell is such that its larger diagonal pseudo unit is a distorted face-centered cube having

$$a_0' = 4.588 \text{ A.}, \quad c_0' = c_0 = 4.938 \text{ A.}$$

Indium is therefore another distorted cubic close-packing, but in this instance there is elongation along one of the cube edges. In indium the 12 equidistant neighbors of the cubic close-packing fall into one group of four at a distance of 3.24 A. and a group of eight not quite so near, at 3.37 A.

The δ' form of *plutonium*, Pu, is described as having this structure with a body-centered diatomic tetragonal cell of the dimensions:

$$a_0 = 3.339 \text{ A.}, \quad c_0 = 4.446 \text{ A.} \quad (477°\text{C.})$$

Protoactinium, Pa, also has been given this type of tetragonal arrangement, with

$$a_0 = 3.925 \text{ A.}, \quad c_0 = 3.238 \text{ A.}$$

The different shape of the cell, involving a compression rather than an elongation along the fourfold axis, gives each Pa atom eight neighbors at 3.212 A. and two more at 3.238 A.

Originally this structure was also ascribed to the gamma modification of manganese (**II,t**).

II,f1. Complete structures have been established for two of the seemingly numerous forms of *boron*, B.

Tetragonal *boron*, B, has a unit containing 50 atoms with cell edges which vary from specimen to specimen around a value of

$$a_0 = 8.74 \text{ A.}, \qquad c_0 = 5.03 \text{ A.}$$

The structure as described most recently is based on D_{4h}^{12} ($P4_2/nnm$) with atoms in the following positions:

(2*b*) $0\,0\,{}^1/_2;\ {}^1/_2\,{}^1/_2\,0$

(8*m*) $uuv;\ \bar{u}\bar{u}v;\ u+{}^1/_2,u+{}^1/_2,{}^1/_2-v;\ {}^1/_2-u,{}^1/_2-u,{}^1/_2-v;$
$\bar{u}u\bar{v};\ u\bar{u}\bar{v};\ {}^1/_2-u,u+{}^1/_2,v+{}^1/_2;\ u+{}^1/_2,{}^1/_2-u,v+{}^1/_2$

(16*n*) $xyz;\ \bar{x}\bar{y}z;\ x+{}^1/_2,y+{}^1/_2,{}^1/_2-z;\ {}^1/_2-x,{}^1/_2-y,{}^1/_2-z;$
$\bar{x}y\bar{z};\ x\bar{y}\bar{z};\ {}^1/_2-x,y+{}^1/_2,z+{}^1/_2;\ x+{}^1/_2,{}^1/_2-y,z+{}^1/_2;$
$\bar{y}x\bar{z};\ y\bar{x}\bar{z};\ {}^1/_2-y,x+{}^1/_2,z+{}^1/_2;\ y+{}^1/_2,{}^1/_2-x,z+{}^1/_2;$
$yxz;\ \bar{y}\bar{x}z;\ y+{}^1/_2,x+{}^1/_2,{}^1/_2-z;\ {}^1/_2-y,{}^1/_2-x,{}^1/_2-z$

The parameters are those of Table II,4.

The resulting structure (Fig. II,7) consists of four nearly regular icosahedra having their centers in the close-packed positions:

(4*e*) ${}^1/_4\,{}^1/_4\,{}^1/_4;\ {}^1/_4\,{}^3/_4\,{}^3/_4;\ {}^3/_4\,{}^1/_4\,{}^3/_4;\ {}^3/_4\,{}^3/_4\,{}^1/_4$

plus the two extra atoms in (2*b*). Within icosahedra the B–B distances vary between 1.74 A. and 1.85 A.; between icosahedra the close B–B approaches are 1.601 A. to 1.861 A. The exact values depend on the crystal supplying the data.

TABLE II,4
Parameters of Atoms in Tetragonal Boron

Atom	Positions	x	y	z
B(1)	(16*n*)	0.325	0.088	0.400
B(2)	(16*n*)	0.227	0.081	0.088
B(3)	(8*m*)	0.121	0.121	0.383
B(4)	(8*m*)	0.245	0.245	0.584

These values are the means of those derived from two different crystals.

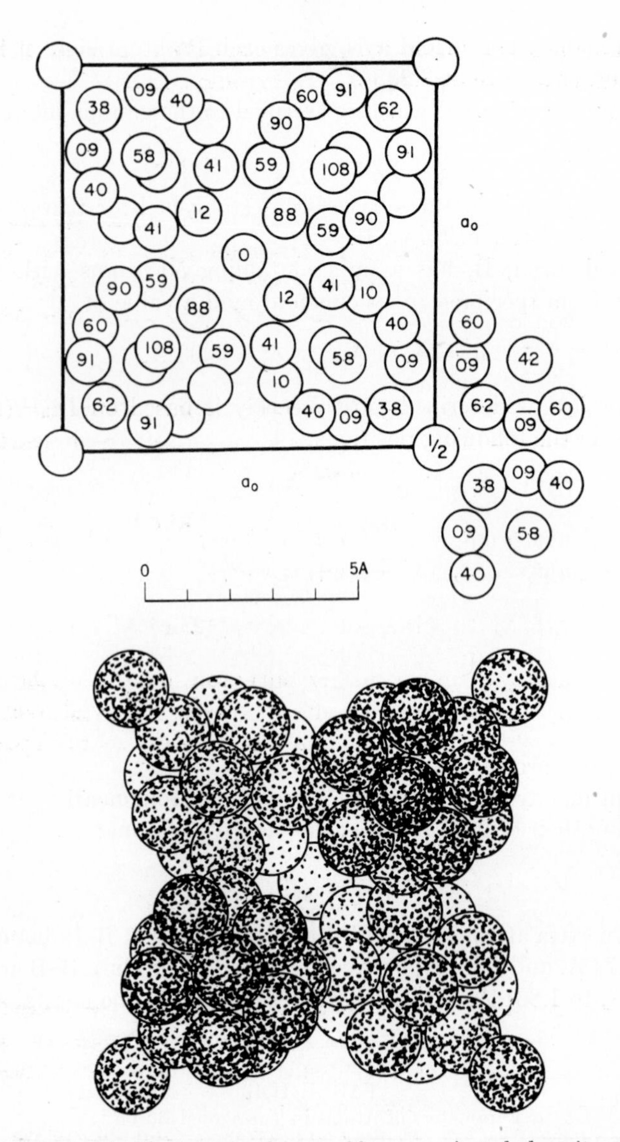

Fig. II,7a (top). The tetragonal form of boron projected along its c_0 axis.
Fig. II,7b (bottom). A packing drawing of the tetragonal form of boron viewed along its $c_0$axis. The four icosahedral groupings are readily distinguished.

When rhombohedral boron is heated with several percent of beryllium, BeB_{12} is formed with the structure of tetragonal boron. It has substantially the same cell:

$$a_0 = 8.80 \text{ A.}, \qquad c_0 = 5.08 \text{ A.}$$

There are four molecules in this cell with boron atoms forming the same icosahedra as in the beta form. The beryllium atoms fill twice as many of the holes between these icosahedra as are filled by the "free" borons in beta-boron; replacing the two borons in (2*b*), the other two Be might be expected to be in (2*a*) 000; $^1/_2\,^1/_2\,^1/_2$.

II,f2. When BI_3 is decomposed by heat, a rhombohedral modification of *boron* is produced. Its clear red crystals have a structure in which the unit is a rhombohedron with the dimensions:

$$a_0 = 5.057 \text{ A.}, \qquad \alpha = 58°4'$$

The corresponding hexagonal cell has the edges

$$a_0' = 4.908 \text{ A.}, \qquad c_0' = 12.567 \text{ A.}$$

There are 12 atoms in the unit rhombohedron, and 36 in the hexagonal prism. All atoms have been found to lie in two sets of special positions (18*h*) of D_{3d}^5 ($R\bar{3}m$), which for the hexagonal cell have the coordinates:

$$(18h) \quad \pm(u\bar{u}v;\ u,2u,v;\ 2\bar{u},\bar{u},v);\ \text{rh}$$

For the two sets of boron atoms:

$$\text{B}(1)\text{: } u = 0.1177;\ v = -0.1073$$

$$\text{B}(2)\text{: } u = 0.1961;\ v = 0.0245$$

The resulting atomic arrangement (Fig. II,8) can be considered as a nearly perfect cubic close-packing of B_{12} icosahedra. Within an icosahedron the B–B distances lie between 1.73 and 1.79 A. Between icosahedra there is a short B–B = 1.71 A.; others are 2.03 A.

There is a close relation between this structure and that of B_4C (Chapter V). It can be imagined as derived from the carbide by the omission of the carbon atoms and their replacement by the three-cornered B–B = 2.03 A. bonds.

A second, high-temperature, rhombohedral form of boron also exists. It has, however, a very large unit (a_0 = 10.944 A.; c_0 = 23.811 A.) and no determination has as yet been made of its atomic arrangement.

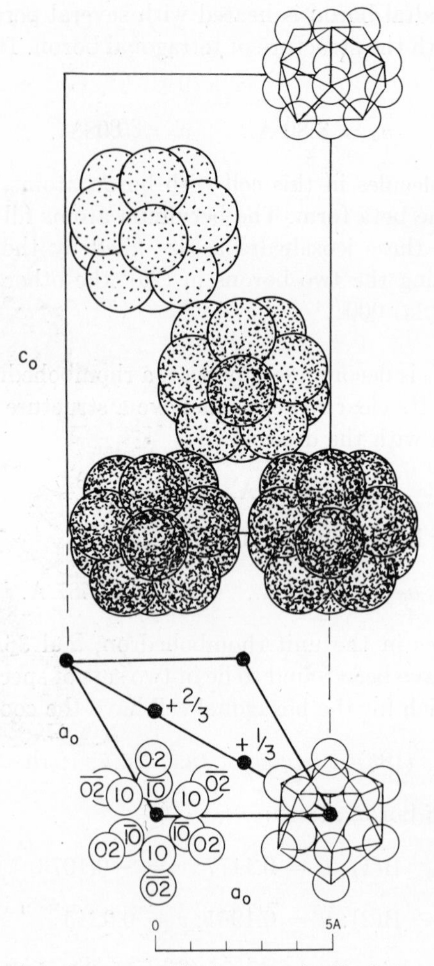

Fig. II,8. Two projections in terms of its hexagonal cell of the rhombohedral form of boron. The icosahedral form of the boron molecule is evident.

II,g1. Ordinary *gallium*, Ga, is orthorhombic with two axes of its pseudotetragonal unit almost identical in length. The assigned structure, containing eight atoms per cell, makes these atoms equivalent and places them in special positions of the space group V_h^{18} (*Bmab*):

$$(8f) \quad \pm(0uv;\ {}^1/_2,u+{}^1/_2,\bar{v};\ {}^1/_2,u,v+{}^1/_2;\ 0,u+{}^1/_2,{}^1/_2-v)$$

These coordinates apply to a unit prism with axes in the sequence:

$$a_0 = 4.5107 \text{ A.};\ b_0 = 4.5167 \text{ A.};\ c_0 = 7.6448 \text{ A.}$$

The established parameters, $u = 0.0785$, $v = 0.1525$, give an arrangement (Fig. II,9) in which a gallium atom has seven neighbors. One of these is especially near, at 2.44 A.; the others are in three sets of two each, at distances between 2.71 and 2.80 A.

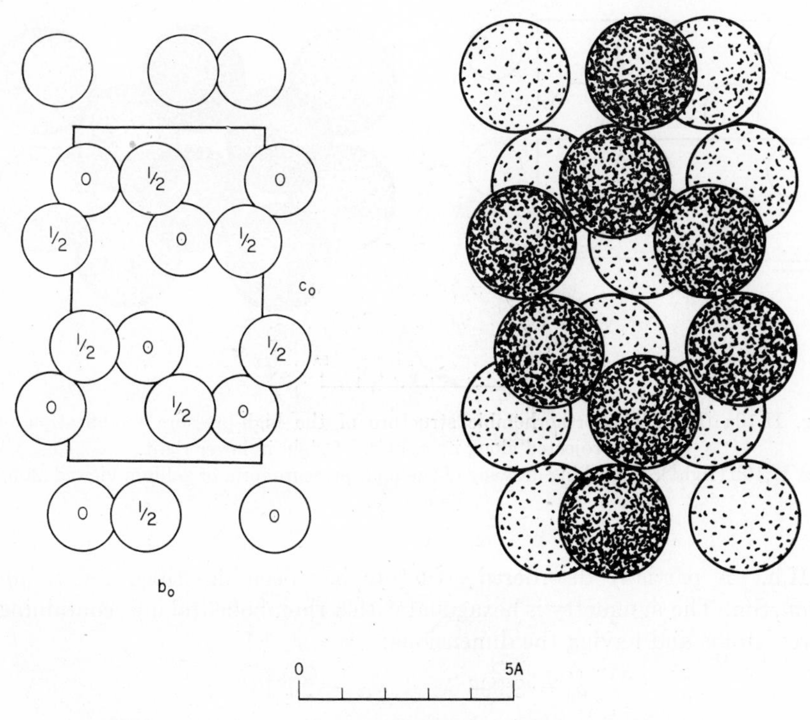

Fig. II,9a (left). The orthorhombic structure of the usual form of gallium projected along its a_0 axis. Origin in lower right.
Fig. II,9b (right). A packing drawing of the structure of ordinary gallium viewed along its a_0 axis.

II,g2. A second form of *gallium*, Ga, unstable under atmospheric pressure, has been obtained by supercooling the liquid metal to −16.3°C.

It is orthorhombic with a tetraatomic cell of the dimensions:

$$a_0 = 2.90 \text{ A.}; \; b_0 = 8.13 \text{ A.}; \; c_0 = 3.17 \text{ A.}$$

The space group is V_h^{17} (*Cmcm*) with the atoms in

$$(4c) \quad \pm(0\, u\, {}^1/_4;\; {}^1/_2, u+{}^1/_2, {}^1/_4), \qquad u = 0.133$$

In this arrangement (Fig. II,10), the gallium atoms form zigzag chains along the c_0 axis. In a chain Ga–Ga = 2.68 A. and Ga–Ga–Ga = 72°30′. Between chains, a gallium atom has four neighbors at 2.87 A. and two more at 2.90 A. There are two others at 3.17 A.

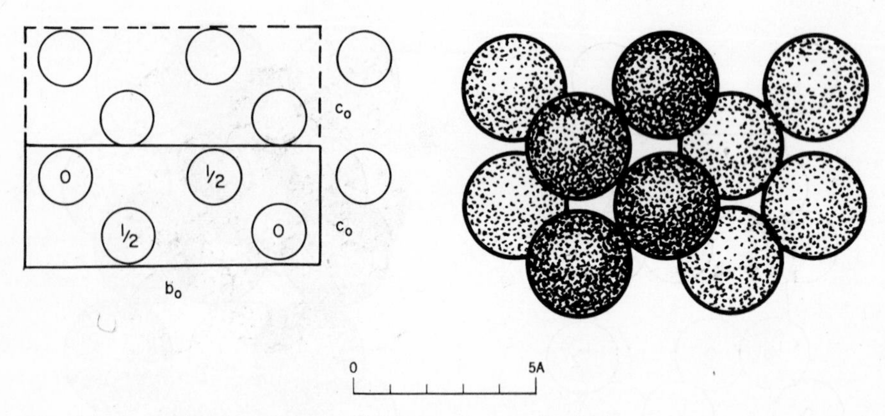

Fig. II,10a (left). The orthorhombic structure of the high-pressure modification of gallium projected along its a_0 axis. Origin in lower right.
Fig. II,10b (right). A packing drawing of the high-pressure form of gallium viewed along its a_0 axis.

II,h. A partially disordered structure has been described for *samarium*, Sm. The symmetry is hexagonal with a rhombohedral unit containing three atoms and having the dimensions:

$$a_0 = 8.996 \text{ A.}, \qquad \alpha = 23°13'$$

The corresponding hexagonal cell containing nine atoms has

$$a_0' = 3.621 \text{ A.}, \qquad c_0' = 26.25 \text{ A.}$$

In a perfectly ordered structure, the three atoms of the rhombohedral cell would be in the following positions of D_{3d}^5 ($R\bar{3}m$):

$$(1a) \quad 000 \qquad \text{and} \qquad (2c) \quad \pm(uuu), \text{ with } u = 0.222.$$

Actual crystals are said to be submicroscopically twinned, or disordered, so that half the units are turned 180° about the principal axis with respect to the others.

In this structure each samarium atom has six neighbors in the same c plane, with Sm–Sm = 3.629 A., and three others slightly nearer in the plane above and the plane below (3.587 A.).

It is thought probable that other rare-earth metals, notably praseodymium and neodymium, show this general type of disorder. The observation that they possess the double hexagonal close-packing of **II,b2,** however, points to the desirability of more work on all metals of the group.

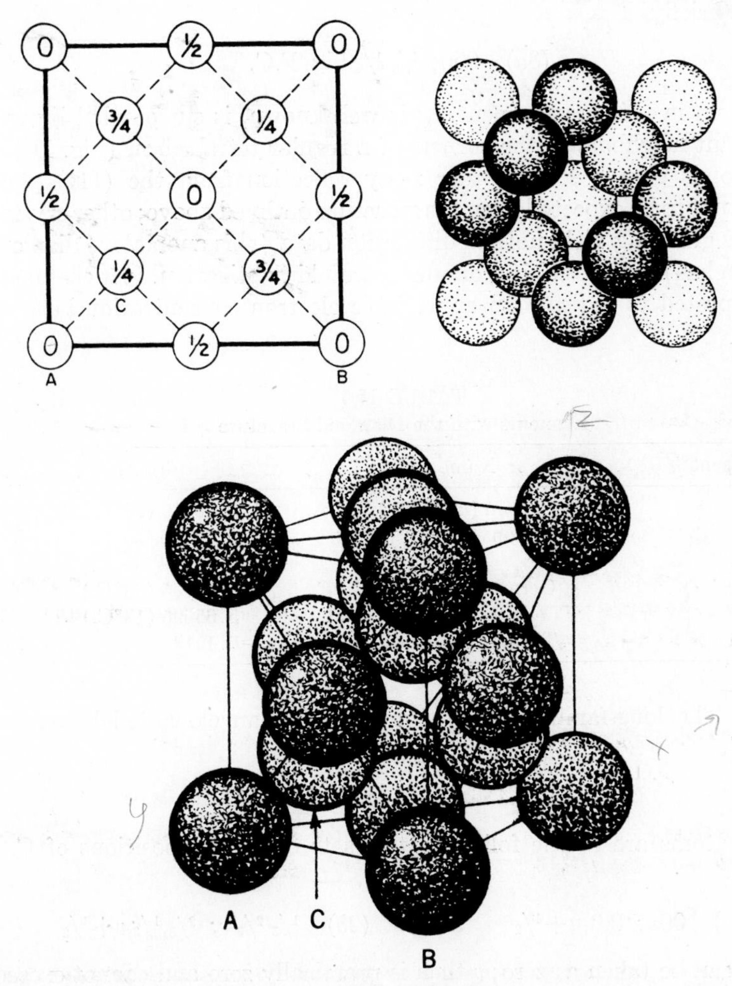

Fig. II,11a (top left). The atomic positions in the unit cell of the diamond arrangement projected on a cube face.

Fig. II,11b (top right). A packing drawing of the atomic distribution in the diamond structure corresponding to the projection of Figure II,11a.

Fig. II,11c (bottom). A perspective drawing of the unit cube of the diamond arrangement showing how its atoms pack within it.

II,i1. The unit cubes of crystals (Table II,5) with the *diamond* structure have eight atoms in the positions of O_h^7 ($Fd3m$):

(8*a*) 000; 0 ½ ½; ½ 0 ½; ½ ½ 0;
¼ ¼ ¼; ¼ ¾ ¾; ¾ ¼ ¾; ¾ ¾ ¼

or more briefly

(8*a*) 000; ¼ ¼ ¼; F.C.

Each atom in this arrangement, as is well known, is surrounded by four equidistant neighbors at the corners of a regular tetrahedron (Fig. II,11).

Diamonds give a second-order x-ray reflection from the (111) plane which is not to be expected from the structure outlined above; other crystals with this grouping yield normal diffraction data. This anomalous diamond reflection has been the subject of much careful investigation. It is customarily interpreted to indicate an appreciable electron concentration between atoms.

TABLE II,5
Elements with the Diamond Structure

Element	Name	a_0, A.
C	Carbon (diamond)	3.56679 (20°C.)
Si	Silicon	5.43070 (25°C.)
		5.445 (1300°C.)
Ge	Germanium	5.65735 (20°C.)
		5.65695 (18°C.)
α-Sn	Tin (gray)	6.4912

II,i2. The long-familiar form of *graphite* has an elongated hexagonal unit with

$$a_0 = 2.456 \text{ A.}, \qquad c_0 = 6.696 \text{ A.}$$

Its four atoms are in the following two sets of special positions of C_{6v}^4 ($C6mc$):

(2*a*) $00u;\ 0,0,u+\frac{1}{2}$ and (2*b*) $\frac{1}{3}\,\frac{2}{3}\,v;\ \frac{2}{3},\frac{1}{3},v+\frac{1}{2}$

where u can be taken as zero; v then is practically zero and cannot exceed 0.05.

The structure thus defined (Fig. II,12) consists of planes of linked hexagons of carbon atoms widely spaced parallel to one another along the principal, c_0, axis. Within these planes the C–C distances are all equal (1.42 A.) and substantially the same as the C–C separations in the benzene ring. The nearest approach of carbon atoms in neighboring planes is 3.40 A. Obviously the forces operating between basal planes are very different from, and far weaker than, the bonds within the planes of carbon; this is associated with the pronounced cleavage and flaky nature of graphite. It is in accord with this difference in bonding character that, when graphite is heated to incandescence, expansion takes place along the c_0 axis, while the dimensions within the carbon planes remain practically unaltered.

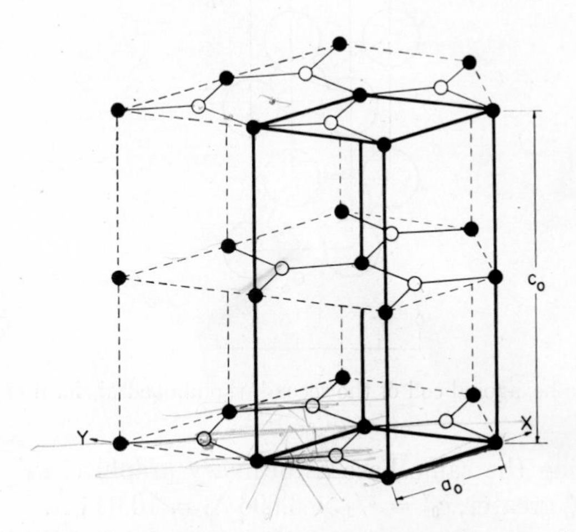

Fig. II,12. A perspective drawing showing positions of the atoms in ordinary graphite. Both the full and the open circles indicate positions of carbon atoms. The hexagonal unit is outlined by the heavy lines.

II,i3. Some *graphites* show weak lines that cannot be explained by the foregoing structure. Treatment with strong acid eliminates them but does not eliminate the usual diffraction lines of graphite. It was demonstrated that they are not due to an impurity but result from a second form of graphite itself. In ordinary graphite, half the carbon atoms, those in (2*a*), lie directly over one another in adjacent layers, while the other half, those in (2*b*), lie over empty centers of the hexagons of neighboring planes. Alternate layers are thus identically placed, as they are in the much more densely packed hexagonal close-packing (**II,b1**).

The structure described for this second, *rhombohedral, graphite* differs from this in that the third vertical layer is turned so that one of its atoms is directly over the $^2/_3,^1/_3,v+^1/_2$ atom and none is over the $0\ 0\ ^1/_2$ atom of the second layer of the usual arrangement (see Fig. II,13). This makes every third instead of every second plane identical and results in a hexagonal

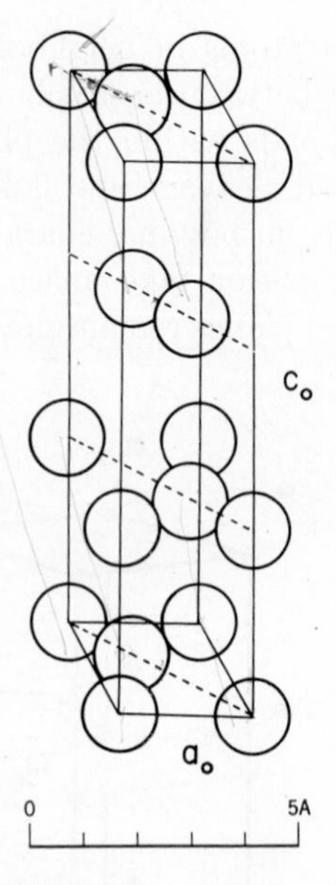

Fig. II,13. The hexagonal cell of the second, rhombohedral, form of graphite.

pseudo-unit having the same base as ordinary graphite, $a_0' = 2.456$ A., and a height 50% greater, $c_0' = {}^3/_2 \times 6.696$ A. $= 10.044$ A. In this cell, atoms are at

$$000;\ {}^1/_3\,{}^2/_3\,0;\ 0\,0\,{}^1/_3;\ {}^2/_3\,{}^1/_3\,{}^1/_3;\ {}^1/_3\,{}^2/_3\,{}^2/_3;\ {}^2/_3\,{}^1/_3\,{}^2/_3$$

The simple unit is a rhombohedron having

$$a_0 = 3.635 \text{ A.}, \qquad \alpha = 39°30'$$

Its two atoms are at

$$uuu;\ \bar{u}\bar{u}\bar{u} \qquad \text{with } u \text{ close to } {}^1/_6$$

Evidently this graphite has the same sort of relation to the ordinary form that the cubic bears to the hexagonal close-packing.

II,i4. *White tin*, Sn, stable at room temperature, has a tetragonal unit containing four atoms in the special positions:

$$(4a)\quad 000;\ 0\,{}^1/_2\,{}^1/_4;\ {}^1/_2\,0\,{}^3/_4;\ {}^1/_2\,{}^1/_2\,{}^1/_2$$

of the space group D_{4h}^{19} ($I4/amd$). The unit, with

$$a_0 = 5.8197 \text{ A.}, \qquad c_0 = 3.17488 \text{ A.} \quad \text{at } 25°\text{C.}$$

is of such a shape that each atom is surrounded by a distorted tetrahedron of neighbors at a distance of 3.02 A. and by two more only slightly farther away (3.17 A.) in the c direction. As Figure II,14 indicates, the structure is a flattened diamond grouping whose base is its inscribed diagonal.

The diamond-like gray tin (**II,i1**) has a Sn–Sn separation of 2.80 A. Change from its fourfold coordination to the sixfold coordination of white tin thus involves increases in interatomic distances of from 8 to 11%.

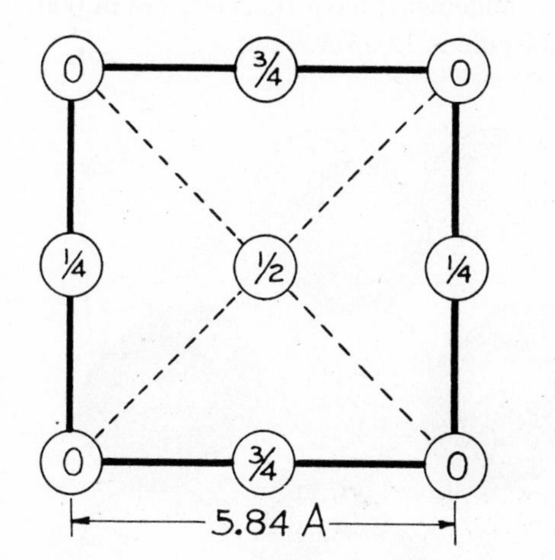

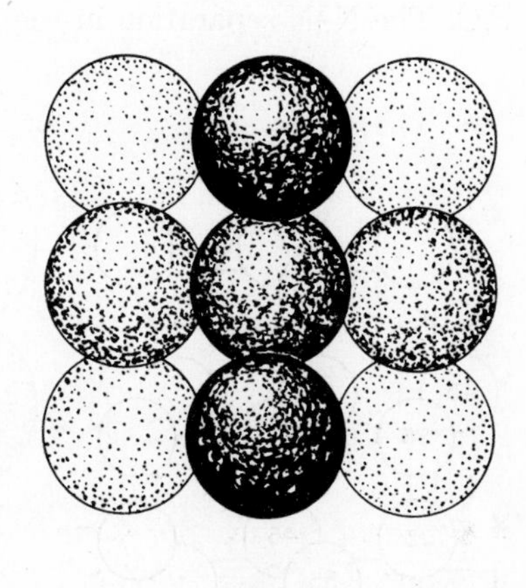

Fig. II,14a (left). The atomic arrangement within the tetragonal unit of white tin projected on the basal, c_0, plane.

Fig. II,14b (right). A packing drawing corresponding to the projection of the structure of white tin shown in Figure II,14a.

II,j1. *Solid nitrogen,* N_2, is dimorphous. The modification existing between 35°K. and the melting point is hexagonal, while the low temperature form is cubic. *Hexagonal nitrogen* has a cell with

$$a_0 = 4.039 \text{ A.}, \qquad c_0 = 6.670 \text{ A.}$$

Its two molecules of N_2 appear to be hexagonally close-packed with centers at 000; $^1/_3\,^2/_3\,^1/_2$, and it has been suggested that the atoms are not in fixed positions but rotate about these centers.

II,j2. The alpha form of *nitrogen,* N_2, is cubic with a tetramolecular unit having

$$a_0 = 5.644 \text{ A.} \qquad (4.2°\text{K.})$$

The structure as originally given was based on T^4 ($P2_13$), but recent work provides no satisfactory evidence for this low symmetry. Instead, the space group appears as T_h^6 ($Pa3$), with atoms in the positions:

$$(8c) \quad \pm(uuu;\ u+{}^1/_2,{}^1/_2-u,\bar{u};\ \bar{u},u+{}^1/_2,{}^1/_2-u;\ {}^1/_2-u,\bar{u},u+{}^1/_2)$$

with u = ca. 0.054.

This structure is a cubic close-packing of N_2 molecules (Fig. II,15) which, in contrast to the earlier proposed arrangement, have their centers in 000; F.C. The N–N separation in the molecule is 1.055 A.

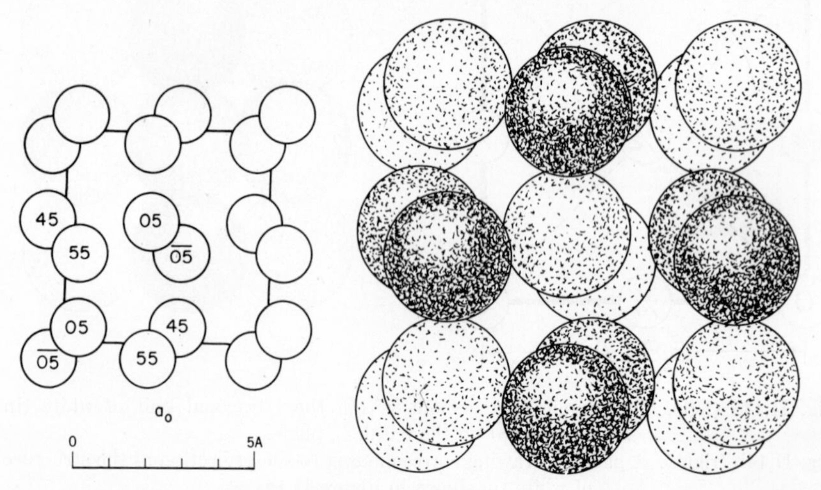

Fig. II,15a (left). A projection down a cubic axis of the low-temperature form of solid nitrogen.
Fig. II,15b (right). A packing drawing of the close-packed cubic structure of solid N_2 viewed along a cube axis.

II,k. A complete orthorhombic structure has been deduced for *black phosphorus,* P. It is like gallium in having V_h^{18} as its space group, and if axes are chosen so that atomic coordinates follow in the same sequence, *Bmab,* its unit prism has the dimensions:

$$a_0 = 3.31 \text{ A.};\ b_0 = 4.38 \text{ A.};\ c_0 = 10.50 \text{ A.}$$

The eight atoms in this cell, in the special positions:

$$(8f) \quad \pm(0uv;\ {}^1/_2,u+{}^1/_2,\bar{v};\ {}^1/_2,u,v+{}^1/_2;\ 0,u+{}^1/_2,{}^1/_2-v)$$

have the parameters u = 0.090, v = 0.098. Though these are not very different from the values for gallium (**II,g1**), the two cells are so dissimilar in shape that the atomic environments in the two crystals are unrelated.

According to this structure, atoms are in double layers which are widely separated from one another (Fig. II,16a). Each atom has two close neighbors in one layer at a P–P distance of 2.17 A. and a third almost as near (2.20 A.). In adjacent double layers (Fig. II,16b) the shortest P–P separation is 3.87 A. Discrete molecules do not exist. In this respect it is not like phosphorus vapor, which at 200°C. is shown by electron diffraction to be composed of P_4 molecules that are regular tetrahedra.

The mineral *arsenolamprite* is said to be an orthorhombic modification of arsenic having the structure of black phosphorus, and to have the cell dimensions:

$$a_0 = 3.63 \text{ A.}; \; b_0 = 4.45 \text{ A.}; \; c_0 = 10.96 \text{ A.}$$

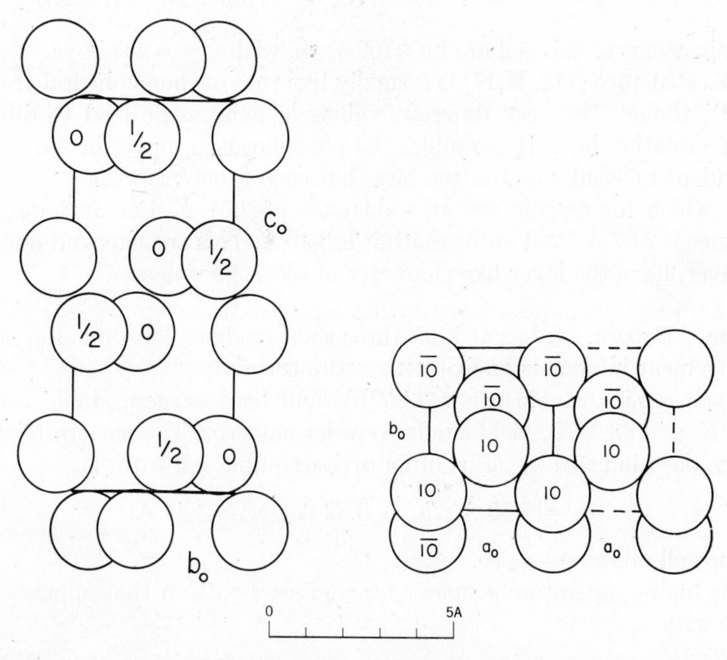

Fig. II,16a (left). The orthorhombic unit of black phosphorus projected along the a_0 axis. Origin in lower right.

Fig. II,16b (right). Two cells of the structure of orthorhombic phosphorus projected along its c_0 axis. Origin in lower middle of the figure.

II,l. *Arsenic*, *antimony*, and *bismuth* have the same rhombohedral structure with cells of the dimensions:

$$\begin{array}{lll} \text{As:} & a_0 = 4.131 \text{ A.}; & \alpha = 54°10' \\ \text{Sb:} & a_0 = 4.50661 \text{ A.}; & \alpha = 57°6.5' \\ \text{Bi:} & a_0 = 4.7459 \text{ A.}; & \alpha = 57°14.2' \end{array}$$

The two atoms in each are in the positions $uuu; \bar{u}\bar{u}\bar{u}$, where u has the values:

$$\begin{array}{ll} \text{As:} & u = 0.226 \\ \text{Sb:} & u = 0.233 \\ \text{Bi:} & u = 0.237 \ (0.23806 \text{ at } 4.2°\text{K.}, 0.2380 \text{ at } 78°\text{K.}) \end{array}$$

The corresponding hexagonal cells containing six atoms have the edges:

$$\begin{array}{llll} \text{As:} & a_0' = 3.760 \text{ A.}; & c_0' = 10.548 \text{ A.} & (26°\text{C.}) \\ \text{Sb:} & a_0' = 4.3083 \text{ A.}; & c_0' = 11.2743 \text{ A.} & (25°\text{C.}) \\ \text{Bi:} & a_0' = 4.54590 \text{ A.}; & c_0' = 11.86225 \text{ A.} & (25°\text{C.}) \\ & a_0' = 4.5345 \text{ A.}; & c_0' = 11.8178 \text{ A.} & (78°\text{K.}) \\ & a_0' = 4.5333 \text{ A.}; & c_0' = 11.8065 \text{ A.} & (4.2°\text{K.}) \end{array}$$

The six atoms of this cell are in $\pm(00v)$; rh, with $v = u$ as above.

This structure (Fig. II,17) is formally like that of rhombohedral graphite (**II,i3**), though the very different values of axial angle lead to different atomic relationships. It resembles the phosphorus grouping in the absence of evident molecules and in the fact that each atom has three close neighbors, which for arsenic are at a distance of 2.51 A. For antimony this distance is 2.87 A., and for bismuth it is 3.10 A. This arrangement does not, however, have the layer-like character of black phosphorus.

II,m. *Oxygen*, O_2, has at least three solid modifications, but the atomic arrangement in none is known with entire certainty.

Alpha oxygen, stable below 23.5°K., and beta oxygen, stable between 23.5°K. and 43.4°K., yield similar powder patterns. The pattern for alpha O_2 has been indexed in terms of an orthorhombic cell with

$$a_0 = 5.50 \text{ A.}; \ b_0 = 3.82 \text{ A.}; \ c_0 = 3.44 \text{ A.}$$

but not all observers agree.

The high-temperature *gamma oxygen* gives a pattern that appears to be cubic with

$$a_0 = 6.83 \text{ A.} \quad (50°\text{K.})$$

It has been proposed that its packing is the same kind of close-packing of O_2 molecules that has been described for alpha nitrogen (**II,j2**).

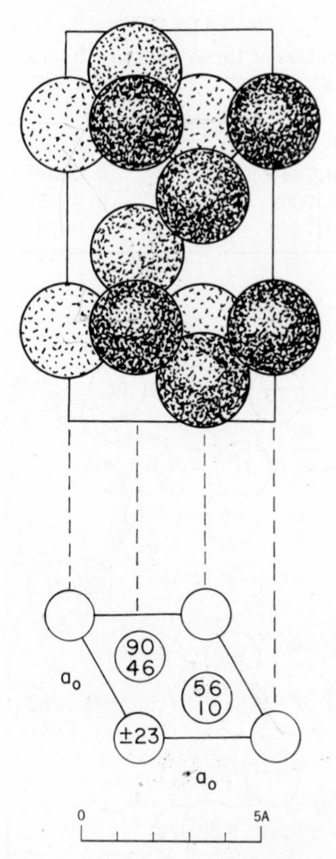

Fig. II,17. Two projections of the rhombohedral structure of arsenic in terms of its hexagonal cell.

II,n1. The unit prism of ordinary *rhombic sulfur*, S, is a very large one containing 128 atoms and having

$$a_0 = 10.467 \text{ A.}; \; b_0 = 12.870 \text{ A.}; \; c_0 = 24.493 \text{ A.} \quad (25°\text{C.})$$

The space group is V_h^{24} (*Fddd*) with all of these atoms in general positions which, placing the origin in a center of symmetry, have the coordinates:

$$(32h) \quad \pm(xyz;\; x,{}^1/_4-y,{}^1/_4-z;\; {}^1/_4-x,y,{}^1/_4-z;\; {}^1/_4-x,{}^1/_4-y,z);\; \text{F.C.}$$

A recent detailed study confirms the original determination and results in parameters, as listed in Table II,6, which though more accurate are not greatly different from those established many years ago. The structure is an assemblage of 16 S_8 molecules per cell (Fig. II,18). Each molecule is a

TABLE II,6
Parameters of the Atoms in Rhombic Sulfur

Atom	x	y	z
S(1)	0.8554	0.9526	0.9516
S(2)	0.7844	0.0301	0.0763
S(3)	0.7069	0.9795	0.0040
S(4)	0.7862	0.9073	0.1290

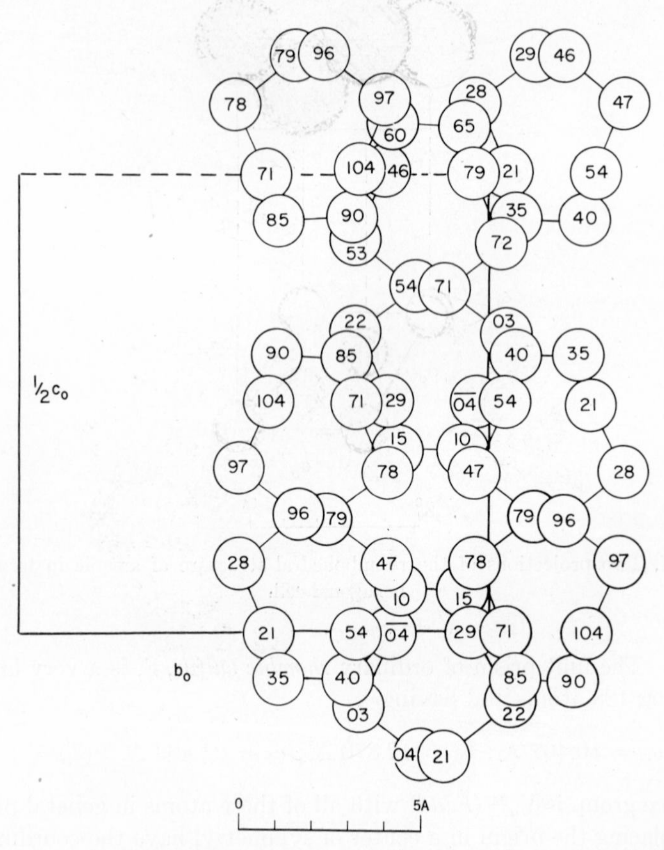

Fig. II,18a. A portion of the structure of orthorhombic sulfur projected along its a_0 axis. Origin in lower right.

closed puckered ring of sulfur atoms in which the S–S separation is 2.048 A. and the S–S–S angle is ca. 107°54′. The nearest approach of the sulfur atoms of adjacent molecules is 3.69 A.

The present description, placing the origin in a center of symmetry, represents a shift of $(-^1/_8, -^1/_8, -^1/_8)$ from the origin of the earlier descriptions.

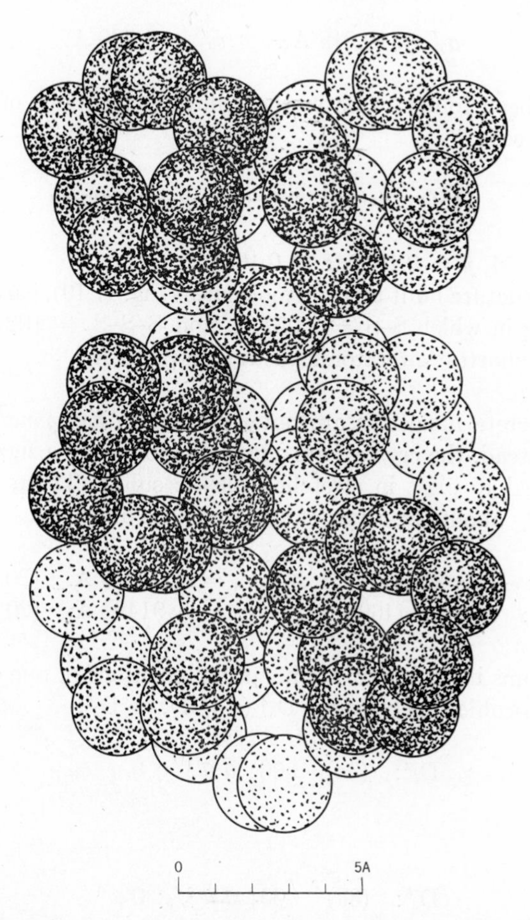

Fig. II,18b. A packing drawing of the portion of the structure of rhombic sulfur shown in Figure II,18a as viewed along its a_0 axis. The nature of the S_8 molecules that are present is apparent in these two drawings.

II,n2. The *rhombohedral* form of *sulfur*, S, has a rhombohedral unit containing six atoms:

$$a_0 = 6.46 \text{ A.}, \qquad \alpha = 115°18'$$

The hexagonal cell containing 18 atoms has the edges:

$$a_0' = 10.818 \text{ A.}, \qquad c_0' = 4.280 \text{ A.}$$

Atoms have been found to be in the following positions of C_{3i}^2 ($R\bar{3}$) referred to this cell:

$$(18f) \quad \pm(xyz;\ \bar{y},x-y,z;\ y-x,\bar{x},z);\ \text{rh}$$

with $x = 0.1454$, $y = 0.1882$, $z = 0.1055$.

This is a structure built up of S_6 molecules (Fig. II,19), each of which is a puckered ring in which S–S = 2.057 A. and S–S–S = 102°12′. Between molecules the shortest interatomic distance is 3.501 A.

II,o1. Discrete molecules are not apparent in *hexagonal selenium and tellurium*. Instead, the crystals are aggregates of endless zigzag chains like those supposed to exist in stretched plastic sulfur. These chains are arranged spirally about the *c* axis of cells with the dimensions:

Se:	$a_0 = 4.35517$ A.;	$c_0 = 4.94945$ A.	(20°C.)
Te:	$a_0 = 4.44693$ A.;	$c_0 = 5.91492$ A.	(20°C.)

The three atoms in each unit are in special positions of one or the other of the enantiomorphic space groups D_3^4 ($C3_121$), or D_3^6:

$$D_3^4: \quad (3a) \quad u00;\ \bar{u}\,\bar{u}\,{}^1/_3;\ 0\,u\,{}^2/_3$$

or

$$D_3^6: \quad (3a) \quad u00;\ \bar{u}\,\bar{u}\,{}^2/_3;\ 0\,u\,{}^1/_3$$

with u(Se) = 0.217 and u(Te) = 0.269 (Fig. II,20). These values of u make Se–Se = 2.32 A. and Te–Te = 2.86 A. along the chains. The angle Se–Se–Se in a chain, presumably between the two primary bonds of a selenium atom, is 105°; the corresponding angle for tellurium is 102°. The nearest approach of selenium atoms in different chains is 3.46 A.; in tellurium the corresponding distance between chains is 3.74 A.

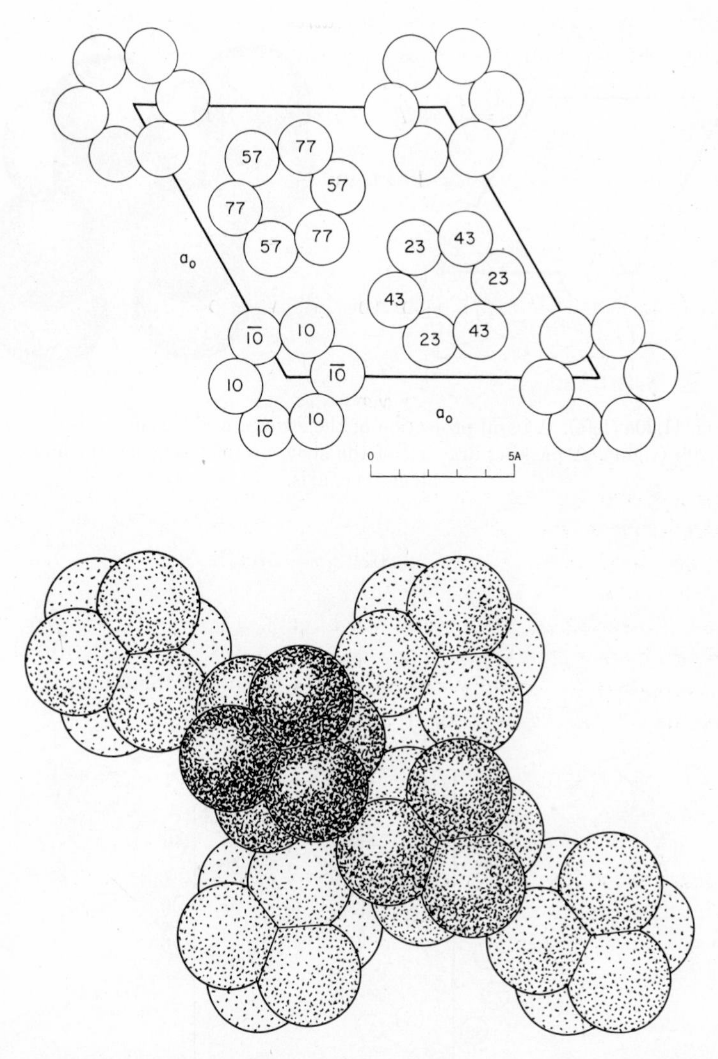

Fig. II,19a (top). The structure of rhombohedral sulfur projected along its c_0 hexagonal axis.

Fig. II,19b (bottom). A packing drawing of the structure of rhombohedral sulfur seen along its hexagonal c_0 axis. The difference between the S_6 molecules in this modification and the S_8 molecules of rhombic sulfur becomes evident on comparing Figures II,18b and II,19b.

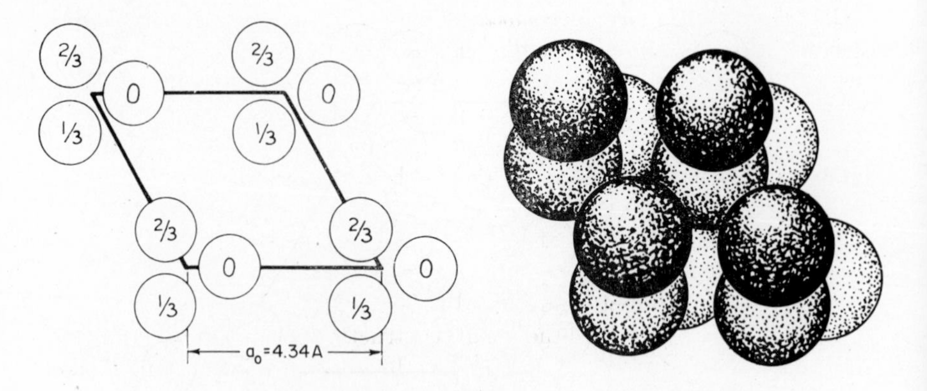

Fig. II,20a (left). A basal projection of the structure of hexagonal selenium.
Fig. II,20b (right). A packing drawing of the structure of hexagonal selenium viewed along its c_0 axis.

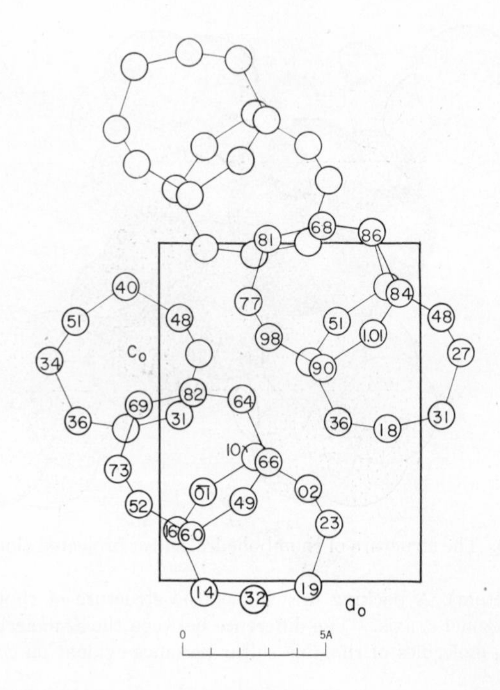

Fig. II,21a. A projection along b_0 of the monoclinic structure of α-selenium. Origin at lower left.

II,o2. The *alpha* form of *monoclinic selenium,* Se, possesses a unit which contains 32 atoms and has the dimensions:

$$a_0 = 9.05 \text{ A.};\ b_0 = 9.07 \text{ A.};\ c_0 = 11.61 \text{ A.};\ \beta = 90°46'$$

Atoms are in general positions of C_{2h}^5 in the orientation $P2_1/n$:

$$(4e) \quad \pm(xyz;\ x+{}^1/_2,{}^1/_2-y,z+{}^1/_2)$$

The determined parameters are listed in Table II,7.

As Figure II,21 indicates, the resulting structure is a grouping of four Se_8 molecules. Each molecule, with the bond dimensions of Figure II,22, is an eight-membered ring in which alternate atoms lie (within 0.05 A.) in one of two parallel planes.

Between molecules the closest Se–Se distance is 3.53 A.

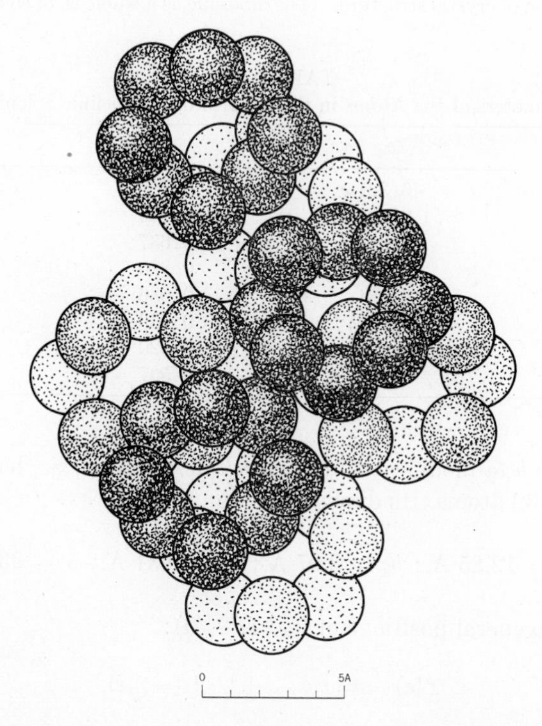

Fig. II,21b. A packing drawing of the molecules of the monoclinic structure of α-selenium shown in Figure II,21a. The puckering of the Se_8 molecules is clear.

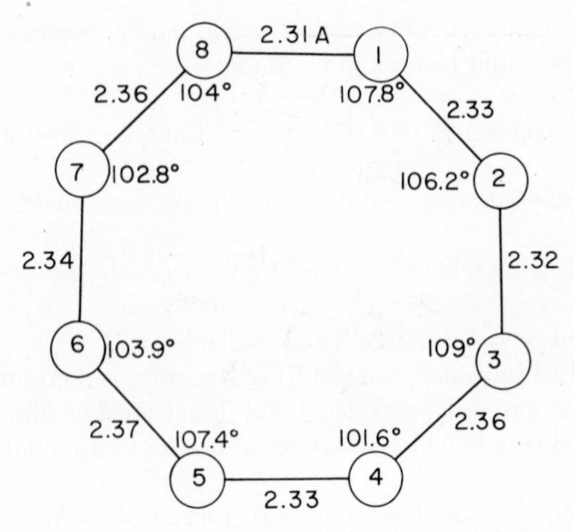

Fig. II,22. Dimensions of the molecule of the α form of monoclinic selenium derived from the determination of crystal structure. The molecule as a whole is, of course, not planar.

TABLE II,7

Parameters of the Atoms in the α Form of Monoclinic Selenium

Atom	x	y	z
Se(1)	0.321	0.486	0.237
Se(2)	0.427	0.664	0.357
Se(3)	0.317	0.637	0.535
Se(4)	0.134	0.820	0.556
Se(5)	−0.081	0.686	0.521
Se(6)	−0.156	0.733	0.328
Se(7)	−0.084	0.520	0.229
Se(8)	0.131	0.597	0.134

II,o3. The *beta* modification of *monoclinic selenium*, Se, has a unit that also contains 32 atoms. Its dimensions are

$$a_0 = 12.85 \text{ A.}; \; b_0 = 8.07 \text{ A.}; \; c_0 = 9.31 \text{ A.}; \; \beta = 93°8'$$

Atoms are in general positions of C_{2h}^5 $(P2_1/a)$:

$$(4e) \quad \pm(xyz;\; x+{}^1/_2, {}^1/_2-y, z)$$

A correction of the original, improbable, parameters gives the set stated in Table II,8.

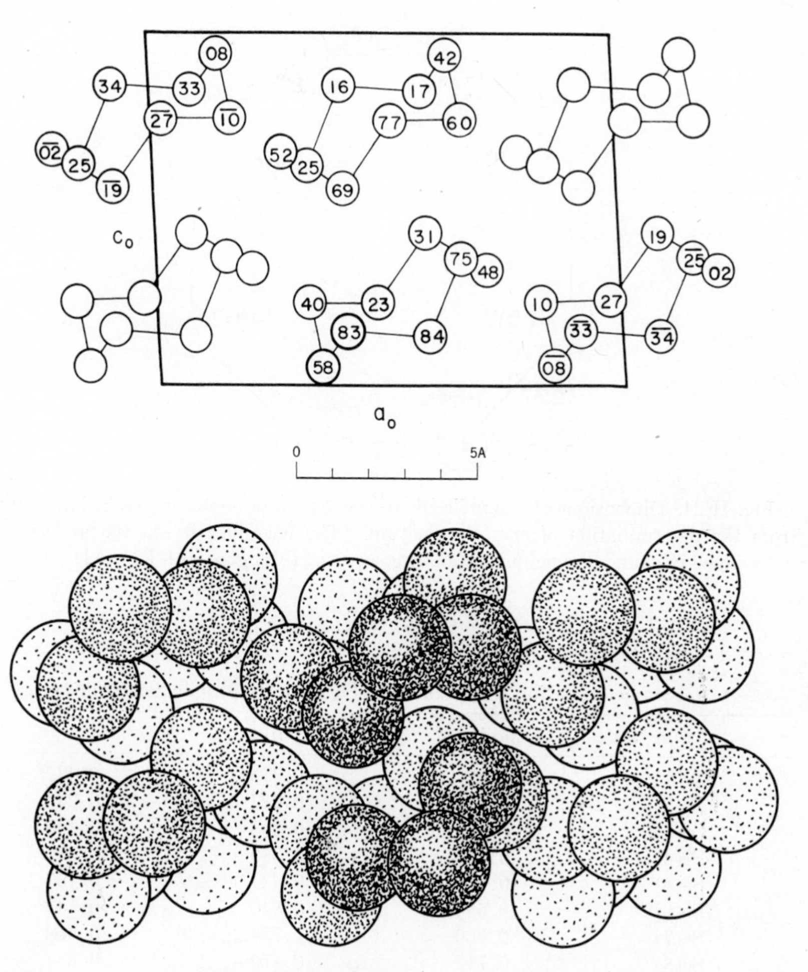

Fig. II,23a (top). A projection along b_0 of the monoclinic β-selenium structure. Origin in lower left.

Fig. II,23b (bottom). A packing drawing of the structure of monoclinic β-selenium as viewed along the b_0 axis. The molecules in this projection are seen nearly edge on.

The structure to which they give rise is illustrated in Figure II,23. Like the alpha form, it is an array of Se_8 molecules which are puckered rings with the bond values shown in Figure II,24. Alternate atoms are planar to 0.04 A. or less, and in these two parallel planes the atoms are at the corners of nearly perfect squares. Thus the same molecule is found in the two forms of monoclinic selenium.

Between molecules, the Se–Se separations range upwards from 3.48 A.

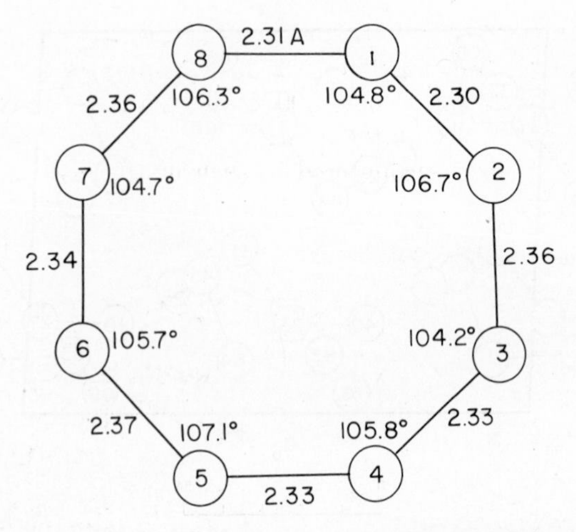

Fig. II,24. Dimensions of the molecule of the β form of monoclinic selenium derived from the determination of crystal structure. The near identity in the molecules as determined in α- and β-selenium is seen by comparing this figure with Figure II,22.

TABLE II,8

Parameters of the Atoms in the β Form of Monoclinic Selenium

Atom	x	y	z
Se(1)	0.584	0.315	0.437
Se(2)	0.477	0.227	0.246
Se(3)	0.328	0.398	0.240
Se(4)	0.352	0.580	0.050
Se(5)	0.410	0.831	0.157
Se(6)	0.590	0.840	0.142
Se(7)	0.660	0.754	0.368
Se(8)	0.710	0.479	0.334

II,p. The body-centered form of *tungsten*, W, has already been described (**II,c**).

A second modification appears when certain fused tungstates are electrolyzed. It is cubic with $a_0 = 5.083$ A. The eight atoms in its cell have been assigned the following two sets of special positions of O_h^3 ($Pm3n$):

W(1): (2a) 000; $\frac{1}{2}\,\frac{1}{2}\,\frac{1}{2}$

W(2): (6c) 0 $\frac{1}{2}\,\frac{1}{4}$; $\frac{1}{4}$ 0 $\frac{1}{2}$; $\frac{1}{2}\,\frac{1}{4}$ 0; 0 $\frac{1}{2}\,\frac{3}{4}$; $\frac{3}{4}$ 0 $\frac{1}{2}$; $\frac{1}{2}\,\frac{3}{4}$ 0

Atoms of the type W(1) have 12 neighbors at a distance of 2.82 A., while each W(2) atom has two neighbors at 2.52 A., four at 2.82 A., and eight at 3.09 A. (Fig. II,25). Heating above 700°C. drives this modification irreversibly into the simple body-centered form.

This structure is also encountered in such intermetallic compounds as $SiCr_3$ and SiV_3 (1939,W).

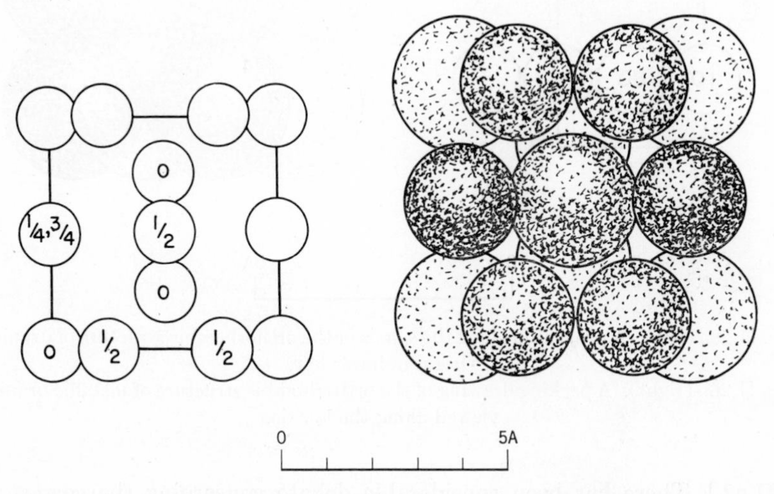

Fig. II,25a (left). A projection along a cubic axis of the second, beta, form of tungsten.
Fig. II,25b (right). A packing drawing of atoms in the second form of tungsten. In intermetallic compounds RX_3 with this structure, the R atoms have the positions indicated by the larger spheres.

II,q1. *Alpha uranium,* U, is orthorhombic, rather than monoclinic as first supposed. The unit cell has

$$a_0 = 2.854 \text{ A.};\ b_0 = 5.869 \text{ A.};\ c_0 = 4.955 \text{ A.}$$

with four atoms in the positions of V_h^{17} (*Cmcm*):

$$(4c)\quad 0\,u\,{}^1/_4;\ 0\,\bar{u}\,{}^3/_4;\ {}^1/_2,u+{}^1/_2,{}^1/_4;\ {}^1/_2,{}^1/_2-u,{}^3/_4$$

with $u = 0.1025$. Each uranium atom has two neighbors at 2.76 A., two at 2.85 A., four at 3.27 A., and four more at 3.36 A. (Fig. II,26).

There are two other modifications. The very complicated beta form, stable between ca. 1200°F. and ca. 1400°F., is described below (**II,q2**). The gamma form is body-centered cubic (**II,c**).

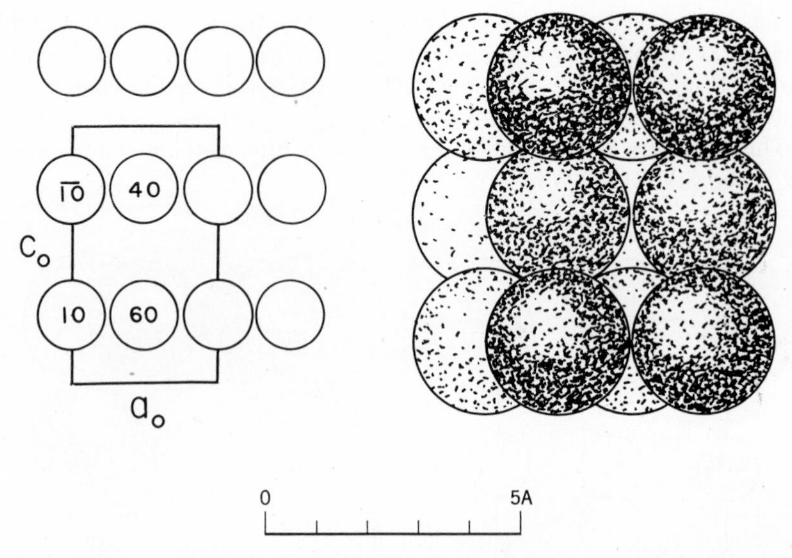

Fig. II,26a (left). A projection along the b_0 axis of the orthorhombic structure of uranium. Origin in lower left.

Fig. II,26b (right). A packing drawing of the orthorhombic structure of metallic uranium viewed along the b_0 axis.

II,q2. There has been considerable debate concerning the correct arrangement in the beta form of uranium stable between 1200°F. and 1400°F. It is tetragonal with a unit that contains 30 atoms and has the dimensions, when quenched to room temperature:

$$a_0 = 10.590 \text{ A.}, \qquad c_0 = 5.634 \text{ A.} \quad (25°\text{C.})$$

Two closely related structures give approximately equal good agreement with existing data. One of these developed from powder data is based on C_{4v}^4 ($P4nm$) with atoms in the following positions:

(2a) $00u;\ {}^1/_2,{}^1/_2,u+{}^1/_2$

(4c) $uuv;\ \bar{u}\bar{u}v;\ u+{}^1/_2,{}^1/_2-u,v+{}^1/_2;\ {}^1/_2-u,u+{}^1/_2,v+{}^1/_2$

(8d) $xyz;\ x+{}^1/_2,{}^1/_2-y,z+{}^1/_2;\ {}^1/_2-x,y+{}^1/_2,z+{}^1/_2;\ \bar{x}\bar{y}z;$
$yxz;\ y+{}^1/_2,{}^1/_2-x,z+{}^1/_2;\ {}^1/_2-y,x+{}^1/_2,z+{}^1/_2;\ \bar{y}\bar{x}z$

The chosen parameters are given in Table II,9.

The structure (Fig. II,27) results in a very complicated atomic environment. There is a close approach of 2.531 A.; others range upwards from 2.831 A.

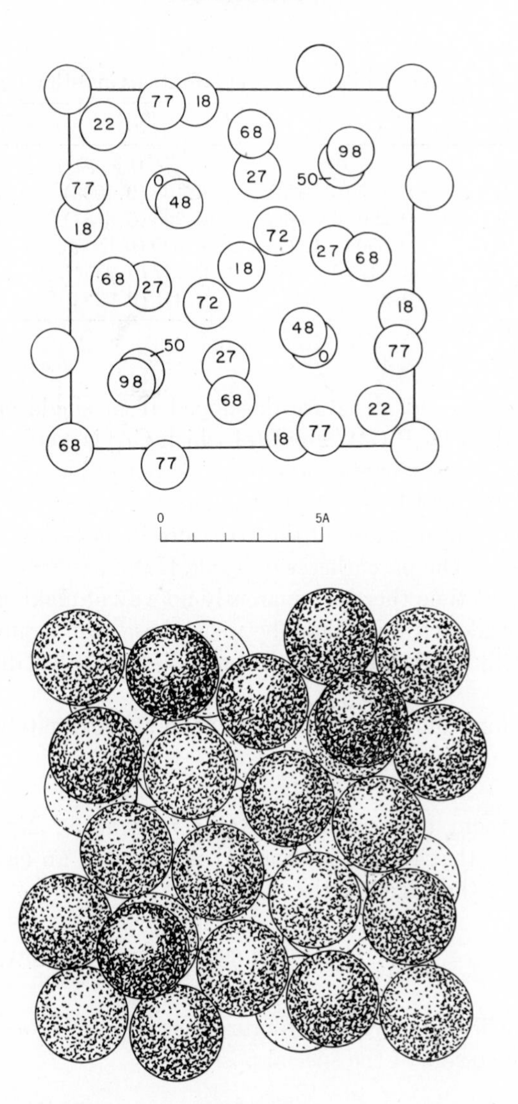

Fig. II,27a (top). A projection along the c_0 axis of the tetragonal structure of the β form of uranium.

Fig. II,27b (bottom). A packing drawing of the complicated tetragonal structure found for β-uranium as viewed along its c_0 axis.

TABLE II,9
Positions and Parameters of the Atoms in β-Uranium

Atom	Position	x	y	z
U(1)	(2*a*)	0	0	0.68 (0.50)
U(2)	(4*c*)	0.105 (0.1033)	0.105 (0.1033)	0.22 (0.25)
U(3)	(4*c*)	0.290 (0.3183)	0.290 (0.3183)	0.00 (0.9800)
U(4)	(4*c*)	0.690 (0.6817)	0.690 (0.6817)	0.48 (0.5200)
U(5)	(8*d*)	0.547 (0.5608)	0.227 (0.2354)	0.27 (0.25)
U(6)	(8*d*)	0.367 (0.3667)	0.041 (0.0383)	0.18 (0.25)

The other proposed structure developed from single crystal measurements is based on D_{4h}^{14} ($P4/mnm$) (of which C_{4v}^{4} is a subgroup). It differs mainly in possessing a center of symmetry and in having z parameters of U(1), U(2), U(5), and U(6) set by the higher symmetry at values that depart appreciably from those of the structure described above. Its parameters are those in the parentheses of Table II,9.

At the present time there is apparently no way of making measurements that would decide with certainty between these two arrangements which, though generally similar, yield interatomic distances that differ by as much as 0.3 A.

Low-chromium alloys with uranium have approximately the beta uranium structure.

II,r1. At room temperature, metallic *neptunium*, Np, gives powder diffraction data that can be interpreted in terms of an eight-atom orthorhombic unit of the edge lengths:

$$a_0 = 4.723 \text{ A.}; \; b_0 = 4.887 \text{ A.}; \; c_0 = 6.663 \text{ A.}$$

These data are in accord with a structure based on V_h^{16} ($Pmcn$) which places the atoms in two sets of the special positions:

$$(4c) \quad \pm(^1/_4\, u\, v;\; ^1/_4, ^1/_2 - u, v + ^1/_2)$$

with $u(1) = 0.208$, $v(1) = 0.036$ and $u(2) = 0.842$, $v(2) = 0.319$.

This is a considerably distorted body-centered arrangement in which each neptunium atom has four especially close neighbors (2.60–2.64 A.), as well as seven or nine others at greater distances (3.06–3.53 A.). The arrangement is shown in Figure II,28.

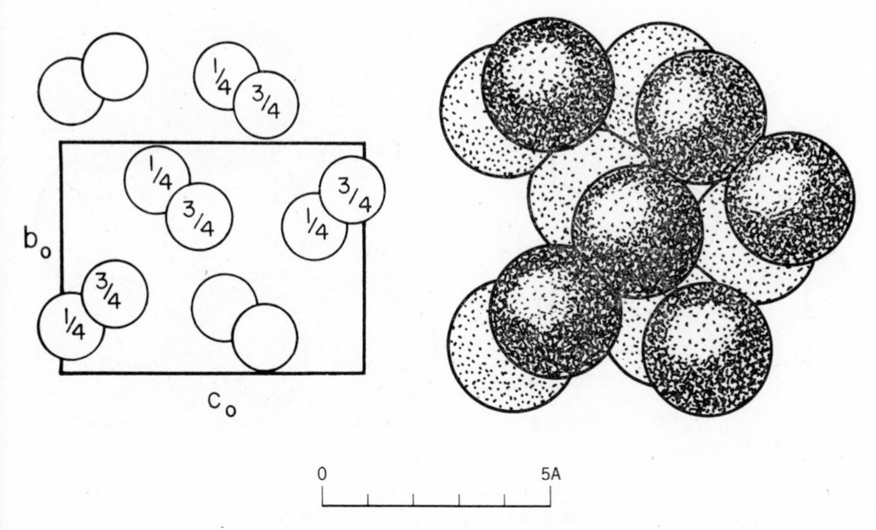

Fig. II,28a (left). A projection along a_0 of the orthorhombic room-temperature, α, neptunium structure. Origin in lower left.

Fig. II,28b (right). A packing drawing of the α-neptunium structure viewed along its a_0 axis.

II,r2. At ca. 278°C., metallic *neptunium*, Np, passes to a higher temperature form that persists up to ca. 570°C. This *beta* modification is tetragonal and has at 313°C. the unit:

$$a_0 = 4.897 \text{ A.}, \qquad c_0 = 3.388 \text{ A.}$$

The four atoms in this cell can be placed in the following special positions of D_4^2 ($P42_1$):

$$(2a) \quad 000; \ {}^1/_2\,{}^1/_2\,0$$
$$(2c) \quad 0\,{}^1/_2\,u; \ {}^1/_2\,0\,\bar{u}, \quad \text{with } u = -0.375.$$

In this structure (Fig. II,29), as in that of the alpha form, each atom has four especially close neighbors (Np–Np = 2.76 A.). Here it has ten others at distances between 3.24 and 3.56 A.

There is evidence of a third, *gamma*, modification of neptunium at temperatures between ca. 570°C. and the melting point (640°C.). The observed x-ray lines from it point to a body-centered cubic structure which, at 600°C., has a_0 = 3.52 A. (**II,c**).

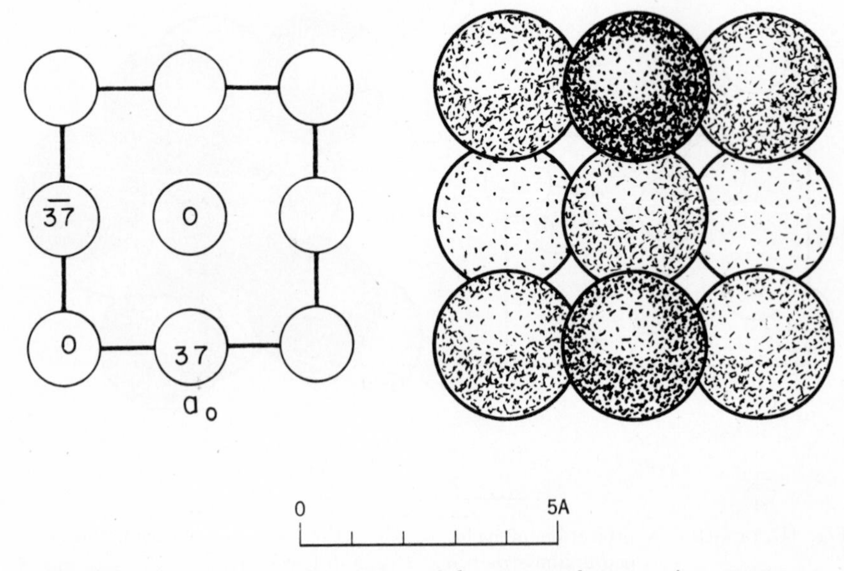

Fig. II,29a (left). A projection along c_0 of the tetragonal β-neptunium structure
Fig. II,29b (right). A packing drawing of the tetragonal β form of neptunium viewed along the c_0 axis.

II,s. Six modifications have been described for metallic *plutonium*, Pu. The lowest temperature, *alpha*, form, stable up to 110°C., is monoclinic with a 16-atom unit cell of the dimensions:

$$a_0 = 6.1835 \text{ A.};\ b_0 = 4.8244 \text{ A.};\ c_0 = 10.973 \text{ A.};\ \beta = 101°48' \text{ (18°C.)}$$

A preliminary structure based on powder data, which are all that are available, places these atoms in special positions:

$$(2e) \quad \pm(u\ ^1/_4\ v)$$

of C_{2h}^2 ($P2_1/m$) with the parameters listed in Table II,10. The atomic arrangement is shown in Figure II,30. Each atom has from three to five nearest neighbors at distances that range between 2.52 and 2.73 A.

Beta plutonium is also monoclinic with 34 atoms in a cell of the dimensions:

$$a_0 = 9.284 \text{ A.};\ b_0 = 10.463 \text{ A.};\ c_0 = 7.859 \text{ A.};\ \beta = 92°8' \text{ (190°C.)}$$

The atomic arrangement has not yet been described.

TABLE II,10
Parameters of the Atoms in α-Plutonium

Atom	u	v
Pu(1)	0.332	0.155
Pu(2)	0.774	0.175
Pu(3)	0.144	0.341
Pu(4)	0.658	0.457
Pu(5)	0.016	0.621
Pu(6)	0.465	0.644
Pu(7)	0.337	0.926
Pu(8)	0.892	0.897

The *gamma* form measured at 235°C. is orthorhombic with eight atoms in a unit of the dimensions:

$$a_0 = 3.1587 \text{ A.}; \; b_0 = 5.7682 \text{ A.}; \; c_0 = 10.162 \text{ A.}$$

In the space group V_h^{24} (*Fddd*), these atoms are in the positions:

$$(8a) \quad 000; \; {}^1/_4\,{}^1/_4\,{}^1/_4; \; \text{F.C.}$$

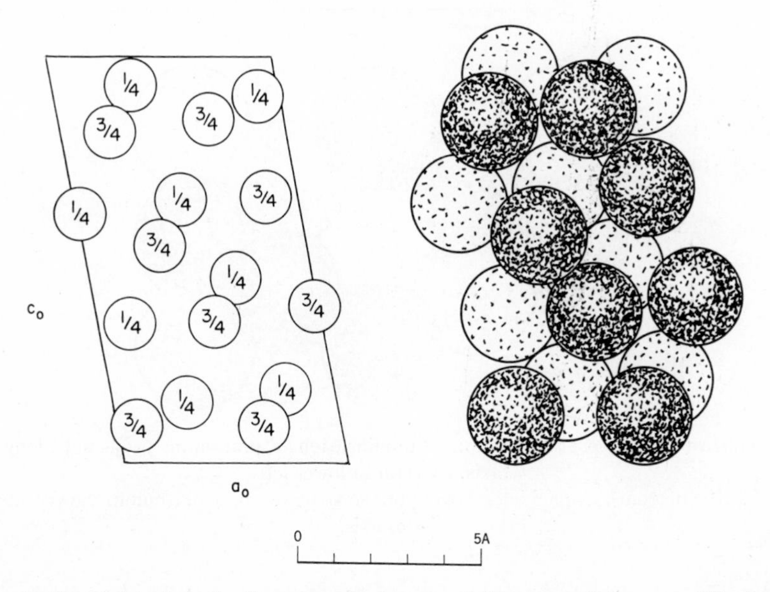

Fig. II,30a (left). A projection along its b_0 axis of the monoclinic α modification of plutonium. Origin in lower left.

Fig. II,30b (right). A packing drawing of the structure of α-plutonium seen along its b_0 axis.

The structure that results (Fig. II,31) surrounds each atom with ten close neighbors, four at 3.026 A., two at 3.159 A. and four more at 3.288 A.

Of the three remaining modifications, *delta*, measured at 320°C., is face-centered cubic (**II,a**) and *epsilon*, measured at 510°C., is body-centered cubic (**II,c**). The intermediate, *delta prime*, form is reported to be tetragonal with the indium structure described in paragraph **II,e3**.

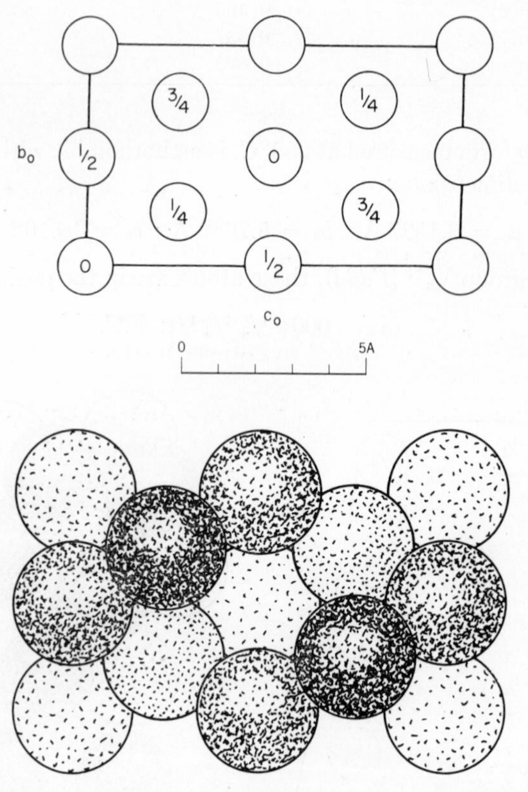

Fig. II,31a (top). The orthorhombic γ modification of plutonium projected along its a_0 axis. Origin in lower left.

Fig. II,31b (bottom). A packing drawing of the structure of γ-plutonium viewed along its a_0 axis.

II,t. The seventh-column metal *manganese*, Mn, has four modifications. Two of these are complicated cubic structures, while the other two, also cubic, are simpler.

The *alpha* form, obtained by fusing or distilling electrolytic manganese, has 58 atoms in a unit cube of edge:

$$a_0 = 8.894 \text{ A.}$$

Its atoms are in the following special positions of T_d^3 ($I\bar{4}3m$):

$$\begin{array}{ll}
(2a) & 000;\ {}^1/_2\,{}^1/_2\,{}^1/_2 \\
(8a) & www;\ w\bar{w}\bar{w};\ \bar{w}w\bar{w};\ \bar{w}\bar{w}w;\ \text{B.C.} \\
(24g) & uuv;\ vuu;\ uvu;\ u\bar{u}\bar{v};\ v\bar{u}\bar{u};\ u\bar{v}\bar{u}; \\
 & \bar{u}u\bar{v};\ \bar{v}u\bar{u};\ \bar{u}v\bar{u};\ \bar{u}\bar{u}v;\ \bar{v}\bar{u}u;\ \bar{u}\bar{v}u;\ \text{B.C.}
\end{array}$$

with $w = 0.317$, $u = 0.356$, $v = 0.042$, $u' = 0.089$, $v' = 0.278$.

In this very complicated grouping, each manganese atom has between 12 and 16 neighbors at distances from 2.24 to 3.00 A. In spite of this complexity, however, the arrangement is not a great departure from the body-centered **II,c,** its cube edge being three times longer and the distortion in atomic positions being such as to permit the introduction of four additional manganese atoms.

Beta manganese, stable between ca. 800°C. and 1100°C. and obtained by quenching from this range, is generally considered to have a unit in which 20 atoms are present in a cube 6.30 A. on a side. A structure has been given which places eight atoms in the following positions of O^6 ($P4_332$):

$$\begin{array}{ll}
(8c) & uuu;\ u+{}^1/_2,{}^1/_2-u,\bar{u};\ \bar{u},u+{}^1/_2,{}^1/_2-u;\ {}^1/_2-u,\bar{u},u+{}^1/_2; \\
 & {}^1/_4-u,{}^1/_4-u,{}^1/_4-u;\ u+{}^3/_4,{}^3/_4-u,u+{}^1/_4; \\
 & {}^3/_4-u,u+{}^1/_4,u+{}^3/_4;\ u+{}^1/_4,u+{}^3/_4,{}^3/_4-u \\
(12d) & {}^1/_4-v,v,{}^1/_8;\ {}^3/_4-v,{}^1/_2-v,{}^7/_8;\ v+{}^3/_4,v+{}^1/_2,{}^3/_8; \\
 & v+{}^1/_4,\bar{v},{}^5/_8;\ {}^1/_8,{}^1/_4-v,v;\ {}^7/_8,{}^3/_4-v,{}^1/_2-v; \\
 & {}^3/_8,v+{}^3/_4,v+{}^1/_2;\ {}^5/_8,v+{}^1/_4,\bar{v};\ v,{}^1/_8,{}^1/_4-v; \\
 & {}^1/_2-v,{}^7/_8,{}^3/_4-v;\ v+{}^1/_2,{}^3/_8,v+{}^3/_4;\ \bar{v},{}^5/_8,v+{}^1/_4
\end{array}$$

with $u = 0.061$ and $v = 0.206$, or in the corresponding enantiomorphic positions of O^7. According to this grouping each atom is surrounded by 12 neighbors at a variety of distances between 2.36 and 2.67 A.

There has also been published another atomic arrangement based on a unit of twice the edge length:

$$a_0' = 12.58 \text{ A.}$$

The 160 atoms in this cube have been said to be in the following special positions of O_h^7: (8*a*) 000; etc., (8*b*) $^1/_2\,^1/_2\,^1/_2$; etc., (48*f*) *w*00; etc., (96*g*) *uuv;* etc., with $w = 0.10$, $u = v = 0.33$. The experimental basis for selecting such a larger unit has not been published and it is clear that additional work on beta manganese is needed.

The *gamma* modification, stable immediately above 1100°C., was at first examined in the quenched condition. It then showed a tetragonal distortion of the cubic close-packing resembling the structure of indium (**II,e3**). When photographed at high temperatures, the gamma phase proves to be precisely cubic close-packed (**II,a**), with:

$$a_0 = 3.863 \text{ A.} \qquad (1100°\text{C.})$$

In the quenched material the face-centered pseudo-cell had the dimensions:

$$a_0' = 3.774 \text{ A.,} \qquad c_0' = 3.533 \text{ A.}$$

It is suggested that this structure, which also has been reported as produced by electrodeposition, arises from gamma manganese by a martensitic transformation.

At 1134°C. the gamma form transforms to a body-centered (**II,c**) δ modification, for which

$$a_0 = 3.081 \text{ A.} \qquad (1140°\text{C.})$$

This *delta* manganese is stable up to the melting point (ca. 1245°C.) just below which $a_0 = 3.093$ A. (1240°C.).

II,u. The solidified halogens *chlorine, bromine,* and *iodine* are orthorhombic with the same structure. All atoms in the tetramolecular units are in the following special positions of V_h^{18} (*Bmab*):

$$(8f) \quad \pm(0uv;\ {}^1/_2,u+{}^1/_2,\bar{v};\ {}^1/_2,u,v+{}^1/_2;\ 0,u+{}^1/_2,{}^1/_2-v)$$

For *chlorine*, Cl_2:

$$a_0 = 6.24 \text{ A.};\ b_0 = 8.26 \text{ A.};\ c_0 = 4.48 \text{ A.} \ (-160°\text{C.})$$

The parameters are $u = 0.100$, $v = 0.130$. In this structure (Fig. II,32), the shortest Cl–Cl = 2.02 A. Between molecules each chlorine atom has two neighbors at a distance of 3.34 A. The molecules thus related yield layers normal to the b_0 axis. Between layers the closest Cl–Cl = 3.69 A.

For *bromine*, Br_2:

$$a_0 = 6.67 \text{ A.};\ b_0 = 8.72 \text{ A.};\ c_0 = 4.48 \text{ A.}\quad (-150°\text{C.})$$
$$a_0 = 6.692 \text{ A.};\ b_0 = 8.737 \text{ A.};\ c_0 = 4.498 \text{ A.}\quad (-106°\text{C.})$$
$$a_0 = 6.700 \text{ A.};\ b_0 = 8.748 \text{ A.};\ c_0 = 4.517 \text{ A.}\quad (-78.5°\text{C.})$$
$$a_0 = 6.737 \text{ A.};\ b_0 = 8.761 \text{ A.};\ c_0 = 4.548 \text{ A.}\quad (-23.5°\text{C.})$$

The parameters are $u = 0.110$, $v = 0.135$. These give an intramolecular Br–Br of 2.27 A.; between molecules the atomic separation is 3.30 A. and more.

For *iodine*, I_2:

$$a_0 = 7.27007 \text{ A.};\ b_0 = 9.79344 \text{ A.};\ c_0 = 4.79004 \text{ A.}\quad (26°\text{C.})$$

The parameters are $u = 0.1156$, $v = 0.1493$. Within the molecule I–I is 2.70 A. and between molecules from 3.54 A. up.

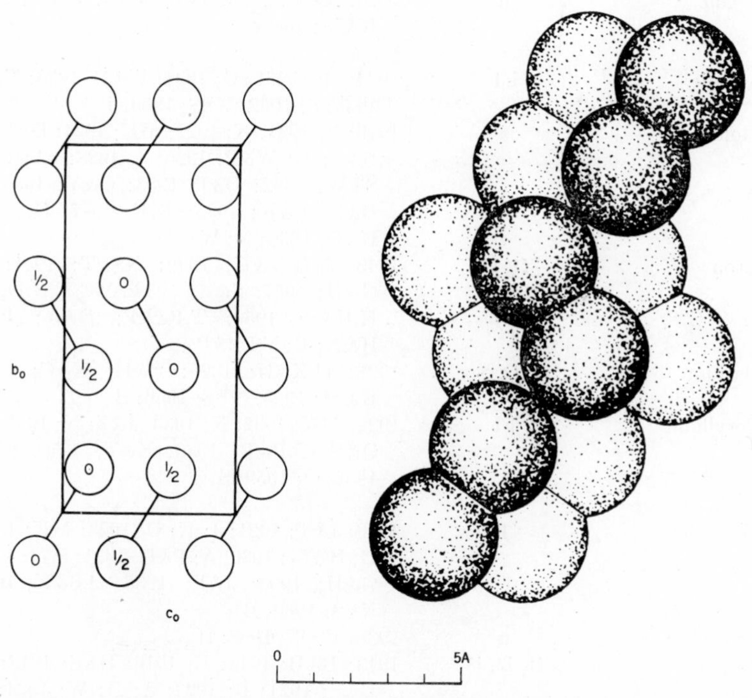

Fig. II,32a (left). The orthorhombic structure of solid iodine projected along its a_0 axis. Origin in lower left.

Fig. II,32b (right). A packing drawing of solid iodine corresponding to the projection of Figure II,32a.

BIBLIOGRAPHY TABLE, CHAPTER II

Element	Paragraph	Literature
Ac, Actinium	a	1961: F,G,B&M
Ag, Silver	a	1916: V; 1921: K; 1922: MK; S&S; T; 1924: D; 1925: B&L; D; W&P; 1926: D&W; E; J; 1928: A; vA; 1930: S&W; 1932: O&I; 1933: O&Y; S; 1934: O&Y; 1935: J&F; 1938: H&R; 1948: H&O; 1953: NBS; 1954: B; 1955: S&K; 1956: N
Al, Aluminum	a	1917: H; 1918: S; 1922: K; 1923: O&P; Y; 1924: D; J,P&W; W; 1925: D; L; vO; P&B; 1929: A; J,B&W; 1931: A; 1932: N; O&I; 1933: O&Y; S; 1935: J&F; 1936: J&S; 1949: Z; 1953: H&A; NBS; 1955: S&K; 1956: G&O; 1960: S&E
Am, Americium	a,b2	1956: G,C,D,W,T&R; 1960: MW,W,C,A, E&Z
Ar, Argon	a	1923: S&vS; 1924: S&vS; 1925: dS&K; 1930: K&M; 1958: H
As, Arsenic	k,l	1924: B; 1925: vO; 1934: W&E; 1938: T&B; 1939: S; 1952: NBS; 1959: J
Au, Gold	a	1916: V; 1921: K; 1922: MK; 1924: D; 1925: B&L; D; W&P; 1926: J; 1928: vA; 1930: S&W; 1932: O&I; 1933: O&Y; 1934: J, B&F; V&K; 1935: J&F; 1947: C; 1948: H&O; 1956: N; W
B, Boron	f1, f2	1943: L,H,N&H; 1949: NS&T; 1951: H, G&H; 1957: S&H; 1958: MC,K,H,D&N; H,H&S; 1959: T,P&LP; D&K; 1960: H&N; B; T,LP&P
Ba, Barium	c	1928: C,K&H; 1929: E&H; K&C; 1941: K&M; 1953: NBS; 1956: B
Be, Beryllium	b1	1922: MK; 1932: N; 1933: J&Z; N; 1935: N; O&P; O,P&R; 1936: K&T; 1950: S&H; 1952: O; 1959: S
Bi, Bismuth	l	1920: J&T; 1921: J; K; O; 1923: MK; 1924: D; H&M; 1930: A; P&C; 1931: S&J; 1932: G&H; 1935: J&F; 1938: J,S&K; 1952: NBS; 1960: B
Br, Bromine	u	1936: V&W; 1959: H
C, Carbon	i1, i2, i3	1913: B&B; 1914: E; 1916: D&S; 1917: H; D&S; 1921: B; 1922: A; H; W; 1924: B; H&M; R; 1925: E; 1926: E; M; 1928: O; 1931: H; 1936: H&W; 1937: T; 1939:

(continued)

BIBLIOGRAPHY TABLE, CHAPTER II (*continued*)

Element	Paragraph	Literature
		B,G,H&P; 1940: T&L; 1941: L&S; T; 1942: L&S; 1943: B; 1944: K; R; 1945: L; R; N&R; 1946: G; 1947: N&R; L; R&S; 1949: H&W; H; 1950: B; L; 1951: B; B&F; L; S&A; 1952: H&W; 1953: J,K&K; K; 1954: B; NBS; 1955: B&M; 1959: G&W; V&K
Ca, Calcium	a, b1, c	1920: H; 1921: H; 1933: E,H&P; G; 1934: G; 1941: K&M; 1956: M,T&B
Cd, Cadmium	b1	1920: H; 1921: H; 1924: vS; 1929: ML&M; 1931: J&P; 1932: S&W; 1934: S; 1935: J&F; K&T; 1936: B; 1952: NBS
Ce, Cerium	a, b1	1921: H; 1924: S&L; 1932: Q; 1937: K&B; 1949: L&T; 1950: S&S
Cl, Chlorine	u	1936: K&T; 1952: C; 1956: C
Co, Cobalt	a, b1	1919: H; 1921: H; 1924: S&L; 1925: M; 1927: S; 1930: H,J&S; O; 1932: K; 1934: S; 1937: M; 1941: E,L&W; 1942: E&L; F&M; 1943: E&L; 1947: H&P; 1954: D,B&M; 1958: A
Cr, Chromium	a, b1, c	1919: H; 1921: H; 1925: P; P&B; 1926: B&O; 1927: S; 1930: S&S; 1931: P; 1932: P; 1934: J,N,Q&F; 1935: W,H&R; 1937: W; 1940: S; 1954: E; NBS; 1955: S&W; 1961: V,K&F
Cs, Cesium	c	1927: S&V; 1928: S&V; 1956: B
Cu, Copper	a	1914: B; 1922: K; 1923: O&P; 1924: D; J, P&W; P; 1925: B&L; D; W&P; 1926: D&W; E; J; Q; 1928: vA; T&W; 1932: O&I; 1933: O&W; O&Y; 1934: V&K; 1935: B&S; 1942: H&A; 1946: R; 1953: NBS
D_2, Deuterium	d	1956: K,L&B; 1959: K,L&B
Dy, Dysprosium	b1	1937: K&B; 1953: B,L&S; 1954: B,L&S
Er, Erbium	b1	1929: ML&M; 1937: K&B; 1953: B,L&S; 1954: B,L&S; 1955: K&W
Eu, Europium	c	1937: K&B; 1956: B
Fe, Iron	a, c	1917: H; 1919: H; 1921: W; W&L; 1922: W&P; 1923: W&P; MK; O&P; 1924: D; H; W; W&P; 1925: B; D; 1926: Y; 1927:

(*continued*)

BIBLIOGRAPHY TABLE, CHAPTER II (*continued*)

Element	Paragraph	Literature
		L; 1928: O&O; 1929: M; B; 1930: R; 1931: vA&B; 1932: B&J; N; P; 1933: E&M; O&Y; 1934: J,B&F; K&S; 1935: J&F; 1936: J&F; 1937: B,J&T; M; 1941: vB; 1948: T; 1949: Z; 1953: NBS; 1954: vB&R; E; 1956: G&O
Ga, Gallium	g1, g2	1926: J,T&W; 1932: L; 1933: L; 1935: B; 1961: C,R&D
Gd, Gadolinium	b1	1937: K&B; 1953: B,L&S; 1954: B,L&S; 1958: B
Ge, Germanium	i1	1922: H; K; 1952: S&A; 1953: NBS; 1955: S&K; 1956: T&S; B&D; 1958: M
H_2, Hydrogen	d	1929: V; 1956: K,L&B; 1959: K,L&B
He, Helium	b1	1938: K&T; T; 1957: G; 1958: H; S,G&M; 1959: D; 1961: S&M
Hf, Hafnium	b1	1925: N&T; 1927: vA; 1952: S&MG; 1953: NBS
Hg, Mercury	e1, e2	1922: A&A; MK&C; 1928: T&W; W; 1929: H; M&B; 1932: H&R; 1933: N; 1957: B; 1959: A,S&S
Ho, Holmium	b1	1939: B
I, Iodine	u	1927: F; 1928: H,M&B; 1943: S&S; 1952: NBS; 1953: K,K&S
In, Indium	e3	1920: H; 1921: H; 1932: D&M; 1933: S; Z&N; 1935: F&O; 1936: A&A; 1938: B; 1952: NBS; 1955: G,M&R
Ir, Iridium	a	1920: H; 1921: H; 1923: W; 1925: B&L; 1932: O&I; 1933: O&Y
K, Potassium	c	1922: MK; 1926: G; 1927: H; S&V; 1928: P; S&V; 1956: B
Kr, Krypton	a	1930: K&M; N&N; 1932: R&S; 1957: C&S
La, Lanthanum	a, b1, b2	1930: ML&MK; 1932: Q; 1933: Z&N; 1934: R; 1937: K&B; 1952: DA; 1953: Z,Y&F; 1961: F,G,B&M
Li, Lithium	a, b1, c	1917: H; 1921: B&K; 1923: B; 1927: S&V; 1928: S&V; 1940: P&A; 1946: L&HR; 1947: B; 1948: B&T; 1956: B; 1957: C&M; 1959: V&K
Lu, Lutetium	b1	1937: K&B; 1956: S,D&H
Mg, Magnesium	b1	1917: H; 1920: B; 1923: O&P; 1928: R,A&W; 1932: S&W; 1934: S; 1935: J&F; O,P&R; 1937: H&F; 1938: B&R; J,S&K; 1939: R&H; 1953: NBS

(*continued*)

BIBLIOGRAPHY TABLE, CHAPTER II (*continued*)

Element	Paragraph	Literature
Mn, Manganese	t	1923: Y; 1925: B; W&P; 1927: B&T; 1928: P; 1929: P&O; S; 1930: O; 1931: S; 1934: W; 1945: P&H; 1954: B&C
Mo, Molybdenum	a, c	1921: H; S; 1924: D; 1925: D; 1926: vA; 1932: O&I; 1935: J&F; 1951: E,S&J; 1953: NBS; 1957: A&G
N_2, Nitrogen	j1, j2	1925: dS&K; 1929: dS&K; V; 1932: R; V; 1934: V; 1937: V; 1959: B,B,M&P
Na, Sodium	b1, c	1917: H; 1927: S&V; 1928: S&V; 1938: A&P; 1956: B
Nb, Niobium (Columbium)	c	1925: vO; 1929: ML&M; 1930: M; 1931: N; 1932: Q; 1934: B&B; 1936: N; 1951: E, S&J; 1953: S&W
Nd, Neodymium	b1, b2	1932: Q; 1937: K&B; 1939: K&B; 1953: E&Z
Ne, Neon	a	1930: K&M; dS,K&M; 1958: H; 1961: K, L&B
Ni, Nickel	a, b1	1917: H; 1919: H; 1920: B; 1921: H; 1922: W; 1923: MK; 1924: D; 1925: C,A&W; C&F; D; L; L&T; P&B; 1929: G; M&N; T; V&B; 1931: B&vB; 1932: B&J; O&I; 1934: J; J,N,Q&F; O&P; 1935: B&S; J&F; 1936: J&F; O&Y; 1937: B,J&T; 1939: L&M; 1944: C; 1953: NBS; 1954: vB&R
Np, Neptunium	c, r1, r2	1952: Z; 1958: MK
O, Oxygen	m	1925: dS&K; 1927: ML&W; 1929: V; 1932: M; R; 1936: K&T; 1937: V
Os, Osmium	b1	1921: H; 1925: B&L; 1926: L&H; 1932: S; 1935: O,P&R; 1937: O&R
P, Phosphorus	k	1925: L&J; 1930: N&P; 1935: H,G&W; 1949: S,S&K; 1952: C&L
Pa, Protoactinium	e3	1952: Z; 1954: S,F,E&Z
Pb, Lead	a	1916: V; 1923: O&P; 1924: D; K; L; 1925: D; P&B; 1927: K&O; 1928: H&S; 1931: S&J; 1932: D; O&I; 1933: O; O&Y; 1934: O; 1939: H; 1941: S&W; 1946: K; 1953: NBS
Pd, Palladium	a	1920: H; 1921: H; 1922: MK; 1923: MK; 1925: B&L; D; G,B&L; 1931: S&W; 1932: O&I; 1933: O&Y; 1953: NBS
Pr, Praseodymium	a, b1, b2	1932: C&R; R; 1937: K&B; 1939: K&B

(*continued*)

BIBLIOGRAPHY TABLE, CHAPTER II (*continued*)

Element	Paragraph	Literature
Pt, Platinum	**a**	1920: H; 1921: H; K; 1924: D; 1925: B&L: D; G,B&L; 1928; vA; 1929: D&T; 1933; O&Y; 1934: O&Y; S; 1951; E,S&J; 1953: NBS
Pu, Plutonium	**a, c, e3, s**	1955: J; Z&E; 1956: E; 1958: B,G,M&R; Z&E; 1959: Z&E; 1961: M&L
Rb, Rubidium	**c**	1927: S&V; 1928: S&V; 1946: HR&L; 1956: B
Re, Rhenium	**b1**	1929: G; 1931: A,A,B,H&M; M; 1932: S&W; 1952: NBS
Rh, Rhodium	**a**	1921: H; 1925: B&L; G,B&L; 1928: vA; 1931: J&R; J&Z; 1932: O&I; 1933: O&Y; 1952: NBS
Ru, Ruthenium	**b1**	1920: H; 1921: H; 1925: B&L; 1926: L&H; 1935: O,P&R; 1937: O&R; 1955: NBS; 1957: R&P
S, Sulfur	**n1, n2**	1914: B; 1924: M&W; 1925: B&B; 1930: H&B; H&M; 1931: T&F; 1932: T&F; 1934: M&G; 1935: W&B; 1937: B; 1950: F&W; 1951: V; 1955: D; A; 1958: D,C&G; 1959: NBS; 1961: D,C&G; A
Sb, Antimony	**l**	1920: J&T; 1921: O; 1925: B&B; 1932: K; 1935: J&F; 1938: T&B; 1952: NBS; 1955: T
Sc, Scandium	**a, b1**	1939: M; 1956: S,D&H
Se, Selenium	**o1, o2, o3**	1923: S; 1924: B; 1925: vO; S; 1929: B; 1931: H,B&M; H,M&B; 1934: K; T; 1937: P&D; 1939: D&G; 1940: S; 1949: K; 1951: B; C,S&O; G; Y; 1952: B; 1953: M,P&MC; 1954: NBS
Si, Silicon	**i1**	1916: D&S; 1917: H; 1921: G; 1922: G; 1923: K&R; 1926: B; 1928: vA; 1929: N&C; 1935: J&F; N; 1944: L&R; 1947: H,K&K; 1952: S&A; 1953: H&J; 1955: S&K; 1956: N; 1959: G&W; V&K; 1960: O&I; 1961: S,B&J
Sm, Samarium	**h**	1953: D,D&S; E&Z; 1954: D,R,S&S
Sn, Tin	**i1, i4**	1918: B&K; 1919: B&K; 1923: vA; M&P; M,P&S; 1925: P&B; 1932: S&W; 1933: S; 1935: J&F; 1936: K&T; 1938: J,S&K; 1950: B; 1956: G&O; 1959: V&K

(continued)

BIBLIOGRAPHY TABLE, CHAPTER II *(continued)*

Element	Paragraph	Literature
Sr, Strontium	a, b1, c	1928: S&V; 1929: E&H; K; 1941: K&M; 1953: S&K
T_2, Tritium	d	1959: K,L&B
Ta, Tantalum	c	1920: H; 1921: H; 1932: O&I; Q; 1934: B&B; 1936: N; 1951: E,S&J; 1953: NBS; 1954: S
Tb, Terbium	b1	1937: K&B
Te, Tellurium	o1	1923: S; 1924: B; 1925: vO; S; 1940: S; 1951: G; 1953: NBS; 1954: Z
Th, Thorium	a, c	1920: B; 1921: H; 1954: C; 1956: J&S; 1959: E&R
Ti, Titanium	b1, c	1920: H; 1921: H; 1925: P; 1936: B&J; 1941: F,K&S; 1950: E&P; 1952: NBS; 1953: B&R; L; 1955: S; 1958: S,A&O; 1960: M&K
Tl, Thallium	b1, c	1924: L; 1926: L; S; T; 1927: B; 1928: P&W; 1929: S; 1931: S; 1941: L&S; 1958: B
Tm, Thulium	b1	1937: K&B
U, Uranium	c, q1, q2	1930: ML&MK; 1933: W; 1937: J&W; 1947: R; 1948: P&E; 1949: L; W&R; 1950: T, 1951: T; 1952: T; 1953: T&S; 1954: T&S; T; 1956: T,S,T&S; 1958: R,B&W; 1959: S; 1960: S&P; 1961: C,H&M
V, Vanadium	c	1922: H; 1936: N; 1961: J&S
W, Tungsten	c, p	1917: D; 1921: H; 1922: B; 1924: D; 1925: D; 1926: vA; B; C; D&W; 1932: O&I; 1933: N; 1934: N; S; 1935: J&F; 1937: M; 1944: P; R; 1953: NBS; 1959: V&K
Xe, Xenon	a	1930: K&M; N&N; 1932: R&S; 1957: C&S
Y, Yttrium	b1	1932: Q; 1939: B; 1956: S,D&H
Yb, Ytterbium	a	1937: K&B; 1953: D,D&S
Zn, Zinc	b1	1920: H; 1921: H; 1922: M,P&S; 1924: vS; 1925: P,A&vD; P&B; 1926: F,S&B; 1927: W&H; 1928: O&O; 1929: ML&M; 1932: B; S&W; 1933: O&I; O&Y; O&P; W; 1934: O&Y; 1935: B&S; J&F; O,P&R; 1936: B; 1937: W&H; 1953: NBS
Zr, Zirconium	b1, c	1921: H; 1925: N&T; 1927: vA; 1932: B; 1934: S; 1941: F,K&S; 1952: NBS; 1958 S,A&O; 1960: L

BIBLIOGRAPHY, CHAPTER II

1913

Bragg, W. H., and Bragg, W. L., "The Structure of the Diamond," *Nature*, **91**, 557; *Proc. Roy. Soc.* (*London*), **89A**, 277.

1914

Bragg, W. H., "X-Ray Spectra Given by Sulfur and Quartz," *Proc. Roy. Soc.* (*London*), **89A**, 575.

Bragg, W. L., "The Crystalline Structure of Copper," *Phil. Mag.*, **28**, 355.

Ewald, P. P., "The Symmetry of Graphite," *Sitz.-Ber. Math.-Phys. Kl. Bayer. Akad. Wiss. Muenchen*, p. 325.

1916

Debye, P., and Scherrer, P., "Interference of X-Rays Using Irregularly Oriented Substances," *Physik. Z.*, **17**, 277.

Vegard, L., "The Structure of Silver Crystals," *Phil. Mag.*, **31**, 83.

Vegard, L., "Results of Crystal Analysis," *Phil. Mag.*, **32**, 65.

1917

Debye, P., "The Atomic Structure of Tungsten," *Physik. Z.*, **18**, 483.

Debye, P., and Scherrer, P., "X-Ray Interference Produced by Irregularly Oriented Particles III; Constitution of Graphite and Amorphous Carbon," *Physik. Z.*, **18**, 291.

Hull, A. W., "The Crystal Structure of Magnesium," *Proc. Natl. Acad. Sci.*, **3**, 470.

Hull, A. W., "The Crystal Structure of Iron," *Phys. Rev.*, **9**, 84.

Hull, A. W., "The Crystal Structure of Al and Si," *Phys. Rev.*, **9**, 564.

Hull, A. W., "A New Method of X-Ray Crystal Analysis," *Phys. Rev.*, **10**, 661.

1918

Bijl, A. J., and Kolkmeijer, N. H., "The Crystal Structure of Gray Tin," *Chem. Weekblad*, **15**, 1264.

Scherrer, P., "The Crystal Structure of Aluminum," *Physik. Z.*, **19**, 23.

1919

Bijl, A. J., and Kolkmeijer, N. H., "Investigation by Means of X-Rays of the Crystal Structure of White and Gray Tin," *Proc. Acad. Sci. Amsterdam*, **21**, 405, 494, 501.

Hull, A. W., "The Crystal Structure of Ferromagnetic Metals," *Phys. Rev.*, **14**, 540.

1920

Bohlin, H., "New Method for the X-Ray Crystallography of Pulverized Substances," *Ann. Physik*, **61**, 421.

Hull, A. W., "Arrangement of the Atoms in Some Common Metals," *Science*, **52**, 227.

James, R. W., and Tunstall, N., "The Crystalline Structure of Antimony," *Phil. Mag.*, **40**, 233.

1921

Bijvoet, J. M., and Karssen, A., "Research by Means of X-Rays on the Structure of Crystals of Lithium and Some of its Compounds with Light Elements. I. Lithium Metal," *Proc. Acad. Sci. Amsterdam*, **23**, 1365; *Verslag Akad. Wetenschap. Amsterdam*, **29**, 1208.

Bragg, W. H., "Intensity of X-Ray Reflection by Diamond," *Proc. Phys. Soc.* (*London*), **33**, 304.

Gerlach, W., "Crystal Lattice Structure Investigations with X-Rays and a Simple X-Ray Tube," *Physik. Z.*, **22**, 557.

Hull, A. W., "The Crystal Structure of Calcium," *Phys. Rev.*, **17**, 42.

Hull, A. W., "X-Ray Crystal Analysis of Thirteen Common Metals," *Phys. Rev.*, **17**, 571.

Hull, A. W., "The Crystal Structures of Ti, Zr, Ce, Th and Os." *Phys Rev.*, **18**, 88.

James, R. W., "Crystalline Structure of Bismuth," *Phil. Mag.*, **42**, 193.

Kahler, H., "The Crystalline Structures of Sputtered and Evaporated Metallic Films," *Phys. Rev.*, **18**, 210.

Ogg, A., "The Crystalline Structure of Antimony and Bismuth," *Phil. Mag.*, **42**, 163.

Stoll, P., "X-Ray Research by the Debye-Scherrer Method; Length of the Side of the Elementary Lattice of Molybdenum," *Arch. Sci. Phys. Mat.*, **3**, 546.

Westgren, A., "X-Ray Spectrographic Investigations of Iron and Steel," *J. Iron Steel Inst.*, **103**, 303; *Engineering*, **111**, 727, 757.

Westgren, A., and Lindh, A. E., "The Crystal Structure of Iron and Steel," *Z. Physik. Chem.* (*Leipzig*), **98**, 181.

1922

Alsen, N., and Aminoff, G., "On the Structure of Crystallized Mercury," *Geol. Foren. Stockholm Forh.*, **44**, 124.

Asahara, G., "The Nature of Graphite and Amorphous Carbon," *Papers Japan. Inst. Phys. Chem. Res.*, **1**, 23; *Japan. J. Chem.*, **1**, 35.

Burger, H. C., "The Determination of the Density of Tungsten by Means of X-Rays," *Physica*, **2**, 114.

Burger, H. C., "The Structure of Tungsten," *Physik. Z.*, **23**, 14.

Gerlach, W., "The *K*-alpha Doublet, Including a New Determination of the Lattice Constants of Some Crystals," *Physik. Z.*, **23**, 114.

Hull, A. W., "The Crystal Structures of Vanadium, Germanium and Graphite," *Phys. Rev.*, **20**, 113.

Kirchner, F., "Experiments on Structure with X-Rays," *Ann. Physik*, **69**, 59.

Kolkmeijer, N. H., "The Crystal Structure of Germanium," *Proc. Acad. Sci. Amsterdam*, **25**, 125; *Verslag Akad. Wetenschap. Amsterdam*, **31**, 155.

McKeehan, L. W., "The Crystal Structure of Potassium," *Proc. Natl. Acad. Sci.*, **8**, 254.

McKeehan, L. W., "The Crystal Structures of Beryllium and Beryllium Oxide," *Proc. Natl. Acad. Sci.*, **8**, 270.

McKeehan, L. W., "The Crystal Structures of Silver–Palladium and Silver–Gold Alloys," *Phys. Rev.*, **19**, 537; **20**, 424.

McKeehan, L. W., "The Effect of Occluded Hydrogen Upon the Crystalline Space Lattice of Palladium," *Phys. Rev.*, **20**, 82.

McKeehan, L. W., and Cioffi, P. P., "The Crystal Structure of Mercury," *Phys. Rev.*, **19**, 444.

Mark, H., Polanyi, M., and Schmid, E., "Phenomena Accompanying the Stretching of Zinc Crystals," *Z. Physik*, **12**, 58, 78, 111.

Scherrer, P., and Stoll, P., "The Determination of the Structure of Inorganic Compounds Derived by Werner," *Z. Anorg. Chem.*, **121**, 319.

Tutton, A. E. H., "Ten Years of X-Ray Crystal Analysis," *Nature*, **110**, 47.

Westgren, A., and Phragmen, G., "X-Ray Studies on the Crystal Structure of Steel," *J. Iron Steel Inst.*, **105**, 241; *Engineering*, **113**, 630.

Westgren, A., and Phragmen, G., "The Crystal Structure of Iron and Steel," *Z. Physik. Chem.*, **102**, 1.

Wever, F., "Atomic Arrangement of Magnetic and Non-magnetic Nickel," *Mitt. Kaiser Wilhelm-Inst. Eisenforsch.*, **3**, No. 2, p. 17.

Wever, F., "The Nature of Graphite and Temper-Carbon," *Mitt. Kaiser Wilhelm-Inst. Eisenforsch.*, **4**, 81.

1923

Arkel, A. E. van, "On the Crystal Structure of White Tin," *Verslag Akad. Wetenschap. Amsterdam*, **32**, 197; *Proc. Acad. Sci. Amsterdam*, **27**, 97 (1924).

Bijvoet, J. M., "X-Ray Investigation of the Crystal Structure of Lithium and Lithium Hydride," *Rec. Trav. Chim.*, **42**, 859.

Küstner, H., and Remy, H., "The Structure of Silicon," *Physik. Z.*, **24**, 25.

McKeehan, L. W., "The Crystal Structures of the System Palladium–Hydrogen," *Phys. Rev.*, **21**, 334.

McKeehan, L. W., "The Crystal Structure of Iron-Nickel Alloys," *Phys. Rev.*, **21**, 402.

McKeehan, L. W., "The Crystal Structure of Bismuth," *J. Franklin Inst.*, **195**, 59.

Mark, H., and Polanyi, M., "The Space Lattice, Glide Directions and Glide Planes in White Tin, *Z. Physik*, **18**, 75; **22**, 200 (1924).

Mark, H., Polanyi, M., and Schmid, E., "Investigations of Unicrystalline Wires of Tin," *Naturwiss.*, **11**, 256.

Owen, E. A., and Preston, G. D., "Modification of the Powder Method of Determining the Structure of Metal Crystals," *Proc. Phys. Soc. (London)*, **35**, 101.

Owen, E. A., and Preston, G. D., "X-Ray Analysis of Solid Solutions," *Proc. Phys. Soc. (London)*, **36**, 14.

Owen, E. A., and Preston, G. D., "X-Ray Analysis of Zinc–Copper Alloys," *Proc. Phys. Soc. (London)*, **36**, 49.

Simon, F., and Simson, C. v., "The Crystal Structure of Argon," *Naturwiss.*, **11**, 1015.

Slattery, M. K., "The Crystal Structure of Metallic Selenium and Tellurium," *Phys. Rev.*, **21**, 378.

Westgren, A., and Phragmen, G., "X-Ray Investigations on the Crystal Structure of Steel," *Jernkontorets Ann.*, **1923**, 449.

Wyckoff, R. W. G., "The Crystal Structure of Metallic Iridium," *Z. Krist.*, **59**, 55.

Young, J. F. T., "The Crystal Structure of Various Heusler Alloys by the Use of X-Rays," *Phil. Mag.*, **46**, 291.

1924

Bernal, J. D., "The Structure of Graphite," *Proc. Roy. Soc. (London)*, **106A**, 749.

Bradley, A. J., "The Crystal Structure of Metallic Arsenic," *Phil. Mag.*, **47**, 657.

Bradley, A. J., "The Crystal Structures of the Rhombohedral Forms of Selenium and Tellurium," *Phil. Mag.*, **48**, 477.
Davey, W. P., "Precision Measurement of the Lattice Constants of Pure Metals," *Phys. Rev.*, **23**, 292.
Davey, W. P., "Application of X-Ray Crystal Analysis to Metallurgy," *Trans. Am. Soc. Steel Treating*, **6**, 375.
Hassel, O., and Mark, H., "The Structure of Bismuth," *Z. Physik*, **23**, 269.
Hassel, O., and Mark, H., "The Crystal Structure of Graphite," *Z. Physik*, **25**, 317.
Heindlhofer, K., "Crystal Structure of Hard Steel," *Phys. Rev.*, **24**, 426.
Jette, E. R., Phragmen, G., and Westgren, A., "X-Ray Studies on the Copper–Aluminum Alloys," *J. Inst. Metals*, **31**, 193.
Kolderup, N.-H., "The Crystal Structure of Lead, Galena, Lead Fluoride and Cadmium Fluoride," *Bergens Museums Aarbok, Naturvidensk. Raek*, No. 2.
Levi, G. R., "On the Structure of Lead and Thallium," *Nuovo Cimento*, **1**, 137.
Mark, H., and Wigner, E., "The Space Lattice of Rhombic Sulfur," *Z. Physik. Chem. (Leipzig)*, **111**, 398.
Patterson, R. A., "Crystal Analysis by the Diffraction of X-Rays," *Ind. Eng. Chem.*, **16**, 689.
Rinne, F., "On the Question of Permanent Structural Deformation of Graphite," *Central. Mineral. Geol.*, **1924**, 513.
Schumacher, E. E., and Lucas, F. F., "Photomicrographic Evidence of the Crystal Structure of Pure Cerium," *J. Am. Chem. Soc.*, **46**, 1167.
Simon, F., and Simson, C. v., "The Crystal Structure of Argon," *Z. Physik*, **25**, 160.
Simson, C. v., "X-Ray Investigation of Amalgams," *Z. Physik. Chem. (Leipzig)*, **109**, 183.
Westgren, A., and Phragmen, G., "X-Ray Studies on the Crystal Structure of Steel," *J. Iron Steel Inst.*, **109**, 159; *Nature*, **114**, 94.
Wever, F., "The Structure of Cubic Metals after Rolling," *Z. Physik*, **28**, 69.

1925

Barth, T., and Lunde, G., "The Influence of the Lanthanide Contraction on the Lattice Dimensions of Cubic Metals of the Platinum Group," *Norsk Geol. Tidsskr.*, **8**, 220; *Z. Physik. Chem. (Leipzig)*, **117**, 478.
Barth, T., and Lunde, G., "X-Ray Investigations on the Platinum Metals, Silver and Gold," *Norsk Geol. Tidsskr.*, **8**, 258; *Z. Physik. Chem. (Leipzig)*, **121**, 78 (1926).
Blake, F. C., "Precision X-Ray Measurements by the Powder Method," *Phys. Rev.*, **26**, 60.
Bradley, A. J., "The Allotropy of Manganese," *Phil. Mag.*, **50**, 1018.
Bragg, W. H., and Bragg, W. L., *X-Rays and Crystal Structure*, 5th Ed., G. Bell & Sons, Ltd., London.
Clark, G. L., Asbury, W. C., and Wick, R. M., "An Application of X-Ray Crystallometry to the Structure of Nickel Catalysts," *J. Am. Chem. Soc.*, **47**, 2661.
Clark, G. L., and Frölich, P. K., "X-Ray Investigation of Electrolytic Nickel," *Z. Elektrochem.*, **31**, 655.
Davey, W. P., "Precision Measurements of the Lattice Constants of Twelve Common Metals," *Phys. Rev.*, **25**, 753; *Z. Krist.*, **63**, 316 (1926).
Davey, W. P., "Lattice Parameter and Density of Pure Tungsten," *Phys. Rev.*, **26**, 736.
Ebert, F., "Abnormal Powder Photographs and the Structure of Graphite," thesis, Greifswald.

Goldschmidt, V. M., Barth, T., and Lunde, G., "Isomorphy and Polymorphy of the Sesquioxides. The Lanthanide Contraction and its Consequences," *Skrifter Norske Videnskaps-Akad. Oslo I: Mat.-Naturv. Kl.*, **1925**, No. 7.

Lange, H., "X-Ray Spectroscopic Studies of Several Metal Alloys by the Method of Seemann and Bohlin," *Ann. Physik*, **76**, 476.

Levi, G. R., and Tacchini, G., "The Non-Existence of Nickel Suboxide," *Gazz. Chim. Ital.*, **55**, 28.

Linck, G., and Jung, H., "X-Ray Investigation of Black (Metallic) Phosphorus," *Z. Anorg. Chem.*, **147**, 288.

Masumoto, H., "A New Allotropy of Cobalt," *Kinzoku no Kenku*, **2**, 877; *Trans. Am. Soc. Steel Treating*, **10**, 489 (1926); *J. Inst. Metals*, **37**, 377 (1927).

Noethling, W., and Tolksdorf, S., "The Crystal Structure of Hafnium," *Z. Krist.*, **62**, 255.

Olshausen, S. v., "Crystal Structure Studies using the Debye-Scherrer Method," *Z. Krist.*, **61**, 463.

Patterson, R. A., "Crystal Structure of Titanium and Chromium," *Phys. Rev.*, **25**, 581; **26**, 56.

Peirce, W. M., Anderson, E. A., and van Dyck, P., "An Investigation of the Alleged Allotropy of Zinc by X-Ray Analysis and a Redetermination of the Zinc Lattice," *J. Franklin Inst.*, **200**, 349.

Phebus, W. C., and Blake, F. C., "The X-Ray Analysis of Certain Alloys," *Phys. Rev.*, **25**, 107.

Slattery, M. K., "The Crystal Structure of Metallic Tellurium and Selenium and of Strontium and Barium Selenide," *Phys. Rev.*, **25**, 333.

Smedt, J. de, and Keesom, W. H., "The Crystal Structure of Argon; Investigation of the Structures of Nitrogen and Oxygen at Liquid Hydrogen Temperature," *Physica*, **5**, 344.

Westgren, A., and Phragmen, G., "X-Ray Analysis of Copper–Zinc, Silver–Zinc and Gold–Zinc Alloys," *Phil. Mag.*, **50**, 311.

Westgren, A., and Phragmen, G., "The Crystal Structure of Manganese," *Z. Physik*, **33**, 777.

1926

Arkel, A. E. van, "The Structure of Mixed Crystals," *Physica*, **6**, 64.

Becker, K., "X-Ray Method of Determining Coefficient of Expansion at High Temperatures," *Z. Physik*, **40**, 37.

Bradley, A. J., and Ollard, E. F., "Allotropy of Chromium," *Nature*, **117**, 122.

Chudoba, K., "A Probable Relation between Twinning and the Atomic Structure in Cubic Heteropolar Compounds," *Z. Krist.*, **65**, 133.

Davey, W. P., and Wilson, T. A., "Lattice Parameters and Densities of Cu, Ag and W," *Phys. Rev.*, **27**, 105.

Ehrenberg, W., "The Dimensions of the Diamond Lattice," *Z. Krist.*, **63**, 320.

Erdal, A., "Contributions to the Analysis of Mixed Crystals and Alloys," *Z. Krist.*, **65**, 69.

Freeman, J. R., Jr., Sillers, F., Jr., and Brandt, P. F., "Pure Zinc at Normal and Elevated Temperatures I and II," *Sci. Papers U. S. Bur. Std.*, **20**, 661, 686.

Goldschmidt, V. M., "The Laws of Crystal Chemistry," *Skrifter Norske Videnskaps-Akad. Oslo I. Mat.-Naturv. Kl.*, **1926**, No. 2; *Naturwiss.*, **14**, 477

Jaeger, F. M., Terpstra, P., and Westenbrink, H. G. K., "The Crystal Structure of Gallium," *Proc. Acad. Sci. Amsterdam*, **29**, 1193; *Verslag Akad. Wetenschap. Amsterdam*, **35**, 832; *Z. Krist.*, **66**, 195 (1927).

Jung, H., "An X-Ray Investigation of Copper, Silver, and Gold," *Z. Krist.*, **64**, 413.

Levi, G. R., "The Crystalline Structure of Thallium," *Nuovo Cimento*, **3**, 297.

Levi, G. R., and Haardt, R., "The Crystal Structures of Ruthenium and Osmium," *Rend. Inst. Lombardo Sci. Lettere*, **59**, 208; *Gazz. Chim. Ital.*, **56**, 369.

Mauguin, C., "The Structure of Graphite," *Bull. Soc. Franc. Mineral.*, **49**, 32.

Quilico, A., "X-Ray Examination of Metallic Hydrides; Hydrides of Copper," *Rend. Accad. Lincei*, **4**, 57.

Sasahara, T., "The Crystal Structure of α-Thallium," *Sci. Papers Inst. Phys. Chem. Res. (Tokyo)*, **5**, 82.

Terpstra, P., "The Crystal Structure of Thallium," *Z. Krist.*, **63**, 318.

Young, J., "The Crystal Structure of Meteoric Iron as Determined by X-Ray Analysis," *Proc. Roy. Soc. (London)*, **112A**, 630.

1927

Arkel, A. E. van, "The Atomic Volume of Zirconium and of Hafnium," *Z. Physik. Chem. (Leipzig)*, **130**, 100.

Becker, K., "The Crystal Structure of Thallium," *Z. Physik*, **42**, 479.

Bradley, A. J., and Thewlis, J., "The Crystal Structure of α-Manganese," *Proc. Roy. Soc. (London)*, **115A**, 456.

Ferrari, A., "The Crystal Structure of Iodine I," *Rend. Accad. Lincei*, **5**, 582.

Horovitz, K., "Investigation of Metal Films by X-Ray Analysis," *Phys. Rev.*, **29**, 352.

Keesom, W. H., and Onnes, H. K., "The Question of the Possibility of a Change of Allotropic State at the Point of Transition to the Super-Conducting State," *Arch. Neerl. Sci., IIIA*, **10**, 221.

Leonhardt, J., "Morphological and Structural Relations of Meteoric Irons in Connection with their Origin," *Fortschr. Mineral. Kryst. Petrog.*, **12**, 52.

McLennan, J. C., and Wilhelm, J. O., "The Crystal Structure of Solid Oxygen," *Phil. Mag.*, **3**, 383.

Sekito, S., "The Lattice Constant of Metallic Cobalt," *Sci. Rept. Tohoku Imp. Univ.*, **16**, 545.

Sillers, F., Jr., "The Crystal Structure of Electrodeposited Chromium," *Trans. Am. Electrochem. Soc.*, **52**, 301.

Simon, F., and Vohsen, E., "The Crystal Structure of Alkali Metals," *Naturwiss.*, **15**, 398.

Wilson, T. A., and Hoyt, S. L., "X-Ray Analysis of Plastic Deformation of Zinc," Am. Inst. Mining Met. Engrs. Tech. Pub. No. 25.

1928

Allard, G., "An Allotropic Form of Silver," *Compt. Rend.*, **187**, 223.

Arkel, A. E. van, "A Method of Increasing the Accuracy of Debye-Scherrer Photographs," *Z. Krist.*, **67**, 235.

Clark, G. L., King, A. J., and Hyde, J. F., "The Crystal Structures of the Alkaline Earth Metals," *Proc. Natl. Acad. Sci.*, **14**, 617.

Halla, F., and Staufer, R., "X-Ray Investigations in the System Lead–Thallium," *Z. Krist.*, **67**, 440; **68**, 299.

Harris, P. M., Mack, E., Jr., and Blake, F. C., "The Atomic Arrangement in the Crystal of Orthorhombic Iodine," *J. Am. Chem. Soc.*, **50**, 1583.

Osawa, A., and Ogawa, Y., "X-Ray Investigation of Iron and Zinc Alloys," *Z. Krist.*, **68**, 177; *Sci. Rept. Tohoku Imp. Univ.*, **18**, 165 (1929).

Ott, H., "The Crystal Structure of Graphite," *Ann. Physik,* **85**, 81.

Persson, E., and Westgren, A., "X-Ray Analysis of the Thallium–Antimony Alloys," *Z. Physik. Chem. (Leipzig)*, **136**, 208.

Posnjak, E., "The Crystal Structure of Potassium," *J. Phys. Chem.*, **32**, 354.

Preston, G. D., "The Crystal Structure of α-Manganese," *Phil. Mag.*, **5**, 1198.

Preston, G. D., "The Crystal Structure of β-Manganese," *Phil. Mag.*, **5**, 1207.

Runquist, A., Arnfelt, H., and Westgren, A., "X-Ray Analysis of the Copper–Magnesium Alloys," *Z. Anorg. Chem.*, **175**, 43.

Simon, F., and Vohsen, E., "The Crystal Structure Determination of the Alkali Metals and Strontium," *Z. Physik. Chem. (Leipzig)*, **133**, 165.

Terrey, H., and Wright, C. M., "The Crystal Structure of Mercury, Copper and Copper Amalgam," *Phil. Mag.*, **6**, 1055.

Wolf, M., "The Crystal Structure of Solid Mercury, *Nature,* **122**, 314; *Z. Physik,* **53**, 72 (1929).

1929

Alichanov, A. I., "X-Ray Investigation of Aluminum at High Temperatures," *Z. Metallk.*, **21**, 127; *Metals and Alloys,* **1**, 30.

Bach, R., "X-Ray Study of the Crystalline State of Iron," *Helv. Phys. Acta,* **2**, 95.

Briegleb, G., "The Dynamical Allotropic States of Selenium," *Naturwiss.*, **17**, 51.

Damianovitch, H., and Trillat, J.-J., "The Action of Helium on Platinum," *Compt. Rend.*, **188**, 991.

Ebert, F., and Hartmann, H., "The Crystal Structures of Strontium and Barium," *Z. Anorg. Chem.*, **179**, 418.

Goldschmidt, V. M., "The Crystal Structure of Rhenium," *Naturwiss.*, **17**, 134; *Z. Physik. Chem. (Leipzig)*, **2B**, 244.

Greenwood, G., "Fiber Texture in Nickel Wires," *Z. Krist.*, **72**, 309.

Horovitz, K., "The Crystal Structure of Solid Mercury," *Phys. Rev.*, **33**, 121.

James, R. W., Brindley, G. W., and Wood, R. G., "A Quantitative Study of the Reflection of X-Rays from Crystals of Aluminum," *Proc. Roy. Soc. (London)*, **125A**, 401.

King, A. J., "The Crystal Structure of Strontium," *Proc. Natl. Acad. Sci.*, **15**, 337; *see also* Simon, F., and Vohsen, E., *ibid.*, **15**, 695.

King, A. J., and Clark, G. L., "The Crystal Structure of Barium," *J. Am. Chem. Soc.*, **51**, 1709.

McLennan, J. C., and Monkman, R. J., "On the Thermal Expansion of Zinc and Cadmium Crystals and on the Crystal Structure of Erbium and Niobium," *Trans. Roy. Soc. Canada III*, **23**, 255.

Mayer, G., "The Lattice Constant of Pure α-Iron," *Z. Krist.*, **70**, 383.

Mazza, L., and Nasini, A. G., "The Crystal Structure of Nickel," *Phil. Mag.*, **7**, 301.

Mehl, R. F., and Barrett, C. S., "System Cadmium–Mercury," Am. Inst. Mining Met. Engrs. Tech. Pub. No. 225.

Nasini, A. G., and Cavallini, A., "Crystal Structure of Silicon," *Atti Congr. Naz. Chim. Pura Appl., 3° May 1929*, p. 463.

Persson, E., and Öhman, E., "A High-Temperature Modification of Manganese," *Nature*, **124**, 333.
Sekito, S., "On the Crystal Structure of Manganese," *Z. Krist.*, **72**, 406.
Sekito, S., "Crystal Structure of Thallium," *J. Study Metals*, **6**, 372; *J. Inst. Metals*, **42**, 516; *Z. Krist.*, **74**, 189 (1930).
Smedt, J. de, Keesom, W. H., and Mooy, H. H., "Crystal Analysis of Solid α-Nitrogen," *Proc. Acad. Sci. Amsterdam*, **32**, 745.
Thomson, G. P., "The Crystal Structure of Nickel Films," *Nature*, **123**, 912.
Valentiner, S., and Becker, G., "The Lattice Structure of Nickel," *Naturwiss.*, **17**, 639.
Vegard, L., "The Structure of Solid Nitrogen Stable Below 35.5°K.," *Z. Physik*, **58**, 497.
Vegard, L., "The Crystal Structure of Solid Nitrogen," *Naturwiss.*, **17**, 543, 672; *Nature*, **124**, 267, 337.
Vegard, L., "Gases Condensed to Solid by Extremely Low Temperature," *Ber. 18th Skand. Naturforsch. Copenhagen*, **1929**, p. 537.

1930

Adinolfi, E., "The Lattice Distance and the Reflecting Power for X-Rays of Bismuth Relative to the Cleavage Planes," *Rend. Accad. Sci. Napoli*, **36**, 69.
Halla, F., and Bosch, F. X., "X-Ray Investigation in the Sulfur–Selenium System I. The Rhombic Mixed Crystals of Sulfur and Selenium," *Z. Physik. Chem.*, **10B**, 149.
Halla, F., and Mehl, E., "Fiber Structure of Plastic Sulfur," *Sitz. Akad. Wiss. Wien, Math.-Naturw. Kl., 15 Mai 1930*, Akad. Anzeiger No. 13.
Hendricks, S. B., Jefferson, M. E., and Shultz, J. F., "The Transition Temperatures of Cobalt and of Nickel. Some Observations on the Oxides of Nickel," *Z. Krist.*, **73**, 376.
Keesom, W. H., and Mooy, H. H., "The Crystal Structure of Krypton," *Nature*, **125**, 889; *Proc. Acad. Sci. Amsterdam*, **33**, 447.
Keesom, W. H., and Mooy, H. H., "Atomic Diameters of the Rare Gases," *Nature*, **126**, 243.
McLennan, J. C., and McKay, R. W., "Crystal Structure of Uranium," *Trans. Roy. Soc. Can.*, **24**, Sect. 3, 1.
McLennan, J. C., and McKay, R. W., "Crystal Structure of Metallic Lanthanum," *Trans. Roy. Soc. Can.*, **24**, Sect. 3, 33.
Meisel, K., "Crystal Structure of Columbium," *Z. Anorg. Chem.*, **190**, 237.
Nasini, A., and Natta, G., "The Crystal Structure of the Inert Gases II. Krypton," *Rend. Accad. Lincei*, **12**, 141.
Natta, G., and Nasini, A. G., "The Crystal Structure of Xenon," *Nature*, **125**, 457.
Natta, G., and Nasini, A., "The Structure of the Inert Gases I. Xenon," *Rend. Accad. Lincei*, **11**, 1009.
Natta, G., and Nasini, A. G., "The Crystal Structure of Krypton," *Nature*, **125**, 889.
Natta, G., and Passerini, L., "The Crystal Structure of White Phosphorus," *Nature*, **125**, 707.
Öhman, E., "X-Ray Investigation of Manganese," *Svensk Kem. Tidskr.*, **42**, 210; *Metallwirtsch.*, **9**, 825.
Osawa, A., "X-Ray Investigation of Alloys of the Nickel–Cobalt and Iron–Cobalt Systems," *Sci. Rept. Tohoku Imp. Univ.*, **19**, 109.

Parravano, N., and Caglioti, V., "Investigation of the System: Bismuth–Selenium," *Gazz. Chim. Ital.*, **60**, 923.

Roberts, O. L., "X-Ray Study of Very Pure Iron," *Phys. Rev.*, **35**, 1426.

Sachs, G., and Weerts, J., "Lattice Constants of Gold–Silver Alloys," *Z. Physik*, **60**, 481.

Sasaki, K., and Sekito, S., "Three Crystalline Modifications of Electrolytic Chromium," *J. Soc. Chem. Ind.* (*Japan*), **33**, Suppl. No. 11, p. 482.

Sekito, S., "Crystal Structure of Thallium," *Z. Krist.*, **74**, 189.

Smedt, J. de, Keesom, W. H., and Mooy, H. H., "The Crystal Structure of Neon," *Proc. Acad. Sci. Amsterdam*, **33**, 255.

1931

Agte, C., Alterthum, H., Becker, K., Heyne, G., and Moers, K., "The Physical Properties of Rhenium," *Naturwiss.*, **19**, 108.

Agte, C., Alterthum, H., Becker, K., Heyne, G., and Moers, K., "Physical and Chemical Properties of Rhenium," *Z. Anorg. Chem.*, **196**, 129.

Alichanov, A. I., "The X-Ray Examination of Aluminum at High Temperatures," *J. Appl. Phys.* (*USSR*), **6**, 19; *Metals and Alloys*, **3**, Abstr. 285.

Arkel, A. E. van, and Burgers, W. G., "X-Rays Suitable for Determination of Small Changes in the Lattice Constant of α-Iron," *Z. Metallk.*, **23**, 149.

Bredig, G., and Bergkampf, E. S. v., "Hexagonal Nickel," *Z. Physik. Chem. Bodenstein-Festband*, 172.

Halla, F., Bosch, F. X., and Mehl, E., "X-Ray Studies in the Sulfur–Selenium System, II. The Space Lattice of Monoclinic Selenium (First Modification)," *Z. Physik. Chem.*, **11B**, 455.

Halla, F., Mehl, E., and Bosch, F. X., "X-Ray Studies in the Sulfur–Selenium System III. The Space Lattice of Mixed Crystals of the γ-Sulfur Type (Type A According to Groth)," *Z. Physik. Chem.*, **12B**, 377.

Hirata, M., "X-Ray Diffraction by Incandescent Carbon," *Sci. Papers Inst. Phys. Chem. Res.* (*Tokyo*), **15**, 219.

Jaeger, F. M., and Rosenbohm, E., "The Exact Measurement of the Specific Heat of Osmium and Rhodium between 0° and 1625°C., *Proc. Acad. Sci. Amsterdam*, **34**, 85.

Jaeger, F. M., and Zanstra, J. E., "The Allotropism of Rhodium and Some Phenomena Observed in the X-Ray Analysis of Heated Metal Wires," *Proc. Acad. Sci. Amsterdam*, **34**, 15.

Jenkins, C. H. M., and Preston, G. D., "Some Properties of Metallic Cadmium," *J. Inst. Metals*, **45**, 307.

Moeller, K., "The Lattice Constants of Rhenium," *Naturwiss.*, **19**, 575.

Neuberger, M. C., "Precision Measurements of the Lattice Constant of Columbium," *Z. Anorg. Chem.*, **197**, 219.

Neuberger, M. C., "The Density, Crystal Structure and Lattice Constant of Columbium," *Z. Krist.*, **78**, 164.

Preston, G. D., "X-Ray Examination of Chromium–Iron Alloys," *J. Iron Steel Inst.*, **124**, 139.

Sekito, S., "X-Ray Investigation of the Allotropic Transformations of Manganese, Thallium and Their Alloys," *Proc. World Engr. Congr. Tokyo 1929*, **36**, 139.

Solomon, D., and Jones, W. M., "An X-Ray Investigation of the Lead–Bismuth and Tin–Bismuth Alloys," *Phil. Mag.*, **11**, 1090.

Stenzel, W., and Weerts, J., "Lattice Constants of Silver–Palladium and Gold–Palladium Alloys," *Siebert Festschrift*, **1931**, 288.

Trillat, J.-J., and Forestier, J., "Study of the Structure of Plastic Sulfur," *Compt. Rend.*, **192**, 559.

1932

Boas, W., "Determination of the Solubility of Cadmium in Zinc by X-Rays," *Metallwirtschaft*, **11**, 603.

Bradley, A. J., and Jay, A. H., "A Method for Deducing Accurate Values of the Lattice Spacing from X-Ray Powder Photographs Taken by the Debye-Scherrer Method," *Proc. Phys. Soc. (London)*, **44**, 563.

Burgers, W. G., "The Crystal Structure of β-Zirconium," *Z. Anorg. Chem.*, **205**, 81.

Burgers, W. G., "Crystal Structure of β-Zirconium," *Nature*, **129**, 281.

Canneri, G., and Rossi, A., "The Preparation of Metallic Praseodymium," *Gazz. Chim. Ital.*, **62**, 1160.

Darbyshire, J. A., "An X-Ray Examination of the Oxides of Lead," *J. Chem. Soc.*, **1932**, 211.

Dwyer, F. P. J., and Mellor, D. P., "Crystal Structure of Indium," *J. Proc. Roy. Soc. N. S. Wales*, **66**, 234.

Goetz, A., and Hergenrother, R. C., "X-Ray Studies on Bismuth Single Crystals," *Phys. Rev.*, **40**, 137.

Hermann, C., and Ruhemann, M., "The Crystal Structure of Mercury," *Z. Krist.*, **83**, 136.

Kersten, H., "Influence of Hydrogen Ion Concentration on the Crystal Structure of Electrodeposited Cobalt," *Physics*, **2**, 274.

Kersten, H., "Influence of Temperature on the Crystal Structure of Electrodeposited Antimony," *Physics*, **2**, 276.

Laves, F., "Crystal Structure of Gallium," *Naturwiss.*, **20**, 472.

Mooy, H. H., "Preliminary Experiments with X-Rays on Oxygen, Acetylene and Ethylene in the Solid State," *Rapp. Commun. No. 24, Congr. Intern. Froid Buenos Aires, Commun. Kamerlingh Onnes Lab. Univ. Leiden* No. 223, p. 1.

Neuburger, M. C., "Precision Determination of Lattice Constants of Beryllium," *Z. Physik. Chem.*, **17B**, 285.

Nishiyama, Z., "On the Corrections for Debye-Scherrer X-Ray Photographs," *Sci. Rept. Tohoku Imp. Univ.*, **21**, 364.

Owen, E. A., and Iball, J., "Precision Measurements of the Crystal Parameters of Some of the Elements," *Phil. Mag.*, **13**, 1020.

Preston, G. D., "An X-Ray Examination of Iron-Chromium Alloys," *Phil. Mag.*, **13**, 419.

Quill, L. L., "The Crystal Structure of Yttrium," *Z. Anorg. Chem.*, **208**, 59.

Quill, L. L., "The Lattice Constants of Columbium, Tantalum and Several Columbates and Tantalates," *Z. Anorg. Chem.*, **208**, 257.

Quill, L. L., "X-Ray Investigation of Metallic Lanthanum, Cerium and Neodymium," *Z. Anorg. Chem.*, **208**, 273.

Rossi, A., "The Crystalline Structure of Praseodymium," *Rend. Accad. Lincei*, **15**, 298.

Ruhemann, M., "X-Ray Investigation of Solid Nitrogen and Oxygen," *Z. Physik*, **76**, 368.

Ruhemann, B., and Simon, F., "Crystal Structures of Krypton, Xenon, Hydrogen Iodide and Hydrogen Bromide in Relation to the Temperature," *Z. Physik. Chem.*, **15B**,389.

Stenzel, W., and Weerts, J., "Precision Determination of Lattice Constants of Non-Cubic Substances," *Z. Krist.*, **84**, 20.

Swjaginzeff, O. E., "Osmiridium I. (with Brunowski, B. K.), II. X-Ray Investigation," *Z. Krist.*, **83**, 172, 187.

Trillat, J.-J., and Forestier, J., "Some Physical Properties of Plastic Sulfur," *Bull. Soc. Chim. France*, **51**, 248.

Vegard, L., "Structure of β-Nitrogen and the Different Behavior of the Two Forms of Solid Nitrogen Regarding Phosphorescence," *Z. Physik*, **79**, 471.

1933

Ebert, F., Hartmann, H., and Peisker, H., "$\alpha \rightleftharpoons \beta$ Transition of Calcium," *Z. Anorg. Chem.*, **213**, 126.

Esser, H., and Mueller, G., "The Lattice Constants of Pure Iron and Iron–Carbon Alloys at Temperatures up to 1100°," *Arch. Eisenhuettenw.*, **7**, 265.

Graf, L., "X-Ray Examination of Calcium at Elevated Temperatures," *Metallwirtschaft*, **12**, 649.

Jaeger, F. M., and Zanstra, J. E., "The Allotropism of Beryllium," *Proc. Acad Sci. Amsterdam*, **36**, 636.

Laves, F., "The Crystal Structure and Morphology of Gallium," *Z. Krist.*, **84**, 256.

Neuburger, M. C., "The Crystal Structure and Lattice Constants of α-(β)-Tungsten," *Z. Krist.*, **85**, 232.

Neuburger, M. C., "Precision Measurements of the Lattice Constants of Beryllium," *Z. Krist.*, **85**, 325.

Neuburger, M. C., "Remarks on Crystal Grating Structure and Grating Constants of Mercury," *Z. Anorg. Chem.*, **212**, 40.

Obinata, I., "X-Ray Examination of Antimony–Lead and Tin–Lead Alloys," *Metallwirtschaft*, **12**, 101.

Obinata, I., and Wassermann, G., "X-Ray Study of the Solubility of Aluminum in Copper," *Naturwiss.*, **21**, 382.

Owen, E. A., and Iball, J., "Thermal Expansion of Zinc by the X-Ray Method," *Phil. Mag.*, **16**, 479.

Owen, E. A., and Pickup, L., "The Relation between Mean Atomic Volume and Composition in Copper–Zinc Alloys," *Proc. Roy. Soc.* (*London*), **140A**, 179.

Owen, E. A., and Yates, E. L., "Precision Measurements of Crystal Parameters," *Phil. Mag.*, **15**, 472.

Owen, E. A., and Yates, E. L., "Crystal Parameters of Four Metals when under Reduced Pressure," *Phil. Mag.*, **16**, 606.

Saini, H., "Thermal Expansion of Ag Measured by X-Rays," *Helv. Phys. Acta*, **6**, 597.

Schulze, A., "Investigations on the Supposed Allotropy of Aluminum," *Metallwirtschaft*, **12**, 667.

Shinoda, G., "X-Ray Investigations on the Thermal Expansions of Solids I," *Mem. Coll. Sci, Kyoto Imp. Univ.*, **16A**, 193.

Weigle, J., "Measurement of a Hexagonal Crystal Lattice: Zinc," *Helv. Phys. Acta*, **7**, 51.

Wilson, T. A., "Crystal Structure of Uranium," *Physics*, **4**, 148.

Zintl, E., and Neumayr, S., "Metals and Alloys VII. Lattice Structure of Indium," *Z. Elektrochem.*, **39**, 81.

Zintl, E., and Neumayr, S., "Metals and Alloys VIII. Crystal Structure of β-Lanthanum," *Z. Elektrochem.*, **39**, 84.

1934

Burgers, W. G., and Basart, J. C. M., "Preparation of Ductile Tantalum by Thermal Dissociation of $TaCl_5$," *Z. Anorg. Chem.*, **216**, 223.

Graf, L., "X-Ray Investigation of Calcium at High Temperatures II," *Physik. Z.*, **35**, 551.

Jesse, W. P., "X-Ray Crystal Measurements of Nickel at High Temperatures," *Physics*, **5**, 147.

Jette, E. R., Bruner, W. L., and Foote, F., "An X-Ray Study of the Gold–Iron Alloys," Am. Inst. Mining Met. Engrs., Inst. Metals Div., Tech. Publ. No. 526.

Jette, E. R., Nordstrom, V. H., Queneau, B., and Foote, F., "X-Ray Studies on the Nickel–Chromium System," Am. Inst. Mining Met. Engrs., Inst. Metals Div., Tech. Publ. No. 522.

Klug, H. P., "The X-Ray Study of Red Monoclinic Selenium. Proof of the Existence of Two Red Monoclinic Varieties of Selenium," *Z. Krist.*, **88**, 128.

Koester, W., and Schmidt, W., "Relation between Lattice Parameter and Ferromagnetism," *Arch. Eisenhuettenw.*, **8**, 25.

Meyer, K. H., and Go, Y., "Stretched Plastic Sulfur and its Structure," *Helv. Chim. Acta*, **17**, 1081.

Neuburger, M. C., "Precision Measurement of the Lattice Constants of Body-Centered Cubic β-Tungsten," *Z. Anorg. Chem.*, **217**, 154.

Ölander, A., "An Electrochemical and X-Ray Study of Solid Thallium–Lead Alloys, *Z. Physik. Chem.*, **168A**, 274.

Owen, E. A., and Pickup, L., "Parameter Values of Copper–Nickel Alloys, *Z. Krist.*, **88**, 116.

Owen, E. A., and Yates, E. L., "The Thermal Expansion of the Crystal Lattices of Silver, Platinum and Zinc," *Phil. Mag.*, **17**, 113.

Rossi, A., "Crystal Structure of Lanthanum, Cerium and Praseodymium Hydrides," *Nature*, **133**, 174.

Shinoda, G., "X-Ray Investigations on the Thermal Expansion of Solids," *Mem. Coll. Sci. Kyoto Imp. Univ.*, **17**, 27; *Proc. Phys. Math. Soc. Japan*, **16**, 436.

Tanaka, K., "The X-Ray Examination of Se Crystals," *Mem. Coll. Sci. Kyoto Imp. Univ.*, **17**, 59.

Vegard, L., "Structure of the β-Form of Solid Carbon Monoxide," *Z. Physik*, **88**, 235.

Vegard, L., and Kloster, A., "Gold–Copper Alloys Especially at High Temperatures," *Z. Krist.*, **89**, 560.

Willott, W. H., and Evans, E. J., "X-Ray Investigation of the Arsenic–Tin System of Alloys," *Phil. Mag.*, **18**, 114.

Wilson, T. A., "The Crystal Structure of beta Mn," *Bull. Am. Phys. Soc.*, **9**, 16.

1935

Bradley, A. J., "The Crystal Structure of Ga," *Z. Krist.*, **91A**, 302.

Brindley, G. W., and Spiers, F. W., "Atomic Scattering Factors of Nickel, Copper and Zinc," *Phil. Mag.*, **20**, 865.

Frevel, L. K., and Ott, E., "The X-Ray Study of In and the In–Ag System," *J. Am. Chem. Soc.*, **57**, 228.

Hultgren, R., Gingrich, N. S., and Warren, B. E., "The Atomic Distribution in Red P. Black P and the Crystal Structure of Black P," *Phys. Rev.*, **47**, 808; *J. Chem. Phys.*, **3**, 351.

Jette, E. R., and Foote, F., "Precision Determination of Lattice Constants," *J. Chem. Phys.*, **3**, 605.

Kossolapow, G. F., and Trapesnikow, A. K., "X-Ray Investigation of the Thermal Expansion of Cd," *Z. Krist.*, **91A**, 410.

Maxwell, L. R., Mosley, V. M., and Hendricks, S. B., "Electron Diffraction by Gas Molecules," *J. Chem. Phys.*, **3**, 698.

Neuburger, M. C., (1) "Lattice Constants and Allotropy of Be," *Z. Krist.*, **92A**, 474; (2) "Precision Measurement of the Lattice Constant of Si," *ibid.*, **92A**, 313.

Owen, E. A., and Pickup, L., "The Lattice Constants of Be," *Phil. Mag.*, **20**, 1155.

Owen, E. A., Pickup, L., and Roberts, J. O., "Lattice Constants of Five Elements Possessing Hexagonal Structure," *Z. Krist.*, **91A**, 70.

Warren, B. E., and Burwell, J. T., "The Structure of Rhombic Sulfur," *J. Chem. Phys.*, **3**, 6.

Wright, L., Hirst, H., and Riley, J., "The Structure of Electrolytic Cr," *Trans. Faraday Soc.*, **31**, 1253.

1936

Ackermann, P., and Mayer, J. E., "Determination of Molecular Structures by Electron Diffraction," *J. Chem. Phys.*, **4**, 377.

Ageev, N. W., and Ageeva, V., "Solid Solutions of Indium and Lead," *J. Inst. Metals*, **59**, Advance copy No. 735, 8 pp.

Brindley, G. W., (1) "X-Ray Examination of Atomic Vibrations in Zinc and Cadmium," *Nature*, **137**, 315; (2) "The Atomic Scattering Factor for Cu $K\alpha$ Radiation," *Proc. Leeds Phil. Lit. Soc., Sci. Sec.*, **3**, 200; (3) "X-Ray Investigations of Atomic Vibrations in Zinc," *Phil. Mag.*, **21**, 790.

Burgers, W. G., and Jacobs, F. M., "Crystal Structure of beta-Titanium," *Z. Krist.*, **94**, 299.

Hofmann, U., and Wilman, D., "On the Crystal Structure of Graphite," *Elektro-chem.*, **42**, 504.

Jette, E. R., and Foote, F., "X-Ray Study of Fe–Ni Alloys," Am. Inst. Mining Met. Engrs. Tech. Publ., No. 670; *Metals Tech.*, **13**, 1.

Jevins, A., and Straumanis, M., "The Lattice Constant of Pure Al," *Z. Physik. Chem.*, **33B**, 265; **34B**, 402.

Keesom, W. H., and Taconis, K. W., (1) "On the Crystal Structure of Cl," *Physica*, **3**, 237; *Proc. Roy. Acad. Amsterdam*, **39**, 314; (2) "On the Crystal Structure of Solid Oxygen," *Proc. Roy. Acad. Amsterdam*, **39**, 149; *Physica*, **3**, 141.

Kosolapov, G. F., and Trapeznikov, A. K., (1) "The Structure of Be," *J. Exptl. Theoret. Phys. USSR*, **6**, 1163; (2) "X-Ray Determination of the Thermal Expansion Coefficients of Be and Sn," *Z. Krist.*, **94A**, 53.

Maxwell, L. R., Hendricks, S. B., and Mosley, V. M., "The Structure of the Sulfur Molecule by Electron Diffraction," *Phys. Rev.*, **49**, 199.

Neuburger, M. C., (1) "Precision Measurement of the Lattice Constants of Very Pure Ta," *Z. Krist.*, **93A**, 312; (2) "Precision Measurement of the Lattice Constants of Pure Cd," *ibid.*, **93A**, 158; (3) "Precision Measurement of the Lattice Constants of Very Pure V," *ibid.*, **93A**, 314.

Owen, E. A., and Yates, E. L., "X-Ray Measurement of the Thermal Expansion of Pure Ni," *Phil. Mag.*, **21**, 809.

Rollier, M. A., "The Crystal Structure of Po by Electron Diffraction," *J. Chem. Phys.*, **4**, 648; *Chim. Ind.* (*Milan*), **18**, 205.

Vonnegut, B., and Warren, B. E., "The Structure of Crystalline Br," *J. Am. Chem. Soc.*, **58**, 2459.

Zener, C., "Theory of the Effect of Temperature on the Reflection of X-Rays by Crystals. II. Anisotropic Crystals," *Phys. Rev.*, **49**, 122.

1937

Bradley, A. J., Jay, A. H., and Taylor, A., "Lattice Spacing of Iron–Nickel Alloys," *Phil. Mag.*, **23**, 545.

Burwell, J. T., II, "The Unit Cell and Space Group of Monoclinic S," *Z. Krist.*, **97**, 123.

Hanawalt, J. D., and Frevel, L. K., "X-Ray Measurement of the Thermal Expansion of Mg," *Z. Krist.*, **98A**, 84.

Howe, J. D., and Lark-Horowitz, K., "Electron Diffraction Patterns of S Molecules and Se Molecules," *Phys. Rev.*, **51**, 380.

Jacob, C. W., and Warren, B. E., "The Crystalline Structure of U," *J. Am. Chem. Soc.*, **59**, 2588.

Klemm, W., and Bommer, H., "Metals of the Rare Earths," *Z. Anorg. Allgem. Chem.*, **231**, 138.

Maxwell, L. R., Hendricks, S. B., and Deming, L. S., "The Molecular Structure of P_4O_6, P_4O_8, P_4O_{10} and As_4O_6 by Electron Diffraction," *J. Chem. Phys.*, **5**, 626.

Meyer, W. F., "Investigations on Co and in the System Co–C," *Z. Krist.*, **97A**, 145.

Moeller, K., "Precision Measurements of Lattice Constants by the Powder Method," *Z. Krist.*, **97A**, 170.

Montoro, V., "Accurate Measurement of the Lattice Constant of Fe," *Met. Ital.*, **29**, 8.

Owen, E. A., and Roberts, E. W., "The Crystal Parameters of Os and Ru at Different Temperatures," *Z. Krist.*, **96A**, 497.

Prins, J. A., and Dekeyser, W., "X-Ray Study of Vitreous Se and its Crystallization," *Physica*, **4**, 900.

Trzebiatowski, W., "On the Structure of Graphite," *Roczniki Chem.*, **17**, 73.

Vegard, L., "Recent Cryogenic Work at Oslo," *Proc. 7th Intern. Cong. Refrigeration, The Hague, Amsterdam*, **1936**, **I**, No. 34, p. 311.

Wollan, E. O., and Harvey, G. G., "Effect of Temperature on the Intensity of Reflection of X-Rays from Zinc Crystals," *Phys. Rev.*, **51**, 1054.

Wood, W. A., "Lattice Dimensions of Electroplated and Normal Cr," *Phil. Mag.*, **23**, 984.

1938

Aruja, E., and Perlitz, H., "New Determination of the Lattice Constant of Na," *Z. Krist.*, **100**, 195.

Betteridge, W., "The Crystal Structure of Cd–In Alloys Rich in In," *Proc. Phys. Soc. (London)*, **50A**, 519.

Brindley, G. W., and Ridley, P., "An X-Ray Investigation of Atomic Vibrations in Mg between 86°K. and 293°K.," *Proc. Phys. Soc. (London)*, **50**, 757.

Brockway, L. O., "The Internuclear Distance in the F_2 Molecule," *J. Am. Chem. Soc.*, **60**, 1348.

Brockway, L. O., and Beach, J. Y., "The Electron Diffraction Investigation of the Molecular Structures of $POCl_3$, $POCl_2F$, $POClF_2$, POF_3, $POFCl_2$, PF_5, PF_3Cl_2, Si_2H_6, $SiHCl_3$, and Si_2Cl_6," *J. Am. Chem. Soc.*, **60**, 1836.

Hampson, P. C., and Stosick, A. J., "The Molecular Structure of As_4O_6, P_4O_6, P_4O_{10}, and $(CH_2)_6N_4$ by Electron Diffraction," *J. Am. Chem. Soc.*, **60**, 1814.

Hume-Rothery, W., and Reynolds, P. W., "A High Temperature Debye-Scherrer Camera and its Application to the Study of the Lattice Spacing of Ag," *Proc. Roy. Soc.* (*London*), **167A**, 25.

Jevins, A., Straumanis, M., and Karlsons, K., "Precision Determinations of the Lattice Constants of Non-Cubic Substances (Bi, Mg, Sn) by the Asymmetric Method," *Z. Physik. Chem.*, **40B**, 347.

Keesom, W. H., and Taconis, K. W., "On the Structure of Solid He," *Physica*, **5**, 161.

Palmer, K. J., "The Electron Diffraction Investigation of $SOCl_2$, SO_3, SO_2Cl_2, S_2Cl_2, SCl_2, $VOCl_3$ and CrO_2Cl_2," *J. Am. Chem. Soc.*, **60**, 2360.

Pauling, L., Laubengayer, A. W., and Hoard, J. L., "The Electron Diffraction Study of Ge_2H_6 and Ge_3H_8," *J. Am. Chem. Soc.*, **60**, 1605.

Stevenson, D. P., and Beach, J. Y., "The Electron Diffraction Investigation of the Molecular Structures of H_2S_2, $(CH_3)_2S_2$ and S_2Cl_2," *J. Am. Chem. Soc.*, **60**, 2872.

Taconis, K. W., "The Structure of Solid and Liquid He," *Ned. Tijdschr. Natuurk.*, **5**, 169.

Trzebiatowski, W., and Bryjak, E., "X-Ray Analysis of the System As–Sb," *Z. Anorg. Allgem. Chem.*, **238**, 255.

1939

Bommer, H., (1) "The Crystal Structure and Magnetic Properties of Metallic Ho," *Z. Anorg. Allgem. Chem.*, **242**, 277; (2) "Magnetochemical Investigations XXXIV. The Magnetic Properties of Sc, Y and La," *Z. Elektrochem.*, **45**, 357.

Brill, R., Grimm, H. G., Hermann, C., and Peters, C., "Application of X-Ray Fourier Analysis to Questions of Chemical Combination," *Ann. Physik*, **34**, 393.

Brindley, G. W., and Ridley, P., "An X-Ray Investigation of Atomic Vibrations in Cadmium," *Proc. Phys. Soc.* (*London*), **51**, 73.

Das, S. R., and Gupta, K. D., "Conversion of Vitreous and Monoclinic Se to the Hexagonal Modification," *Nature*, **143**, 165.

Fordham, S., and Khalsa, R. G., "Single Crystal Pd Films and the Interaction with Gases," *J. Chem. Soc.*, **1939**, 406.

Hayasi, M., "X-Ray Determination of the Solid Solubilities of Bi in Pb," *Nippon Kinzoku Gakkai-Si*, **3**, 123.

Klemm, W., and Bommer, H., "The Rare-Earth Metals," *Z. Anorg. Allgem. Chem.*, **241**, 264.

LeClerc, G., and Michel, A., "Hexagonal Ni," *Compt. Rend.*, **208**, 1583.

Maxwell, L. R., and Mosley, V. M., "Internuclear Distances in the Gas Molecules Se_2, HgCl, Cu_2Cl_2, Cu_2Br_2 and Cu_2I_2 by Electron Diffraction," *Phys. Rev.*, **55**, 238.

Meisel, K., "The Crystal Structure of Metallic Sc," *Naturwiss.*, **27**, 230.

Raynor, G. V., and Hume-Rothery, W., "A Technique for the X-Ray Powder Photography of Reactive Metals and Alloys with Special Reference to the Lattice Spacings of Mg at High Temperatures," *J. Inst. Metals*, **65**, 477.

Stöhr, H., "The Allotropy of As," *Z. Anorg. Allgem. Chem.*, **242**, 138.

Stosick, A. J., "The Electron Diffraction Investigation of Phosphorus Sulfoxide, $P_4S_4O_6$," *J. Am. Chem. Soc.*, **61**, 1130.

Wallbaum, H. J., "On Vanadium Silicide," *Z. Metallk.*, **31**, 362.

1940

Perlitz, H., and Aruja, E., "A Redetermination of the Crystal Structure of Li," *Phil. Mag.*, **30**, 55.

Söchtig, H., "Investigations of Pure Cr in the Region of Anomaly," *Ann. Physik*, **38,** 97.
Straumanis, M., "Lattice Constants and Coefficients of Expansion of Selenium and Tellurium," *Z. Krist.*, **102,** 432.
Taylor, A., and Laidler, D., "Anomalous Diffractions in the Hull-Debye-Scherrer Spectrum of Graphite," *Nature* **146,** 130.

1941

Bergen, H. van, "Precision Measurement of Lattice Constants with the Compensation Method II," *Ann. Physik*, **39,** 553.
Edwards, O. S., Lipson, H., and Wilson, A. J. C., "The Structure of Co," *Nature*, **148,** 165.
Fitzwilliam, J., Kaufmann, A. R., and Squire, C. F., "Magnetic and X-Ray Studies on Ti and Zr," *J. Chem. Phys.*, **9,** 678.
Klemm, W., and Mika, G., "Interactions of the Alkaline Earth Metals," *Z. Anorg. Allgem. Chem.*, **248,** 155.
Lipson, H., and Stokes, A. R., "The Structure of Tl," *Nature*, **148,** 437.
Lonsdale, K., and Smith, H., (1) "Diffuse X-Ray Diffraction from the Two Types of Diamonds," *Nature*, **148,** 112; (2) "X-Ray Diffuse Reflections from Li and Na in Relation to Elastic Anisotropy," *ibid.*, **148,** 628.
Stokes, A. R., and Wilson, A. J. C., "The Thermal Expansion of Pb from 0°C.–320°C.," *Proc. Phys. Soc.* (*London*), **53,** 658.
Taylor, A., "Study of C by the Powder Method," *J. Sci. Instr.*, **18,** 90.

1942

Edwards, O. S., and Lipson, H., "Imperfections in the Structure of Co," *Proc. Roy Soc.* (*London*), **180A,** 268.
Fricke, R., and Müller, H., "The Allotropic Transformation of Finely Divided Metals in Inert Substances," *Naturwiss.*, **30,** 439.
Hume-Rothery, W., and Andrews, K. W., "The Lattice Spacing and Thermal Expansion of Cu," *J. Inst. Metals*, **63,** Pt. 2, p. 19.
Lipson, H., and Stokes, A. R., (1) "A New Structure for Graphite," *Nature*, **149,** 328; (2) "The Structure of Graphite," *Proc. Roy. Soc.* (*London*), **181A,** 101.

1943

Bhagavantam, S., "Normal Oscillations of the Diamond Structure," *Proc. Indian Acad. Sci.*, **18A,** 251.
Edwards, O. S., and Lipson, H., "An X-Ray Study of the Transformation of Co," *J. Inst. Metals*, **69,** Pt. 4, 177.
Laubengayer, A. W., Hurd, D. T., Newkirk, A. E., and Hoard, J. L., "Boron I. Preparation and Properties of Pure Crystalline Boron," *J. Am. Chem. Soc.*, **65,** 1924.
Straumanis, M., and Sauka, J., "Lattice Constants and the Expansion Coefficient of Iodine," *Z. Physik. Chem.*, **53B,** 320.

1944

Colombani, A., "Properties of Ionoplastic Nickel," *Ann. Phys.*, **19,** 272.
Krishnan, R. S., "Thermal Expansion of Diamond," *Nature*, **154,** 486.
Lipson, H., and Rogers, L. E. R., "The Measurement of X-Ray Wave Lengths by the Powder Method. Cr $K\beta_1$ and Mn $K\beta_1$," *Phil. Mag.*, **35,** 544.
Petch, N. J., "α-Tungsten," *Nature*, **154,** 337.

Riley, D. P., "Lattice Constant of Diamond and the C–C Single Bond," *Nature*, **153**, 587.

Rooksby, H. P., "α-Tungsten," *Nature*, **154**, 337.

1945

Lonsdale, K., "Extra X-Ray Reflections from Diamonds," *Nature*, **155**, 572.

Nelson, J. B., and Riley, D. P., "The Thermal Expansion of Graphite from 15° to 800°. I. Experimental," *Proc. Phys. Soc.* (*London*), **57**, 477.

Potter, E. V., and Huber, R. W., "Effect of Hydrogen on the X-Ray Parameter and Structure of Electrolytic Manganese," *Phys. Rev.*, **68**, 24.

Raman, C. V., "Allotropic Modifications of Diamond," *Nature*, **156**, 22.

1946

Beamer, W. H., and Maxwell, C. R., "The Crystal Structure of Polonium," *J. Chem. Phys.*, **14**, 569.

Gibson, J., "Structure of Graphite," *Nature*, **158**, 752.

Hume-Rothery, W., and Lonsdale, K., "The Lattice Spacing of Rubidium Between −183° and +19°," *Phil. Mag.*, **36**, 842.

Klug, H. P., "A Redetermination of the Lattice Constant of Lead," *J. Am. Chem. Soc.*, **68**, 1493.

Lonsdale, K., and Hume-Rothery, W., "The Lattice Spacing of Lithium at 20° and −183°," *Phil. Mag.*, **36**, 799.

Rose, A. J., "X-Ray Spectra Using Strictly Monochromatic Radiation," *Compt. Rend.*, **222**, 805.

1947

Barrett, C. S., "A Low-Temperature Transformation in Lithium," *Phys. Rev.*, **72**, 245.

Cullity, B. D., "The Crystal Structure of Gold and Beryllium," *Metals Technol.*, **14**, No. 3; Am. Inst. Mining Met. Engrs., Inst. Metals Div., Tech. Publ. No. 2152, 5 pp.

Heyd, F., Khol, F., and Kochanovska, A., "Existence of a New Noncubic Form of Silicon," *Coll. Czechoslov. Chem. Commun.*, **12**, 502.

Hofer, L. J. E., and Peebles, W. C., "Preparation and X-Ray Diffraction Studies of a New Cobalt Carbide," *J. Am. Chem. Soc.*, **69**, 893.

Lonsdale, K., "Divergent-Beam X-Ray Photography of Crystals," *Trans. Roy. Soc.*, (*London*), **A240**, 219.

Mooney, R. C. L., "The Crystal Structure of Element 43," *Phys. Rev.*, **72**, 1269.

Nelson, J. B., and Riley, D. P., "'Structure of Graphite," *Nature*, **159**, 637.

Rooksby, H. P., and Stewart, E. G., "Structure of Graphite," *Nature*, **159**, 638.

Rundle, R. E., "The Structure of Uranium Hydride and Deutride," *J. Am. Chem. Soc.* **69**, 1719.

1948

Barrett, C. S., and Trautz, O. R., "Low Temperature Transformations in Lithium and Lithium-Magnesium Alloys," *Metals Technol.* **15**, No. 3; Am. Inst. Mining Met. Engrs., Inst. Metals Div., Tech. Bull. No. 2346, 23 pp.

Hauk, V., and Osswald, E., "The Effect of Temperature on X-Ray Measurements," *Z. Metallk.*, **39**, 190.

Mooney, R. C. L., "Crystal Structure of Element 43," *Acta Cryst.*, **1**, 161

Pauling, L., and Ewing, F. J., "The Structure of Uranium Hydride," *J. Am. Chem. Soc.*, **70**, 1660.
Thomas, D. E., "Precision Measurement of Crystal-Lattice Parameters," *J. Sci. Instr.*, **25**, 440.

1949

Beamer, W. H., and Maxwell, C. R., "X-Ray Studies and Crystal Structure (of Polonium)," *J. Chem. Phys.*, **17**, 1293.
Hoerni, J., "Diffraction of Electrons in Graphite," *Nature*, **164**, 1045.
Hoerni, J., and Weigle, J., "Structure of Graphite," *Nature*, **164**, 1088.
Krebs, H., "X-Ray Investigations of Black Selenium," *Z. Metallk.*, **40**, 29.
Lawson, A. W., and Tang, T. Y., "Concerning the High-Pressure Allotropic Modification of Cerium," *Phys. Rev.*, **76**, 301.
Lukesh, J. S., "Note on the Structure of Uranium," *Acta Cryst.*, **2**, 420.
Naray-Szabo, St. v., and Tobais, C. W., "X-Ray Powder Patterns of Boron-Coated Molybdenum and Tungsten Filaments," *J. Am. Chem. Soc.*, **71**, 1882.
Sugawara, T., Sakamoto, Y., and Kanda, E., "On the Transition of Yellow Phosphorus at Low Temperatures," *Sci. Rept. Res. Inst. Tokyo, 1A*, **29**, 153.
Wilson, A. S., and Rundle, R. E., "The Structures of Uranium Metal," *Acta Cryst.*, **2**, 126.
Zhmudskii, A. Z., "Accurate Measurement of Lattice Constants," *Zavodskaya Lab.*, **15**, 1055.

1950

Bacon, G. E., "Unit Cell Dimensions of Graphite," *Acta Cryst.*, **3**, 137.
Bacon, G. E., "A Note on the Rhombohedral Modification of Graphite," *Acta Cryst.*, **3**, 320.
Bacon, G. E., "X-Ray and Neutron Diffraction by Graphite Layers," *Nature*, **166**, 794.
Brownlee, L. D., "Lattice Constant of Gray Tin," *Nature*, **166**, 482.
Eppelsheimer, D. S., and Penman, R. R., "Accurate Determination of the Lattice of β-Titanium at 900°C.," *Nature*, **166**, 960.
Frondel, C., and Whitfield, R. E., "Crystallography of Rhombohedral Sulfur," *Acta Cryst.*, **3**, 242.
Lukesh, J. S., "On the Symmetry of Graphite," *Phys. Rev.*, **80**, 226.
Schuch, A. F., and Sturdivant, J. H., "The Structure of Cerium at the Temperature of Liquid Air," *J. Chem. Phys.*, **18**, 145.
Sidhu, S. S., and Henry, C. O., "Allotropy of Beryllium," *J. Appl. Phys.*, **21**, 1036.
Tucker, C. W., Jr., "The Crystal Structure of Metallic Uranium," *Trans. Am. Soc. Metals*, **42**, 762.

1951

Bacon, G. E., "Interlayer Spacing of Graphite," *Acta Cryst.*, **4**, 558.
Bacon, G. E., and Franklin, R. E., "The a-Dimension of Graphite," *Acta Cryst.*, **4**, 561.
Burbank, R. D., "Crystal Structure of α-Monoclinic Selenium," *Acta Cryst.*, **4**, 140.
Chihaya, T., Shiota, N., and Onozaki, C., "Fundamental Studies on Selenium, III. X-Ray Studies of the Growth of Selenium Crystals," *Nippon Kinzoku Gakkai-Shi*, **B15**, 86.
Edwards, J. W., Speiser, R., and Johnston, H. L., "High-Temperature Structure and Thermal Expansion of Some Metals Determined by X-Ray Diffraction. Platinum, Tantalum, Niobium and Molybdenum," *J. Appl. Phys.*, **22**, 424.

Grison, E., "Tellurium-Selenium Alloys," *J. Chem. Phys.*, **19**, 1109.

Hoard, J. L., Geller, S., and Hughes, R. E., "The Structure of Elementary Boron," *J. Am. Chem. Soc.*, **73**, 1892.

Lukesh, J. S., "The Symmetry of Graphite," *Phys. Rev.*, **84**, 1068.

Straumanis, M. E., and Aka, E. Z., "Precision Determination of Lattice Parameter, Coefficient of Thermal Expansion, and Atomic Weight of Carbon in Diamond," *J. Am. Chem. Soc.*, **73**, 5643.

Thewlis, J., "Structure of Uranium," *Nature*, **168**, 198.

Ventriglia, U., "Structure of Orthorhombic Sulfur," *Periodico Mineral.* (*Rome*), **20**, 237.

Yamamori, S., "Metallic Selenium Converted from the Vitreous State," *Nippon Kinzoku Gakkai-Shi*, **B15**, 274.

1952

Burbank, R. D., "The Crystal Structure of β-Monoclinic Selenium," *Acta Cryst.*, **5**, 236.

Collin, R. L., "The Crystal Structure of Solid Chlorine," *Acta Cryst.*, **5**, 431.

Corbridge, D. E. C., and Lowe, E. J., "Structure of White Phosphorus: Single-Crystal X-Ray Examination," *Nature*, **170**, 629.

Dreyfus-Alain, B., "Radio Crystallographic Study of the Hydrogenation of Lanthanum," *Compt. Rend.*, **235**, 540.

Hoerni, J., and Wooster, W. A., "The Diffuse X-Ray Reflection from Diamond," *Experientia*, **8**, 297.

Owen, E. A., "The Crystal Parameters of Beryllium," *Proc. Phys. Soc.* (*London*), **65A**, 294.

Sidhu, S. S., and McGuire, J. C., "An X-Ray Diffraction Study of the Hafnium–Hydrogen System," *J. Appl. Phys.*, **23**, 1257.

Straumanis, M. E., and Aka, E. Z., "Lattice Parameters, Coefficients of Thermal Expansion, and Atomic Weights of Purest Silicon and Germanium," *J. Appl. Phys.*, **23**, 330.

Thewlis, J., "X-Ray Powder Study of β-Uranium," *Acta Cryst.*, **5**, 790.

Tucker, C. W., Jr., "A Supplementary Note on the Crystal Structure of β-Uranium," *Acta Cryst.*, **5**, 389, 395.

Zachariasen, W. H., "Identification and Crystal Structure of Protoactinium Metal and of Protoactinium Monoxide," *Acta Cryst.*, **5**, 19.

Zachariasen, W. H., "The Crystal Structure of Neptunium Metal," *Acta Cryst.*, **5**, 660.

Zachariasen, W. H., "Crystal Structure Studies of Neptunium Metal at Elevated Temperatures," *Acta Cryst.*, **5**, 664.

1953

Banister, J. R., Legvold, S., and Spedding, F. H., "Structure of the Rare Earth Metals at Low Temperatures," U.S. At. Energy Comm. ISC-342, 44 pp.

Berry, R. L., and Raynor, G. V., "Lattice Spacings of Titanium at Elevated Temperatures," *Research* (*London*), **6**, 21 S.

Daane, A. H., Dennison, D. H., and Spedding, F. H., "The Preparation of Samarium and Ytterbium Metals," *J. Am. Chem. Soc.*, **75**, 2272.

Ellinger, F. H., and Zachariasen, W. H., "Crystal Structure of Samarium Metal and of Samarium Monoxide," *J. Am. Chem. Soc.*, **75**, 5650.

Hill, R. B., and Axon, H. J., "Lattice Spacings of Two Samples of Super-Purity Aluminum at 25°C.," *Research* (*London*), **6**, 23 S.

Hoch, M., and Johnston, H. L., "Formation and Crystal Structure of Solid Silicon Monoxide," *J. Am. Chem. Soc.*, **75**, 5224.

Jumpertz, E., Kircher, H., and Kleber, W., "Morphology and Structure of Diamond," *Naturwiss.*, **40**, 409.

Kitaigorodskii, A. I., Khotsyanova, T. L., and Struchkov, Y. T., "Crystal Structure of Iodine," *Zh. Fiz. Khim.*, **27**, 780.

Kochanovska, A., "Rhombohedral Modification of Graphite," *Czech. J. Phys.*, **3**, 193.

Levinger, B. W., "Lattice Parameter of β-Titanium at Room Temperature," *J. Metals, Trans. Sec.*, **5**, 195.

Marsh, R. E., Pauling, L., and McCullough, J. D., "The Crystal Structure of Selenium," *Acta Cryst.*, **6**, 71.

Sheldon, E. A., and King, A. J., "Structure of the Allotropic Forms of Strontium," *Acta Cryst.*, **6**, 100.

Tucker, C. W., Jr., and Senio, P., "An Improved Determination of the Crystal Structure of β-Uranium," *Acta Cryst.*, **6**, 753.

Ziegler, W. T., Young, R. A., and Floyd, A. L., Jr., "The Crystal Structure and Superconductivity of Lanthanum," *J. Am. Chem. Soc.*, **75**, 1215.

1954

Bacon, G. E., "The (*hki*0) Reflection of Graphite," *Acta Cryst.*, **7**, 359.

Bagnall, K. W., and D'Eye, R. W. M., "The Preparation of Polonium Metal and Polonium Dioxide," *J. Chem. Soc.*, **1954**, 4295.

Banister, J. R., Legvold, S., and Spedding, F. H., "Structure of Gadolinium, Dysprosium, and Erbium at Low Temperatures," *Phys. Rev.*, **94**, 1140.

Basinski, Z. S., and Christian, J. W., "A Pressurized High-Temperature Debye-Scherrer Camera, and its Use to Determine the Structures and Coefficients of Expansion of γ and δ Manganese," *Proc. Roy. Soc.* (*London*), **A223**, 554.

Batchelder, F. W. von, and Raeuchle, R. F., "Re-examination of the Symmetries of Iron and Nickel by the Powder Method," *Acta Cryst.*, **7**, 464.

Bublik, A. I., "Electron Diffraction Study of Thin Silver Films," *Dokl. Akad. Nauk SSSR*, **95**, 521.

Chiotti, P., "High-Temperature Crystal Structure of Thorium," *J. Electrochem. Soc.*, **101**, 567.

Daane, A. H., Rundle, R. E., Smith, H. G., and Spedding, F. H., "The Crystal Structure of Samarium," *Acta Cryst.*, **7**, 532.

Drain, J., Bridelle, R., and Michel, A., "Effect of Foreign Elements on the Allotropic Transformation of Cobalt," *Bull. Soc. Chim. France*, **1954**, 828.

Ebert, F., "Rapid Determination of Lattice Constants," *Z. Metallk.*, **45**, 436.

Pines, B. Y., and Kaluzhinova, N. V., "Exact Determination of Crystal Lattice Constants from X-Ray Pictures of Polycrystals," *Zh. Tekhn. Fiz.*, **24**, 320.

Schönberg, N., "An X-Ray Investigation of the Tantalum–Oxygen System," *Acta Chem. Scand.*, **8**, 240.

Sellers, P. A., Fried, S., Elson, R. E., and Zachariasen, W. H., "Preparation of Some Protoactinium Compounds and the Metal," *J. Am. Chem. Soc.*, **76**, 5935.

Thewlis, J., and Steeple, H., "The β-Uranium Structure," *Acta Cryst.*, **7**, 323.

Tucker, C. W., Jr., "The Status of the β-Uranium Structure," *Acta Cryst.*, **7**, 752.

Zorll, U., "Lattice Constants of Tellurium, Mercury Telluride and Mercury Selenide," *Z. Physik*, **138**, 167.

1955

Abrahams, S. C., "The Crystal and Molecular Structure of Orthorhombic Sulfur," *Acta Cryst.*, **8**, 661.

Baskin, Y., and Meyer, L., "Lattice Constants of Graphite at Low Temperatures," *Phys. Rev.*, **100**, 544.

Donnay, J. D. H., "The Lattice of Rhombohedral Sulfur," *Acta Cryst.*, **8**, 245.

Graham, J., Moore, A., and Raynor, G. V., "The Effect of Temperature on the Lattice Spacings of Indium," *J. Inst. Metals*, **84**, 86.

Jette, E. R., "Physical Properties of Plutonium Metal," *J. Chem. Phys.*, **23**, 365.

Koehler, W. C., and Wollan, E. O., "Neutron Diffraction by Metallic Erbium," *Phys. Rev.*, **97**, 1177.

Smakula, A., and Kalnajs, J., "Precision Determination of Lattice Constants with a Geiger-Counter X-Ray Diffractometer," *Phys. Rev.*, **99**, 1737.

Straumanis, M. E., and Weng, C. C., "The Precise Lattice Constant and the Expansion Coefficient of Chromium between 10 and 60°," *Acta Cryst.*, **8**, 367.

Szántó, I., "The Determination of High-Purity α-Titanium Lattice Parameters," *Acta Tech. Acad. Sci. Hung.*, **13**, 363.

Tatarinova, L. I., "Electron Diffraction Investigation of Crystalline Antimony," *Tr. Inst. Kristallogr., Akad. Nauk SSSR*, **1955**, No. 11, 101.

Zachariasen, W. H., and Ellinger, F. H., "Crystal Chemical Studies of the 5*f*-Series of Elements. XXIV. The Crystal Structure and Thermal Expansion of γ-Plutonium," *Acta Cryst.*, **8**, 431.

1956

Abrahamson, E. P., II, and Grant, N. J., "β-Chromium," *J. Metals*, **8**; *AIME Trans.*, **206**, 975.

Barrett, C. S., "Crystal Structure of Barium and Europium at 293, 78 and 5°K.," *J. Chem. Phys.*, **25**, 1123.

Barrett, C. S., "X-Ray Study of the Alkali Metals at Low Temperatures," *Acta Cryst.*, **9**, 671.

Blum, P., and Durif, A., "Comparison of Two Methods for the Determination of Lattice Constants with the Help of an X-Ray Diffractometer," *Acta Cryst.*, **9**, 829.

Collin, R. L., "The Crystal Structure of Solid Chlorine: Correction," *Acta Cryst.*, **9**, 537.

Ellinger, F. H., "Crystal Structure of δ' Plutonium and the Thermal Expansion Characteristics of δ, δ' and ϵ Plutonium," *J. Metals*, **8**; *AIME Trans.*, **206**, 1256.

Graf, P., Cunningham, B. B., Dauben, C. H., Wallmann, J. C., Templeton, D. H., and Ruben, H., "Crystal Structure and Magnetic Susceptibility of Americium Metal," *J. Am. Chem. Soc.*, **78**, 2340.

Gurevich, M. A., and Ormont, B. F., "Precision Determination of the Periods of Identity of Polycrystals in X-Ray Cameras with Back-Reflection Exposure of High Resolving Power," *Zh. Tekhn. Fiz.*, **26**, 1106.

James, W. J., and Straumanis, M. E., "Lattice Parameter and Coefficient of Thermal Expansion of Thorium," *Acta Cryst.*, **9**, 376.

Kogan, V. S., Lazarev, B. G., and Bulatova, R. F., "The Crystalline Structure of Hydrogen and Deuterium," *Zh. Eksperim. Teoret. Fiz.*, **31**, 541.

Melsert, H., Tiedema, T. J., and Burgers, W. G., "X-Ray Study of the Allotropic Modifications of Calcium Metal," *Acta Cryst.*, **9**, 525.

Neff, H., "Precision Measurements of Lattice Constants with the Geiger-Counter Diffraction Goniometer," *Z. Angew. Phys.*, **8**, 505.

Spedding, F. H., Daane, A. H., and Herrmann, K. W., "The Crystal Structures and Lattice Parameters of High-Purity Scandium, Yttrium and the Rare-Earth Metals," *Acta Cryst.*, **9**, 559.

Tucker, C. W., Jr., and Sénio, P., "Rapid Lattice-Parameter Measurements of Moderate Precision on Single Crystals," U.S. At. Energy Comm. Rept. AECU-3428, 5 pp.

Tucker, C. W., Jr., Sénio, P., Thewlis, J., and Steeple, H., "Joint Note on the β-Uranium Structure," *Acta Cryst.*, **9**, 472.

Weyerer, H., "Precision Determination of Lattice Constants," *Z. Angew. Phys.*, **8**, 202.

Weyerer, H., "Precision Measurements of Lattice Constants. II," *Z. Angew. Phys.*, **8**, 297.

1957

Aggarwal, P. S., and Goswami, A., "A New Phase Structure of Molybdenum," *Proc. Phys. Soc. (London)*, **70B**, 708.

Barrett, C. S., "The Structure of Mercury at Low Temperatures," *Acta Cryst.*, **10**, 58.

Cheesman, G. H., and Soane, C. M., "The Lattice Constants of the Inert Elements," *Proc. Phys. Soc. (London)*, **70B**, 700.

Covington, E. J., and Montgomery, D. J., "Lattice Constants of Separated Lithium Isotopes," *J. Chem. Phys.*, **27**, 1030.

Grassmann, P., "Properties of Liquid and Solid Helium," *Vierteljahrsschr. Naturforsch. Ges. Zurich*, **102**, No. 3, 61.

Rudnitskii, A. A., and Polyakova, R. S., "Physical Properties of Ruthenium," *Zh. Neorgan. Khim.*, **2**, 2758.

Sands, D. E., and Hoard, J. L., "Rhombohedral Elemental Boron," *J. Am. Chem. Soc.*, **79**, 5582.

Worsham, J. E., Jr., Wilkinson, M. K., and Shull, C. G., "Neutron-Diffraction Observations on the Palladium–Hydrogen and Palladium–Deuterium Systems," *J. Phys. Chem. Solids*, **3**, 303.

Zachariasen, W. H., and Ellinger, F., "Crystal Structure of α-Plutonium Metal," *J. Chem. Phys.*, **27**, 811.

1958

Anatharaman, T. R., "Lattice Parameters and Crystallographic Angles of Hexagonal Cobalt," *Current Sci. (India)*, **27**, 51.

Ball, J. G., Greenfield, P., Mardon, P. G., and Robertson, J. A. L., "Crystal Structure of Plutonium, δ and ϵ Phases," At. Energy Research Estab. (Gt. Brit.) M/R 2416, 11 pp.

Barrett, C. S., "Structure of Thallium and Gadolinium at Low Temperatures," *Phys. Rev.*, **110**, 1071.

Donohue, J., Caron, A., and Goldish, E., "Crystal Structure of Rhombohedral Sulfur," *Nature*, **182**, 518.

Henshaw, D. G., "Structure of Solid Helium by Neutron Diffraction," *Phys. Rev.*, **109**, 328.

Henshaw, D. G., "Atomic Distribution in Liquid and Solid Neon and Solid Argon by Neutron Diffraction," *Phys. Rev.*, **111**, 1470.

Hoard, J. L., Hughes, R. E., and Sands, D. E., "The Structure of Tetragonal Boron," *J. Am. Chem. Soc.*, **80**, 4507.

McCarty, L. V., Kasper, J. S., Horn, F. H., Decker, B. F., and Newkirk, A. E., "A New Crystalline Modification of Boron," *J. Am. Chem. Soc.*, **80**, 2592.

Mack, G., "Precision Measurements of the Lattice Constants of Germanium Single Crystals by the Method of Kossel and van Bergen," *Z. Physik*, **152**, 19.

McKay, H. A. C., et al., "Chemistry and Metallurgy of Neptunium," *Proc. U.N. Intern. Conf. Peaceful Uses At. Energy, 2nd, Geneva, 1958*, **28**, 299.

Rundle, R. E., Baenziger, N. C., and Wilson, A. S., "X-Ray Study of the Uranium–Oxygen System," U. S. At. Energy Comm. TID-5290, Bk. 1, 131.

Schuch, A. F., Grilly, E. R., and Mills, R. L., "Structure of the α and β Forms of Solid Helium-3," *Phys. Rev.*, **110**, 775.

Sof'ina, V. V., Azarkh, Z. M., and Orlova, N. N., "X-Ray Phase Analysis of the Zirconium–Hydrogen and Titanium–Hydrogen Systems," *Kristallografiya*, **3**, 539.

1959

Atoji, M., Schirber, J. E., and Swenson, C. A., "Crystal Structure of β-Hg," *J. Chem. Phys.*, **31**, 1628.

Bolz, L. H., Boyd, M. E., Mauer, F. A., and Peiser, H. S., "A Reexamination of the Crystal Structure of α- and β-Nitrogen," *Acta Cryst.*, **12**, 247.

Decker, B. F., and Kasper, J. S., "The Crystal Structure of a Simple Rhombohedral Form of Boron," *Acta Cryst.*, **12**, 503.

Donohue, J., "Crystal Structure of Helium Isotopes," *Phys. Rev.*, **114**, 1009.

Evans, D. S., and Raynor, G. V., "The Lattice Spacing of Thorium with Reference to Contamination," *J. Nuclear Materials*, **1**, 281.

Göttlicher, S., and Wölfel, E., "X-Ray Determination of Electron Distribution in Crystals. VII. The Electron Density in the Diamond and Silicon Lattices," *Z. Elektrochem.*, **63**, 891.

Hawes, L. L., "The Thermal Expansion of Solid Bromine," *Acta Cryst.*, **12**, 34.

Johan, Z., "Arsenolamprite: The Orthorhombic Modification of Arsenic from Černý Důl (Schwarzental) in the Riesengebirge," *Chem. Erde*, **20**, 71.

Kogan, V. S., Lazarev, B. G., and Bulatova, R. F., "X-Ray Diffraction in Polycrystalline Samples of Hydrogen Isotopes," *Zh. Eksperim. Teoret. Fiz.*, **37**, 678.

Schwarzenberger, D. R., "Accurate Determination of the Lattice Parameters of Beryllium," *Phil. Mag.*, [8], **4**, 1242.

Sturcken, E. F., "Determination of Theoretical Diffraction Intensities for α-Uranium," U.S. At. Energy Comm. NLCO-804, 49.

Talley, C. P., Post, B., and LaPlaca, S., "A New Modification of Elemental Boron," *Boron Syn., Struct., Properties, Proc. Conf. Asbury Park, N.J.* **1959**, p. 83 (pub. 1960).

Vogel, R. E., and Kempter, C. P., "Mathematical Technique for the Precision Determination of Lattice Constants," U.S. At. Energy Comm. LA-2317, 30 pp.

Zachariasen, W. H., and Ellinger, F. H., "Unit Cell and Thermal Expansion of β-Plutonium Metal," *Acta Cryst.*, **12**, 175.

1960

Barrett, C. S., "The Structure of Bismuth at Low Temperatures," *Australian J. Phys.*, **13**, 209.

Becher, H. J., "Beryllium Boride, BeB_{12}, with β-Boron Structure," *Z. Anorg. Allgem. Chem.*, **306**, 266.

Hoard, J. L., and Newkirk, A. E., "An Analysis of Polymorphism in Boron Based upon X-Ray Diffraction Results," *J. Am. Chem. Soc.*, **82**, 70.

Lichter, B. D., "Precision Lattice-Parameter Determination of Zirconium–Oxygen Solid Solutions," *Trans. AIME*, **218**, 1015.

McWhan, D. B., Wallmann, J. C., Cunningham, B. B., Asprey, L. B., Ellinger, F. H., and Zachariasen, W. H., "Preparation and Crystal Structure of Americium Metal," *J. Inorg. Nuclear Chem.*, **15**, 185.

Makarov, E. S., and Kuznetsov, L. M., "Crystal Structure and Chemical Nature of the Lower Oxides of Titanium $TiO_{0 \rightarrow 0.48}$," *Zh. Strukt. Khim.*, **1**, No. 2, 170.

Ozolins, G., and Ievins, A., "The Use of Extrapolation in the Pictures from the Asymmetric method. Determination of the Lattice Constant of Silicon," *Latvijas PSR Zinatnu Akad. Vestis*, **1960**, No. 12, 61.

Straumanis, M. E., and Ejima, T., "Perfection of the Aluminum Lattice," *Z. Physik. Chem. (Frankfurt)*, **23**, 440.

Sturcken, E. F., and Post, B., "The Atomic Position Parameter in α-Uranium," *Acta Cryst.*, **13**, 852.

Talley, C. P., LaPlaca, S., and Post, B., "A New Polymorph of Boron," *Acta Cryst.*, **13**, 271.

1961

Abrahams, S. C., "Scale Factors, Form Factors and Bond Lengths in Orthorhombic Sulfur," *Acta Cryst.*, **14**, 311.

Caron, A., and Donohue, J., "The Lattice Constants of Orthorhombic Sulfur and Revision of the Interatomic Distances," *Acta Cryst.*, **14**, 548.

Cash, A. H., Hughes, E. W., and Murdock, C. C., "The Structure of Crystalline Uranium," *Acta Cryst.*, **14**, 313.

Curien, H., Rimsky, A., and Defrain, A., "Atomic Structure of a Crystalline Phase of Gallium, Unstable at Atmospheric Pressure," *Bull. Soc. Franc. Mineral. Crist.*, **84**, 260.

Donohue, J., Caron, A., and Goldish, E., "The Crystal and Molecular Structure of S_6 (Sulfur-6)," *J. Am. Chem. Soc.*, **83**, 3748.

Farr, J. D., Giorgi, A. L., Bowman, M. G., and Money, R. K., "Crystal Structure of Actinium Metal and Actinium Hydride," *J. Inorg. Nuclear Chem.*, **18**, 42.

James, W. J., and Straumanis, M. E., "Lattice Parameter and Coefficient of Thermal Expansion of Vanadium," *Z. Physik. Chem. (Frankfurt)*, **29**, 134.

Kogan, V. S., Lazarev, B. G., and Bulatova, R. F., "Difference in the Lattice Constants of Neon Isotopes," *Zh. Eksperim. Teoret. Fiz.*, **40**, 29.

McWhan, D. B., "Crystal Structure and Physical Properties of Americium Metal," U.S. At. Energy Comm. UCRL-9695.

Marples, J. A. C., and Lee, J. A., "Structure of δ-Plutonium," *Nature*, **191**, 1391.

Mills, R. L., and Schuch, A. F., "Crystal Structure of the β-Form of He^4," *Phys. Rev. Letters*, **6**, 263.

Schuch, A. F., and Mills, R. L., "New Allotropic Form of He^3," *Phys. Rev. Letters*, **6**, 596.

Straumanis, M. E., Borgeaud, P., and James, W. J., "Perfection of the Lattice of Dislocation-Free Silicon Studied by the Lattice Constant and Density methods," *J. Appl. Phys.*, **32**, 1382.

Vasyutinskii, B. M., Kartmazov, G. N., and Finkel, V. A., "Structure of Chromium at 700–1700°," *Fiz. Metal. Metalloved., Akad. Nauk SSSR*, **12**, 771.

Chapter III

STRUCTURES OF THE COMPOUNDS RX

Structures have now been determined for more than 400 compounds RX. Most belong to one or another of a few simple types which thus provide the natural basis for their description. Some of these, as exemplified by NaCl and CsCl, are strongly ionic in the type of chemical bonding that prevails; others, such as NiAs, are essentially homopolar in bonding and often exhibit chainlike or sheetlike structures. A few are composed of well-defined molecules and some show more or less clearly the influence of hydrogen bonding in holding the crystals together. Advantage has been taken of such relationships to group together the various structures for description. Nevertheless, a steadily growing number among even these simple RX compounds are being found to give rather complicated structures. These do not fit readily into the kind of classification used here, but thus far no more rational alternative has been apparent. Consequently it may be necessary to refer to the bibliographic master table at the end of the chapter or to the index at the end of the volume to find a desired compound.

NaCl-Like Structures

III,a1. The largest group of RX-type crystals have the structure of *sodium chloride*, NaCl. The unit cube of this arrangement contains four molecules with atoms in the positions:

R: (4*a*) 000; $\frac{1}{2}\frac{1}{2}0$; $\frac{1}{2}0\frac{1}{2}$; $0\frac{1}{2}\frac{1}{2}$, or 000; F.C.

X: (4*b*) $\frac{1}{2}\frac{1}{2}\frac{1}{2}$; $\frac{1}{2}00$; $0\frac{1}{2}0$; $00\frac{1}{2}$, or $\frac{1}{2}\frac{1}{2}\frac{1}{2}$; F.C.

As Figure III,1 indicates, each R atom has six equidistant X atoms as nearest neighbors, and vice versa. Since these neighbors are at the corners of a regular octahedron, the association is customarily said to involve an octahedral coordination.

Compounds of the following chemical types crystallize with the NaCl arrangement:

A. Alkali and silver halides and alkali pseudohalides, meaning by pseudohalides such radicals as the cyanide, CN, and the hydrosulfide, SH, ions.

B. Alkaline earth oxides, sulfides, selenides and tellurides, together with similar compounds involving such non-rare gas shell metals as iron and cadmium.

C. Nitrides, phosphides and other fifth column binary compounds with trivalent metals, chiefly rare earths.

D. Carbides of various metals.

The lengths of edge of the unit cells of these substances are listed in Table III,1.

TABLE III,1
Crystals with the NaCl Arrangement (**III,a1**)

Crystal	a_0, A.
AgBr	5.7745
AgCl	5.547
AgF	4.92
AmO	5.05
BaO	5.523
BaNH	5.84
BaS	6.3875 (21°C.)
BaSe	6.600
BaTe	6.986
BiSe	5.99
BiTe	6.47
CaO	4.8105
CaNH	5.006
CaS	5.6903 (21°C.)
CaSe	5.91
CaTe	6.345
CdO	4.6953 (27°C.)
CeAs	6.060
CeBi	6.487
CeN	5.011
CeP	5.897
CeS	5.778
CeSb	6.399
CeSe	5.982
CeTe	6.346
CoO	4.2667 (22°C.)
CrN	4.140
CsCl	7.02 (450°C.)
CsF	6.008
CsH	6.376
DyAs	5.780
DyN	4.905
DySb	6.153

(continued)

TABLE III,1 (*continued*)

Crystal	a_0, A.
DyTe	6.092
ErAs	5.732
ErN	4.839
ErSb	6.106
ErTe	6.063
EuN	5.014
EuS	5.957
EuO	5.1439
EuSe	6.191
EuTe	6.584
FeO	4.2774 (on Fe_3O_4 surface)
FeO	4.3108 (on Fe boundary)
GdAs	5.854
GdN	4.999
GdSb	6.217
GdSe	5.771
HfC	4.4578
HoAs	5.771
HoBi	6.228
HoN	4.874
HoP	5.626
HoS	5.465
HoSb	6.130
HoSe	5.680
HoTe	6.049
KBr	6.6000 (25°C.)
KCN[a]	6.527 (25°C.)
KCl	6.29294 (25°C.)
KF	5.347
KH	5.700
KI	7.06555 (25°C.)
KOH	5.78 (130°C.)
KSH	6.60 (170°C.)
KSeH	6.92 (180°C.)
LaAs	6.125
LaBi	6.565
LaN	5.301
LaP	6.013
LaS	5.842
LaSb	6.475

(*continued*)

TABLE III,1 (*continued*)

Crystal	a_0, A
LaSe	6.060
LaTe	6.409
LiBr	5.5013 (26°C.)
LiCl	5.12954
LiD	4.065
LiF	4.0173
Li^6F	4.0271
Li^7F	4.0263
LiH	4.085
LiI	6.000
LuN	4.766
MgO	4.2112 (21°C.)
MgS	5.2033 (21°C.)
MgSe	5.451
MnO	4.4448 (26°C.)
MnS	5.2236 (26°C.)
MnSe	5.448
NH_4Br	6.90 (250°C.)
NH_4Cl	6.52 (250°C.)
NH_4I	7.2613 (26°C.)
NaBr	5.97324 (26°C.)
NaCN	5.893
NaCl	5.62779 (18°C.)
	5.63978 (18°C.)
	5.64056 (26°C.)
NaF	4.620
NaH	4.880
NaI	6.4728 (26°C.)
NaSH	6.06 (150°C.)
NaSeH	6.30 (150°C.)
NbC	4.4691 (20°C.)
$NbC_{0.7}$	4.432
$NbN_{0.98}$	4.702 (25°C.)
NbO	4.2097
NdAs	5.958
NdBi	6.424
NdN	5.151
NdP	5.826
NdS	5.681

(*continued*)

TABLE III,1 (*continued*)

Crystal	a_0, A.
NdSb	6.309
NdSe	5.879
NdTe	6.249
NiO	4.1684
NpN	4.897
NpO	5.01
PaO	4.961
PbS	5.9362 (26°C.)
PbSe	6.1243 (26°C.)
PbTe	6.454
PdH[b]	4.02
PrAs	5.997
PrBi	6.448
PrN	5.155
PrP	5.860
PrS	5.727
PrSb	6.353
PrSe	5.952
PrTe	6.309
PuAs	5.855
PuB	4.92
PuC	4.920
PuN	4.905
PuO	4.959
PuP	5.644
PuS	5.536
PuTe	6.183
RbBr	6.854
RbCN	6.82
RbCl	6.5810 (27°C.)
RbF	5.64
RbH	6.037
RbI	7.342 (27°C.)
$RbNH_2$	6.395 (50°C.)
RbSH	6.93 (200°C.)
RbSeH	7.21 (180°C.)
ScAs	5.487
ScN	4.44
ScSb	5.859

(*continued*)

TABLE III,1 (*continued*)

Crystal	a_0, A.
SmAs	5.921
SmBi	6.362
SmN	5.0481
SmO	4.9883
SmP	5.760
SmS	5.970
SmSb	6.271
SmSe	6.200
SmTe	6.594
SnAs	5.681
SnSb	6.130
SnSe	6.020
SnTe	6.313
SrNH	5.45
SrO	5.1602 (25°C.)
SrS	6.0198 (20°C.)
SrSe	6.23
SrTe	6.47
TaC	4.4540 (23°C.)
TaO	4.422–4.439
TbAs	5.827
TbBi	6.280
TbN	4.933
TbP	5.688
TbS	5.516
TbSb	6.181
TbSe	5.741
TbTe	6.102
ThAs	5.972
ThC	5.338
ThP	5.818
ThS	5.682
ThSb	6.318
ThSe	5.875
TiC	4.3186
TiN	4.235
TiO[c]	4.1766 (25°C.)
TmAs	5.711

(*continued*)

TABLE III,1 (*continued*)

Crystal	a_0, A.
TmN	4.809
TmSb	6.083
TmTe	6.042
UAs	5.766
UBi	6.364
UC	4.9591 (24°C.)
UN	4.884
UO	4.92
UP	5.589
US	5.484
USb	6.191
USe	5.750
UTe	6.163
VC	4.182
$VC_{0.75}$	4.136
VN	4.128
VO[c]	4.062 (800°C.)
YAs	5.786
YN	4.877 (23°C.)
$YSb_{1.10}$	5.658
YTe	6.095
YbAs	5.698
YbN	4.7852
YbO	4.86
YbSb	5.922
YbSe	5.867
YbTe	6.353
ZrB	4.65
ZrC	4.6828 (23°C.)
ZrN	4.61
ZrO	4.62
ZrP	5.27
ZrS	5.250

[a] Recent neutron diffraction data on KCN support the idea that the CN radicals are freely rotating.

[b] The beta phase of the Pd–H system has, by neutron diffraction, an incomplete NaCl structure with up to 70% of the H, or D, positions filled.

[c] a_0 varies with composition between $TiO_{0.86}$ and $TiO_{1.204}$ and between $VO_{0.75}$ and $VO_{1.32}$.

Under certain conditions, monoxides of transition metals have shown slight distortions of the NaCl arrangement. Thus at low temperatures (90°K.) FeO becomes rhombohedral. Similarly, NiO when made by heating $Ni(NO_3)_2$ in air at 1350°C. is rhombohedral with

$$a_0 = 4.16768 \text{ A.}, \qquad \alpha = 90°4'$$

Cobalt monoxide, CoO, on the other hand, appears tetragonal below 280°K., with

$$a_0 = 4.2638 \text{ A.}, \qquad c_0 = 4.2143 \text{ A.}$$

If it is assumed that in the NaCl structure one is dealing with spherical ions and that stability of an ionic structure is furthered by contact between ions of opposite charge, it is simple to calculate from a knowledge of ionic radii the range of substances RX which may be expected to have this atomic arrangement. Inspection of Figure III,1 shows that large X ions will be in contact with one another along the diagonals of a cube face if the diameter of the R ions is smaller than ca. 0.41 that of X. To have cation–anion contact with such smaller R ions, each would require fewer than six (i.e., four) X neighbors. In other words for octahedral, sixfold, coordination and the NaCl structure, the *radius ratio* $r(\mathrm{R})/r(\mathrm{X}) \geqq 0.41$. On the other hand, if the R ion is especially large it may have more than six X neighbors without their making contact with one another, and this is the case with CsCl (**III,b1**) and the crystals isomorphous with it. There each R atom has eight X neighbors at the corners of an enveloping cube and it is easy to see that these X atoms must be in contact with one another if the radius of R is less than 0.73 that of X. If the simple assumptions made here were valid, therefore, NaCl-like structures are to be expected for ionic RX compounds if $0.73 \geqq r(\mathrm{R})/r(\mathrm{X}) \geqq 0.41$. In a general sort of way this relation prevails, though there are exceptions, notably LiI which is NaCl-like even though the lithium ion is small enough to allow I–I contacts; and many crystals with the NaCl structure exist in which the bonding is certainly not strongly ionic.

For many years it was more or less taken for granted that the NaCl arrangement was the ideal example of a structure held together by ionic bonds and that with it departures from an exact 1:1 ratio of the R and X atoms were scarcely to be expected. It is now apparent that in many NaCl-type compounds the atoms are not strongly ionized and that, perhaps as a corollary of this, wide variations in composition are often encountered. These are particularly evident among compounds belonging to groups *C* and *D*, above.

Fig. III,1. A perspective drawing of the unit cube of NaCl with atoms given their accepted ionic radii. The large spheres are the chloride ions.

One of the earliest known and most thoroughly studied examples of a non-1:1 compound is *ferrous oxide*, FeO. Crystals of it commonly contain only about 48 atomic per cent iron and the edge length of the unit depends on both the iron–oxygen ratio and the temperature of preparation. When the compound is heated with an excess of iron and quenched from a series of temperatures, a_0 is greater the higher the quenching temperature. This was early interpreted to mean that at elevated temperatures iron dissolves in and swells the FeO structure, or as an attempt to complete the structure by approaching the 1:1 composition ratio. It now appears clear that Fe_xO should be considered as a close-packing of the oxygen atoms with some of the metal positions of the NaCl arrangement empty. Recent neutron diffraction studies further show that some iron atoms lie in the tetrahedrally coordinated positions occupied by metal atoms in the spinel (Chapter VIII) structure instead of the octahedrally coordinated positions characteristic of an NaCl-type crystal. Thus FeO can be thought of as the stabilization at room temperatures of a somewhat defective and distorted approach at high temperatures towards spinel-like Fe_3O_4.

Similar oxides of variable composition are CdO which shows a real though less marked expansion when heated with excess of cadmium, and vanadium monoxide which is said to vary by 10% on either side of a 1:1 atomic ratio without a significant change in unit cell size.

Several examples have been found of NaCl-like structures of much more variable composition. Thus the composition of Ti_xC rises from only 29 at.-% of carbon to 50 at.-% (when there are equal numbers of titanium and carbon atoms present) with only an increase of a_0 from 4.294 to 4.321 A.; that is to say, doubling the number of carbon atoms in the structure increased the size of the unit by only 0.03 A., or less than one per cent. Similarly, a niobium carbide containing only 2.5% of carbon by weight was found to have a_0 = 4.395 A.; and this was very little expanded (to 4.462 A.) when the carbon content was raised to 10.9%. Other compounds, such as SnAs and SnSb have shown large variations in composition without change in structure. As another example, ThS_x appears NaCl-like with x anything between $^1/_2$ and $^3/_4$ (for $ThS_{0.75}$, a_0 = 5.674 A.).

It has become customary to explain all of these deficit structures by saying that the element in short supply is distributed haphazardly among one set of the four equivalent positions of the NaCl arrangement. This undoubtedly is often true, though certainly it is not always the case (see, for instance, **III,a2**).

Another somewhat different example of a deficit structure is provided by the solid solutions that exist between some monosulfides and sesquisulfides. Thus at high temperatures (ca. 1200°C.) MgS, MnS, and CaS dissolve up to 20% Y_2S_3 without change of structure type (1961: F, D, and P). These solutions and their dimensional variations are readily understood if the crystals are viewed as cubic close-packings of the sulfur atoms with some voids unfilled by the smaller numbers of cations required for electrical neutrality: the structure swells with Y_2S_3 addition when Y is bigger than the bivalent cations, as with MgS and MnS, and it shrinks when the Y cation is the smaller, as with CaS.

III,a2. The compound *niobium monoxide*, NbO, has a simple structure that has commonly been described as a defect NaCl arrangement. There are but three molecules in a unit cube of the edge length:

$$a_0 = 4.2103 \text{ A.} \qquad (27°\text{C.})$$

Atomic positions, described in terms of O_h^1 ($Pm3m$) are:

$$\text{Nb: } (3c) \quad 0\,^1/_2\,^1/_2;\ ^1/_2\,0\,^1/_2;\ ^1/_2\,^1/_2\,0$$
$$\text{O: } (3d) \quad ^1/_2\,0\,0;\ 0\,^1/_2\,0;\ 0\,0\,^1/_2$$

The structure is shown in Figure III,2. Existing data demonstrate that this is indeed an ordered atomic grouping and not a defective NaCl arrangement in which three molecules are randomly distributed over two sets of four equivalent positions of NaCl.

III,a3. At room temperature the *hydrosulfides* and *hydroselenides* of *sodium*, *potassium*, and *rubidium* are rhombohedral; above 150° to 180°C. the distortion this expresses disappears and they have the exact NaCl grouping discussed in **III,a1**.

The unit rhombohedra of the room temperature form containing a single molecule have the dimensions listed in columns 2 and 3 of Table III,2.

TABLE III,2
Rhombohedral Hydrosulfides and Hydroselenides (**III, a3**)

	Unit rhombohedron (unimolecular)		NaCl-like pseudocell	
Crystal	a_0, A.	α	a_0', A.	α'
NaSH	3.99	67°56′	5.99	96°21′
NaSeH	4.16	68°4′	6.24	96°27′
KSH	4.37	69°2′	6.61	97°9′
KSeH	4.51	69°18′	6.84	97°21′
RbSH	4.56	69°8′	6.89	97°13′
RbSeH	4.66	70°24′	7.12	98°7′

	Hexagonal cell (trimolecular)	
	a_0, A.	c_0, A.
NaSH	4.46	9.14
NaSeH	4.66	9.52
KSH	4.95	9.91
KSeH	5.13	10.20
RbSH	5.17	10.33
RbSeH	5.37	10.43

Alkali atoms can be considered to be at the origin in 000, and sulfur or selenium atoms at uuu with u = ca. 0.50. The NaCl-like character is apparent when this grouping is described in terms of the larger four-molecule pseudo-unit that has as axes the face diagonals of the true unit. The dimensions of these larger face-centered rhombohedra (columns 4 and 5) show them to be cubes somewhat flattened along their body diagonals. The reason for this flattening, which presumably is linked with the positions of the hydrogen atoms, i.e., with the shape of the SH and SeH ions, is not supplied by the x-ray data. It would be helpful in this connection to determine if u for S and Se is exactly $^1/_2$.

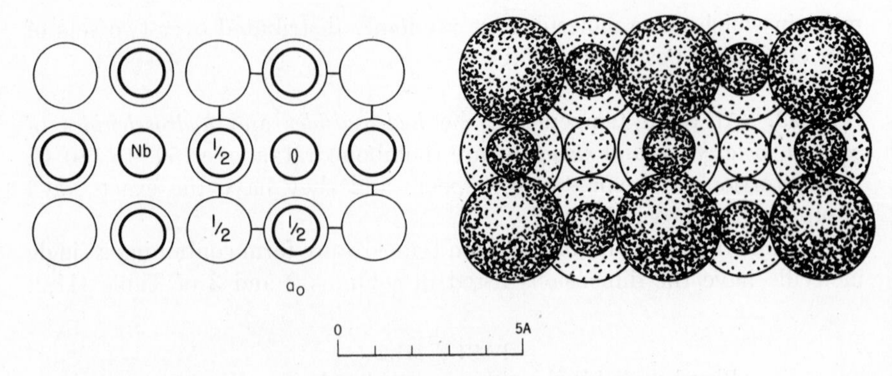

Fig. III,2a (left). The contents of two unit cubes of NbO projected along a cube axis.
Fig. III,2b (right). A packing drawing of the NbO arrangement viewed along a cubic axis. The small circles are the niobium atoms.

III,a4. Crystals of *thallous fluoride*, TlF, provide an orthorhombic distortion of the NaCl structure. The unit prism contains four molecules and has the approximately equilateral dimensions:

$$a_0 = 5.180 \text{ A.}; \quad b_0 = 5.495 \text{ A.}; \quad c_0 = 6.080 \text{ A.}$$

Thallium atoms are at points of a face-centered lattice with the coordinates,

$$000; \; {}^1/_2\,{}^1/_2\,0; \; {}^1/_2\,0\,{}^1/_2; \; 0\,{}^1/_2\,{}^1/_2$$

and in the structure as described fluorine atoms are at

$${}^1/_2\,{}^1/_2\,{}^1/_2; \; {}^1/_2\,0\,0; \; 0\,{}^1/_2\,0; \; 0\,0\,{}^1/_2$$

Thus the distortion is expressed by the different lengths of the unit axes. The short Tl–F separation along a_0 (2.59 A.) is ascribed to the known high polarizability of thallium ions.

III,a5. At room temperatures the *cyanides* of *sodium* and *potassium*, NaCN and KCN, have the cubic NaCl structure (see Table III,1). The separate atoms of their cyanide ions apparently do not have fixed positions; instead these ions have a spherical symmetry that is commonly considered to be an expression of their "rotation."

This rotation cannot be maintained at lowered temperatures and then these salts invert to an orthorhombic modification in which the carbon and nitrogen atoms have definite positions. Their bimolecular units have the following dimensions:

NaCN: $a_0 = 3.774$ A.; $b_0 = 4.719$ A.; $c_0 = 5.640$ A. (6°C.)
KCN: $a_0 = 4.24$ A.; $b_0 = 5.14$ A.; $c_0 = 6.16$ A. (−60°C.)

In both, atoms are in the following special positions of C_{2v}^{20} (Imm):

Na or K: (2a) 000; $\frac{1}{2}\frac{1}{2}\frac{1}{2}$
C and N: (2b) $0\,u\,\frac{1}{2}$; $\frac{1}{2},u+\frac{1}{2},0$

The intensities of the reflections observed from the sodium salt are well explained if u(C) = 0.111 and u(N) = −0.111, or vice versa. This corresponds to a C–N separation of 1.05 A. within the cyanide ion and leads to the grouping shown in Figure III,3. Intensities observed for the potassium salt are satisfactorily accounted for by cyanide ions of the same size, corresponding to a u of 0.105.

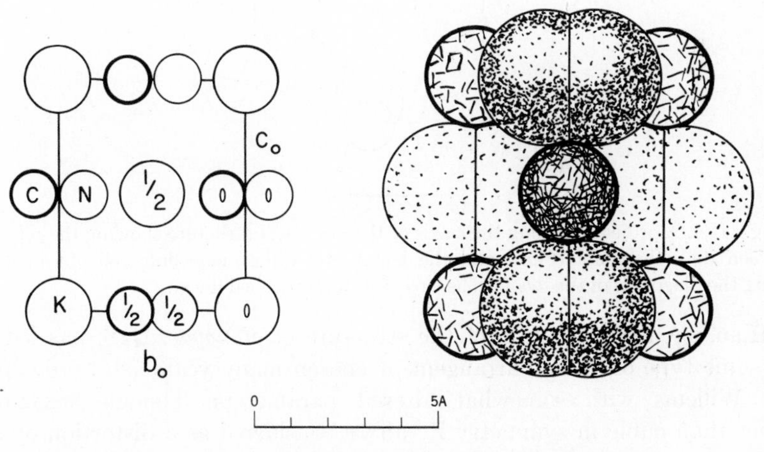

Fig. III,3a (left). A projection along the a_0 axis of the orthorhombic, low-temperature form of KCN. Origin in lower right.

Fig. III,3b (right). A packing drawing of the low-temperature, orthorhombic structure of KCN viewed along the a_0 axis. The line-shaded atoms are potassium; nitrogen and carbon are not distinguished.

The relation of this structure to that of NaCl is clear if it is described in terms of the larger four-molecule monoclinic pseudo-cell whose base is the parallelogram that can be circumscribed about the rectangular a_0b_0 face of the true unit (Fig. III,4). This larger pseudo-unit has for NaCN $a_0' = b_0' = 6.01$ A., $c_0' = c_0 = 5.61$ A., and the angle between a_0' and b_0' equal to 104°. The distortion which produces this departure from high symmetry is conveniently pictured by considering the cyanide ions in this case as ellipsoids of revolution having radii of 2.15 A. and 1.78 A. It should be noted that the major axis is appreciably greater than the diameter of the "rotating" ions in the high-temperature forms of these salts.

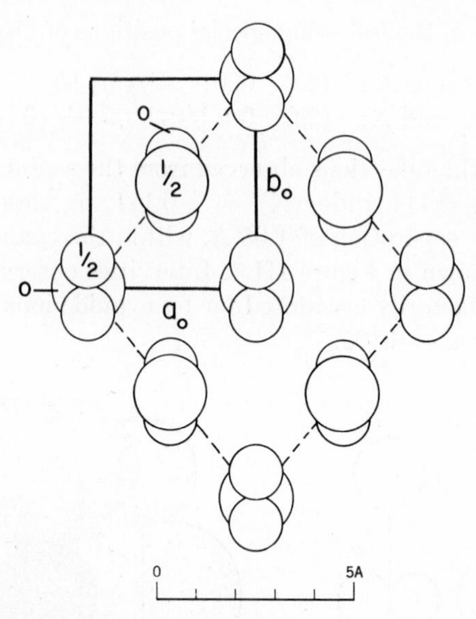

Fig. III,4. A projection along the c axis of the low KCN structure showing the relation between its orthorhombic unit and the larger NaCl-like monoclinic cell obtained by taking the diagonals of the a_0b_0 face as two of its axes (dashed lines).

III,a6. A reexamination of the structure of *cinnabar*, HgS, has led to the same type of atomic arrangement chosen many years ago by de Jong and Willems with somewhat altered parameters. Though hexagonal rather than cubic in symmetry it can be considered as a distortion of the NaCl arrangement. The trimolecular hexagonal unit has the dimensions:

$$a_0 = 4.149 \text{ A.}, \quad c_0 = 9.495 \text{ A.} \quad (26°\text{C.})$$

Atoms are placed in the following special positions of D_3^4 ($C3_12$) (or in the corresponding positions of the enantiomorphic D_3^6):

Hg: (3*a*) $u\,0\,{}^1/_3;\ 0\,u\,{}^2/_3;\ \bar{u}\bar{u}0$ with $u = 0.720$
S: (3*b*) $u\,0\,{}^5/_6;\ 0\,u\,{}^1/_6;\ \bar{u}\,\bar{u}\,{}^1/_2$ with $u = 0.485$

This is a structure (Fig. III,5) composed of Hg–S chains running along the c_0 axis. In the chain Hg–S = 2.36 A.; between chains it is ca. 3.2 A.

Recently, *mercuric oxide*, HgO, has been shown to have a modification possessing this structure. For it:

$$a_0 = 3.577 \text{ A.}, \quad c_0 = 8.681 \text{ A.}$$

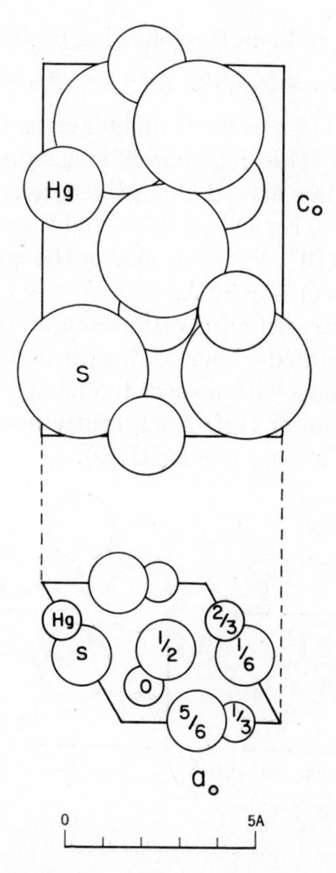

Fig. III,5. Two projections of the hexagonal structure of cinnabar, HgS.

The parameter of the mercury atoms in $(3a)$ is $u = 0.745$ and of the oxygen atoms in $(3b)$ 0.46. In the Hg–O chains the separation is 2.03 A. and the angles Hg–O–Hg and O–Hg–O are 108° and 176°. Between chains the shortest Hg–O distance is 2.79 A. These are substantially the same separations that prevail in orthorhombic HgO (**III,a7**).

III,a7. Conflicting structures based on both x-ray and neutron diffraction have been published for the usual form of *mercuric oxide*, HgO. Apparently that based on the larger orthorhombic cell containing four molecules (1956:A) is to be preferred. It has the edges:

$$a_0 = 6.6121 \text{ A.}; \; b_0 = 5.5201 \text{ A.}; \; c_0 = 3.5213 \text{ A.}$$

Atoms have been found to be in the following positions of V_h^{16} (*Pnma*)

$$(4c) \quad \pm(u\,{}^1/_4\,v;\; u+{}^1/_2,{}^1/_4,{}^1/_2-v)$$

with for mercury $u = 0.115$, $v = 0.245$ and for oxygen $u = 0.365$, $v = 0.585$.

In this structure (Fig. III,6) indefinitely long zigzag but planar chains of mercury and oxygen atoms extend along the a_0 axis; their plane is normal to b_0. Within these chains, Hg–O = 2.03 A. and the angles are Hg–O–Hg = 109° and O–Hg–O = 179°. Between chains the shortest separations are Hg–O = 2.82 A. and O–O = 3.39 A.

The other proposed structure for orthorhombic HgO had a bimolecular cell with a_0 half that stated above and with a different distribution of mercury and oxygen atoms with respect to one another.

A new, hexagonal, form of HgO has recently been prepared. It has the cinnabar (HgS) type of arrangement (**III,a6**).

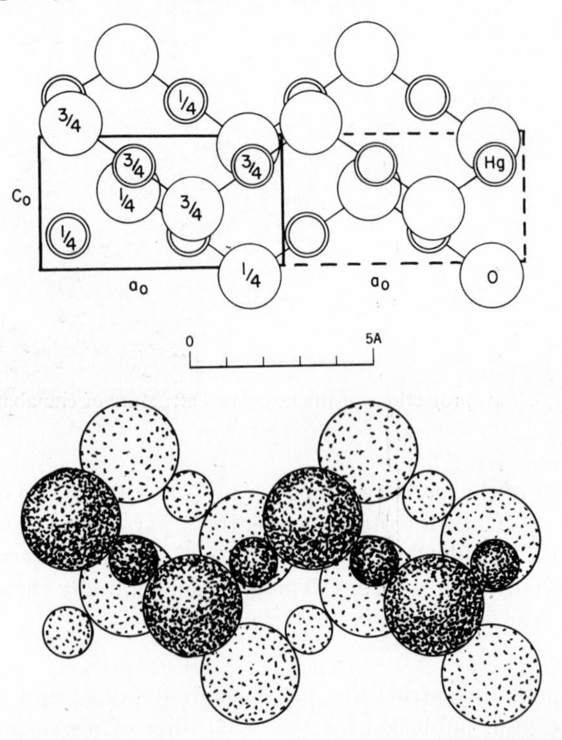

Fig. III,6a (top). The contents of two cells of orthorhombic HgO projected along its b_0 axis. Origin in lower left.

Fig. III,6b (bottom). A packing drawing of the orthorhombic HgO structure seen along its b_0 axis. The small circles are the mercury atoms. The atomic chains running parallel to the a_0 axis are apparent.

III,a8. Crystals of *potassium hydroxide*, KOH, were several years ago given the TlI structure (**III,g2**). Recent work using single crystals leads to a different, monoclinic, arrangement. The unit cell, containing two molecules, has the dimensions:

$$a_0 = 3.95 \text{ A.};\ b_0 = 4.00 \text{ A.};\ c_0 = 5.73 \text{ A.};\ \beta = 103°36'$$

at room temperature, and

$$a_0 = 3.96 \text{ A.};\ b_0 = 3.94 \text{ A.};\ c_0 = 5.67 \text{ A.};\ \beta = 105°54'$$

at -133°C. The space group has been chosen as C_2^2 ($P2_1$), partly on the basis of infrared evidence. All atoms would then be in the general positions:

$$(2a)\quad xyz;\ \bar{x},y+{}^1/_2,\bar{z}$$

The finally chosen parameters are:

For K: $x = 0.175$; $y = 0.25$; $z = 0.288$
For O: $x = 0.318$; $y = 0.25$; $z = 0.770$

In the resulting structure (Fig. III,7), each atom is surrounded by a distorted octahedron of atoms of the opposite sort with K–O = 2.69–3.15 A.

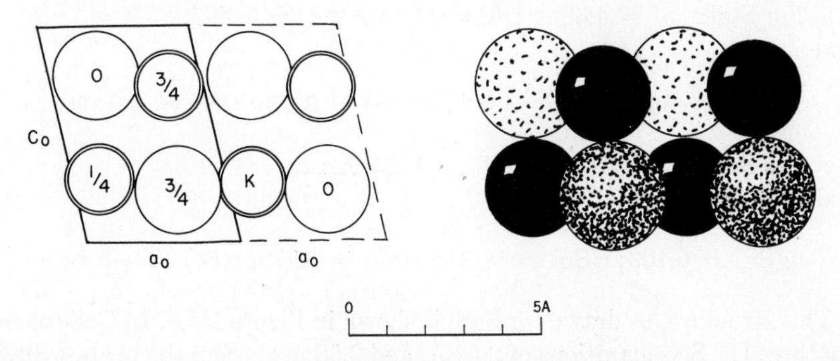

Fig. III,7a (left). Two unit cells of monoclinic KOH projected along the b_0 axis. Origin in lower left.

Fig. III,7b (right). A packing drawing of the very simple monoclinic KOH viewed along its b_0 axis. The potassium atoms are black.

The shortest O–O separation of 3.35 A. exceeds the customary O–H–O bond and the x-ray data do not decide between imaginable positions of the hydrogen atoms.

This structure is a considerable distortion of the NaCl (**III,a1**) arrangement. Above 248°C. KOH is reported to be cubic with exactly this structure (Table III,1).

III,a9. *Sulfides and selenides* of the fourth group metals *germanium and tin* have structures which also can be described as distortions of that of NaCl. They are orthorhombic with unit cells similar in size and shape. Each has four molecules in the unit with atoms in the coordinate positions of V_h^{16} (*Pbnm*)

$$(4c) \quad \pm (u\, v\, {}^1/_4;\ {}^1/_2-u, v+{}^1/_2, {}^1/_4)$$

In *germanium selenide*, GeSe, which has most recently been studied, the cell dimensions and parameters are:

$$a_0 = 4.40 \text{ A.};\ b_0 = 10.82 \text{ A.};\ c_0 = 3.85 \text{ A.}$$

and

$$u(\text{Ge}) = 0.111,\ v(\text{Ge}) = 0.121,\ u(\text{Se}) = 0.500,\ v(\text{Se}) = -0.146.$$

For GeS: $a_0 = 4.30$ A.; $b_0 = 10.44$ A.; $c_0 = 3.65$ A.

and

$$u(\text{Ge}) = 0.106;\ v(\text{Ge}) = 0.121;\ u(\text{S}) = 0.503;\ v(\text{S}) = -0.148$$

For SnS: $a_0 = 4.33$ A.; $b_0 = 11.18$ A.; $c_0 = 3.98$ A.

and

$$u(\text{Sn}) = 0.115;\ v(\text{Sn}) = 0.118;\ u(\text{S}) = 0.478;\ v(\text{S}) = -0.150$$

For SnSe: $a_0 = 4.46$ A.; $b_0 = 11.57$ A.; $c_0 = 4.19$ A.

and

$$u(\text{Sn}) = 0.103;\ v(\text{Sn}) = 0.118;\ u(\text{Se}) = 0.479;\ v(\text{Se}) = -0.145$$

This structure, as drawn for SnS, is shown in Figure III,8. In GeSe there are three Ge–Se separations of 2.54 A. and 2.58 A.; in SnS the nearest Sn–S distance is 2.62 A.

In spite of the fact that it has a unit of very much the same dimensions as SnS, crystals ot *indium sulfide*, InS, have been described in terms of a different space group V_h^{12} (*Pnnm*). This is, however, not necessary. Instead, in terms of the structure given above, that of InS is

$$a_0 = 4.442 \text{ A.};\ b_0 = 10.642 \text{ A.};\ c_0 = 3.940 \text{ A.}$$

and

$$u(\text{In}) = 0.125;\ v(\text{In}) = 0.121;\ u(\text{S}) = 0.505;\ v(\text{S}) = -0.145$$

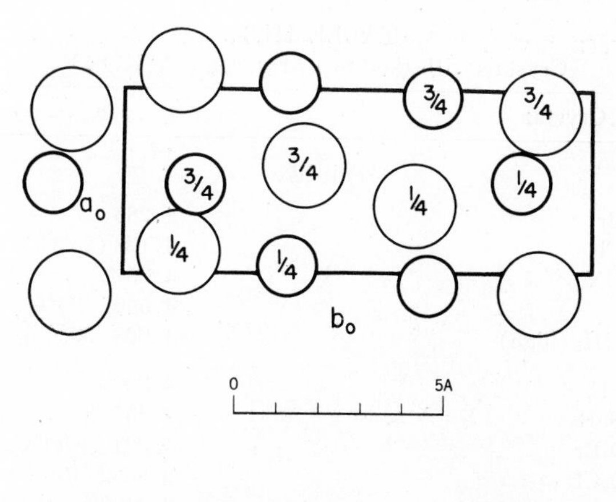

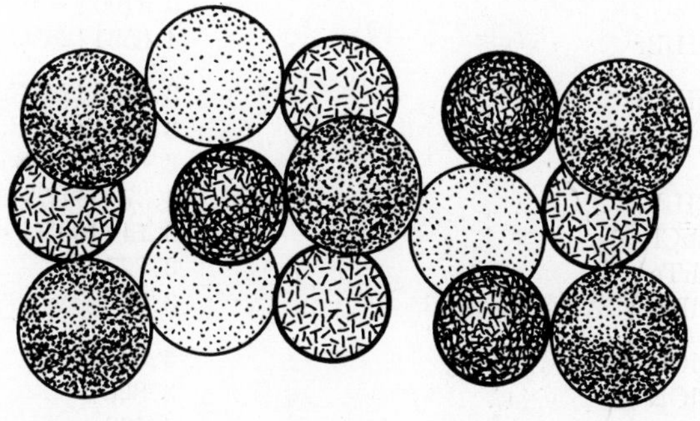

Fig. III,8a (top). A projection along c_0 of the orthorhombic structure of SnS. The tin are the larger circles. Origin in lower left.

Fig. III,8b (bottom). A packing drawing of the orthorhombic SnS structure viewed along the c_0 axis. The sulfur atoms are line-shaded.

CsCl-Like Structures

III,b1. The very simple *cesium chloride*, CsCl, grouping places a single molecule in a unit cube having atoms in the body-centered positions:

$$R:\ 000 \qquad \text{and} \qquad X:\ {}^1/_2\,{}^1/_2\,{}^1/_2$$

Each atom is at the center of a cube of atoms of the opposite sort (Fig. III,9), and thus has the coordination number eight.

Two different kinds of substances crystallize with this arrangement. In one group are halides of the largest univalent ions—cesium, thallium, and ammonium; in the other are intermetallic compounds of which beta-brass, CuZn, is representative (Table III,3).

TABLE III,3
Crystals with the CsCl Arrangement (III,b1)

Crystal	a_0, A.
Salt-like	
CsBr	4.286
CsCl	4.123 (25°C.)
CsCN	4.25
CsI	4.5667 (20°C.)
$CsNH_2$ (high)	4.063 (50°C.)
CsSH	4.30
CsSeH	4.437
ND_4Br	4.034 (−30°C.)
ND_4Cl	3.8682 (18°C.)
	3.8190 (−185°C.)
NH_4Br	4.0594 (26°C.)
NH_4Cl	3.8756 (26°C.)
	3.8684 (18°C.)
	3.8200 (−185°C.)
NH_4I	4.37 (−17°C.)
RbCl	3.74 (190°C.)
ThTe	3.827
TlBr	3.97
TlCl	3.8340
TlCN	3.82
TlI	4.198
Some intermetallic examples	
AgCd	3.33
AgCe	3.731
AgLa	3.760
AgMg	3.28
AgZn	3.156

(continued)

In crystals of the ammonium halides which have this structure the hydrogen atoms presumably are in (4*e*) of T_d^1 ($P\bar{4}3m$):

$$(4e) \quad uuu;\ u\bar{u}\bar{u};\ \bar{u}\bar{u}u;\ \bar{u}u\bar{u}$$

if they are in positions dictated by symmetry. X-ray diffraction did not settle this but it can be established through electron or neutron diffraction. From a recent study at −40°C. and at room temperature using electrons it has been decided that in NH_4Cl, $u = 0.153$, corresponding to N–H = 1.03 A.; in another investigation u was found to be 0.146, leading to N–H = 0.98 A.

TABLE III,3 (*continued*)

Crystal	a_0, A.
AlNd	3.73
AlNi	2.881
AuCd	3.34 (400°C.)
AuMg	3.259
AuZn	3.19
BeCo	2.606
BeCu	2.698
BePd	2.813
CaTl	3.847
CdCe	3.86
CdLa	3.90
CdPr	3.82
CuPd	2.988
CuZn	2.945
LiAg	3.168
LiHg	3.287
LiTl	3.424
MgCe	3.898
MgHg	3.44
MgLa	3.965
MgPr	3.88
MgSr	3.900
MgTl	3.628
SrTl	4.024
TlBi	3.98
TlSb (excess Tl)	3.84
ZnCe	3.70
ZnLa	3.75
ZnPr	3.67

The eightfold coordination that characterizes this CsCl arrangement is, as stated above, encountered in compounds having especially large cations. The radius ratio of the ions involved conforms to the $r(\text{R})/r(\text{X}) \geqq 0.73$ that must prevail if ions of opposite signs are to make contact.

Under high pressures, crystals with the NaCl arrangement are known to transform to the CsCl structure. Thus at 7500 kg./cm.^{2} and room temperature, RbCl is CsCl-like with $a_0 = 3.82$ A. Similarly, at 11,000 atmospheres and 25°C., RbI is CsCl-like with $a_0 = 4.29$ A.

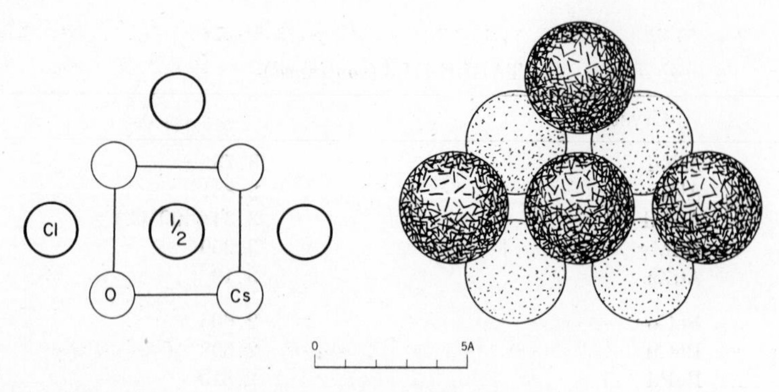

Fig. III,9a (left). A projection along a cubic axis of the CsCl arrangement.
Fig. III,9b (right). A packing drawing of the simple CsCl structure viewed along a cube axis.

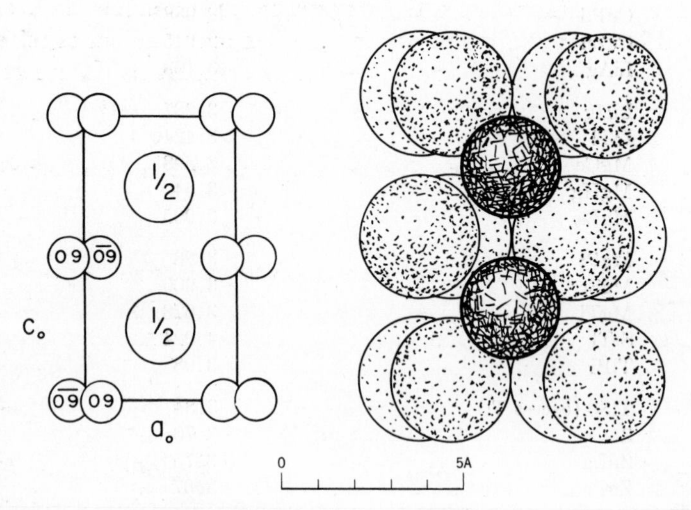

Fig. III,10a (left). A projection along an a_0 axis of the tetragonal structure of NH_4CN. The large circles are NH_4 ions; the carbon and nitrogen atoms are not distinguished from one another. Origin in lower left.

Fig. III,10b (right). A packing drawing of the tetragonal NH_4CN structure viewed along an a_0 axis. The dotted carbon and nitrogen atoms are not distinguished from one another.

III,b2. *Ammonium cyanide*, NH_4CN, resembles NH_4Cl in having a structure like CsCl, but its symmetry is lower because its cyanide groups are not freely rotating. It is tetragonal without an inversion down to at least -80°C. The unit cell contains two molecules and has the peculiarity of shrinking only along the c_0 axis as the temperature is lowered:

$$\text{At } 35°\text{C.:} \quad a_0 = 4.16 \text{ A.}, \; c_0 = 7.64 \text{ A.}$$
$$\text{At } -80°\text{C.:} \quad a_0 = 4.16 \text{ A.}, \; c_0 = 7.45 \text{ A.}$$

The ammonium ions are in the positions:

$$(2c) \quad {}^1/_2\,{}^1/_2\,{}^1/_4; \; {}^1/_2\,{}^1/_2\,{}^3/_4$$

of D_{4h}^{10} ($P4/mcm$), and the carbon and nitrogen atoms together seem to be in the fourfold positions:

$$(4i) \quad \pm(uu0; \; \bar{u}\,u\,{}^1/_2)$$

with $u = 0.093$ (Fig. III,10).

This structure explains both the anisotropic expansion and the x-ray intensities at high and low temperatures if one assumes that the cyanide ions vibrate preferentially along the c_0 direction and normal to their own axis. Such a favored vibration can be imagined as a first step towards rotation.

III,b3. Crystals of *phosphonium iodide*, PH_4I, and related substances are tetragonal with the bimolecular units:

Crystal	a_0,A.	c_0,A.	u
$CsNH_2$ (low)	5.641	4.194	—
PH_4Br	6.042	4.378	0.42
PH_4I	6.34	4.62	0.40
NH_4Br	5.697 (-145°C.)	4.046	0.53
	5.713 (-71°C.)	4.055	
NH_4SH	6.011	4.009	0.34
NH_4I	6.18 (-100°C.)	4.37	0.51
$N(CH_3)_4Cl$	7.78	5.53	0.35
$N(CH_3)_4Br$	7.76	5.53	0.37
$N(CH_3)_4I$	7.96	5.75	0.39
$N(CH_3)_4ClO_4$	8.290	6.006	—
$N(CH_3)_4MnO_4$	8.439	6.019	—

The atoms of PH_4I are in the following special positions of D_{4h}^7 (P/nmm):

$$\text{P:} \quad (2a) \quad 000; \; {}^1/_2\,{}^1/_2\,0$$
$$\text{I:} \quad (2c) \quad 0\,{}^1/_2\,u; \; {}^1/_2\,0\,\bar{u}$$

with $u = 0.40$. If the axial ratio c_0/a_0 were $1:2^{1/2}$ and if u were exactly 0.50, this structure would be an undistorted CsCl arrangement.

In tetramethyl ammonium chloride, $N(CH_3)_4Cl$, the carbon atoms are tetrahedrally arranged about the nitrogen atoms in the special positions:

$$(8i) \quad w0v; \quad 0w\bar{v}; \quad \bar{w}0v; \quad 0\bar{w}\bar{v};$$

$$w+{}^1/_2,{}^1/_2,\bar{v}; \quad {}^1/_2,w+{}^1/_2,v; \quad {}^1/_2-w,{}^1/_2,\bar{v}; \quad {}^1/_2,{}^1/_2-w,v$$

with $w = 0.15$ and v somewhere between 0.10 and 0.17. The lower symmetry of the structure as a whole is readily understood in terms of the adjustments necessary to pack together these large tetrahedral methyl ammonium ions with the halide anions (Fig. III,11).

Tetrahedrally Coordinated Structures

III,c1. The *zinc sulfide* arrangement, like that of NaCl, contains four molecules in its unit cube and is developed on a face-centered lattice. Its atoms have the coordinates (of T_d^2—$F\bar{4}3m$):

R: (4a) 000; 0 ½ ½; ½ 0 ½; ½ ½ 0, or 000; F.C.

X: (4c) ¼ ¼ ¼; ¼ ¾ ¾; ¾ ¼ ¾; ¾ ¾ ¼, or ¼ ¼ ¼; F.C.

As can be seen from Figure III,12, each atom has about it four equally distant atoms of the opposite sort arranged at the corners of a regular tetrahedron. If all the atoms were alike this would, of course, be the diamond arrangement (**II,i1**).

The cell dimensions of crystals with this structure are given in Table III,4. As these indicate, preparations are sometimes found which yield the general diffraction pattern of ZnS but depart far from a 1:1 composition. It has become usual to explain these as defect structures with statistically distributed "holes"; but especially when they appear to be definite compounds, more thorough studies are needed.

A number of polymorphs of ZnS, analogous to those studied in great detail for SiC, have now been discovered. These are discussed under SiC in paragraph **III,c3**.

The kinds of compounds which have the tetrahedrally coordinated ZnS and the closely related ZnO structure of paragraph **III,c2** are in general of two sorts:

1. A few compounds, such as BeO and BeS, containing especially small cations for which $r(R)/r(X) < 0.41$.

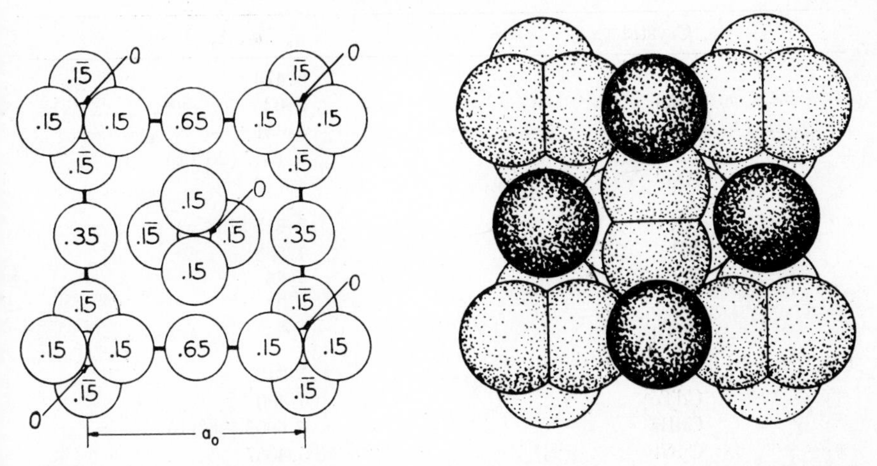

Fig. III,11a (left). A projection of the atomic arrangement in the tetragonal unit cell of $N(CH_3)_4Cl$ upon its base. Smallest circles, lying within groups of four large circles, are nitrogen atoms; the surrounding large circles are carbon atoms as centers of methyl radicals. Isolated circles, of intermediate size, are chlorine atoms.

Fig. III,11b (right). A packing drawing showing the way the $N(CH_3)_4^+$ and Cl^- ions in $N(CH_3)_4Cl$ pack together.

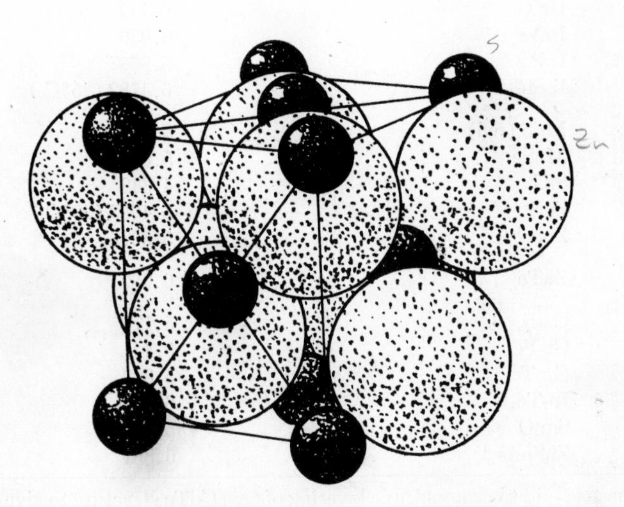

Fig. III,12. A perspective drawing showing the way zinc and sulfur atoms would pack together in the ZnS arrangement if they had the accepted ionic sizes.

TABLE III,4
Crystals with the ZnS Arrangement (**III,c1**)

Crystal	a_0, A.
AgI	6.473
AlAs	5.62
AlP	5.451
AlSb	6.1347 (26°C.)
BAs	4.777
BN	3.615 (25°C.)
BP	4.538
BeS	4.85
BeSe	5.07
BeTe	5.54
CdS	5.818
CdTe[a]	6.480
CuBr	5.6905 (26°C.)
CuCl	5.4057
CuF	4.255
CuI	6.0427
GaAs[b]	5.6537
GaP	5.4505
GaSb	6.118
HgS	5.8517 (26°C.)
HgSe	6.084
HgTe	6.429
InAs	6.036
InP	5.8687
InSb	6.4782 (25°C.)
MnS (red)	5.600
MnSe	5.82
SiC	4.348
ZnS[c]	5.4093 (26°C.)
ZnSe[b]	5.6676
ZnTe	6.089
Ga_2S_3	5.171
Ga_2Se_3	5.441
Ga_2Te_3	5.899
In_2Te_3	6.158
Sm_2O	5.3761
$ZnSnAs_2$[d]	5.851

[a] There is also said to be a multiple layering of this CdTe structure to give a hexagonal cell with $a_0 = 4.60$ A., $c_0 = 45.1$ A.

[b] GaAs and ZnSe form a continuous series of solid solutions.

[c] If 1% ZnO present, 5.4065 A.

[d] Zn and Sn statistically distributed.

2. Halides, oxides, nitrides, etc. of transitional and less electropositive metals. With many it cannot be inferred that ionization is sufficiently complete to give meaning to a calculated radius ratio. An additivity of interatomic distances has, however, been observed and this is usually expressed through the use of nonionic, "tetrahedral radii." In many of the compounds and structures for which such radii are approximately valid there are X–X as well as R–X close contacts and it is these that give the basis for the atomic radii chosen.

III,c2. The atoms in the two-molecule hexagonal unit of the *zincite*, ZnO, arrangement are in the positions:

$$\text{R:}\quad 000;\ \tfrac{1}{3}\,\tfrac{2}{3}\,\tfrac{1}{2}$$
$$\text{X:}\quad 00u;\ \tfrac{1}{3},\tfrac{2}{3},u+\tfrac{1}{2}$$

which are derived from two sets of special positions of C_{6v}^4 ($C6mc$):

$$(2b)\quad \tfrac{1}{3}\,\tfrac{2}{3}\,v;\ \tfrac{2}{3},\tfrac{1}{3},v+\tfrac{1}{2}$$

by a change of origin to the point $\tfrac{2}{3}\,\tfrac{1}{3}\,0$. The axial ratios of crystals with this structure have always been close to $c/a = 1.63$ and the parameter u to 0.375. Under these circumstances (Fig. III,13), each atom has about it a tetrahedron of atoms of the opposite sort just as in the cubic ZnS arrangement.

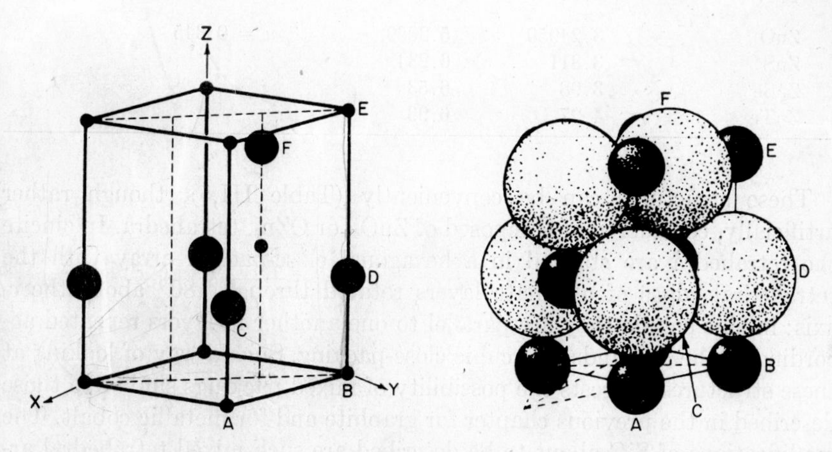

Fig. III,13a (left). A drawing that shows the positions of the atoms in the hexagonal unit cell of ZnO. The zinc atoms are represented by the small black circles.

Fig. III,13b (right). A perspective drawing showing how atoms of ZnO would pack together if they were ions. The small black spheres are zinc. Letters in this drawing refer to atoms similarly designated in Figure III,13a.

TABLE III,5
Crystals with the ZnO Arrangement (III,c2)

Crystal	a_0, A.	c_0, A.	Remarks
AgI	4.580	7.494	
Al_2CO	3.17	5.06	A long range, somewhat disordered structure
AlN	3.111	4.978	$u = 0.385$
BeO	2.698	4.380	$u = 0.378$
CdS	4.1348	6.7490	
CdSe	4.30	7.02	
CuBr	4.06	6.66	391–470°C.
CuCl	3.91	6.42	410°C.
CuH	2.893	4.614	
CuI	4.31	7.09	402–440°C.
GaN	3.180	5.166	
InN	3.533	5.693	
MgTe	4.52	7.33	
MnS (pink)	3.976	6.432	
MnSe	4.12	6.72	
MnTe	4.087	6.701	
NH_4F	4.39	7.02	$u = 0.365$
NbN	3.017	5.580	
SiC	3.076	5.048	
TaN	3.05	4.94	
ZnO	3.24950	5.2069	$u = 0.345$
ZnS	3.811	6.234	
ZnSe	3.98	6.53	
ZnTe	4.27	6.99	

These structures can be conveniently (Table III,5), though rather artificially, considered as composed of ZnO_4, or OZn_4, tetrahedra. In zincite the tetrahedra are stacked in a hexagonal close-packed array with the tetrahedral edges of alternate layers rotated through 180° about the c_0 axis; in zinc sulfide they are parallel to one another in layers repeated according to the demands of a cubic close-packing. Such a way of looking at these structures suggests the possibility of mixed packings similar to those described in the previous chapter for graphite and for metallic cobalt. The modifications of SiC about to be described are such mixed tetrahedral arrangements.

No clear distinction can be drawn between the types of compounds which crystallize with the ZnO and ZnS structures. The two arrangements are so much alike in their spatial distributions that one would not expect important energy differences between them. It is therefore not surprising to find that a number of compounds appear with both structures.

III,c3. *Silicon carbide*, SiC, occurs in several modifications which, in spite of a formal complexity, are similar in basic principle and closely related to one another and to the tetrahedral structures just described. One form is cubic with the ZnS arrangement (**III,c1**), the others are either rhombohedral or hexagonal. Their close relationship to one another is best expressed in terms of hexagonal units and pseudocells; when this is done they prove to have the same value of a_0 and a value of c_0 which is ca. 2.51 A. times the number of molecules in the hexagonal cell (Table III,6).

TABLE III,6
Cell Dimensions of SiC

	True unit			Hexagonal cell		
Modification	a_0, A.	α	M	a_0', A.	c_0', A.	M'
β	4.439	—	4	3.073	7.57	3
2*H*	—	—	—	3.076	5.048	2
4*H*-III	—	—	—	3.073	10.053	4
6*H*-II	—	—	—	3.073	15.08	6
8*H*	—	—	—	3.079	20.1470	8
10*H*	—	—	—	3.079	25.183	10
19*H*	—	—	—	3.079	47.849	19
27*H*	—	—	—	3.079	67.996	27
15*R*-I	12.69	13°55′	5	3.073	37.70	15
21*R*-IV	17.68	9°58′	7	3.073	52.78	21
27*R*	22.735	7°46′	9	3.079	67.996	27
33*R*-VI	27.70	6°21.5′	11	3.073	82.94	33
51*R*(1)-V	42.763	4°7′	17	3.073	129.03	51
51*R*(2)	42.849	4°7′	17	3.079	128.437	51
75*R*	62.984	2°48′	25	3.079	188.878	75
84*R*	70.537	2°30′	28	3.079	211.543	84
87*R*-VII	72.907	2°25′	29	3.073	218.657	87
141*R*	118.359	1°30′	47	3.079	355.04	141
174*R*	—	—	58	3.079	436.7	174
393*R*	329.87	0°32′	131	3.079	989.60	393
Related forms of ZnS						
4*H*-III	—	—	—	3.806	12.44	4
6*H*-II	—	—	—	3.813	18.69	6
8*H*	—	—	—	3.82	24.96	8
10*H*	—	—	—	3.824	31.20	10
15*R*-I	—	—	5	3.822	46.79	15

The form originally designated as beta has the cubic ZnS arrangement (**III,c1**) with $a_0 = 4.348$ A. In it the shortest interatomic distances are C–Si = 1.89 A. and C–C = Si–Si = 3.09 A. Recently an analogous, ZnO-like, modification has been observed. Its two molecules are in a hexagonal cell of the dimensions:

$$a_0 = 3.076 \text{ A.,} \qquad c_0 = 5.048 \text{ A.}$$

The two simplest forms of SiC after these have hexagonal units. Their atomic arrangements, in the units of Table III,6, are as follows:

Type III: The atoms of its four molecules are all on trigonal axes in the special positions of C_{6v}^4 ($C6mc$):

$$(2a) \quad 00u; \qquad 0,0,u+{}^1/_2$$
$$(2b) \quad {}^1/_3\,{}^2/_3\,v; \; {}^2/_3,{}^1/_3,v+{}^1/_2$$

with u(C) = 0, v(C) = $^1/_4$, and u(Si) = $^3/_{16}$, v(Si) = $^7/_{16}$ (i.e., the Si parameter = C parameter plus $^3/_{16}$).

Type II: The atoms of its six molecules are in the same special positions of C_{6v}^4 with u(C) = 0, v(C) = $^1/_6$ and $^5/_6$, and u(Si) = $^1/_8$, v(Si) = $^7/_{24}$ and $^{23}/_{24}$ (the Si parameter = the C parameter plus $^1/_8$).

Other early recognized modifications have rhombohedral units. In each, the atoms are on trigonal axes with the singly equivalent positions uuu corresponding presumably to the space group C_{3v}^5 ($R3m$). The dimensions of these rhombohedra and the number of molecules in each are stated in columns 2, 3, and 4 of Table III,6; the values of u are given below. These structures are, however, best compared with one another by stating positions within the three-times bigger hexagonal pseudocells (columns 5, 6, and 7). The atoms in these can be described as occurring in triplets with the coordinates:

$$00u; \; {}^1/_3,{}^2/_3,{}^1/_3+u; \; {}^2/_3,{}^1/_3,{}^2/_3+u$$

and the same values of the parameters u that apply to the rhombohedral unit.

Type I: The atoms in the 15-molecule hexagonal cell have the parameters: u(C) = 0, $^2/_{15}$, $^6/_{15}$, $^9/_{15}$, and $^{13}/_{15}$, and u_n (Si) = u_n(C) plus $^1/_{20}$.

Type IV: The atoms in the 21-molecule cell have the parameters: u(Si) = 0, $^4/_{21}$, $^6/_{21}$, $^9/_{21}$, $^{12}/_{21}$, $^{15}/_{21}$, and $^{17}/_{21}$, and u_n(C) = u_n(Si) plus $^1/_{28}$.

Type VI: The atoms in the 33-molecule cell have the parameters: u(Si) = 0, $^2/_{33}$, $^6/_{33}$, $^8/_{33}$, $^{12}/_{33}$, $^{15}/_{33}$, $^{18}/_{33}$, $^{21}/_{33}$, $^{25}/_{33}$, $^{27}/_{33}$, and $^{31}/_{33}$, and u_n(C) = u_n(Si) plus $^1/_{44}$.

The cubic beta form is easily expressed in terms of a similar hexagonal pseudocell. It has one molecule in the rhombohedral and three in the hexagonal pseudocell. For both u(C) = 0 and u(Si) = $^1/_4$.

Each of these structures can be considered as having planes of carbon atoms (or of silicon atoms) stacked one above another in different ways. For the simple beta modification, the second of these layers is obviously displaced with respect to one chosen as first by $1/3\ 2/3\ 1/3$ and the third layer by $2/3\ 1/3\ 2/3$; the fourth layer at $0\ 0\ 3/3$ is then directly over the first. If as was done in Chapter II when discussing cobalt and the two forms of graphite, layers in these three positions are designated as *0, 1, 2,* then the vertical sequence in the beta form, that of a cubic close-packing, is

0,1,2,0,1,2,0,1,2, . . .

and has a non-identically placed plane above and below any chosen one.

Expressing planar displacements in the vertical sequence in this fashion, it follows directly from the coordinates listed above that the other modifications already discussed can be represented as:

For III: *0,1,0,2, 0,1,0,2, . . .*
For II: *0,1,2,0,2,1, 0,1,2,0,2,1, . . .*
For I: *0,2,0,1,2,1,0,1,2,0,2,1,2,0,1, 0,2,0,1,2, . . .*
For IV: *0,2,1,2,0,1,0,2,1,0,1,2,0,2,1,0,2,0,1,2,1, . . .*
For VI: *0,2,0,1,2,1,0,2,0,1,2,1,0,1,2,0,2,1,0,1,2,0,*
2,1,2,0,1,0,2,1,2,0,1, ...

Continuing to follow Chapter II by designating as *H* a plane that has the same, and by *C* a plane that has different displacements above and below it (corresponding to the hexagonal and cubic close-packed distributions), the sequences in these forms of SiC can also be written:

For beta: *0,1,2, 0,1,2, 0,1,2,*
C C C C C C C C C
For III: *0,1,0,2, 0,1,0,2, 0,1,0,2,*
H C H C H C H C H C H
For II: *0,1,2,0,2,1, 0,1,2,0,2,1,*
C C H C C H C C H C C
For I: *0,2,0,1,2,1,0,1,2, 0,2,1,2,0,1, 0,2,0,*
H C C H C H C C H C H C C H C H C
For IV: *0,2,1,2,0,1,0,2,1,0,1,2,0,2,1,0,2,0,1,2,1,0,2,*
C H C C H C C C H C C H C C C H C C H C C C
For VI: *0,2,0,1, 2,1,0,2, 0,1,0,2, 0,2,2,0, 2,1,0,1,*
C H C C H C C H C C H C H C C H C C H C
2,0,2,1, 2,0,1,0 2,1,2, 0 1,0,2,0,1
C H C H C C H C C H C C H C H C

From this standpoint the various modifications of SiC appear as regularly repeated mixed sequences of hexagonal and cubic close-packed layers. It is noteworthy that, except for the ZnO-like form, which is a purely *H H H H* sequence, these modifications do not have adjacent *H H* components. Considering this naturally imposed restriction, they are amongst the simplest imaginable sequences of *C* and *H* packings. There are others, however, now to be discussed, which are very complicated in their patterns of repetition.

The two early studied among these are rhombohedral. One was designated V, the other VII. This arbitrary notation using roman numerals was acceptable as long as the known modifications were few in number, but they are now so numerous that a better one is needed. We shall follow the system which uses *H* and *R* to indicate whether the fundamental lattice is hexagonal or rhombohedral and a numeral giving the number of molecules in the hexagonal units. As can be seen from the Table, this is usually, though not always, enough to identify a modification.

The structures found for V and VII are (using the same group of coordinates as above):

Type V: The parameters of the atoms in the 51-molecule hexagonal cell are: $u(\mathrm{Si}) = 0, {}^{2}/_{51}, {}^{6}/_{51}, {}^{8}/_{51}, {}^{12}/_{51}, {}^{14}/_{51}, {}^{18}/_{51}, {}^{21}/_{51}, {}^{24}/_{51}, {}^{27}/_{51}, {}^{30}/_{51}, {}^{33}/_{51}, {}^{37}/_{51}, {}^{39}/_{51}, {}^{43}/_{51}, {}^{45}/_{51}, {}^{49}/_{51}$ and $u(\mathrm{C}) = u(\mathrm{Si})$ plus ${}^{1}/_{68}$.

Type VII: The parameters of the atoms in the 87-molecule cell are: $u(\mathrm{Si}) = 0, {}^{2}/_{87}, {}^{6}/_{87}, {}^{8}/_{87}, {}^{12}/_{87}, {}^{14}/_{87}, {}^{18}/_{87}, {}^{20}/_{87}, {}^{24}/_{87}, {}^{26}/_{87}, {}^{30}/_{87}, {}^{33}/_{87}, {}^{36}/_{87}, {}^{39}/_{87}, {}^{42}/_{87}, {}^{45}/_{87}, {}^{48}/_{87}, {}^{51}/_{87}, {}^{54}/_{87}, {}^{57}/_{87}, {}^{61}/_{87}, {}^{63}/_{87}, {}^{67}/_{87}, {}^{69}/_{87}, {}^{73}/_{87}, {}^{75}/_{87}, {}^{79}/_{87}, {}^{81}/_{87}, {}^{85}/_{87}$ and $u(\mathrm{C}) = u(\mathrm{Si})$ plus ${}^{1}/_{116}$.

Listing as before the vertical planar displacements for these two forms, their sequence can be written as:

For V:	*0,2,0,1,*	*2,1,0,2,*	*0,1,2,1,*	*0,2,0,1,*	*2,1,0,1,*	*2,0,2,1*
	H C C	*H C C H*	*C C H C*	*C H C C*	*H C H C*	*C H C C*
For VII:	*0,2,0,1,*	*2,1,0,2,*	*0,1,2,1,*	*0,2,0,1,*	*2,1,0,2,*	
	H C C	*H C C H*	*C C H C*	*C H C C*	*H C C H*	
	0,1,2,1,	*0,2,0,1,*	*2,1,0,1,*	*2,1,2,0*		
	C C H C	*C H C C*	*H C H C*	*C H C C*		

In an abbreviated form, this series for V can be written as

$$(HCC)_5HC(HCC)_5 \ . \ . \ .$$

For VII it would be

$$(HCC)_9HC(HCC)_9 \ . \ . \ .$$

Writing the other two rhombohedral forms in the same fashion:

$$\text{For I: } (HCC)_1HC(HCC)_1 \ldots$$
$$\text{For VI: } (HCC)_3HC(HCC)_3 \ldots$$

This brings out clearly the fundamental relationship that exists between these four modifications.

In recent years additional rhombohedral modifications have been described. The simplest, $R27$, has nine molecules in the unit rhombohedron:

$$a_0' = 22.735 \text{ A.}, \qquad \alpha = 7°46'$$

and 27 in the hexagonal cell:

$$a_0 = 3.079 \text{ A.}, \qquad c_0 = 67.996 \text{ A.}$$

Parameters describing the positions of the atoms of these 27 molecules are $u(\text{Si}) = 0, {}^4/_{27}, {}^8/_{27}, {}^{10}/_{27}, {}^{12}/_{27}, {}^{14}/_{27}, {}^{16}/_{27}, {}^{20}/_{27}, {}^{24}/_{27}$, and for C, $u(\text{C}) = u(\text{Si})$ plus ${}^1/_{36}$.
These result in the sequence:

$$\begin{array}{llllll} 0,2,1,2, & 0,2,1,2, & 0,1,0,2, & 0,1,0,2, & 0,1,2,1, & 0,1,2,1, \\ \;\;C\,H\,C & H\,C\,H\,C & C\,H\,C\,H & C\,H\,C\,H & C\,C\,H\,C & H\,C\,H\,C \\ 0,1,2,0, & 2,1 & & & & \\ H\,C\,C\,H & C & & & & \end{array}$$

$= (CH)_3 \cdot CCH \cdot (CH)_3 \cdot CCH \cdot (CH)_3 \cdot CCH$ or $HCC \cdot (HC)_3 \cdot HCC \cdot (HC)_3$

There is a second form, $51R(2)$, with 51 molecules in the hexagonal cell (17 in the unit rhombohedron) and the dimensions:

$$a_0' = 42.849 \text{ A.}, \qquad \alpha = 4°7' \quad \text{for the rhombohedron}$$

and

$$a_0 = 3.079 \text{ A.}, \qquad c_0 = 128.437 \text{ A.} \quad \text{for the hexagonal unit}$$

The atomic parameters are: $u(\text{Si}) = 0, {}^3/_{51}, {}^7/_{51}, {}^{11}/_{51}, {}^{15}/_{51}, {}^{19}/_{51}, {}^{21}/_{51}, {}^{23}/_{51}, {}^{25}/_{51}, {}^{27}/_{51}, {}^{29}/_{51}, {}^{31}/_{51}, {}^{33}/_{51}, {}^{35}/_{51}, {}^{39}/_{51}, {}^{43}/_{51}, {}^{47}/_{51}$; $u(\text{C}) = u(\text{Si})$ plus ${}^1/_{68}$.
These result in:

$$\begin{array}{llllll} 0,1,2,0, & 2,1,2,0, & 2,1,2,0, & 2,1,2,0, & 2,1,2,0, & 1,0,2,0, \\ 1,0,2,0, & 1,0,2,0, & 1,0,2,0, & 1,2,1,0, & 1,2,1,0, & 1,2,1,0, \\ 1,2,1, & & & & & \end{array}$$

$= (CH)_8 \cdot C \cdot (CH)_8 \cdot C \cdot (CH)_8$, or $(HC)_7 \cdot HCC \cdot (HC)_7 \cdot HCC$.

Another more complicated form ($75R$) has 75 molecules in the hexagonal cell (25 in the unit rhombohedron). Its dimensions are:

$$a_0' = 62.984 \text{ A.}, \quad \alpha = 2°48' \quad \text{for the rhombohedron}$$
$$a_0 = 3.079 \text{ A.}, \quad c_0 = 188.878 \text{ A.} \quad \text{for the hexagonal unit}$$

Its atomic parameters are: $u(\text{Si}) = 0, {}^3/_{75}, {}^7/_{75}, {}^9/_{75}, {}^{11}/_{75}, {}^{15}/_{75}, {}^{19}/_{75}, {}^{21}/_{75}, {}^{23}/_{75}, {}^{27}/_{75}, {}^{29}/_{75}, {}^{31}/_{75}, {}^{35}/_{75}, {}^{38}/_{75}, {}^{42}/_{75}, {}^{45}/_{75}, {}^{49}/_{75}, {}^{51}/_{75}, {}^{55}/_{75}, {}^{58}/_{75}, {}^{62}/_{75}, {}^{64}/_{75}, {}^{66}/_{75}, {}^{68}/_{75}, {}^{72}/_{75}$, and $u(\text{C}) = u(\text{Si})$ plus ${}^1/_{100}$. This leads to the sequence: $[HCC \cdot HC \cdot HCC \cdot (HC)_2 \cdot HCC \cdot HC \cdot (HCC)_2 \cdot HC]_{12}$.

The next more complicated form has 84 molecules in the hexagonal unit (28 in the unit rhombohedron) with the cell dimensions:

$$a_0' = 70.537 \text{ A.}, \quad \alpha = 2°30' \quad \text{for the rhombohedron}$$
$$a_0 = 3.079 \text{ A.}, \quad c_0 = 211.543 \text{ A.} \quad \text{for the hexagonal unit}$$

The parameters of its atoms are: $u(\text{Si}) = 0, {}^3/_{84}, {}^6/_{84}, {}^9/_{84}, {}^{12}/_{84}, {}^{15}/_{84}, {}^{18}/_{84}, {}^{21}/_{84}, {}^{25}/_{84}, {}^{27}/_{84}, {}^{29}/_{84}, {}^{33}/_{84}, {}^{35}/_{84}, {}^{39}/_{84}, {}^{41}/_{84}, {}^{45}/_{84}, {}^{47}/_{84}, {}^{51}/_{84}, {}^{54}/_{84}, {}^{58}/_{84}, {}^{60}/_{84}, {}^{64}/_{84}, {}^{66}/_{84}, {}^{70}/_{84}, {}^{72}/_{84}, {}^{76}/_{84}, {}^{78}/_{84}, {}^{80}/_{84}$, and $u(\text{C}) = u(\text{Si})$ plus ${}^1/_{112}$. This gives rise to: $[(HCC)_7 \cdot HC \cdot HCC \cdot HC \cdot]_n$.

A structure has been described for the form $141R$ which has 141 molecules in the hexagonal and 47 in the rhombohedral unit of the dimensions:

$$a_0' = 118.359 \text{ A.}, \quad \alpha = 1°30' \quad \text{for the rhombohedral unit}$$
$$a_0 = 3.079 \text{ A.}, \quad c_0 = 355.04 \text{ A.} \quad \text{for the hexagonal unit}$$

The parameters of its atoms are: $u(\text{Si}) = 0, {}^2/_{141}, {}^6/_{141}, {}^8/_{141}, {}^{12}/_{141}, {}^{14}/_{141}, {}^{18}/_{141}, \ldots {}^{139}/_{141}$, and $u(\text{C}) = u(\text{Si})$ plus ${}^1/_{188}$.
This gives rise to: $(HCC)_{15} \cdot HC(HCC)_{15} \cdot HC(HCC)_{15} \ldots$

The most recent form to be analyzed in detail is $174R$. It has 58 molecules in a rhombohedral unit and 174 in the hexagonal cell having:

$$a_0 = 3.08 \text{ A.}, \quad c_0 = 436.7 \text{ A.}$$

The layer sequence was found to be: $[C \cdot (CCH)_7 \cdot CCC \cdot (CCH)_{11}]_3$.

There is still another modification, $393R$, for which the atomic sequence has not, however, yet been deduced. Its units have the dimensions:

$$a_0' = 329.87 \text{ A.}, \quad \alpha = 0°32'$$
$$a_0 = 3.079 \text{ A.}, \quad c_0 = 989.60 \text{ A.}$$

Beyond these there is an indication of a form that repeats with $c_0 = 594 \times 2.51$ A. = ca. 1500 A.

Several additional hexagonal forms have also been analyzed recently. The simplest of these has eight molecules in its unit hexagonal prism with the cell dimensions:

$$a_0 = 3.079 \text{ A.}, \quad c_0 = 20.1470 \text{ A.}$$

Atoms are in the same special positions used to describe the other modifications based on C_{6v}^4 ($C6mc$):

$$(2a)\quad 00u;\qquad 0,0,u+1/2$$
$$(2b)\quad 1/3\,2/3\,v;\qquad 2/3,1/3,v+1/2$$

with u(Si) $= 0$, v(Si) $= 2/8, 5/8, 7/8$, and with the C parameters equal to the Si parameters plus $3/32$.

This corresponds to the sequence:

$$0,2,1,2,\ \ 0,1,2,1,\ \ 0,2,1,2,$$
$$(CC)\cdot HC\cdot(CC) = HCC\cdot C\cdot HCC\cdot C = (HCCC)_n$$

Another new hexagonal form ($10H$) contains ten molecules in a cell of the dimensions:

$$a_0 = 3.079 \text{ A.},\qquad c_0 = 25.183 \text{ A.}$$

In this case atoms have been placed in the following special positions of C_{3v}^1 ($C3m$):

$$(1a)\quad 00u;\qquad (1b)\quad 1/3\,2/3\,v,\qquad (1c)\quad 2/3\,1/3\,w$$

with u(Si) $= 0, 3/10, 7/10$; v(Si) $= 2/10, 4/10, 6/10, 8/10$; w(Si) $= 1/10, 5/10, 9/10$, and with the C parameters equal to the Si parameters plus $3/40$.

These lead to the sequence:

$$0,2,1,0,\ \ 1,2,1,0,\ \ 1,2,\ \ 0,2,1,0, = (HCC)_2\cdot(HC)_2$$

Still another hexagonal form contains 19 molecules in a unit of the dimensions:

$$a_0 = 3.079 \text{ A.},\qquad c_0 = 47.849 \text{ A.}$$

Atoms are in the positions of C_{3v}^1 just stated, with u(Si) $= 0, 2/19, 4/19, 8/19, 11/19, 15/19, 17/19$; v(Si) $= 1/19, 5/19, 7/19, 9/19, 13/19, 16/19$; w(Si) $= 3/19, 6/19, 10/19, 12/19, 14/19, 18/19$, and with the C parameters equal to the Si parameters plus $3/76$. The resulting sequence is $(HCC\cdot HC)_3(HC)_2$.

A hexagonal form containing 27 molecules in the unit has also been described. Its dimensions, the same as those of $27R$, are:

$$a_0 = 3.079 \text{ A.},\qquad c_0 = 67.996 \text{ A.}$$

Atoms have been placed in the following positions of D_{3d}^3 ($P\bar{3}m1$):

$$(1a)\quad 000;\qquad (2c)\quad \pm(00u);\qquad (2d)\quad \pm(1/3\,2/3\,v)$$

with u(Si) = $^2/_{27}$, $^6/_{27}$, $^9/_{27}$, $^{12}/_{27}$; v(Si) = $^1/_{27}$, $^4/_{27}$, $^8/_{27}$, $^{10}/_{27}$, $^{14}/_{27}$, $^{16}/_{27}$, $^{20}/_{27}$, $^{22}/_{27}$, $^{24}/_{27}$, and with the C parameters equal to the Si parameters plus $^1/_{36}$. The planar sequence to which this corresponds is: $CH\cdot(CCH)\cdot CH\cdot$-$(CCH)_5\cdot CH\cdot(CCH)$.

As already indicated in paragraph **III,c1**, ZnS shows the same kind of polymorphism as SiC. Fewer forms have been found, but detailed structures have been described for at least two.

For the 8*H* form the atomic arrangement is the same as that for the corresponding polymorph of SiC with one pair of zinc atoms in (2*a*), u = 0, and three pairs in (2*b*) with v = $^2/_8$, $^5/_8$, $^7/_8$. The sulfur atoms have the same coordinates with the S parameters equal to the Zn parameters plus $^3/_{32}$.

The structure given the 10*H* modification is different from that of the 10*H* type SiC. Like the 8*H* form it is based on C_{6v}^4 ($P6_3mc$). Two pairs of zinc atoms are in (2*a*) with u = 0 and $^3/_{10}$; three pairs are in (2*b*) with v = $^1/_{10}$, $^4/_{10}$, $^7/_{10}$. The sulfur atoms have the same coordinates with the same parameters plus $^3/_{40}$. This gives rise to the sequence:

$$0,1,2,0,\quad 1,0,2,1,\quad 0,2,\quad 0,1,2,0,\quad = (CCCCH)_n$$

This is much simpler than that described for 10*H* silicon carbide and is in line with the observed 6*H* and 8*H* types. Perhaps it would be worthwhile to reexamine the structure of the 10*H* form of SiC.

III,c4. The two compounds *cadmium antimonide*, CdSb, and *zinc antimonide*, ZnSb, have a structure that is essentially intermetallic. Their orthorhombic units containing eight molecules have the dimensions:

CdSb: a_0 = 6.471 A.; b_0 = 8.253 A.; c_0 = 8.526 A.
ZnSb: a_0 = 6.218 A.; b_0 = 7.741 A.; c_0 = 8.115 A.

Atoms have been found to be in general positions of V_h^{15} ($Pbca$):

(4*c*) $\pm(xyz;\ x+{}^1/_2,{}^1/_2-y,\bar{z};\ \bar{x},y+{}^1/_2,{}^1/_2-z;\ {}^1/_2-x,\bar{y},z+{}^1/_2)$

with the parameters:

	Atom	x	y	z
For CdSb				
	Sb	0.136	0.072	0.108
	Cd	0.456	0.119	−0.128
For ZnSb				
	Sb	0.1420	0.0841	0.1083
	Zn	0.4539	0.1115	−0.1327

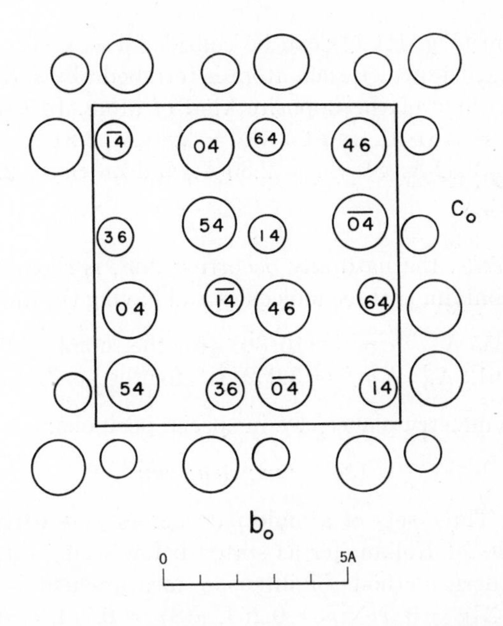

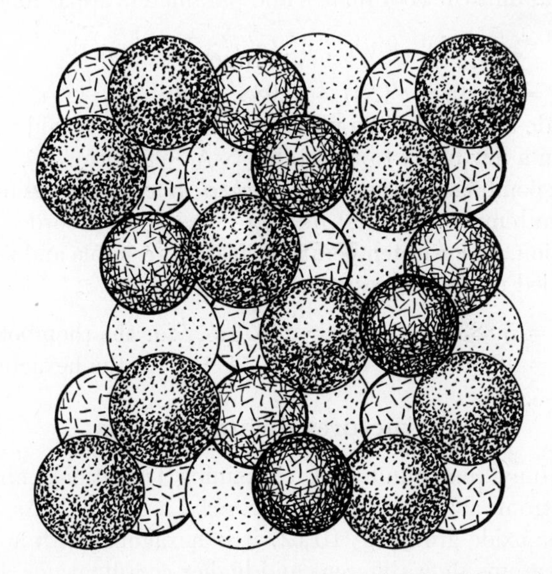

Fig. III,14a (top). A projection along the a_0 axis of the orthorhombic structure of CdSb. The smaller circles are antimony. Origin in the lower right.

Fig. III,14b (bottom). A packing drawing of the orthorhombic CdSb structure viewed along the a_0 axis. The antimony atoms are line-shaded.

This structure (Fig. III,14) can be considered as a very deformed diamond arrangement in which each atom is tetrahedrally surrounded by one of its own and three of the opposite kind of atom. In CdSb, Cd–Cd = 2.99 A., Sb–Sb = 2.81 A., and Cd–Sb = 2.80 A., 2.81 A., and 2.91 A. In ZnSb, Zn–Zn = 2.82 A., Sb–Sb = 2.80 A., and Zn–Sb = 2.64 A., 2.64 A., and 2.76 A.

III,c5. *Millerite*, the naturally occurring NiS, is rhombohedral with a flattened unit containing three molecules and having the dimensions:

$$a_0 = 5.655 \text{ A.}, \quad \alpha = 116°36' \quad \text{for the rhombohedral unit}$$
$$a_0' = 9.612 \text{ A.}, \quad c_0' = 3.259 \text{ A.} \quad \text{for the hexagonal unit}$$

Its atoms, in symmetry planes, are in special positions:

$$(3b) \quad uuv;\ uvu;\ vuu$$

of C_{3v}^5 ($R3m$). Three sets of atomic parameters were early proposed. Of these, the results of Kolkmeijer as stated below seem best, but a further study with modern methods of intensity measurement is to be desired. Parameters: u(Ni) = 0, v(Ni) = 0.264, u(S) = 0.714, v(S) = 0.361. For the hexagonal unit the coordinates and parameters are as follows:

$$(9b) \quad u'\bar{u}'v';\ u',2u',v';\ 2\bar{u}',\bar{u}',v';\ \text{rh}$$

with u'(Ni) = −0.088, v'(Ni) = 0.088, u'(S) = 0.114, v'(S) = 0.596.

When made artificially, the millerite form of NiS is said to be unstable, going over in a few hours to the stable NiAs-like form.

The corresponding millerite-like *nickel selenide*, NiSe, was found so short-lived at room temperature that its pattern could be recorded only by electron diffraction. At lower temperatures it is more stable and x-ray measurements have led to the dimensions:

$$\text{NiSe: } a_0 = 5.8834 \text{ A.}, \quad \alpha = 116°31' \quad \text{for the rhombohedral unit}$$
$$a_0' = 10.005 \text{ A.}, \quad c_0' = 3.351 \text{ A.} \quad \text{for the hexagonal unit}$$

Structures Related to NiAs

III,d1. Many compounds RX containing transitional and other non-rare gas electronic shell metals have the *nickel arsenide*, NiAs, arrangement. Like the zinc oxide grouping (**III,c2**) it is hexagonal with a two-molecule unit, but its atoms show different and higher coordinations. In contrast to most of the structures thus far enumerated, its atoms R and X have different environments and non-interchangeable positions. Atoms are in the same special positions of C_{6v}^4 ($C6mc$) that are used in graphite, for instance, but the axial ratios and values of u are so different that the resulting atomic groupings are not similar.

These positions for the NiAs arrangement are:

$$\begin{array}{lll} \text{R:} & (2a) & 000;\ 0\,0\,{}^1/_2 \\ \text{X:} & (2b) & {}^1/_3\,{}^2/_3\,u;\ {}^2/_3,{}^1/_3,u+{}^1/_2 \end{array}$$

where R is the metallic and X the metalloid element (Fig. III,15).

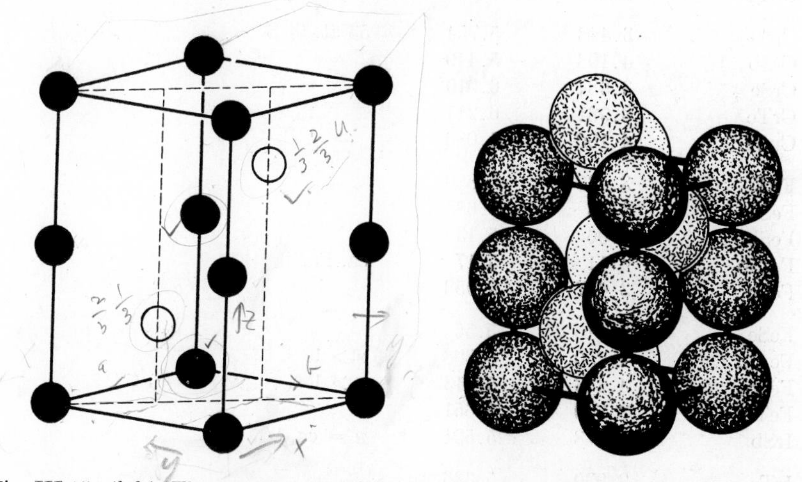

Fig. III,15a (left). The arrangement of the atoms in the hexagonal cell of NiAs. Black circles represent the nickel, open circles the arsenic atoms.

Fig. III,15b (right). A perspective packing drawing of the NiAs arrangement. The nickel atoms are dot-shaded, the arsenic atoms are line-shaded.

With $u = {}^1/_4$, each X atom is surrounded by six equidistant R atoms situated at the corners of a right trigonal prism. Each R atom on the other hand has eight close neighbors, six of which are X atoms while the other two are those R atoms immediately above and below it. It is usual to find the more metallic atom of a compound in the R position, but this is not necessary and thus there arises the possibility of an anti as well as the normal NiAs structure. A few of these anti arrangements are now known.

In contrast with the ZnO arrangement (**III,c2**) with its very constant axial ratio, c/a for NiAs-like crystals ranges between 1.2 and 1.7. The composition of many of these crystals is highly variable, and this excess of one or the other kind of element leads to large variations in the observed cell dimensions. Both these factors must be borne in mind when evaluating the data of Table III,7, which lists the various compounds known to have this structure.

TABLE III,7
Crystals with the NiAs Arrangement (**III,d1**)

Crystal	a_0, A.	c_0, A.	Remarks
AuSn	4.314	5.512	
CoS	3.367	5.160	
CoSb	3.866	5.188	
CoSe	3.6294	5.3006	
CoTe	3.886	5.360	
CrS	3.448	5.754	53 at.-% S
CrSb	4.108	5.440	
CrSe	3.684	6.019	
CrTe	3.981	6.211	
CuSn	4.190	5.086	
FeS	3.438	5.880	
FeS_x	3.43	5.68	
FeSb	4.06	5.13	
Fe_xSb	4.11	5.17	ca. Fe_3Sb_2
FeSe	3.637	5.958	
$FeSe_x$	3.51	5.55	
Fe_xSn	4.230	5.208	$x > 1$
$FeSn_x$	4.233	5.213	$x > 1$
FeTe	3.800	5.651	
IrSb	3.978	5.521	u = ca. $^1/_4$
IrTe	3.930	5.386	
MnAs	3.710	5.691	20°C.
MnBi	4.30	6.12	
MnSb	4.120	5.784	
Mn_xSb	4.13	5.74	ca. Mn_3Sb_2
MnTe	4.1429	6.7031	20°C.
δ′-NbN	2.968	5.549	Anti-NiAs
NbS	3.32	6.46	Excess S
NiAs	3.602	5.009	
NiS	3.4392	5.3484	Stable above 485°C. Excess S reduces cell dimensions
NiSb	3.94	5.14	
NiSe	3.6613	5.3562	Can accept up to 20% excess Se
NiSn	4.048	5.123	42 at.-% Sn
NiTe	3.957	5.354	
PdSb	4.070	5.582	
PdSn	4.378	5.627	36–42 at.-% Sn
PdTe	4.1521	5.6719	
PtB	3.358	4.058	Anti-NiAs

(continued)

The large differences in composition shown by so many of these crystals have given rise to difficult problems. Such variations could be due to additional atoms occupying holes in the normal structure (excess structures) or to vacancies in the normal atomic positions (deficit structures). It should be possible to distinguish between these two types of grouping, since one should be more, the other less, dense than the perfect 1:1 compound. Application of this criterion has indicated that in general they are of the deficit type. Further evidence of this is provided by the observation, for example, that $FeS_{1.x}$ and $FeSe_{1.x}$ have unit cells smaller than those of FeS and FeSe. Apparently, metallic deficits when they occur do not exceed ca. 10%, but deficits in the metalloid atoms run much higher in such materials as $Fe_{1.5}Sb$ and $Mn_{1.5}Sb$. With metalloid-deficient substances, cell dimensions vary little with the composition.

A complete understanding of NiAs-like crystals is made still harder by the "superlattice" that is often associated with them. The apparent cell dimensions of the superlattice structures may be very great, thus indicating an unusually complicated atomic arrangement. Only gradually can we expect to acquire an adequate knowledge of these complications and the reasons for them.

TABLE III,7 (*continued*)

Crystal	a_0, A.	c_0, A.	Remarks
PtBi	4.315	5.490	
PtSb	4.130	5.472	
PtSn	4.103	5.428	
$RhB_{1.1}$	3.309	4.224	Anti-NiAs
RhBi	4.094	5.663	Accepts large amounts of excess Bi with little cell change
RhSn	4.340	5.555	40 at.-% Sn
RhTe	3.99	5.66	
ScTe	4.122	6.735	
TiS[a] (low)	3.30	6.44	
TiSe[a]	3.559	6.220	Form continuous solid solutions with diselenide and ditelluride
TiTe[a]	3.834	6.390	(same as above)
VP	3.18	6.22	
VS	3.360	5.813	
VSe	3.580	5.977	
VTe	3.942	6.126	53 at.-% Te
ZrTe	3.953	6.647	Forms continuous solid solutions with $ZrTe_2$ from $ZrTe_{0.8} \rightarrow ZrTe_{2.0}$

[a] Also reported to have larger cells due to regular stacking disorders or to deficit structures.

A detailed study has been made of stoichiometric *ferrous sulfide*, FeS, in an attempt to define the departures from the NiAs arrangement which account for the lower symmetry of pure FeS and analogous compounds. This has resulted in the choice of a large hexagonal unit containing 12 molecules and having the cell edges

$$a_0 = 5.968 \text{ A.}, \qquad c_0 = 11.74 \text{ A.}$$

Atoms have been placed in the following positions of D_{3h}^4 ($C\bar{6}2c$):

S(1): (2a) $000;\ 00\,{}^1/_2$

S(2): (4f) $\pm({}^1/_3\,{}^2/_3\,u;\ {}^1/_3,{}^2/_3,{}^1/_2-u)$ with $u = 0.016$,

S(3): (6h) $u\,v\,{}^1/_4;\ \bar{v},u-v,{}^1/_4;\ v-u,\bar{u},{}^1/_4;$
$v\,u\,{}^3/_4;\ \bar{u},v-u,{}^3/_4;\ u-v,\bar{v},{}^3/_4$ with $u = {}^2/_3$, $v = 0$

Fe: (12i) $xyz;\quad \bar{y},x-y,z;\quad y-x,\bar{x},z;$
$x,y,{}^1/_2-z;\ \bar{y},x-y,{}^1/_2-z;\ y-x,\bar{x},{}^1/_2-z;$
$y,x,z+{}^1/_2;\ \bar{x},y-x,z+{}^1/_2;\ x-y,\bar{y},z+{}^1/_2;$
$yx\bar{z};\quad \bar{x},y-x,\bar{z};\quad x-y,\bar{y},\bar{z}$

with $x = 0.360$, $y = 0.040$, $z = 0.125$. In this "superlattice" on the NiAs structure there is the same succession of metallic and non-metallic atomic layers along the principal axis. The metallic layers, however, show a certain displacement normal to the axis, while the sulfur atoms are displaced along it.

III,d2. Low-temperature *titanium monosulfide*, TiS, has, as stated in paragraph **III,d1**, the nickel arsenide structure. Its high-temperature form has a structure that can be considered as a relatively simple "superlattice" on NiAs. The probable arrangement is described as one in which three molecules are in a unit rhombohedron of the dimensions:

$$a_0 = 9.04 \text{ A.}, \qquad \alpha = 21°48'$$

The corresponding hexagonal cell containing nine molecules has the edges:

$$a_0' = 3.417 \text{ A.}, \qquad c_0' = 26.4 \text{ A.}$$

The space group was chosen as D_{3d}^5 ($R\bar{3}m$), with atoms in the following positions (for the hexagonal cell):

Ti(1): (3b) $0\,0\,{}^1/_2$; rh

Ti(2): (6c) $\pm(00u)$; rh with $u = 0.378$

S(1): (3a) 000; rh

S(2): (6c) with $u' = 0.226$

This is a stacking of NiAs sub-units turned about the c_0 axis with respect to one another. The shortest Ti–S = 2.32 A. and Ti–Ti = 3.18 A.

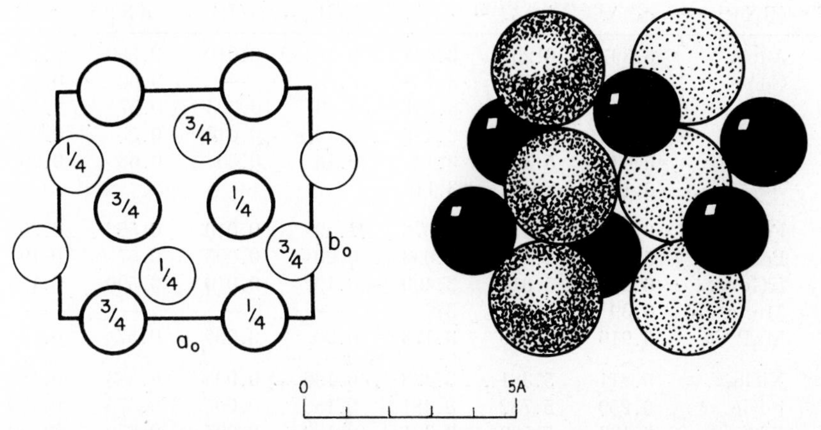

Fig. III,16a (left). A projection along c_0 of the orthorhombic structure of MnP. The manganese atoms are the larger circles. Origin in lower right.

Fig. III,16b (right). A packing drawing of the orthorhombic MnP structure viewed along the c_0 axis. The dotted atoms are manganese.

III,d3. The structure assigned to *manganese phosphide,* MnP, and isomorphous compounds is a distortion of the NiAs arrangement (**III,d1**). It is orthorhombic with four molecules in unit prisms having the dimensions given in Table III,8. All atoms are in the following special positions of V_h^{16} (*Pbnm*):

$$(4c) \quad \pm(u\,v\,{}^1/_4;\ {}^1/_2-u,v+{}^1/_2,{}^1/_4)$$

with the parameters of Table III,8.

In the structure thus defined (Fig. III,16), each phosphorus or other metalloid atom is surrounded by six metal atoms at the corners of a trigonal prism, but this is not regular as in NiAs. Instead, one base is much larger than the other. Metal atoms have six metalloid atoms at the corners of a distorted octahedron and four (rather than two as in NiAs) metal atoms which here are tetrahedrally distributed in four of the octahedral interstices.

III,d4. The compound *ferric boride,* FeB, is also orthorhombic with a unit cell somewhat similar to that of FeP in size and shape. Its four-molecular unit has:

$$a_0 = 4.053 \text{ A.};\ b_0 = 5.495 \text{ A.};\ c_0 = 2.946 \text{ A.}$$

TABLE III,8
Crystals with the Orthorhombic MnP Arrangement (**III,d3**)

Crystal (RX)	a_0, A.	b_0, A.	c_0, A.	u(R)	v(R)	u(X)	v(X)
AuGa	6.397	6.267	3.421	0.184	0.010	0.590	0.195
CoAs	5.869	5.292	3.458	0.25	—	0.60	0.25
CoP	5.599	5.076	3.281	0.20	0.005	0.57	0.19
CrAs	6.222	5.741	3.486	0.19	0.010	0.58	0.20
CrP	5.94	5.366	3.13	0.16	0.00	0.63	0.20
	6.108	5.362	3.113				
FeAs	6.028	5.439	3.373	0.19	0.010	0.58	0.20
FeP	5.793	5.187	3.093	0.20	0.005	0.57	0.19
IrGe	6.281	5.611	3.490	0.192	0.010	0.590	0.185
MnAs	6.39	5.64	3.63	—	—	—	—
MnP	5.916	5.260	3.173	0.20	0.005	0.57	0.19
NiGe	5.811	5.381	3.428	0.190	0.005	0.583	0.188
PdGe	6.259	5.782	3.481	0.188	0.005	0.575	0.190
PdSi	6.133	5.599	3.381	0.190	0.007	0.570	0.190
PdSn	6.32	6.13	3.87	0.182	0.007	0.590	0.182
PtGe	6.088	5.732	3.701	0.195	0.010	0.590	0.195
PtSi	5.932	5.595	3.603	0.195	0.010	0.590	0.195
RhGe	6.48	5.70	3.25	0.202	0.007	0.564	0.191
RhSb	6.333	5.952	3.876	0.192	0.010	0.590	0.195
RuP	6.120	5.520	3.168	—	—	—	—
VAs	6.317	5.879	3.334	—	—	—	—
WP	6.219	5.717	3.238	0.175	0.00	0.62	0.20

Like FeP it has an atomic arrangement based on V_h^{16} (*Pbnm*) with all atoms in symmetry planes and defined by the coordinates:

$$(4c) \quad \pm(u\, v\, {}^1/_4;\ {}^1/_2-u, v+{}^1/_2, {}^1/_4)$$

The parameters found for iron and those assigned to boron are, however, very different from those given the atoms in FeP and they result in a very different marshalling. In FeB, u(Fe) has been found to be 0.125 and v(Fe) = 0.180. In the structure pictured in Figure III,17, u(B) = 0.61 and v(B) = 0.04.

This arrangement differs from those of NiAs and FeP in that its metalloid atoms are in close contact with one another as well as with surrounding metal atoms. Each boron atom has two boron neighbors 1.77 A. away and thus can be thought of as forming part of a zigzag chain extending indefinitely through the crystal. It also has seven iron neighbors at distances of 2.11 A. and 2.17 A. These separations are in acceptable agreement with

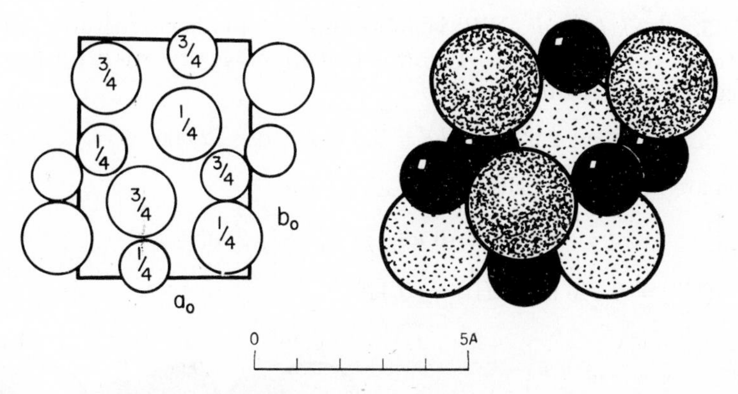

Fig. III,17a (left). A projection along c_0 of the orthorhombic structure of FeB. The iron atoms are the larger circles. Origin in lower right.
Fig. III,17b (right). A packing drawing of the orthorhombic FeB structure viewed along the c_0 axis. The iron atoms are dotted.

the sum of the neutral boron radius (0.88 A.) and the neutral trivalent iron radius (1.28 A.). The iron atoms have a high coordination number inasmuch as each has seven boron neighbors at the foregoing distances as well as an equilateral trigonal right prism of six iron atoms at 2.72 A. and four more at 2.95 A.

Crystals of *cobalt boride*, CoB, were early given this structure with:

$$a_0 = 3.948 \text{ A.};\ b_0 = 5.243 \text{ A.};\ c_0 = 3.037 \text{ A.}$$

The assigned parameters, very similar to those for FeB, are: $u(\text{Co}) = 0.125$, $v(\text{Co}) = 0.180$, $u(\text{B}) = 0.64$, $v(\text{B}) = 0.037$.

The essential correctness of this FeB structure has been confirmed by a more recent study of *titanium boride*, TiB. For it:

$$a_0 = 4.56 \text{ A.};\ b_0 = 6.12 \text{ A.};\ c_0 = 3.06 \text{ A.}$$

The determined atomic parameters are in close agreement with those found and assumed for FeB: $u(\text{Ti}) = 0.123$, $v(\text{Ti}) = 0.177$, $u(\text{B}) = 0.603$, $v(\text{B}) = 0.029$.

It is instructive to compare the three structures described in **III,d1**, **d3**, and **d4** with those of ZnS (**III,c1**) and ZnO (**III,c2**). In the groupings involving the zinc atoms, contact is maintained only between unlike atoms. In the NiAs structure there is close contact between both unlike atoms and the metal atoms among themselves, and this is also true of MnP; in FeB there is contact between both of these and between the metalloid atoms as well.

III,d5. *Nickel boride*, NiB, is an example of another structure containing strings of boron atoms. It forms orthorhombic crystals with a tetramolecular cell of the dimensions:

$$a_0 = 2.925 \text{ A.};\ b_0 = 7.396 \text{ A.};\ c_0 = 2.966 \text{ A.}$$

Atoms are in special positions of V_h^{17} (*Cmcm*):

$$(4c)\quad \pm(0\ u\ {}^1/_4;\ {}^1/_2, u+{}^1/_2, {}^1/_4)$$

with u(Ni) = 0.146 and u(B) = 0.440.

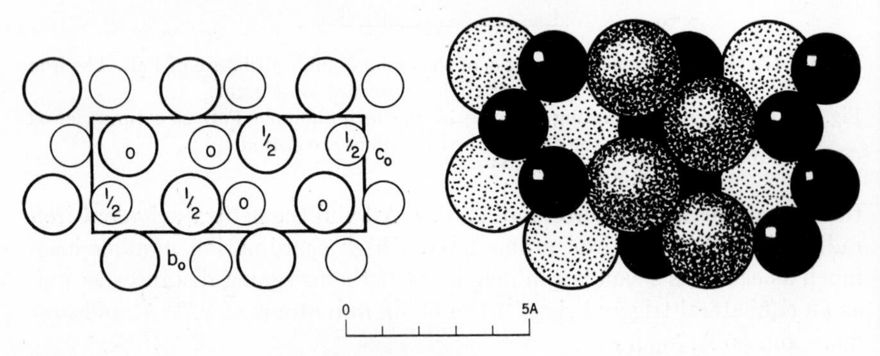

Fig. III,18a (left). A projection along a_0 of the orthorhombic structure of NiB. The nickel are the larger circles. Origin in lower right.

Fig. III,18b (right). A packing drawing of the orthorhombic NiB structure viewed along the a_0 axis. The smaller, black atoms are boron.

In this arrangement (Fig. III,18), a nickel atom has six nickel neighbors, four at 2.584 A. and two at 2.619 A. The boron atoms, making contact with one another as well as with nickel, form chains running in the direction of the c_0 axis. The B–B separation is 1.72 A. and in addition each boron is 2.177 A. away from four nickel atoms and 2.632 A. from two more.

Another compound with this structure is *chromium boride*, CrB. Its cell dimensions are:

$$a_0 = 2.969 \text{ A.};\ b_0 = 7.858 \text{ A.};\ c_0 = 2.932 \text{ A.}$$

In two different determinations u(Cr) has been chosen as 0.146 and 0.143 and u(B) as 0.440 and 0.43. The B–B separation within the chains is 1.72 A., as in NiB; the shortest Cr–B = 2.19 A.

Three other borides are known to have this structure. Their unit cells have been given the dimensions:

VB: a_0 = 3.10 A.; b_0 = 8.17 A.; c_0 = 2.98 A.
TaB: a_0 = 3.276 A.; b_0 = 8.669 A.; c_0 = 3.157 A.
WB: a_0 = 3.19 A.; b_0 = 8.40 A.; c_0 = 3.07 A. (1850°C.)

The a_0 and c_0 axes of this structure are so nearly equal that special care must be taken not to confuse them. It is to be noted that their relative lengths are reversed in some of the foregoing compounds. This probably should be checked through further work.

III,d6. Crystals of *lithium arsenide*, LiAs, are monoclinic with an eight-molecule cell of the dimensions:

$$a_0 = 5.79 \text{ A.};\ b_0 = 5.24 \text{ A.};\ c_0 = 10.70 \text{ A.};\ \beta = 117°24'$$

Atoms are in general positions of C_{2h}^5 ($P2_1/c$):

$$(4e) \quad \pm(xyz;\ x,{}^1/_2-y,z+{}^1/_2)$$

with the parameters of Table III,9.

The same structure is possessed by *sodium antimonide*, NaSb. Its unit has the dimensions:

$$a_0 = 6.80 \text{ A.};\ b_0 = 6.34 \text{ A.};\ c_0 = 12.48 \text{ A.};\ \beta = 117°36'$$

The atomic parameters are stated in Table III,9.

TABLE III,9
Parameters of the Atoms in LiAs and Related Compounds

Atom	x	y	z
		LiAs	
Li(1)	0.235	0.402	0.329
Li(2)	0.232	0.669	0.045
As(1)	0.3042	0.9143	0.2992
As(2)	0.2891	0.1626	0.1011
		NaSb	
Na(1)	0.2108	0.3892	0.3251
Na(2)	0.2179	0.6725	0.0409
Sb(1)	0.3076	0.8994	0.2954
Sb(2)	0.2932	0.1599	0.1051
		KSb	
K(1)	0.226	0.408	0.330
K(2)	0.219	0.668	0.032
Sb(1)	0.325	0.898	0.287
Sb(2)	0.319	0.170	0.124

This atomic arrangement is shown in Figure III,19. It can be thought of as composed of two spiralling chains of the nonmetallic elements with axes parallel to the b_0 axis, interspersed with similar chains of the metallic elements. In LiAs the As–As distance alternates between 2.454 and 2.472 A.; in NaSb the Sb–Sb separations are 2.845 and 2.857 A. Each lithium atom has six arsenic neighbors at distances between 2.63 and 2.89 A.; in NaSb the corresponding Na–Sb distances range between 3.13 and 3.52 A. In LiAs there are three short Li–Li = 2.99 and 3.00 A.; in NaSb one short Na–Na = 3.44 A. The other metal-to-metal contacts lie between 3.22 and 3.78 A. for LiAs and 3.71 and 4.45 A. for NaSb.

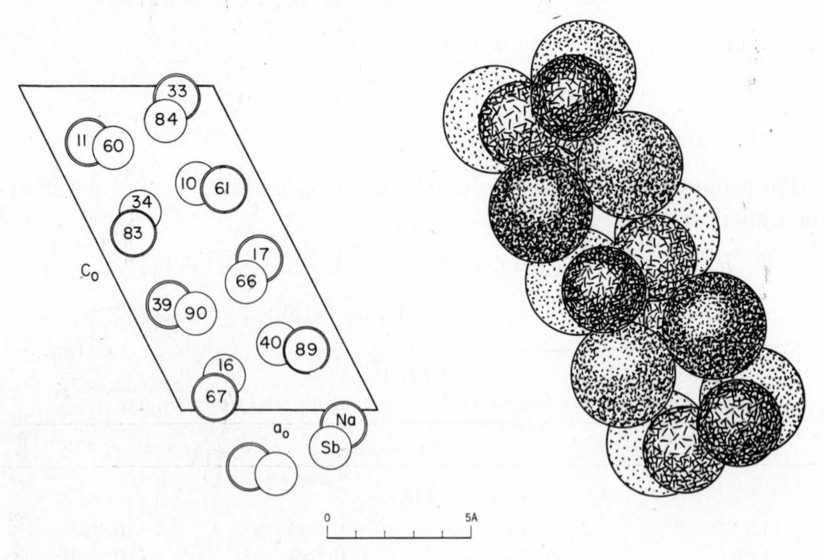

Fig. III,19a (left). The monoclinic structure of NaSb projected along its b_0 axis. Origin in lower left.

Fig. III,19b (right). A packing drawing of the NaSb arrangement seen along its b_0 axis. Atoms have their metallic radii, the smaller antimony atoms being line-shaded.

Potassium monantimonide, KSb, also has this structure with:

$$a_0 = 7.18 \text{ A.};\ b_0 = 6.97 \text{ A.};\ c_0 = 13.40 \text{ A.};\ \beta = 115°6'$$

The atomic parameters are stated in Table III,9. In this crystal the six potassium atoms about an antimony atom are at a mean distance of ca. 3.64 A. In an antimony chain, Sb–Sb = 2.81 and 2.88 A.

III,d7. *Thallium selenide,* TlSe, provides an example of another structure in which the metalloid atoms occur in strings. There are eight molecules in a tetragonal unit of the dimensions:

$$a_0 = 8.02 \text{ A.}, \qquad c_0 = 7.00 \text{ A.}$$

In the structure described for this crystal, based on the body-centered space group D_{4h}^{18} ($I4/mcm$), thallium atoms are of two sorts in positions fixed by symmetry:

$$\text{Tl(1)}: \quad (4a) \quad \pm(0\,0\,{}^1/_4);\ \text{B.C.}$$
$$\text{Tl(2)}: \quad (4b) \quad \pm(0\,{}^1/_2\,{}^1/_4);\ \text{B.C.}$$

Selenium positions are given by the single set of coordinates:

$$\text{Se}: \quad (8h) \quad \pm(u,u+{}^1/_2,0;\ u+{}^1/_2,\bar{u},0);\ \text{B.C.}$$

with $u = 0.179$. In this grouping (Fig. III,20), Tl(2) shows an eightfold coordination with Tl–Se = 3.42 A. and Tl(1) a fourfold coordination with Tl–Se = 2.68 A. The Se–Se separation along the c_0 axis is ${}^1/_2c_0 = 3.50$ A.

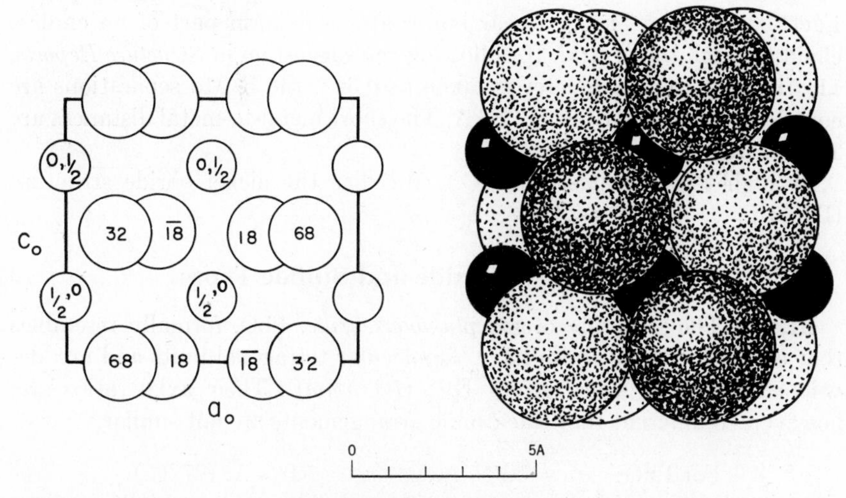

Fig. III,20a (left). The tetragonal structure of TlSe projected along an a_0 axis. The thallium atoms are the smaller circles. Origin in lower left.

Fig. III,20b (right). A packing drawing of the tetragonal TlSe arrangement seen along an a_0 axis. Atoms have been given their approximate ionic radii, the thallium atoms being black.

Two other crystals are known to have this structure. They are:

$$\text{TlS:}\quad a_0 = 7.77\text{ A.},\qquad c_0 = 6.79\text{ A.}$$
$$\text{InTe:}\quad a_0 = 8.437\text{ A.},\qquad c_0 = 7.139\text{ A.}$$

For InTe, $u = 0.180$.

III,d8. A structure has been proposed for the low-temperature forms of *molybdenum monoboride*, MoB, and the corresponding WB. The symmetry is tetragonal with eight molecules in cells of the dimensions:

$$\text{MoB:}\quad a_0 = 3.105\text{ A.},\qquad c_0 = 16.97\text{ A.}$$
$$\text{WB:}\quad a_0 = 3.115\text{ A.},\qquad c_0 = 16.93\text{ A.}$$

Atoms have been placed in:

$$(8e)\quad 00u;\ 00\bar{u};\ 0,{}^1/_2,u+{}^1/_4;\ 0,{}^1/_2,{}^1/_4-u;\ \text{B.C.}$$

of D_{4h}^{19} ($I4/amd$). For molybdenum (and tungsten) $u = 0.197$. The boron atoms have been given the same coordinates $(8e)$ with $u(\text{B}) = 0.352$ (for both compounds).

In the atomic arrangement thus described (Fig. III,21) each metal atom has six boron neighbors at the corners of an enveloping triangular prism. Each boron atom has two near boron atoms to form part of an endless chain through the crystal. If, following the suggestion in *Structure Reports*, **11**, 51, the boron parameter is taken as 0.344, the B–Mo separations are equal to 2.32 A. and B–B = 1.87 A. The short metal-to-metal distances are 2.83–2.89 A.

At high temperatures (1850°C.), WB has the nickel boride structure (**III,d5**).

Miscellaneous Oxide and Sulfide Types

III,e1. The structure of *red plumbous oxide*, PbO, formally resembles that of PH_4I (**III,b3**); both have bimolecular tetragonal units and are developed from the space group D_{4h}^7 ($P4/nmm$). Their axial ratios are, however, so different that the atomic arrangements are not similar.

$$\text{For PbO:}\quad a_0 = 3.975\text{ A.},\qquad c_0 = 5.023\text{ A. (27°C.)}$$

Lead atoms are in:

$$(2c)\quad 0\,{}^1/_2\,u;\ {}^1/_2\,0\,\bar{u}\qquad \text{with } u = 0.2385$$

Their great scattering power has made it hard to be certain of the oxygen

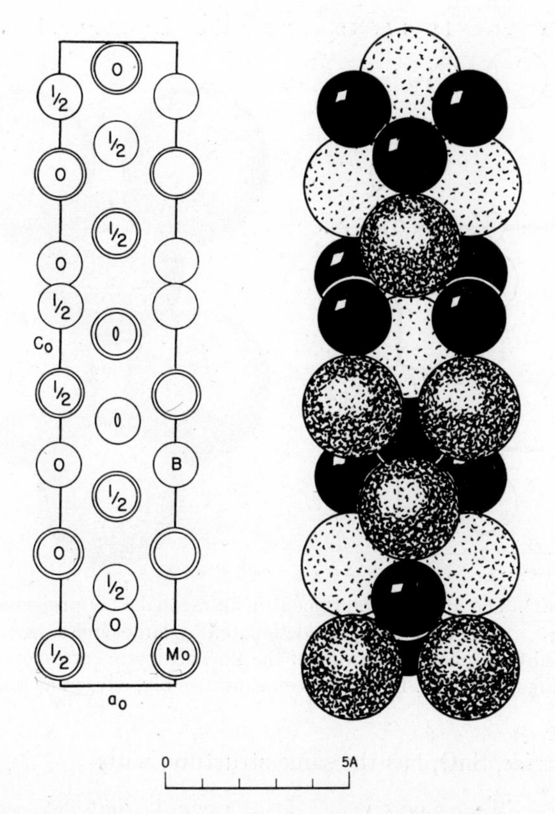

Fig. III,21a (left). The tetragonal structure of MoB projected along an a_0 axis. Origin in lower left.
Fig. III,21b (right). A packing drawing of the MoB structure seen along an a_0 axis. The molybdenum atoms are the large dotted circles.

positions, but the best available evidence indicates that these atoms probably are in:

$$(2a) \quad 000; \ {}^1/_2\,{}^1/_2\,0$$

If this is correct the significant atomic separations are Pb–O = 2.33 A. with four lead atoms about each oxygen, and Pb–Pb = 3.70–3.98 A., with each lead atom having 12 lead as well as four oxygen neighbors.

The lead atoms are in a distorted cubic close-packing that would be perfect if u were $^1/_4$ and if c/a were 1.41 instead of 1.25. This structure (Fig. III,22) can accordingly be described as a metallic lead array swelled and distorted by the introduction of oxygen atoms in sheets.

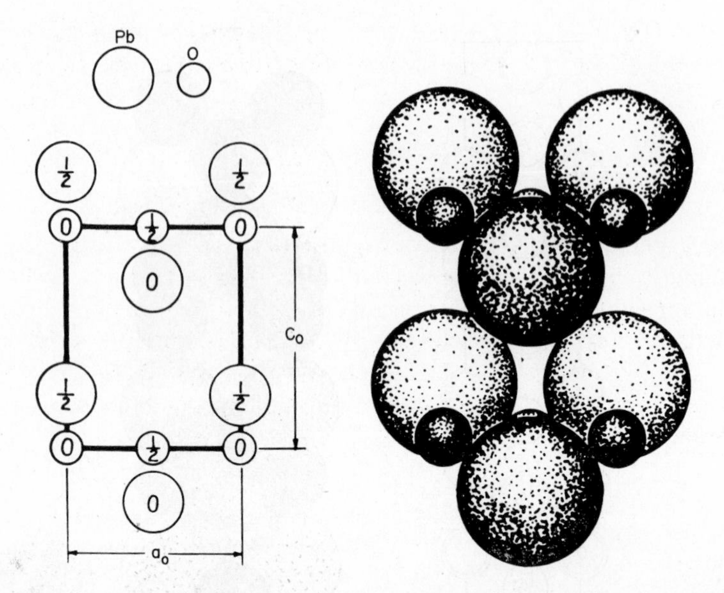

Fig. III,22a (left). The atomic arrangement in tetragonal PbO projected on an *a* face. The lead atoms are designated by large circles.

Fig. III,22b (right). A packing drawing of the PbO structure corresponding to the projection of Figure III,22a. Lead has been shown as big, oxygen as small spheres.

Stannous oxide, SnO, has the same structure, with

$$a_0 = 3.802 \text{ A.}, \qquad c_0 = 4.836 \text{ A. } (26° \text{ C.})$$

Indium bismuthide, InBi, has been given an atomic arrangement "anti" to the foregoing, with

$$a_0 = 5.005 \text{ A.}, \qquad c_0 = 4.771 \text{ A.}$$

This places the indium atoms in $(2a)$ and the bismuth atoms in $(2c)$, but with the rather different parameter $u = 0.393$. In the resulting arrangement, the shortest In–Bi = 3.13 A.

The structure given *lithium hydroxide*, LiOH, is like that of PbO. Its OH ions have been put in $(2c)$ with u between 0.18 and 0.22. For it

$$a_0 = 3.546 \text{ A.}, \qquad c_0 = 4.334 \text{ A.}$$

A recent neutron diffraction study has shown that $u(\text{O}) = 0.1941$ and $u(\text{H}) = 0.4072$. This yields the separations O–H = 0.981 A., a shortest Li–O = 1.96 A., and O–O = 3.06 A.

III,e2. *Zirconium monosulfide*, ZrS, has the type of structure initially ascribed to red PbO. Its bimolecular tetragonal cell has the edge lengths:

$$a_0 = 3.55 \text{ A.}, \qquad c_0 = 6.31 \text{ A.}$$

Both types of atom are in the special positions of D_{4h}^7 ($P4/nmm$):

$$(2c) \quad 0\,{}^1/_2\,u;\ {}^1/_2\,0\,\bar{u} \text{ with } u(\text{Zr}) = 0.19 \text{ and } u(\text{S}) = 0.69$$

This arrangement is illustrated in Figure III,23. It is built up of double zirconium–sulfur sheets perpendicular to the c_0 axis. Within a sheet each zirconium atom has four sulfur neighbors with Zr–S = 2.67 A.; and one more with Zr–S = 3.15 A. Each sulfur has four sulfur atoms at a distance of 3.47 A. and two more at 3.55 A., as well as four zirconium atoms at 2.67 A. The zirconium environment of a zirconium atom is the same as the sulfur environment of a sulfur atom.

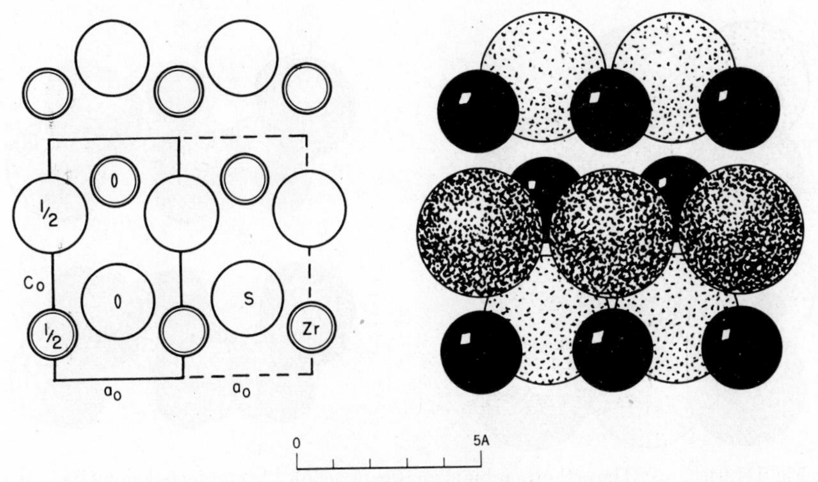

Fig. III,23a (left). Two cells of the tetragonal ZrS structure projected along an a_0 axis. Origin in the lower left.

Fig. III,23b (right). A packing drawing of the ZrS arrangement seen along an a_0 axis. The large dotted atoms of sulfur have their ionic size.

III,e3. The structure of the *yellow* form of *plumbous oxide*, PbO, including an accurate determination of the positions of its oxygen atoms, has been established by neutron diffraction.

There are four molecules in an orthorhombic unit having the dimensions:

$$a_0 = 5.891 \text{ A.};\ b_0 = 4.775 \text{ A.};\ c_0 = 5.489 \text{ A.} \quad (27°\text{C.})$$

The chosen space group is the high symmetry V_h^{11} ($Pbma$) and in this more recent study the a_0 and c_0 axes have been interchanged with respect to those

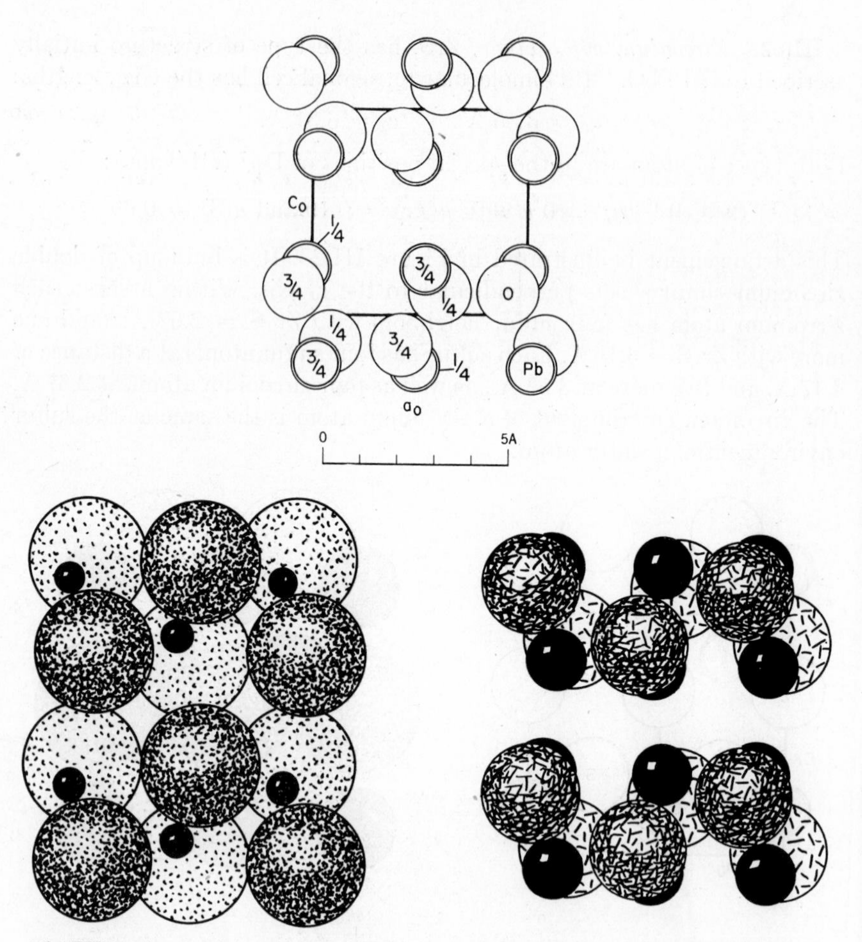

Fig. III,24a (top). The orthorhombic structure of yellow PbO projected along its b_0 axis. Origin in lower left.

Fig. III,24b (bottom, left). A packing drawing of the yellow PbO arrangement, viewed along its b_0 axis. The lead atoms have been given their metallic radius, the oxygen atoms being then the small black circles.

Fig. III,24c (bottom, right). A packing drawing of PbO seen, as in Figure III,24b, along the b_0 axis. Here the line-shaded oxygen atoms have been given their ionic radii.

selected in the earlier x-ray work. The lead atoms have nearly the positions found by x rays, and they and the oxygen atoms occur in the special positions:

$$(4d) \quad \pm(u\ {}^1/_4\ v;\ u+{}^1/_2,{}^1/_4,\bar{v})$$

For lead, $u = -0.0208$, $v = 0.2309$; for oxygen, $u = 0.0886$, $v = -0.1309$.

This is a structure (Fig. III,24) which consists of layers of atoms normal to the c_0 axis. In each layer a lead atom is surrounded by four oxygen atoms, at distances of 2.21, 2.22, 2.49, and 2.49 A. All are on the same side of the lead, thus forming a flat four-sided pyramid. The angle O–Pb–O involving the short bonds is 90°24′. The shortest Pb–Pb separations are 3.47 A. within a sheet and 3.63 A. between sheets.

As in the case of red PbO, the arrangement can be thought of as a packing of metallic lead atoms with small oxygen atoms in the interstices (Fig. III,24b); if the structure is drawn giving the atoms their approximate ionic radii (Fig. III,24c), the sheets normal to c_0 appear wide apart.

III,e4. The mineral *cooperite*, PtS, has been given a structure resembling that of red PbO. Platinum atoms have been put in

$$0\ {}^1/_2\ 0;\quad {}^1/_2\ 0\ {}^1/_2$$

and sulfur in

$$0\ 0\ {}^1/_4;\quad 0\ 0\ {}^3/_4$$

of the bimolecular tetragonal cell:

$$a_0 = 3.47\ \text{A.},\qquad c_0 = 6.10\ \text{A.}$$

An alternative way of expressing these atomic positions brings out the formal analogy between the structures of PtS and PbO. This places the metal atoms in the same positions (2*c*) $0\ {}^1/_2\ u;\ {}^1/_2\ 0\ \bar{u}$ as the lead atoms in PbO with $u = {}^1/_4$. The sulfur positions, at 000 and $0\ 0\ {}^1/_2$, are different from those of the oxygen atoms in PbO.

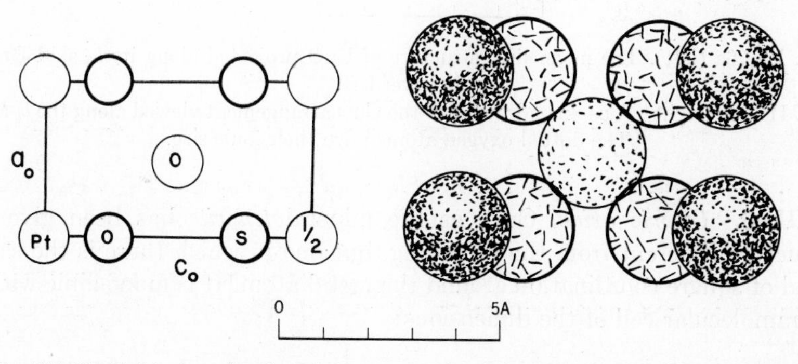

Fig. III,25a (left). A projection along an a_0 axis of the tetragonal structure of PtS
Fig. III,25b (right). A packing drawing of the tetragonal PtS structure viewed along an a_0 axis. The sulfur atoms are line-shaded.

In cooperite each metal atom has about it four metalloid atoms at the corners of a square and each sulfur is surrounded by a tetrahedron of metal atoms (Fig. III,25). This structure, too, can be considered as a distorted cubic close-packing of the metal atoms, but in this case the metalloid atoms are in interstices different from those occupied in PbO.

Two oxides of the platinum metals have also been given this structure. They are:

$$\text{PdO:}\quad a_0 = 3.043 \text{ A.},\quad c_0 = 5.337 \text{ A.}\quad (26°\text{C.})$$
$$\text{PtO:}\quad a_0 = 3.04 \text{ A.},\quad c_0 = 5.34 \text{ A.}$$

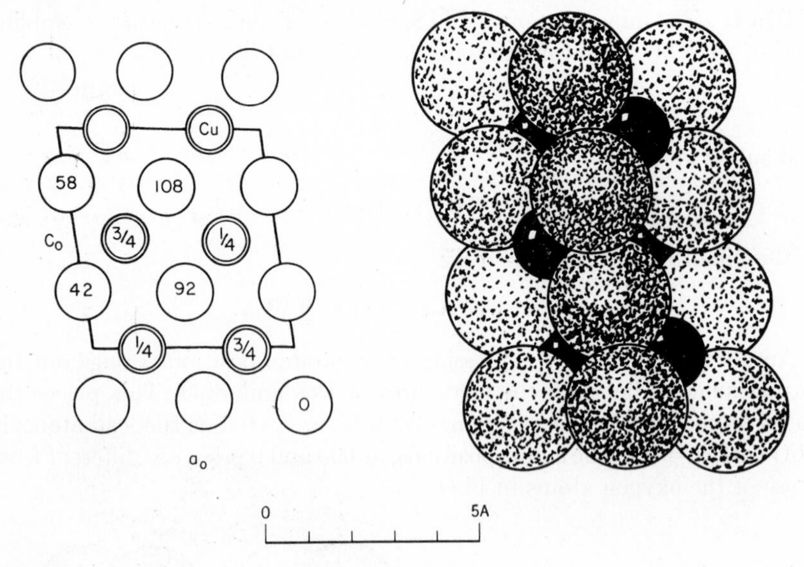

Fig. III,26a (left). The monoclinic structure of CuO projected along its b_0 axis. Origin in lower left.

Fig. III,26b (right). A packing drawing of the CuO arrangement viewed along the b_0 axis. The dotted oxygen atoms have their ionic size.

III,e5. *Cupric oxide*, CuO, as the mineral *tenorite* has been given a structure different from the preceding, but one in which there is the same kind of square coordination around the metal atom. It is monoclinic with a tetramolecular cell of the dimensions:

$$a_0 = 4.653 \text{ A.};\ b_0 = 3.410 \text{ A.};\ c_0 = 5.108 \text{ A.};\ \beta = 99°29'$$

The proposed atomic arrangement based on C_{2h}^6 ($C2/c$) has its atoms in the following positions:

$$\text{Cu: } (4c) \quad \pm(\tfrac{1}{4}\,\tfrac{1}{4}\,0;\ \tfrac{3}{4}\,\tfrac{1}{4}\,\tfrac{1}{2})$$
$$\text{O: } (4e) \quad \pm(0\,u\,\tfrac{1}{4};\ \tfrac{1}{2},u+\tfrac{1}{2},\tfrac{1}{4})$$

If $u = -0.584$, rather than the usually stated 0.584, each atom of copper in the resulting arrangement (Fig. III,26) is surrounded by an approximate square of oxygens, with Cu–O = ca. 1.88 A. and 1.96 A. Each oxygen has about it a distorted tetrahedron of copper atoms.

III,e6. *Silver monoxide*, AgO, as investigated with x rays, was described as having the same structure as CuO though with a cell of the somewhat different shape:

$$a_0 = 5.852 \text{ A.};\ b_0 = 3.478 \text{ A.};\ c_0 = 5.495 \text{ A.};\ \beta = 107°30'$$

A recent neutron diffraction study in giving more weight to the contributions of oxygen shows that the correct space group is C_{2h}^5 ($P2_1/c$) rather than C_{2h}^6. As in the earlier work, the metal atoms are face-centered (origin shifted by $\tfrac{1}{4}\,\tfrac{1}{4}\,0$); the oxygens are in general positions:

$$\text{Ag(1): } (2a) \quad 000;\ 0\,\tfrac{1}{2}\,\tfrac{1}{2}$$
$$\text{Ag(2): } (2d) \quad \tfrac{1}{2}\,0\,\tfrac{1}{2};\ \tfrac{1}{2}\,\tfrac{1}{2}\,0$$
$$\text{O: } (4e) \quad \pm(xyz;\ x,\tfrac{1}{2}-y,z+\tfrac{1}{2})$$

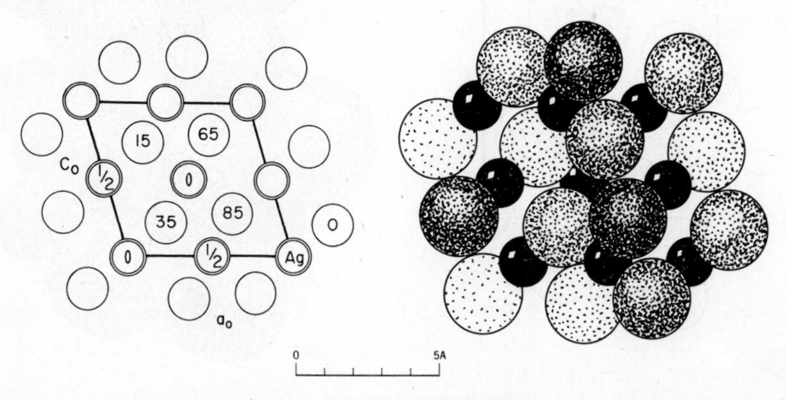

Fig. III,27a (left). The monoclinic AgO arrangement projected along its b_0 axis. The difference between this and the CuO structure becomes evident by comparing Figures 26 and 27. Origin in lower left.

Fig. III,27b (right). A packing drawing of the structure of AgO seen along its b_0 axis. The dotted oxygen circles have been given their ionic size.

with $x = 0.295$, $y = 0.350$, $z = 0.230$. This represents a small though important displacement of the oxygen atoms from the positions given them under C_{2h}^6.

In this structure (Fig. III,27), the two kinds of silver atoms are differently bound to oxygen. About Ag(1) are two linearly distributed oxygens with Ag(1)–O = 2.18 A. There are four oxygens around Ag(2) at the corners of an approximate square (O–Ag–O = 91°30′) with Ag(2)–O = 2.01–2.05 A.

It probably would be worthwhile to reexamine the structure of CuO in the light of these results to see if perhaps it too has this arrangement rather than that of paragraph **III,e5**.

In the neutron study of AgO, no evidence was found to substantiate the existence of a previously reported ZnS form (**III,c1**).

III,e7. The structure found for *palladous sulfide*, PdS, has tetragonal symmetry, with a unit containing eight molecules. Cell dimensions are:

$$a_0 = 6.4287 \text{ A.,} \qquad c_0 = 6.6082 \text{ A.}$$

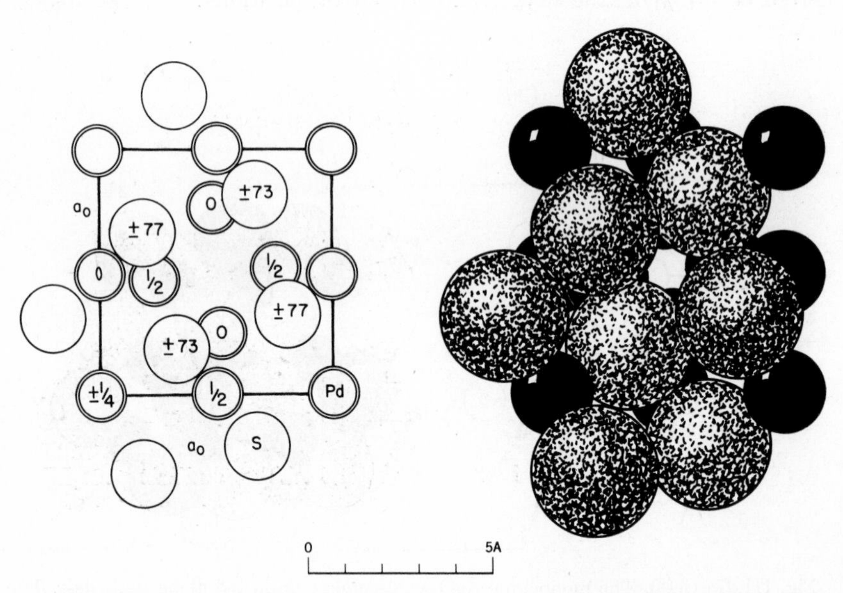

Fig. III,28a (left). The tetragonal PdS structure projected along its c_0 axis. Origin in lower right.

Fig. III,28b (right). A packing drawing of the PdS structure seen along its principal, c_0, axis. The large dotted sulfur atoms have their ionic size.

Atoms have been placed in the following positions of C_{4h}^2 ($P4_2/m$):

Pd(1): (2e) $\pm(0\,0\,{}^1/_4)$
Pd(2): (2c) $0\,{}^1/_2\,0;\ {}^1/_2\,0\,{}^1/_2$
Pd(3): (4j) $\pm(uv0;\ v\,\bar{u}\,{}^1/_2)$ with $u = 0.48$, $v = 0.25$
S: (8k) $\pm(xyz;\ xy\bar{z};\ \bar{y},x,z+{}^1/_2;\ y,\bar{x},z+{}^1/_2)$

with $x = 0.19$, $y = 0.32$, $z = 0.23$.

In this arrangement (Fig. III,28), the palladium atoms, though of three different sorts, are each surrounded by four sulfur atoms at the corners of deformed squares. The four palladium atoms around each sulfur atom are at corners of distorted tetrahedra.

In one investigation the analogous compound PdSe has been found, with

$$a_0 = 6.72 \text{ A.}, \qquad c_0 = 6.90 \text{ A.}$$

For it the variable parameters have been stated to be: Pd(3): $u = 0.455$, $v = 0.235$, and Se: $x = 0.20$, $y = 0.325$, $z = 0.230$.

III,e8. Crystals of a composition close to $CrS_{0.97}$ have a structure that has been described as intermediate between those of NiAs (**III,d1**) and PtS (**III,e4**). They are monoclinic with a tetramolecular cell of the dimensions:

$$a_0 = 3.826 \text{ A.};\ b_0 = 5.913 \text{ A.};\ c_0 = 6.089 \text{ A.};\ \beta = 101°36'$$

Atoms have been placed in the following positions of C_{2h}^6 ($C2/c$):

Cr: (4a) $000;\ 0\,0\,{}^1/_2;\ {}^1/_2\,{}^1/_2\,0;\ {}^1/_2\,{}^1/_2\,{}^1/_2$
S: (4e) $\pm(0\,u\,{}^1/_4;\ {}^1/_2,u+{}^1/_2,{}^1/_4)$ with $u = 0.320$

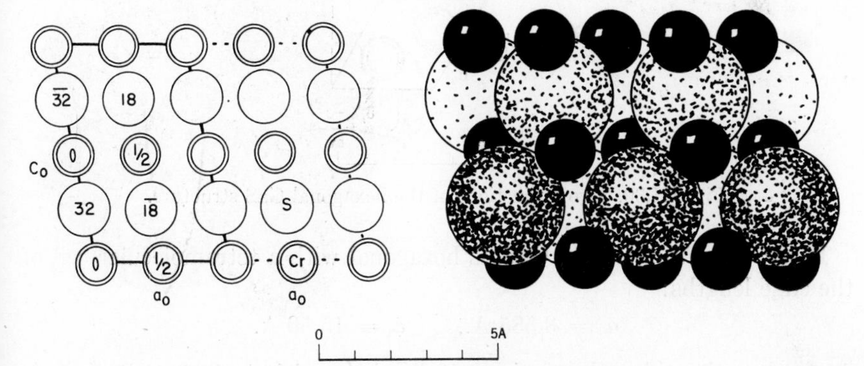

Fig. III,29a (left). A projection, along b_0, of two cells of the monoclinic CrS arrangement. Origin in lower left.

Fig. III,29b (right). A packing drawing of the CrS structure seen along the b_0 axis. The small black circles are the chromium atoms.

The resulting arrangement is shown in Figure III,29. The chromium atoms are surrounded by a tetragonal bipyramid of sulfur atoms with Cr–S = 2.429, 2.437, and 2.878 A; the closest Cr–Cr = 3.044 A. and the closest S–S = 3.357 A.

In earlier work this substance was described as having a structure that was a "superlattice" on the NiAs arrangement. This was found not to be the case and the present atomic arrangement was derived from material carefully selected to give a single-crystal pattern.

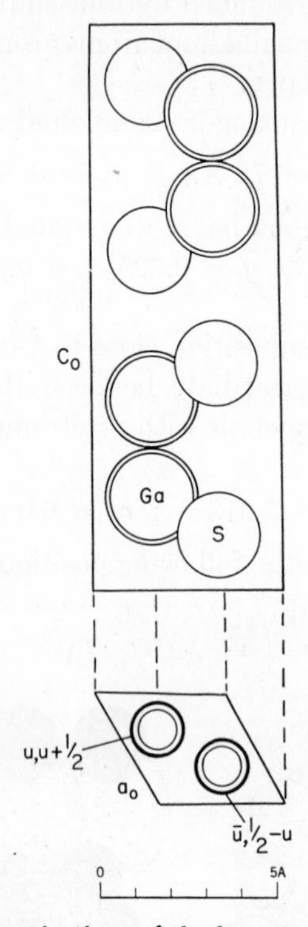

Fig. III,30. Two projections of the hexagonal GaS structure.

III,e9. *Gallium sulfide*, GaS, is hexagonal with a tetramolecular cell of the edge lengths:

$$a_0 = 3.585 \text{ A.}, \qquad c_0 = 15.50 \text{ A.}$$

Atoms have been found to be in the following positions of D_{6h}^4 ($P6_3/mmc$):

$$(4f) \quad \pm(1/3\ 2/3\ u;\ 2/3,1/3,u+1/2)$$

with u = 0.17 for gallium and 0.60 for sulfur.

In this structure (Fig. III,30), double layers of gallium atoms succeed double layers of sulfur along the c_0 axis. The Ga–Ga separation is 2.52 A., and Ga–S = 2.30 A.

Indium selenide, InSe, has the same structure, with

$$a_0 = 4.05 \text{ A.}, \qquad c_0 = 16.93 \text{ A.}$$

The parameters as established by electron diffraction are $u(\text{In}) = 0.157$ and $u(\text{Se}) = 0.602$. Here the shortest interatomic distances are In–In = 3.16 A., In–Se = 2.51 A., and Se–Se = 4.17 A.

Though there appears to be some disagreement in the literature, the best indications to date are that *gallium selenide*, GaSe, has this structure with:

$$a_0 = 3.742 \text{ A.}, \qquad c_0 = 15.919 \text{ A.}$$

For gallium, $u = 0.177$; for selenium, $u = 0.60$.

It has been reported that there is also a rhombohedral form of GaSe. The six-molecule hexagonal cell has been given the dimensions:

$$a_0' = 3.746 \text{ A.}, \qquad c_0' = 23.910 \text{ A.}$$

An atomic arrangement has been stated, but further work seems desirable.

III,e10. *Cupric sulfide*, CuS, as the mineral *covellite*, is hexagonal with the elongated six-molecule cell:

$$a_0 = 3.796 \text{ A.}, \qquad c_0 = 16.36 \text{ A.}$$

The most probable of three suggested structures is based on D_{6h}^4 ($P6_3/mmc$) and has atoms in the following special positions:

$$\begin{aligned} \text{Cu(1)}&: \ (2d) \quad \pm(^2/_3\ ^1/_3\ ^1/_4) \\ \text{Cu(2)}&: \ (4j) \quad \pm(^1/_3\ ^2/_3\ u;\ ^1/_3,^2/_3,^1/_2-u) \\ \text{S(1)}&: \ (2c) \quad \pm(^1/_3\ ^2/_3\ ^1/_4) \\ \text{S(2)}&: \ (4e) \quad \pm(00v;\ 0,0,^1/_2-v) \end{aligned}$$

with $u = 0.107$ and $v = 0.064$. This arrangement (Fig. III,31) consists of two half cells along the c_0 axis related to one another as are two hexagonal close-packed layers and joined by pairs of S–S atoms suggestive of the sulfur pairs in pyrite (Chapter IV,**g1**). The various sulfur and copper atoms have different interatomic distances and environments that are in some ways unexpected. Atom Cu(1) is surrounded by three sulfur atoms at a distance of 2.19 A., and each Cu(2) atom by four tetrahedrally arranged sulfur atoms 2.31 A. away. S(1) atoms have five copper neighbors at distances of 2.19 and 2.35 A., while each S(2) atom has one sulfur atom at 2.05 A. and three copper atoms at 2.31 A.

The selenide mineral *klockmannite*, CuSe, has the same structure with

$$a_0 = 3.938 \text{ A.}, \qquad c_0 = 17.25 \text{ A.}$$

For copper, $u = 0.10$; for selenium, $v = 0.066$.

Miscellaneous Phosphides, Nitrides, and Carbides

III,f1. The compound *titanium phosphide*, TiP, is hexagonal, with a tetramolecular cell of the dimensions:

$$a_0 = 3.513 \text{ A.}, \qquad c_0 = 11.75 \text{ A.}$$

Atoms are in the following special positions of D_{6h}^4 ($P6_3/mmc$):

Ti: (4*f*) $\pm(^1/_3, ^2/_3\, u;\ ^1/_3, ^2/_3, ^1/_2 - u)$ with $u = 0.125$

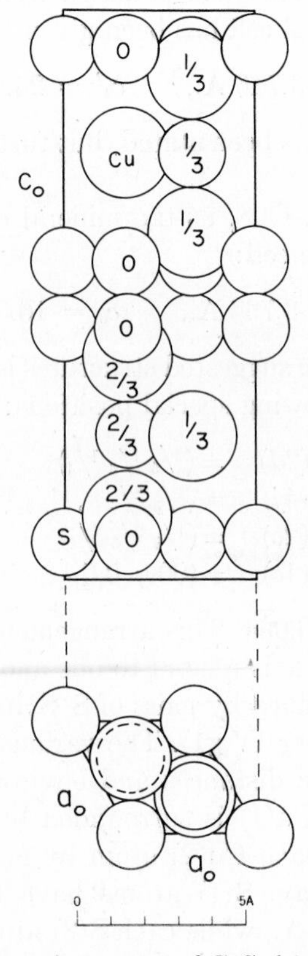

Fig. III,31. The hexagonal structure of CuS shown in two projections.

$$\text{P}(1)\text{:}\quad (2a)\quad 000;\ 0\,0\,{}^1/_2$$
$$\text{P}(2)\text{:}\quad (2d)\quad \pm({}^2/_3\,{}^1/_3\,{}^1/_4)$$

This arrangement is illustrated in Figure III,32. The two kinds of phosphorus atoms are arranged as are the atoms in NiAs.

Zirconium arsenide, ZrAs, has this structure, with

$$a_0 = 3.804 \text{ A.,} \qquad c_0 = 12.867 \text{ A.}$$

The parameter of the zirconium atoms is 0.117.

The following substances, considered to possess this atomic arrangement, have the dimensions:

$$\text{MoC:}\ a_0 = 2.932 \text{ A.,} \qquad c_0 = 10.97 \text{ A.}$$

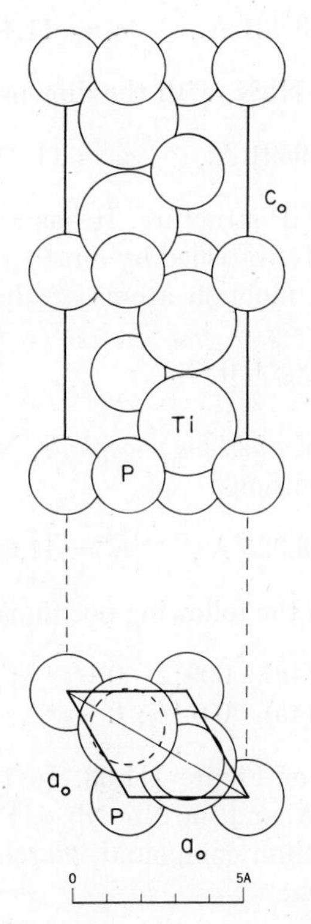

Fig. III,32. Two projections of the hexagonal structure of TiP.

The composition of this gamma-prime phase could not be established by chemical analysis. It has approximately the same a_0 and four times the c_0 of gamma MoC (a_0 = 2.898 A., c_0 = 2.809 A.).

$$\text{TiAs:}\quad a_0 = 3.642 \text{ A.},\qquad c_0 = 12.064 \text{ A.}$$
$$\text{ZrP:}\quad a_0 = 3.686 \text{ A.},\qquad c_0 = 12.48 \text{ A.}$$

Also with this arrangement are the *carbo-sulfides* of *titanium* and *zirconium* (with two molecules per cell):

$$Ti_2CS:\quad a_0 = 3.210 \text{ A.},\qquad c_0 = 11.20 \text{ A.}$$
$$Zr_2CS:\quad a_0 = 3.396 \text{ A.},\qquad c_0 = 12.11 \text{ A.}$$

The carbon atoms are in (2*a*) and the sulfur atoms in (2*d*). The metallic atoms are in (4*f*), with u = 0.099 for both compounds.

It is also reported that Ti_3S_4 is an isomorphous defect structure, with

$$a_0 = 3.431 \text{ A.},\qquad c_0 = 11.44 \text{ A.}$$

Epsilon *niobium nitride*, NbN, with the dimensions

$$a_0 = 2.9591 \text{ A.},\qquad c_0 = 11.2714 \text{ A.}$$

was originally given this TiP structure. It has recently been stated, however, that the data are better satisfied by a rather improbable mixed structure in which 90% of the niobium atoms are in $\pm(^1/_3\,^2/_3\,^1/_8;\ ^1/_3\,^2/_3\,^3/_8)$ and 10% in $^2/_3\,^1/_3\,n$, where $n = {}^1/_8, {}^3/_8, {}^5/_8, {}^7/_8$. The nitrogen positions are given as 000; $0\,0\,^1/_4$; $0\,0\,^1/_2$; $0\,0\,^3/_4$.

III,f2. The beta form of *niobium phosphide*, NbP, has a tetramolecular tetragonal unit of the dimensions:

$$a_0 = 3.325 \text{ A.},\qquad c_0 = 11.38 \text{ A.}$$

Atoms have been placed in the following positions of D_4^{10} ($I4_12$):

Nb: (4*a*) 000; $0\,^1/_2\,^1/_4$; B.C.
P: (4*b*) $0\,0\,^1/_2$; $0\,^1/_2\,^3/_4$; B.C.

The arrangement is that of Figure III,33. In this simple structure the shortest Nb–P = ca. 2.35 A. and the Nb–Nb = P–P = 3.32 A.

The corresponding tantalum compound, *tantalum phosphide*, beta TaP, has the same structure with:

$$a_0 = 3.330 \text{ A.},\qquad c_0 = 11.39 \text{ A.}$$

Each of these compounds has an alpha form with unit cells half as tall:

For alpha NbP: $a_0 = 3.32$ A., $c_0 = 5.69$ A.
For alpha TaP: $a_0 = 3.320$ A., $c_0 = 5.69$ A.

A disorder in the distribution of the two kinds of atoms is said to account for the simpler cells.

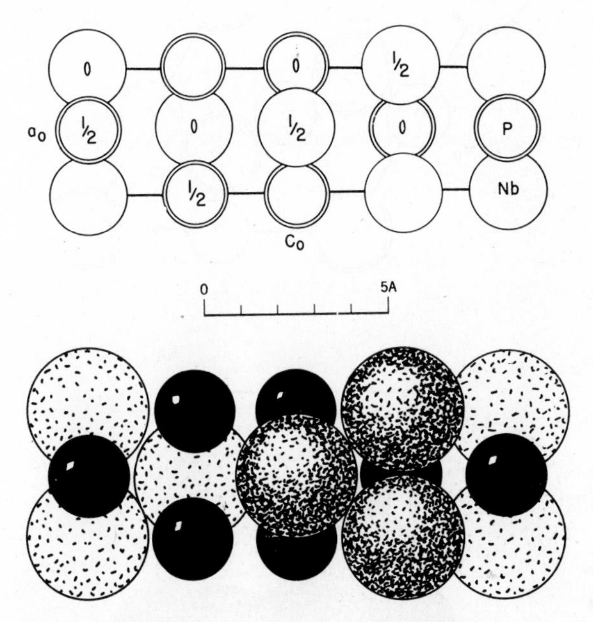

Fig. III,33a (top). The tetragonal NbP structure projected along an a_0 axis.
Fig. III,33b (bottom). A packing drawing of the NbP structure viewed along an a_0 axis. The sulfur atoms are the black circles.

III,f3. According to a recent study, one form of *tantalum nitride*, epsilon TaN, is hexagonal with a trimolecular unit of the dimensions:

$$a_0 = 5.191 \text{ A.}, \qquad c_0 = 2.908 \text{ A.}$$

Atoms have been put in the following positions of D_{6h}^1 ($P6/mmm$):

Ta: (1*a*) 000 and (2*d*) $\frac{1}{3}\frac{2}{3}\frac{1}{2}$; $\frac{2}{3}\frac{1}{3}\frac{1}{2}$
N: (3*f*) $\frac{1}{2}00$; $0\frac{1}{2}0$; $\frac{1}{2}\frac{1}{2}0$.

In this structure (Fig. III,34) the shortest Ta–N = 2.08 A. and the shortest Ta–Ta = 2.90 A.

The same arrangement has been chosen for delta TiO_x with, however, only enough of the (3*f*) positions filled to give a composition in the neighborhood of Ti_2O. The cell dimensions are:

$$a_0 = 4.991 \text{ A.,} \qquad c_0 = 2.879 \text{ A.}$$

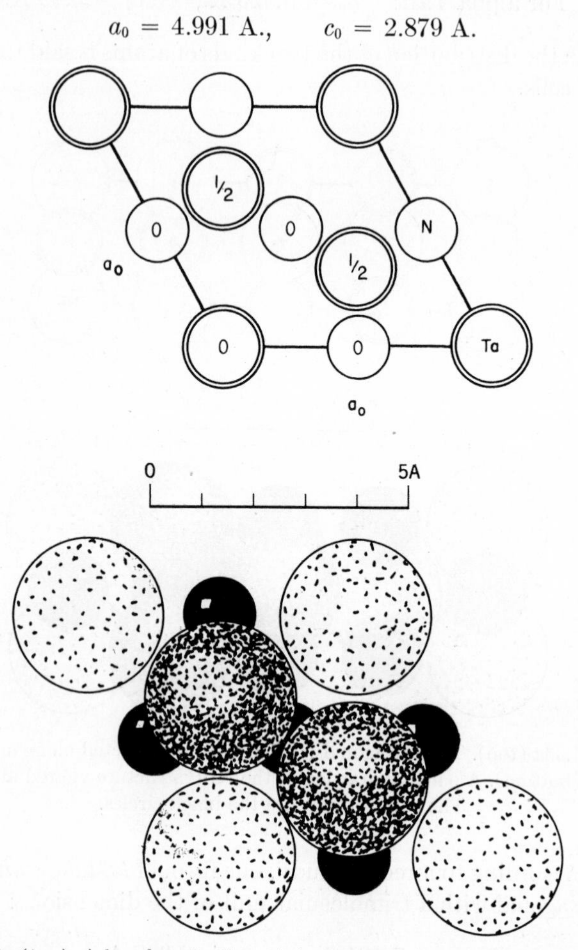

Fig. III,34a (top). A basal projection of the hexagonal structure of epsilon TaN.
Fig. III,34b (bottom). A packing drawing of the ε-TaN arrangement seen along its c_0 axis. The dotted tantalum atoms have their metallic radius.

III,f4. *Tungsten carbide*, WC, is hexagonal with an especially simple unimolecular cell of the dimensions:

$$a_0 = 2.9065 \text{ A.,} \qquad c_0 = 2.8366 \text{ A.}$$

The tungsten atom is at the origin, 000, but x-ray methods have not been able to place the light carbon atom with any definiteness. It has variously been said to be in $^1/_3\,^2/_3\,^1/_2$ (or $^2/_3\,^1/_3\,^1/_2$) or randomly distributed between these two. Recently a neutron diffraction study has decided between these possibilities in showing that the more highly ordered one is correct, the carbon atom being in $^2/_3\,^1/_3\,^1/_2$. The structure is shown in Figure III,35.

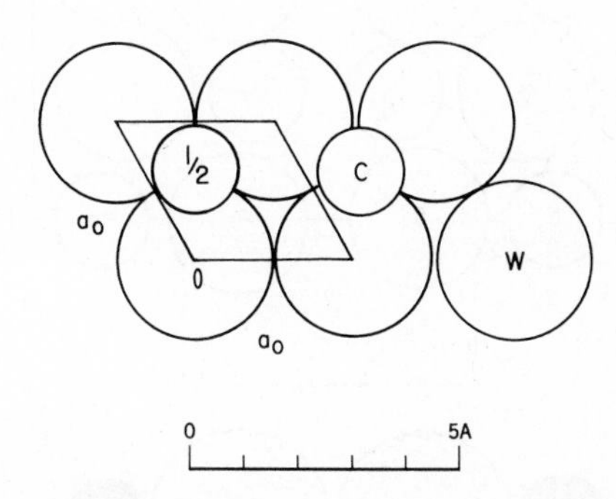

Fig. III,35. A basal projection of the very simple hexagonal structure of WC.

Delta *tungsten nitride*, WN, has a unit of the same shape:

$$a_0 = 2.893 \text{ A.}, \qquad c_0 = 2.826 \text{ A.},$$

with the tungsten atom at 000. It is very possible that the nitrogen atom has the same position as the carbon atom in WC, namely, $^2/_3\,^1/_3\,^1/_2$.

Other substances with similar units are:

$$\begin{array}{llll} \text{RuC}: & a_0 = 2.90785 \text{ A.}, & c_0 = 2.82186 \text{ A.} & (25°\text{C.}) \\ \text{OsC}: & a_0 = 2.90769 \text{ A.}, & c_0 = 2.82182 \text{ A.} & (25°\text{C.}) \\ \text{MoP}: & a_0 = 3.230 \text{ A.}, & c_0 = 3.207 \text{ A.} & \end{array}$$

Miscellaneous Halide and Other Ionic Types

III,g1. The *mercurous halides*, Hg_2X_2, were early given structures which have since been confirmed. They have tetragonal units (Table III,10) containing two Hg_2X_2 molecules. In the established atomic arrangement (Fig. III,36) all atoms are in the following special positions of D_{4h}^{17} ($I4/mmm$):

$$(4e) \quad \pm(00u;\ ^1/_2,^1/_2,u+^1/_2)$$

with the parameters of columns 4 and 5 of Table III,10. The interatomic distances that result are those of columns 6, 7, and 8.

It is noteworthy that there is a real diminution in the Hg–Hg separation as the halogen becomes lighter. The fluoride stands apart from the others by reason of its far larger X–X distance.

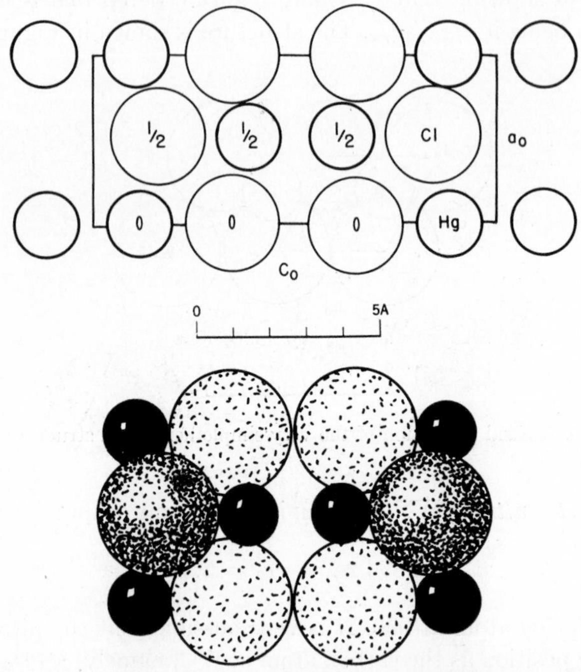

Fig. III,36a (top). The tetragonal structure of Hg_2Cl_2 projected along an a_0 axis.
Fig. III,36b (bottom). A packing drawing of the Hg_2Cl_2 arrangement viewed along an a_0 axis. The large chloride ions are dotted.

TABLE III,10
The Tetragonal Mercurous Halides (**III,g1**)

Crystal	a_0, A.	c_0, A.	u(Hg)	u'(X)	Hg–Hg, A.	Hg–X, A.	X–X, A.
Hg_2F_2	3.66	10.9	0.111	0.323	2.43	2.31	3.85
Hg_2Cl_2	4.478	10.91	0.116	0.347	2.53	2.52	3.33
Hg_2Br_2	4.65	11.10	0.116	0.347	2.58	2.57	3.40
Hg_2I_2	4.933	11.633 (26°C.)	0.116	0.347	2.69	2.68	3.55

III,g2. The form of *thallous iodide*, TlI, stable at temperatures below 175°C. is orthorhombic with the elongated four-molecule cell:

$$a_0 = 4.582 \text{ A.}; \; b_0 = 12.92 \text{ A.}; \; c_0 = 5.251 \text{ A.} \quad (25°\text{C.})$$

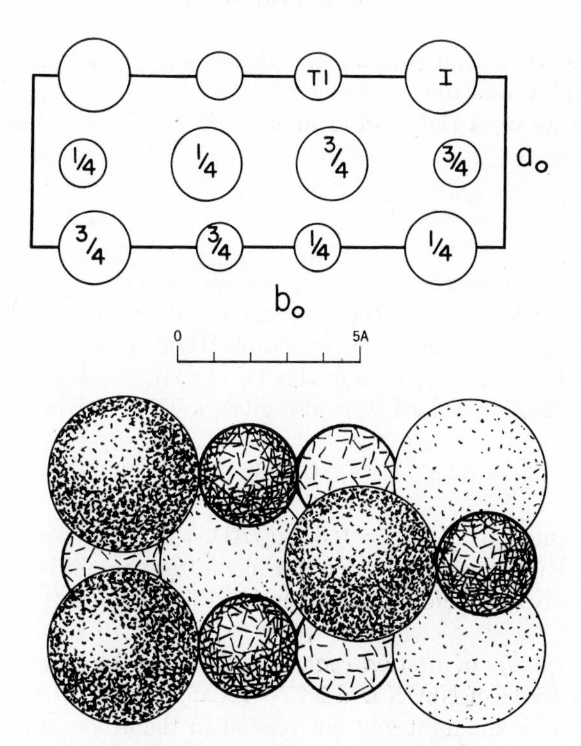

Fig. III,37a (top). A projection along the c_0 axis of the orthorhombic structure of TlI. Origin in upper right.

Fig. III,37b (bottom). A packing drawing of the orthorhombic TlI arrangement viewed along the c_0 axis. The smaller, line-shaded atoms are thallium.

Its atoms have been given the positions:

$$(4c)\quad 0\,u\,{}^1/_4;\ 0\,\bar{u}\,{}^3/_4;\ {}^1/_2,u+{}^1/_2,{}^1/_4;\ {}^1/_2,{}^1/_2-u,{}^3/_4$$

of V_h^{17} ($Cmcm$) with u(Tl) = 0.392 and u(I) = 0.1333. The resulting structure (Fig. III,37) with its sevenfold coordination bears no obvious relationships to other ionic arrangements.

Two alkali hydroxides and two halides of monovalent indium are reported to have this atomic distribution in their crystals. They are

InBr: a_0 = 4.46 A.; b_0 = 12.39 A.; c_0 = 4.73 A.

The atomic parameters are: for indium, u = 0.386; for bromine, u = 0.160.

InI: a_0 = 4.75 A.; b_0 = 12.76 A.; c_0 = 4.91 A.

The atomic parameters are: for indium, u = 0.398; for iodine, u = 0.145.

In InI there is one In–I = 3.23 A. and four In–I = 3.46 A. The shortest In–In = 3.58 A. and the shortest I–I = 4.44 A.

In the earlier work the delta form of KOH was described as having this structure, with

$$a_0 = 3.95 \text{ A.};\ b_0 = 11.4 \text{ A.};\ c_0 = 4.03 \text{ A.}$$

The assigned parameters were $u(\text{K}) = 0.36$, $u(\text{OH}) = 0.11$. More recent studies of what is probably the same modification have led to the monoclinic arrangement described in paragraph **III,a8**. It would appear, therefore, that this initial ascription to the TlI structure was an error.

Sodium hydroxide, NaOH, was also given a TlI-like arrangement, with

$$a_0 = 3.397 \text{ A.};\ b_0 = 11.32 \text{ A.};\ c_0 = 3.397 \text{ A.}$$

and the parameters: $u(\text{Na}) = 0.34$, $u(\text{OH}) = 0.13$. In view of the probability that KOH does not have a TlI-like form it would be worthwhile to reinvestigate NaOH before fully accepting this assignment of structure.

III,g3. Crystals of *lithium cyanide*, LiCN, are like those of the low-temperature forms of NaCN and KCN (**III,a5**) in being orthorhombic, but their structure is different and not related to the others thus far studied. The space group is V_h^{16} (*Pbnm*) with all atoms in the special positions:

$$(4c) \quad \pm(u\, v\, {}^1/_4;\ {}^1/_2-u, v+{}^1/_2, {}^1/_4)$$

The unit cell containing four molecules has the dimensions:

$$a_0 = 6.52 \text{ A.};\ b_0 = 8.73 \text{ A.};\ c_0 = 3.73 \text{ A.}$$

Because of the low scattering power of all the atoms involved, it was possible to place them with special accuracy, to distinguish between the carbon and the nitrogen atoms of the CN radical and thus to determine that it is the nitrogen atom which accepts the extra lithium electron.

The chosen parameters are: for lithium, $u = 0.471$, $v = 0.374$; for carbon, $u = 0.261$, $v = 0.190$; for nitrogen $u = 0.142$, $v = 0.094$.

This structure (Fig. III,38) is a very open one with large empty spaces and channels between atoms. Presumably on account of its small size, lithium has only four CN neighbors, three of which are nitrogen, the fourth carbon, at the corners of a distorted tetrahedron. The C–N separation, 1.15 A., is considerably greater than the 1.05 A. determined for NaCN.

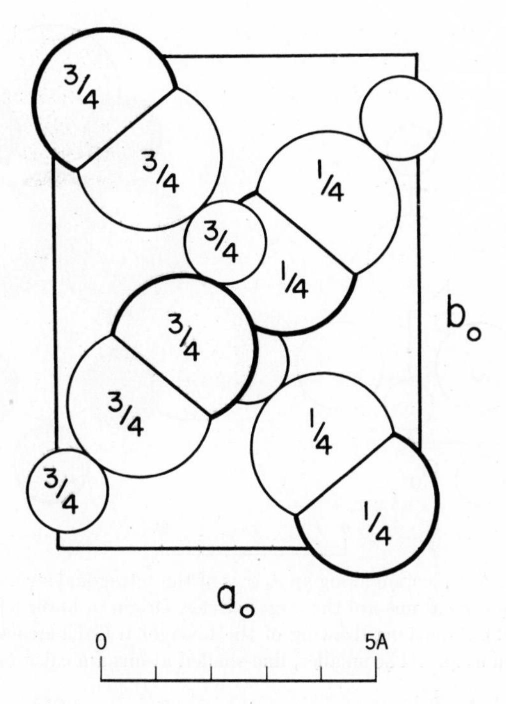

Fig. III,38. A projection along the c_0 axis of the orthorhombic structure of LiCN. The small circles are lithium, the more heavily ringed, large circles are nitrogen. Origin in lower right.

III,g4. The unusual compound *calcium monochloride*, CaCl, is tetragonal with a bimolecular cell of the dimensions:

$$a_0 = 3.851 \text{ A.}, \qquad c_0 = 6.861 \text{ A.}$$

All atoms have been placed in the following special positions of D_{4h}^7 ($P4/nmm$):

$$(2c) \quad 0\,{}^1/_2\,u; \;\; {}^1/_2\,0\,\bar{u},$$

with $u(\text{Ca}) = 0.146$ and $u(\text{Cl}) = 0.695$.

This leads to a structure layered parallel to the c face (Fig. III,39) in which double calcium layers alternate with double layers of chlorine. In these layers the shortest Ca–Ca = 3.38 A. and the shortest Cl–Cl = 3.81 A.; between the layers Ca–Cl = 3.77 A.

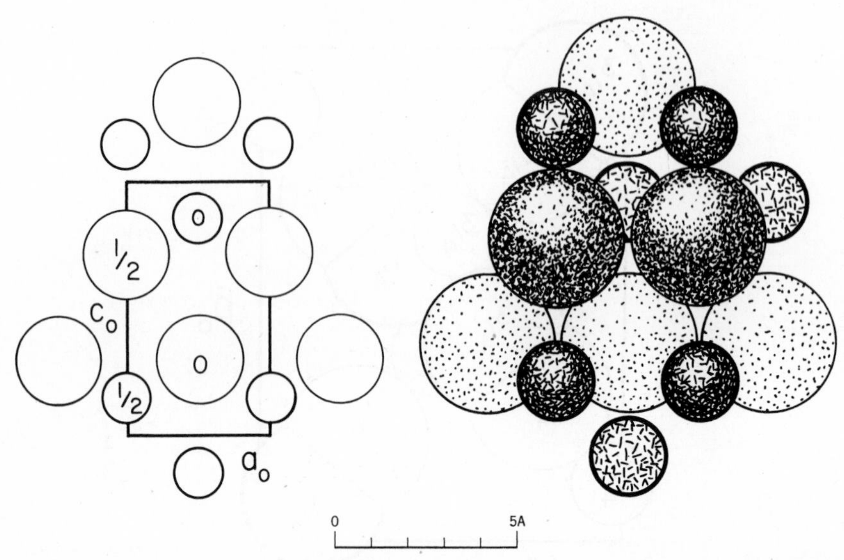

Fig. III,39a (left). A projection along an a_0 axis of the tetragonal structure of CaCl. The chlorine atoms are the larger circles. Origin in lower left.

Fig. III,39b (right). A packing drawing of the tetragonal CaCl structure viewed along an a_0 axis. The smaller, line-shaded atoms are calcium.

III,g5. *Hydrazinium chloride*, N_2H_5Cl, forms orthorhombic crystals that have a large 16-molecular cell of the dimensions:

$$a_0 = 12.491 \text{ A.}; \ b_0 = 21.854 \text{ A.}; \ c_0 = 4.41 \text{ A.}$$

Atoms have been found to be in general positions of C_{2v}^{19} (*Fdd*):

(16*b*) $xyz;\ \bar{x}\bar{y}z;\ {}^1/_4-x,y+{}^1/_4,z+{}^1/_4;\ x+{}^1/_4,{}^1/_4-y,z+{}^1/_4;$ F.C.

with the following parameters.

Atom	x	y	z
Cl	0.0410	0.0835	0.000
N(1)	0.1032	0.2363	0.776
N(2)	0.0687	0.1922	0.549

This and the bromide of the following paragraph have cations which apparently are linked together through hydrogen bonds. In this chloride each N_2H_5 cation has six chlorine neighbors with N–Cl = 3.12–3.49 A. and is hydrogen-bonded to two other cations which thus spiral in chains along the c_0 axis (Fig. III,40). The N–N separation within the cation is 1.45 A.;

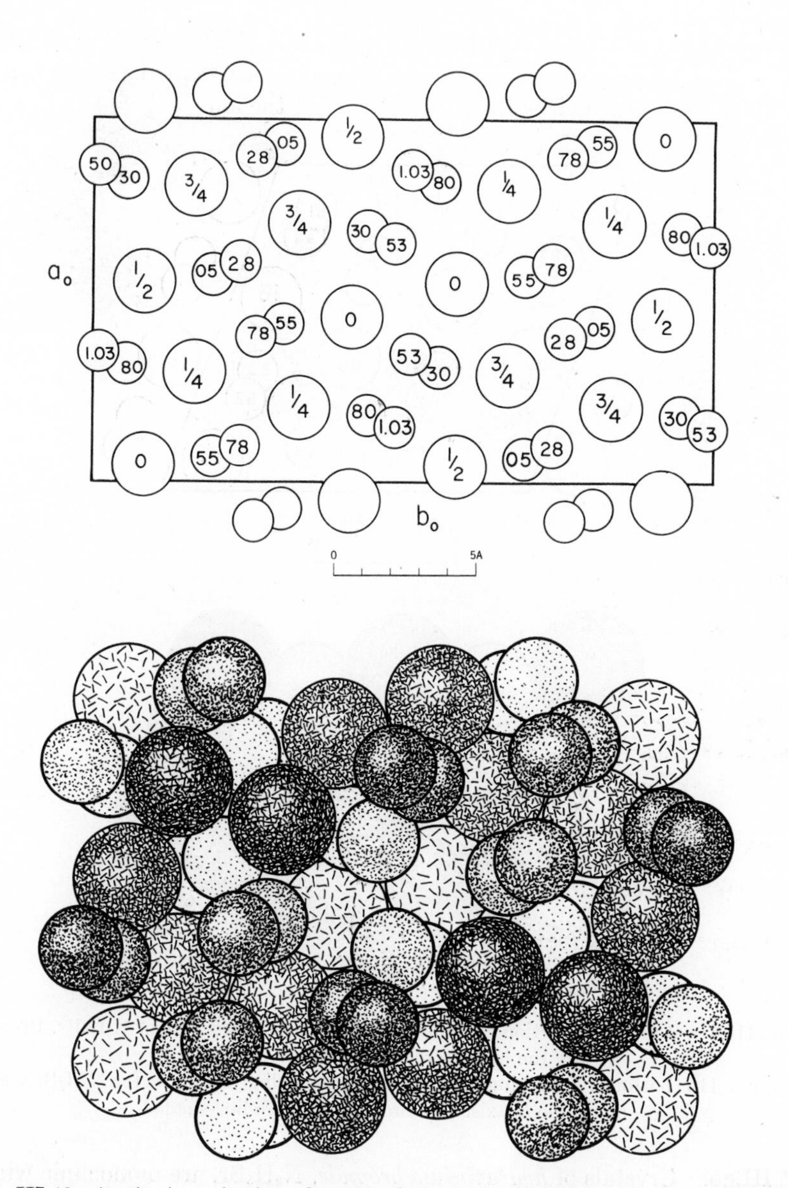

Fig. III,40a (top). A projection along c_0 of the orthorhombic structure of N_2H_5Cl. Chlorine are the larger circles. Origin in lower right.

Fig. III,40b (bottom). A packing drawing of the orthorhombic structure for N_2H_5Cl viewed along the c_0 axis. The chloride ions are line-shaded.

the hydrogen bond distance between cations is N(2)–H–N(1) = 2.95 A. Since N(2) approaches closer to the chloride ions than does N(1), it is thought that N(2) marks the $—NH_3^+$ part of the ion.

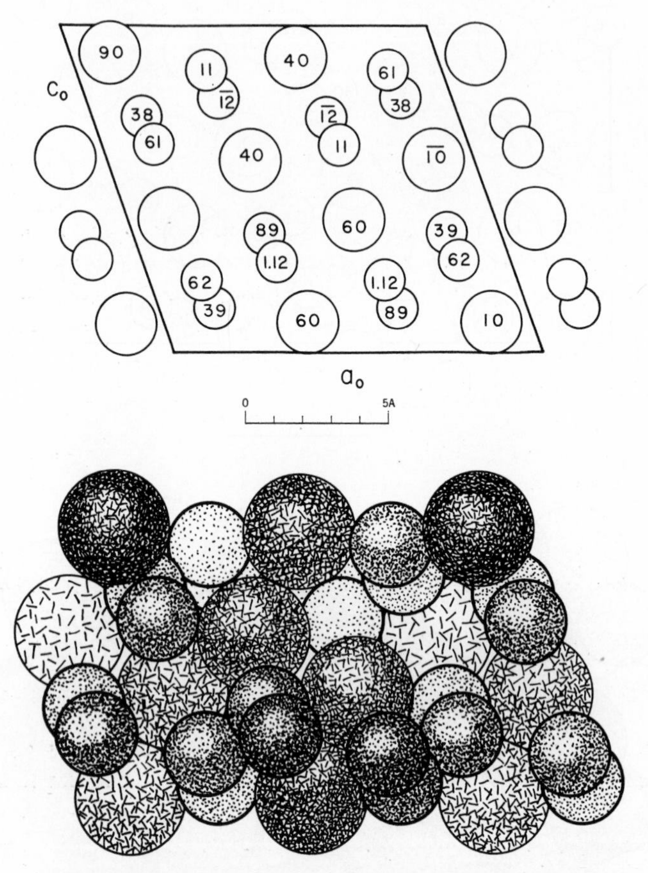

Fig. III,41a (top). A projection along b_0 of the monoclinic structure of N_2H_5Br. Bromine are the larger circles. Origin in lower right.

Fig. III,41b (bottom). A packing drawing of the monoclinic structure for N_2H_5Br viewed along the b_0 axis. The bromide ions are line-shaded.

III,g6. Crystals of *hydrazinium bromide*, N_2H_5Br, are monoclinic with a unit cell of the following dimensions:

$$a_0 = 12.85 \text{ A.};\ b_0 = 4.54 \text{ A.};\ c_0 = 11.94 \text{ A.};\ \beta = 110°16'$$

The space group is C_{2h}^6 ($C2/c$) with all atoms of the eight molecules per cell in its general positions:

$$(8f) \quad \pm(xyz;\ \bar{x},y,{}^1/_2-z;\ x+{}^1/_2,y+{}^1/_2,z;\ {}^1/_2-x,y+{}^1/_2,{}^1/_2-z)$$

The chosen parameters are the following:

Atom	x	y	z
Br	−0.1037	0.097	0.0802
N(1)	0.1492	0.390	0.1350
N(2)	0.1410	0.620	0.2165

The structure is shown in Figure III,41. Within the N_2H_5 cation, N–N = 1.45 A. As in the chloride there is considered to be a linking of cations through hydrogen bonds to form chains spiraling, in this case, along the b_0 axis. These N(2)–H–N(1) distances are 2.93 A. Of the eight neighbors of a N_2H_5 cation, six are bromine with N–Br = 3.29–3.70 A. and two are through the hydrogen bonds.

III,g7. *Aurous iodide,* AuI, is tetragonal with a tetramolecular cell having the edge lengths:

$$a_0 = 4.35 \text{ A.}, \qquad c_0 = 13.73 \text{ A.}$$

Atoms have been placed in the following special positions of D_{4h}^{16} ($P4_2/ncm$):

$$\text{Au: } (4d) \quad 000;\ {}^1/_2\,{}^1/_2\,0;\ {}^1/_2\,0\,{}^1/_2;\ 0\,{}^1/_2\,{}^1/_2$$

$$\text{I: } (4e) \quad \pm({}^1/_4\,{}^1/_4\,u;\ {}^1/_4,{}^1/_4,u+{}^1/_2) \qquad \text{with } u = 0.153$$

This is a structure (Fig. III,42) built up of Au–I chains in which the Au–I separations are 2.62 A. Each gold atom has two iodine neighbors and in addition four gold atoms at a distance of 3.08 A. Each iodine atom has, besides the two gold atoms, eight iodine atoms at the distances I–4I = 4.1 A. and 4.35 A.

III,g8. *Aurous cyanide,* AuCN, is hexagonal with a unimolecular cell having the edges:

$$a_0 = 3.40 \text{ A.}, \qquad c_0 = 5.09 \text{ A.}$$

Placing the gold atom in the origin, 000, the carbon and nitrogen atoms would be in similar positions $00u$. These light atoms could not be placed experimentally but if C–N is assumed to be 1.15 A., Au–C (or N) = 1.97 A. and u = ca. 0.4 and 0.6.

III,g9. Between the melting point (−13°C.) and −103°C. *hydrogen cyanide*, HCN, is tetragonal. As measured at ca. −80°C., its bimolecular unit has the edges:

$$a_0 = 4.6 \text{ A3.}, \qquad c_0 = 4.34 \text{ A.}$$

Atoms are in special positions:

$$(2a) \quad 00u;\ {}^1/_2,{}^1/_2,u+{}^1/_2$$

of C_{4v}^9 (*I4mm*), with u(C) = 0 (arbitrary) and u(N) = ca. 0.267.

The structure thus obtained is, considering the cell dimensions, only a small distortion of a cubic body-centering.

The modification existing below the transition at −103°C. is very little different from the higher temperature form. Its bimolecular orthorhombic unit, measured at −120°C., has the dimensions:

$$a_0 = 4.13 \text{ A.};\ b_0 = 4.85 \text{ A.};\ c_0 = 4.34 \text{ A.}$$

The x-ray reflections from the two forms have the same relative intensities and therefore their structures must differ only in the displacements in the a_0b_0 planes that correspond to the change in cell dimensions. The low form is pyroelectric and its space group therefore probably is C_{2v}^{20} (*Imm*).

Considering the linear character of the molecules, there obviously is for each form a C–H–N distance of 3.18 A. The intermolecular N–C separation is 3.43 A. for the high and 3.35 A. for the low form.

In addition there is a tetramer of hydrogen cyanide, $(HCN)_4$, whose structure shows it to be in fact diaminomaleonitrile (see Chapter XIII).

III,g10. Crystals of *hydrofluoric acid*, HF, are orthorhombic, and not tetragonal as earlier assumed. They have a tetramolecular cell with the edge lengths:

$$a_0 = 3.42 \text{ A.};\ b_0 = 4.32 \text{ A.};\ c_0 = 5.41 \text{ A.}$$

The fluorine atoms are in special positions:

$$(4c) \quad \pm(0\ {}^1/_4\ u;\ {}^1/_2,{}^1/_4,u+{}^1/_2)$$

of V_h^{17} (*Bmmb*), with u = 0.115. In this structure (Fig. III,43), fluorine atoms that are 2.49 A. apart are undoubtedly held together by hydrogen bonds. It was not possible to decide how symmetrical these bonds are in terms of the hydrogen positions. The evidence as a whole, however, favors asymmetric bonding and a preferred direction of thermal motion of the fluorine atoms in the crystal.

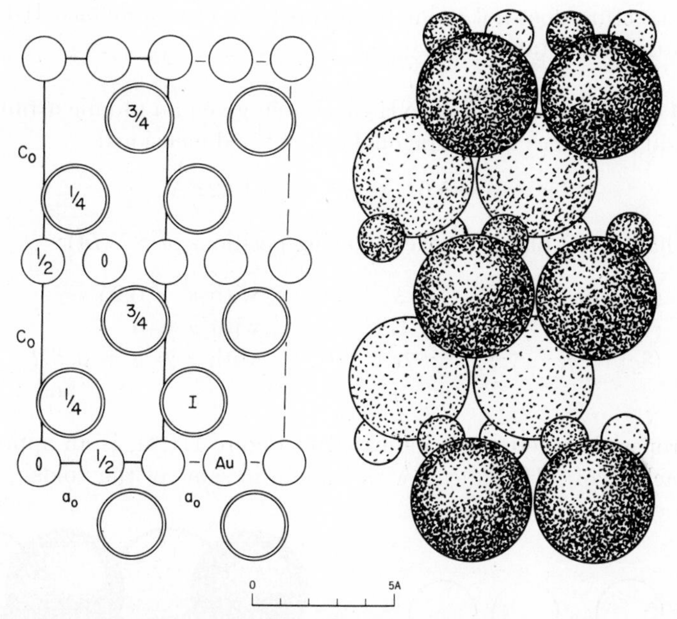

Fig. III,42a (left). The contents of two cells of the tetragonal AuI structure projected along an a_0 axis. Origin in lower left.

Fig. III,42b (right). A packing drawing of the AuI structure seen along an a_0 axis. The large circles are the iodine atoms bearing their iodide radius.

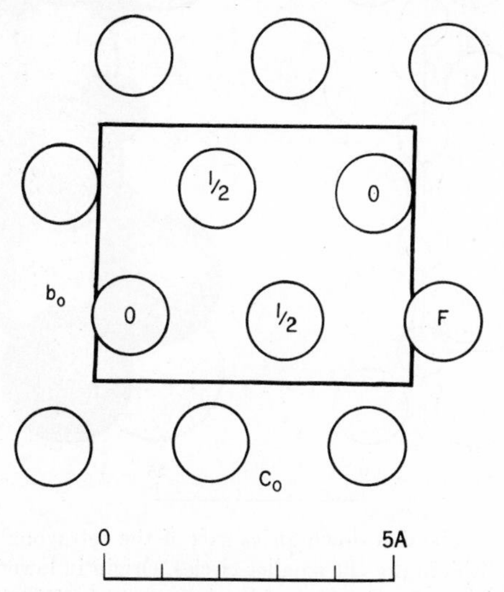

Fig. III,43. The orthorhombic cell of HF projected along its a_0 axis to show the distribution of its fluorine atoms. Origin in lower left.

The structure does not point to the presence of well-defined H_2F_2 molecules in the solid state.

III,g11. *Lithium amide*, $LiNH_2$, has been given an atomic arrangement with eight molecules in a tetragonal cell of the dimensions:

$$a_0 = 5.01 \text{ A.}, \quad c_0 = 10.22 \text{ A.}$$

Atoms have been placed in the following positions of S_4^2 ($I\bar{4}$):

Li(1): (4*e*) $\pm(00u)$; B.C. with $u = 0.25$
Li(2): (4*f*) $0\,{}^1/_2\,u;\ {}^1/_2\,0\,\bar{u}$; B.C. with $u = 0$
NH_2: (8*g*) $xyz;\ \bar{x}\bar{y}z;\ y\bar{x}\bar{z};\ \bar{y}x\bar{z}$; B.C. with $x = y = 0.232$ and $z = 0.116$

This proposed structure, as indicated in Figure III,44, is an approximate close-packing of NH_2 radicals with lithium in some of the voids.

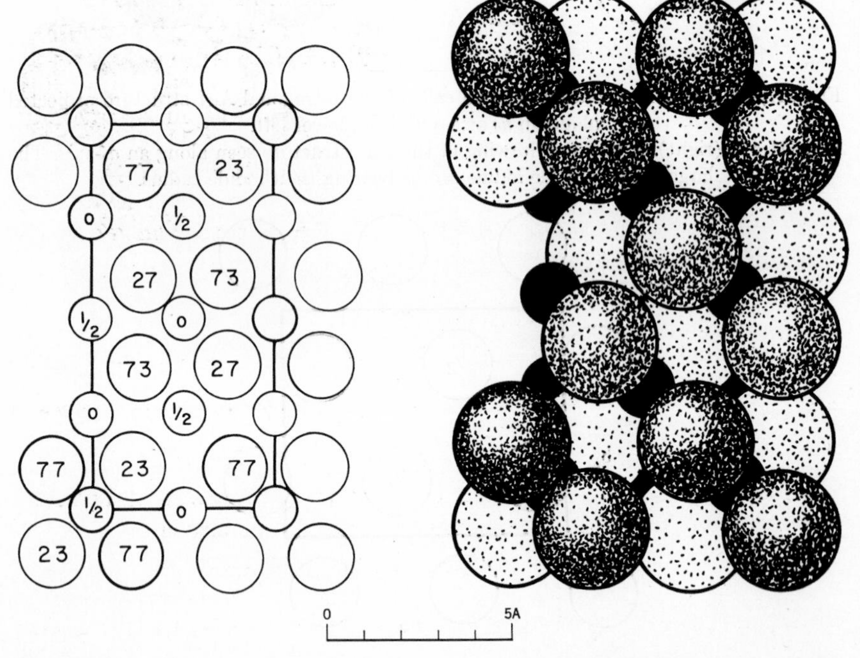

Fig. III,44a (left). A projection down an a_0 axis of the tetragonal structure of $LiNH_2$. The lithium are the smaller circles. Origin in lower left.

Fig. III,44b (right). A packing drawing of the tetragonal $LiNH_2$ structure viewed along an a_0 axis. The large dotted circles are nitrogen, as centers of the NH_2 groups.

Lithium hydrosulfide, LiSH, has the same structure, with

$$a_0 = 5.554 \text{ A.}, \qquad c_0 = 12.340 \text{ A.}$$

The sulfur atoms have been put in (8*g*) with the same parameters as those of the nitrogen atoms in the amide. The lithium atoms have the same positions and the same parameters in the two crystals. In the resulting arrangement each lithium atom has four SH neighbors with Li–SH = 2.43–2.68 A.

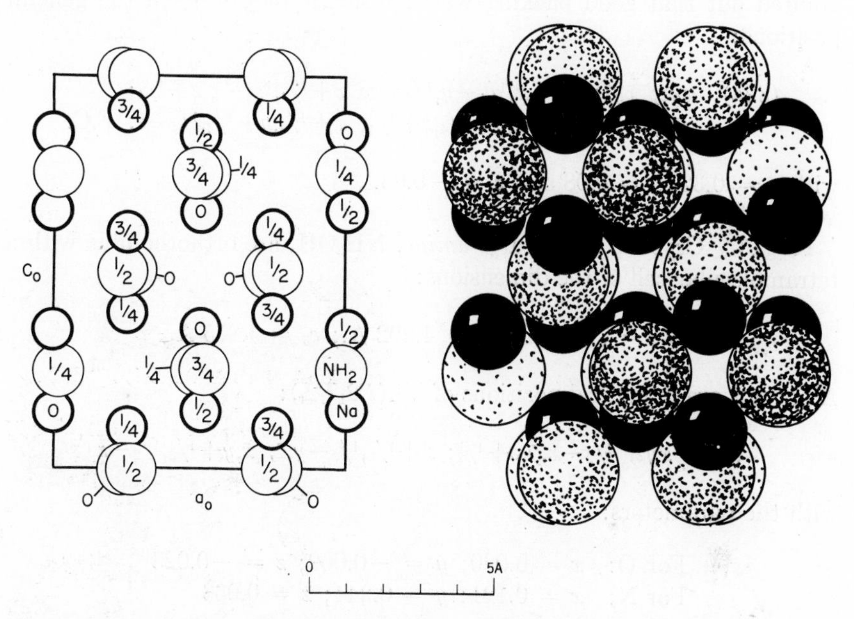

Fig. III,45a (left). The orthorhombic $NaNH_2$ arrangement projected along its b_0 axis. Origin in lower left.

Fig. III,45b (right). A packing drawing of the $NaNH_2$ structure seen along its b_0 axis. The sodium atoms are black.

III,g12. Crystals of *sodium amide*, $NaNH_2$, are orthorhombic with a large unit containing 16 molecules and having the dimensions:

$$a_0 = 8.073 \text{ A.}; \; b_0 = 8.964 \text{ A.}; \; c_0 = 10.456 \text{ A.}$$

The space group is V_h^{24} (*Fddd*), with atoms in the following special positions:

$$\text{Na: } (16g) \quad 00u;\; 00\bar{u};\; {}^1/_4,{}^1/_4,u+{}^1/_4;\; {}^1/_4,{}^1/_4,{}^1/_4-u;\; \text{F.C.}$$

$$\text{N: } (16e) \quad u00;\; \bar{u}00;\; u+{}^1/_4,{}^1/_4,{}^1/_4;\; {}^1/_4-u,{}^1/_4,{}^1/_4;\; \text{F.C.}$$

In two independent determinations u(Na) has been found as 0.146 and 0.142, u(N) as 0.236 and 0.233, respectively.

This structure (Fig. III,45) has been described as a deformed cubic close-packing of the NH_2 groups with the sodium atoms in tetrahedral holes. Each sodium atom thus has four tetrahedrally distributed NH_2 neighbors with Na–2N = 2.39 A. and two more at 2.51 A. The four sodium atoms associated with each NH_2 group are clustered around one side of the group.

The positions of the hydrogen atoms could not be determined, but it was pointed out that good packing would result if they were in the general positions:

$$(32h)\quad xyz;\ \bar{x}y\bar{z};\ {}^1/_4-x,{}^1/_4-y,{}^1/_4-z;\ x+{}^1/_4,{}^1/_4-y,z+{}^1/_4;$$
$$x\bar{y}\bar{z};\ \bar{x}\bar{y}z;\ {}^1/_4-x,y+{}^1/_4,z+{}^1/_4;\ x+{}^1/_4,y+{}^1/_4,{}^1/_4-z;\ \text{F.C.}$$

with x = 0.32, y = 0.08 and z = −0.03.

III,g13. Crystals of *hydroxyl amine*, NH_2OH, are orthorhombic with a tetramolecular cell of the dimensions:

$$a_0 = 7.292\text{ A.};\ b_0 = 4.392\text{ A.};\ c_0 = 4.875\text{ A.}$$

All atoms are in general positions of V^4 ($P2_12_12_1$):

$$(4a)\quad xyz;\ {}^1/_2-x,\bar{y},z+{}^1/_2;\ x+{}^1/_2,{}^1/_2-y,\bar{z};\ \bar{x},y+{}^1/_2,{}^1/_2-z$$

with the parameters:

For O: x = 0.060; y = −0.062; z = −0.023
For N: x = 0.121; y = 0.244; z = 0.063

It was not possible to derive the hydrogen positions from the x-ray data.

The structure is shown in Figure III,46. The N–O separation is 1.47 A., with the large error of ±0.03 A. due in part to an inability to locate the hydrogens. Between molecules there are N–O distances of 2.74, 3.07, 3.11, 3.18, and 3.24 A. and others greater than 3.5 A. It was noted that if the first two are considered to involve hydrogen bondings there are close relationships between this structure and those of H_2O_2 (**III,g14**) and N_2H_4 (Chapter IV,**i3**).

III,g14. *Hydrogen peroxide*, H_2O_2, is solid below −0.9°C. The symmetry of its crystals is tetragonal with a tetramolecular unit of the edge lengths:

$$a_0 = 4.06\text{ A.},\qquad c_0 = 8.00\text{ A.}$$

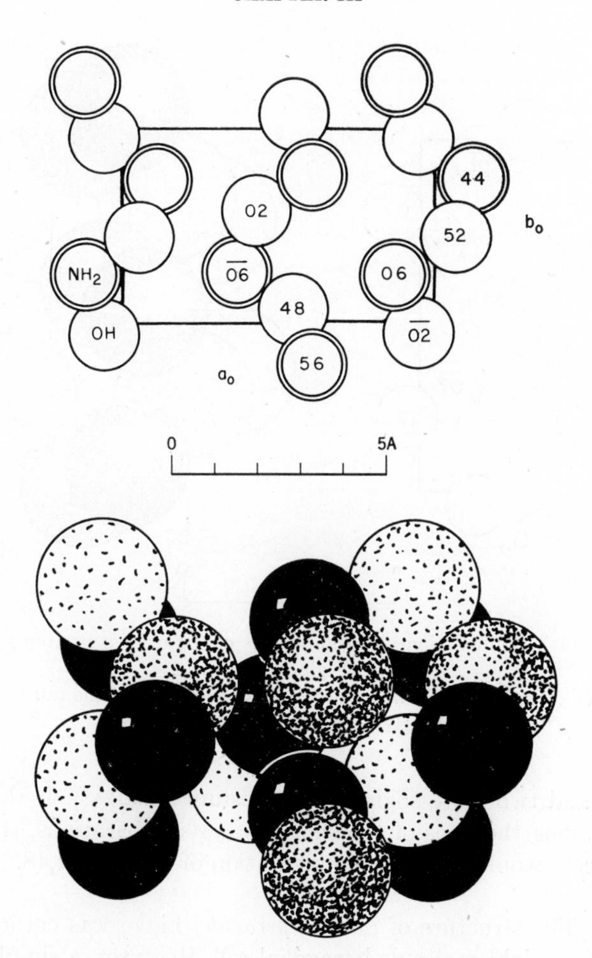

Fig. III,46a. The orthorhombic NH_2OH arrangement projected along its c_0 axis. Origin in lower right.

Fig. III,46b. A packing drawing of the NH_2OH structure seen along its c_0 axis. The hydroxyl groups are black.

Atoms are in general positions of D_4^4 ($P4_12_1$) (or its enantiomorph D_4^8):

(8*b*) $xyz;\ \bar{x},\bar{y},z+{}^1/_2;\ \bar{y},\bar{x},{}^1/_2-z;\ {}^1/_2-x,y+{}^1/_2,{}^1/_4-z;$
$yx\bar{z};\ x+{}^1/_2,{}^1/_2-y,{}^3/_4-z;\ {}^1/_2-y,x+{}^1/_2,z+{}^1/_4;\ y+{}^1/_2,{}^1/_2-x,z+{}^3/_4$

For oxygen, $x = 0.071$, $y = 0.172$, $z = 0.217$.

In the resulting structure (Fig. III,47), O–O = 1.49 A. within the molecule. Between molecules an oxygen has two oxygen neighbors at a distance

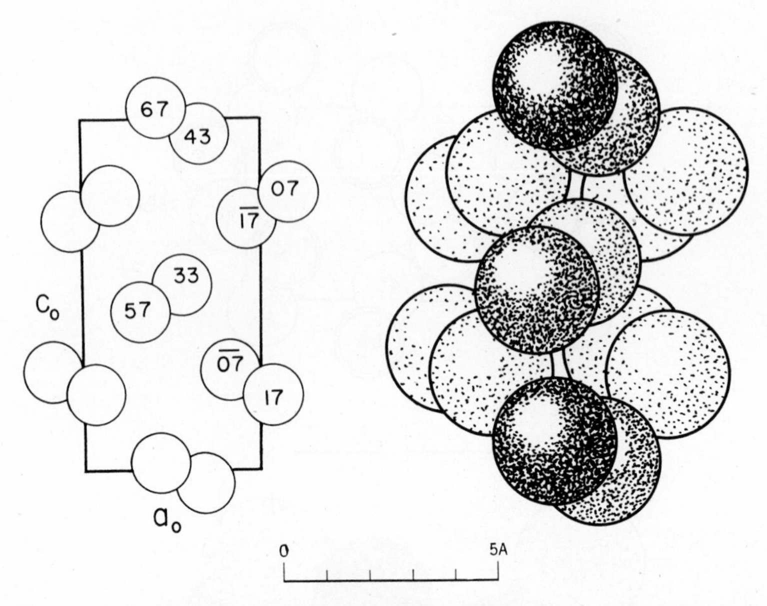

Fig. III,47a (left). A projection along an a_0 axis of the oxygen positions in tetragonal H_2O_2. Origin in lower left.

Fig. III,47b (right). A packing drawing of the oxygen atoms in the tetragonal structure of solid H_2O_2, as viewed along an a_0 axis.

of 2.78 A. and two more slightly more distant (2.90 A.). If the hydrogen atoms are along the lines of the shorter oxygen separations, the complete H_2O_2 molecule would have the configuration of Figure III,48.

III,g15. The structure of *lithium peroxide*, Li_2O_2, was earlier described in terms of an eight-molecule hexagonal cell. Recently, a simpler cell containing two molecules has been found adequate to interpret the data. It has the dimensions:

$$a_0 = 3.142 \text{ A.}, \qquad c_0 = 7.650 \text{ A.}$$

The selected space group is C_{3h}^1 ($P\bar{6}$) with atoms in the positions:

Li(1):	(1*a*)	000	
Li(2):	(1*d*)	$^1/_3\ ^2/_3\ ^1/_2$	
Li(3):	(2*i*)	$^2/_3\ ^1/_3\ u;\ ^2/_3\ ^1/_3\ \bar{u}$	with $u = ^1/_4$
O(1):	(2*g*)	$\pm(00u)$	with $u = 0.401$
O(2):	(2*h*)	$^1/_3\ ^2/_3\ u;\ ^1/_3\ ^2/_3\ \bar{u}$	with $u = 0.10$

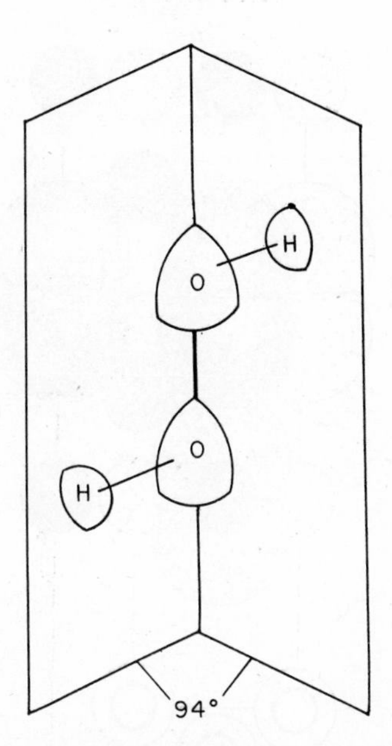

Fig. III,48. The apparent configuration of the H_2O_2 molecule.

In this somewhat unusual arrangement (Fig. III,49) the oxygen atoms of a pair are separated by a distance of 1.51 A. Atoms of lithium have six oxygen neighbors with Li(1)–O = 1.96 A. and Li(2)–O = 2.14 A.

III,g16. The symmetry of *sodium peroxide*, Na_2O_2, was at one time thought to be tetragonal but it is in fact hexagonal. The unit cell has the edges:

$$a_0 = 6.208 \text{ A.}, \qquad c_0 = 4.460 \text{ A.}$$

and contains three molecules. The structure as described, based on D_{3h}^3 ($C\,\bar{6}2m$), has atoms in the positions:

Na(1):	(3f)	$u00;\ 0u0;\ \bar{u}\bar{u}0$	with $u = 0.295$
Na(2):	(3g)	$u\,0\,^1/_2;\ 0\,u\,^1/_2;\ \bar{u}\,\bar{u}\,^1/_2$	with $u = 0.632$
O(1):	(2e)	$\pm(00v)$	with $v = 0.668$
O(2):	(4h)	$\pm(^1/_3\,^2/_3\,v;\ ^1/_3\,^2/_3\,\bar{v})$	with $v = 0.168$

In this arrangement (Fig. III,50) the separation O–O in the peroxide group is 1.495 A. Each sodium atom is surrounded by six oxygen atoms at distances between 2.32 and 2.46 A.

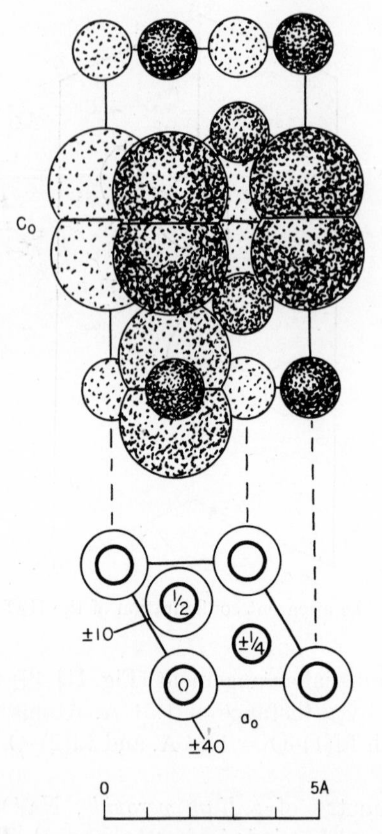

Fig. III,49. Two projections of the hexagonal Li_2O_2 arrangement. The small circles are lithium.

It is pointed out that compounds of the type beta-K_2UF_6 and beta-$KLaF_4$ (Chapters IX and VIII) have cells of the same shape as Na_2O_2 and structures anti to it with fluorine atoms in the same positions as the sodium atoms and metal atoms in the centers of the O_2 pairs.

III,g17. Crystals of *potassium peroxide*, K_2O_2, are orthorhombic with a tetramolecular unit of the dimensions:

$$a_0 = 6.736 \text{ A.}; \; b_0 = 7.001 \text{ A.}; \; c_0 = 6.479 \text{ A.}$$

The chosen space group is V_h^{18} (*Cmca*) with atoms in the positions:

K: (8*e*) $\pm(^1/_4\, u\, ^1/_4;\ ^3/_4\, u\, ^1/_4;\ ^3/_4,u+^1/_2,^1/_4;\ ^1/_4,u+^1/_2,^1/_4)$
O: (8*f*) $\pm(0uv;\ ^1/_2,u,^1/_2-v;\ ^1/_2,u+^1/_2,v;\ 0,u+^1/_2,^1/_2-v)$

with $u(\text{K}) = 0.160$, $u(\text{O}) = 0.088$, $v(\text{O}) = 0.934$.

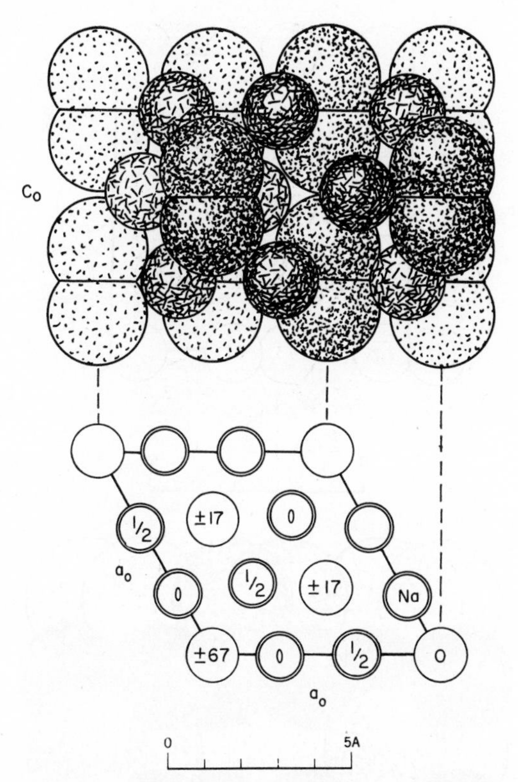

Fig. III,50. Two projections of the hexagonal Na_2O_2 structure. The sodium atoms are line-shaded.

The structure that results is shown in Figure III,51. In the peroxide group O–O = 1.50 A. The K–O separations lie between 2.66 and 2.74 A.

III,g18. Crystals of *rubidium peroxide*, Rb_2O_2, and of *cesium peroxide*, Cs_2O_2, are orthorhombic like the potassium compound (**III,g17**) but with a different atomic arrangement. The bimolecular cells have the dimensions:

$$Rb_2O_2: \quad a_0 = 4.201 \text{ A.}; \; b_0 = 7.075 \text{ A.}; \; c_0 = 5.983 \text{ A.}$$
$$Cs_2O_2: \quad a_0 = 4.322 \text{ A.}; \; b_0 = 7.517 \text{ A.}; \; c_0 = 6.430 \text{ A.}$$

The space group is V_h^{25} (*Immm*) with atoms in the positions:

$$\text{Rb (or Cs)}: \quad (4g) \quad \pm(0u0; \; {}^1/_2 \, u \, {}^1/_2) \qquad \text{with u} = {}^1/_4$$
$$\text{O}: \quad (4i) \quad \pm(00u; \; {}^1/_2, {}^1/_2, u+{}^1/_2).$$

For Rb_2O_2, u(O) = 0.375; for Cs_2O_2, u = 0.38. These parameters give the same O–O = 1.50 A. for the peroxide group. In Rb_2O_2 each oxygen atom has

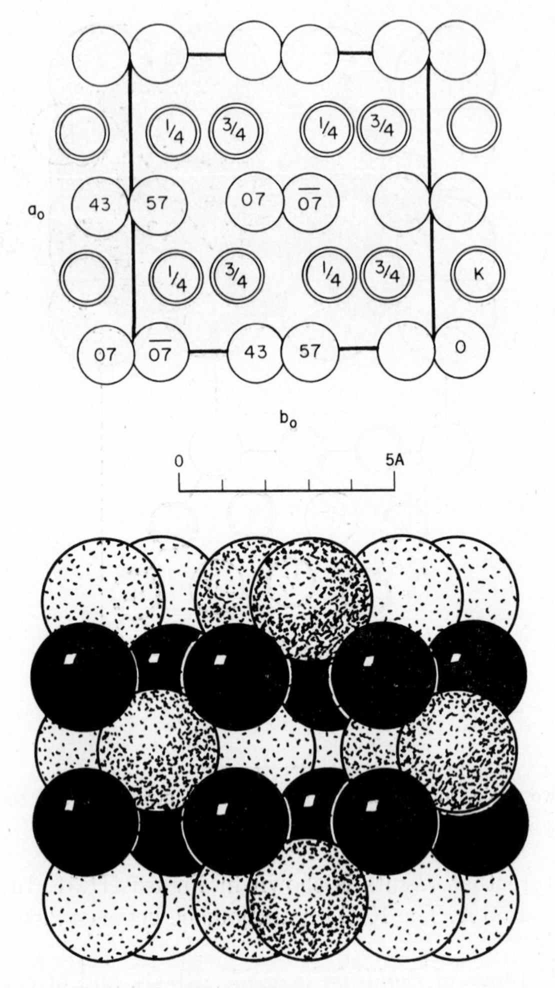

Fig. III,51a (top). The orthorhombic K_2O_2 structure projected along its c_0 axis. Origin in lower left.

Fig. III,51b (bottom). A packing drawing of the K_2O_2 structure seen along its c_0 axis. The potassium atoms are black.

four rubidium neighbors at a distance of 2.84 A. and two more almost equally close (2.85 A.). For Cs_2O_2 the corresponding Cs–O distances are 2.95 and 3.09 A.

The structure is illustrated in Figure III,52. Both it and the K_2O_2 arrangement can be considered as distortions of the fluorite, CaF_2, structure (Chapter IV,**a1**) with O_2 groups replacing calcium.

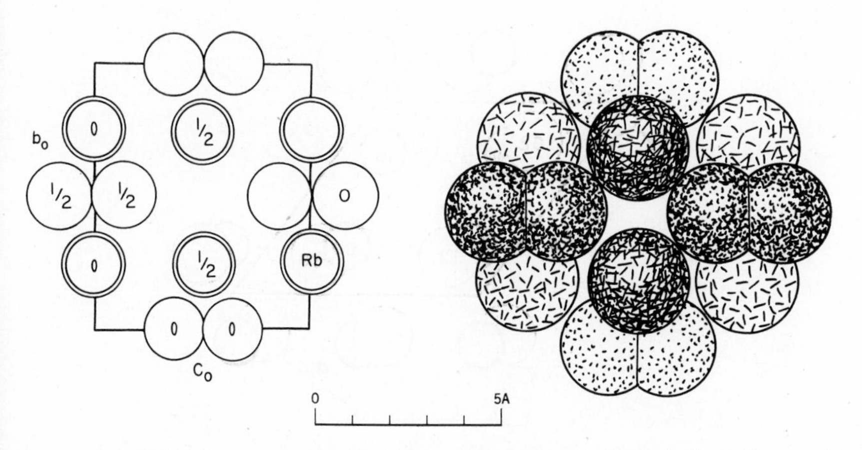

Fig. III,52a (left). The orthorhombic Rb_2O_2 arrangement projected along its a_0 axis. Origin in lower left.

Fig. III,52b (right). A packing drawing of the Rb_2O_2 structure seen along its a_0 axis. The rubidium atoms are line-shaded.

Molecular Structures

III,h1. The structure of solid *nitric oxide*, N_2O_2, at -175°C. is monoclinic with two molecules in the unit:

$$a_0 = 6.68 \text{ A.}; \; b_0 = 3.96 \text{ A.}; \; c_0 = 6.55 \text{ A.}; \; \beta = 127°54'$$

Atoms are in general positions of C_{2h}^5 $(P2_1/a)$:

$$(4e) \quad \pm(xyz; \; x+{}^1/_2,{}^1/_2-y,z)$$

It is impossible to distinguish by means of x rays between nitrogen and oxygen; one has the parameters $x = 0.228$, $y = 0.121$, $z = 0.194$; the other has the parameters $x' = 0.160$, $y' = -0.101$, $z' = 0.241$.

In the resulting structure (Fig. III,53) there is a short N–O = 1.12 A. Between the two halves of the dimers centered around the origin and other symmetry centers (if these dimers have physical reality in the solid) the N–O separation is a far longer 2.40 A.

III,h2. As is well known, the halogens form a series of inter-compounds with one another. *Cyanogen iodide*, CNI, can be considered as one of these, but it is considerably more stable than most and therefore especially well suited for x-ray examination. Its crystals are rhombohedral with a unimolecular cell having the dimensions:

$$a_0 = 4.44 \text{ A.}, \qquad \alpha = 101°24'$$

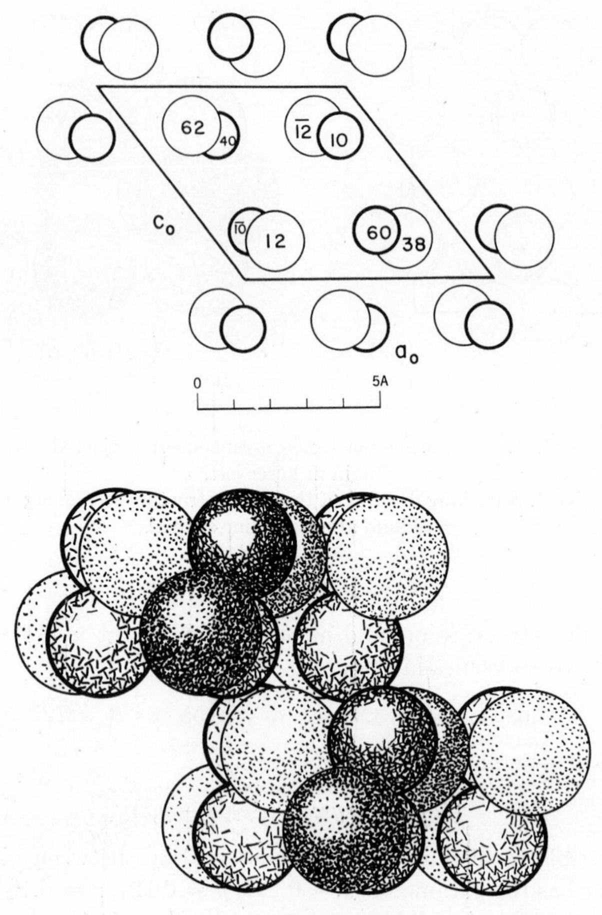

Fig. III,53a (top). A projection along b_0 of the monoclinic structure of N_2O_2. Nitrogen and oxygen atoms cannot be distinguished experimentally. Origin in lower left.
Fig. III,53b (bottom). A packing drawing of the monoclinic N_2O_2 structure viewed along the b_0 axis. It is not known whether the dotted or the line-shaded atoms are oxygen.

The probable space group is C_{3v}^5 ($R3m$). If both the carbon and the nitrogen have positions fixed by symmetry, all atoms are on trigonal axes with the coordinates uuu. The nitrogen and carbon are so light compared with iodine that their positions cannot be established from the intensity data, but it has been shown that a molecular structure with u(C) = 0.34 and u(N) = 0.54, u being taken as zero for iodine, is compatible with the data and gives the acceptable interatomic distances: C–I = 2.03 A., C–N = 1.18 A.

The corresponding trimolecular hexagonal unit has the dimensions:

$$a_0' = 6.87 \text{ A.,} \qquad c_0' = 5.98 \text{ A.}$$

In this cell the atomic parameters are, of course, the same as the foregoing.

III,h3. Crystals of *cyanogen chloride*, CNCl, and of the corresponding *bromide*, CNBr, have a structure somewhat different from the iodide, CNI. They are orthorhombic with bimolecular units of the dimensions:

$$\text{CNCl:} \quad a_0 = 5.684 \text{ A.}; \ b_0 = 3.977 \text{ A.}; \ c_0 = 5.740 \text{ A.}$$
$$\text{CNBr:} \quad a_0 = 6.02 \text{ A.}; \ b_0 = 4.12 \text{ A.}; \ c_0 = 5.80 \text{ A.}$$

The space group is V_h^{13} (*Pmmn*) with all atoms in the special positions:

$$(2a) \quad {}^1/_4\, {}^1/_4\, u; \ {}^3/_4\, {}^3/_4\, \bar{u}$$

For CNCl the parameters have been found to have the values: $u(\text{C}) = 0.4239$, $u(\text{N}) = 0.6258$, $u(\text{Cl}) = 0.1499$. For CNBr they are: $u(\text{C}) = 0.445$, $u(\text{N}) = 0.614$, and $u(\text{Br}) = 0.143$.

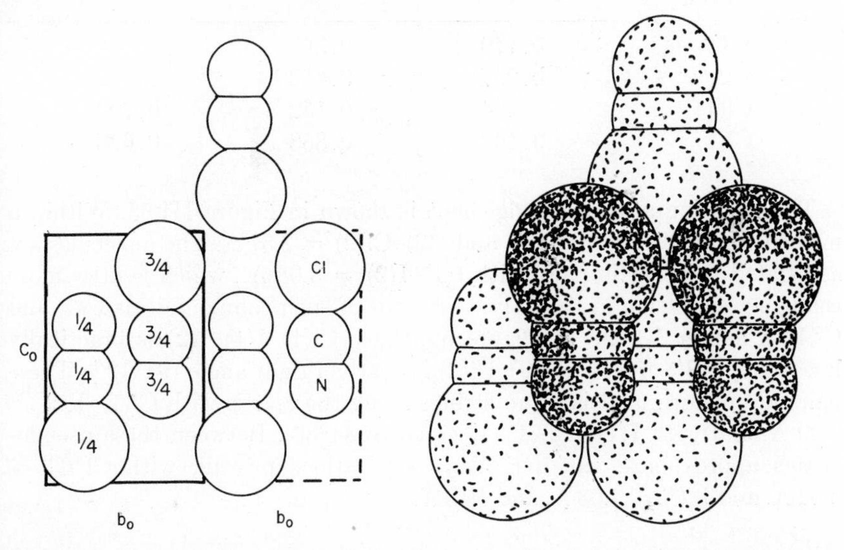

Fig. III,54a (left). Two cells of the orthorhombic structure of CNCl projected along its a_0 axis. In this case the axes were chosen as right-handed (rather than the usual left-handed). Origin in lower left.

Fig. III,54b (right). A packing drawing of the CNCl structure. The large circles are chlorine; the carbon and nitrogen are not distinguished. The direction of the a_0 axis in this case is downwards.

The simple structure that results is shown in Figure III,54. The molecules are strictly linear with C–N = 1.16 A. and C–Cl = 1.57 A. for the chloride. The molecules themselves are stacked in chains and these chains are closely packed in one direction within the crystal. Within a chain the Cl–N between adjacent molecules is 3.01 A. The distance between parallel chains is 3.98 A.; between those pointing in opposite directions it is 3.47 A.

III,h4. Crystals of the alpha form of *iodine monochloride*, ICl, are monoclinic with a unit containing eight molecules and with

$$a_0 = 12.60 \text{ A.};\ b_0 = 4.38 \text{ A.};\ c_0 = 11.90 \text{ A.};\ \beta = 119°30'$$

presumably at room temperature. All atoms are in general positions of C_{2h}^5 ($P2_1/c$):

$$(4e)\quad \pm(xyz;\ \bar{x},y+{}^1/_2,{}^1/_2-z)$$

with the following parameters:

Atom	x	y	z
I(1)	0.179	0.366	0.588
I(2)	0.297	0.632	0.436
Cl(1)	0.084	0.152	0.706
Cl(2)	0.462	0.858	0.620

The resulting atomic arrangement is shown in Figure III,55. Within a molecule I(1)–Cl(1) = 2.44 A. and I(2)–Cl(2) = 2.37 A. The molecules are associated together in pairs with I(1)–I(2) = 3.08 A., which is little more than the separation in the polyiodide ions of such compounds as CsI_3 and Cs_2I_8 (Chapter V). As the figure indicates, Cl(1)–I(1)–I(2) is practically linear (179°18′) and I(1)–I(2)–Cl(2) is nearly a right angle (94°12′). These pairs are in turn associated in strings along the c_0 axis with Cl(1)–I(2) = 3.00 A. and with I(1)–I(2)–Cl(1) equal to 84°36′. Between these pseudo-chains approximate van der Waals separations prevail, with Cl–Cl = 3.52 A. and I–Cl = 4.03 A. and 4.13 A.

III,h5. Considering its chemical simplicity, *realgar*, AsS, has a crystal structure that has been surprisingly hard to establish. It is monoclinic with a 16-molecular unit of the dimensions:

$$a_0 = 9.27 \text{ A.};\ b_0 = 13.50 \text{ A.};\ c_0 = 6.56 \text{ A.};\ \beta = 106°37'$$

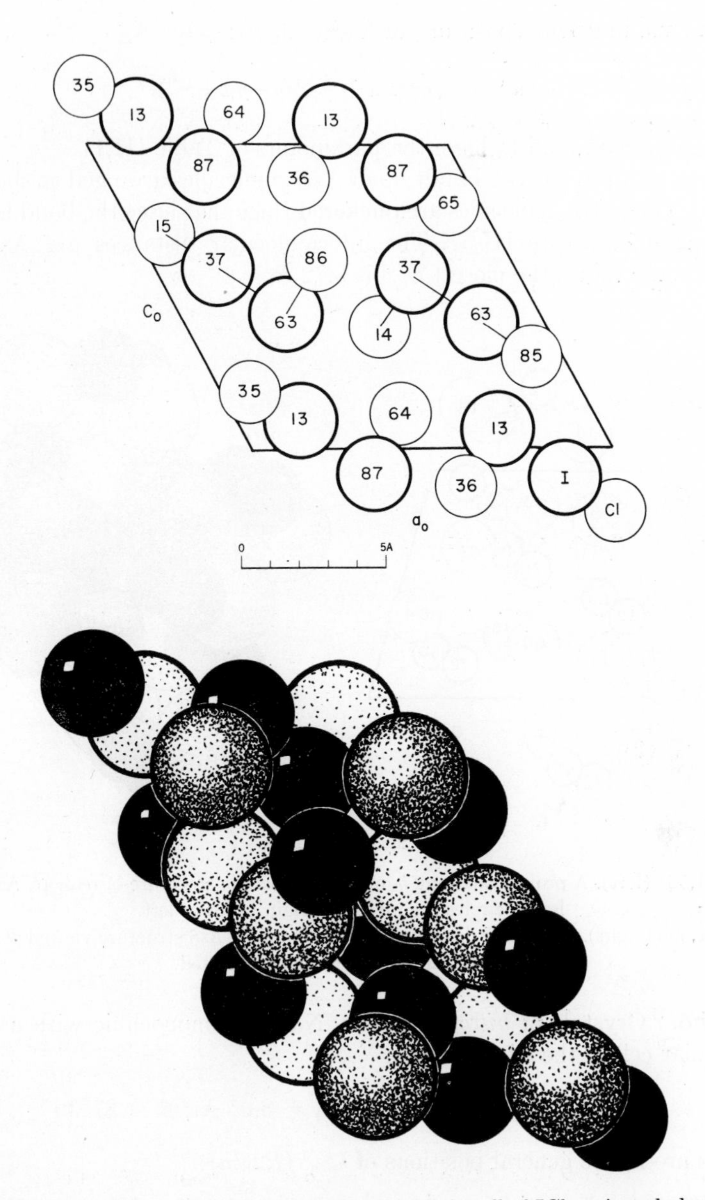

Fig. III,55a (top). The contents of the large monoclinic cell of ICl projected along its b_0 axis. Origin in lower left.

Fig. III,55b (bottom). A packing drawing of the ICl arrangement viewed along its b_0 axis. The chlorine atoms are black.

The atoms, in general positions of C_{2h}^5 ($P2_1$/n):

$$(4e) \quad \pm(xyz;\ x+{}^1/_2,{}^1/_2-y,z+{}^1/_2)$$

have now been found to have the parameters of Table III,11.

These result in a crystal built up of As_4S_4 molecules arranged as shown in Figure III,56. The molecules are puckered rings that have the bond lengths and angles of Figure III,57. The intermolecular distances are As–S = 3.39 A., and more; the closest S–S is 3.72 A.

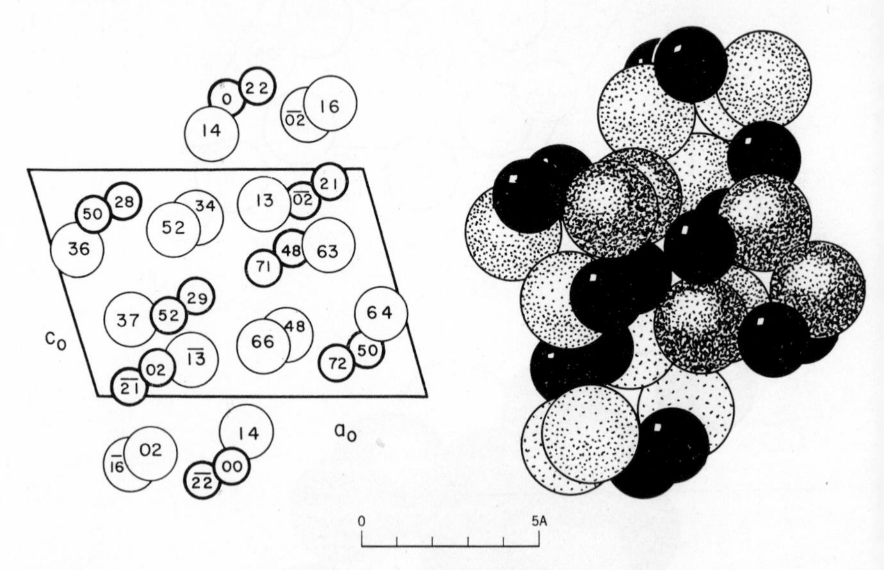

Fig. III,56a (left). A projection along b_0 of the monoclinic structure of realgar, AsS. The larger circles are arsenic. Origin in lower left.

Fig. III,56b (right). A packing drawing of the monoclinic AsS structure viewed along the b_0 axis. The arsenic atoms are dotted.

III,h6. Crystals of *nitrogen sulfide*, N_4S_4, are monoclinic with a tetramolecular cell of the edge lengths:

$$a_0 = 8.75 \text{ A.};\ b_0 = 7.16 \text{ A.};\ c_0 = 8.65 \text{ A.};\ \beta = 87°30'$$

Atoms are in the general positions of C_{2h}^5 ($P2_1/n$):

$$(4e) \quad \pm(xyz;\ x+{}^1/_2,{}^1/_2-y,z+{}^1/_2)$$

with the parameters of Table III,12.

TABLE III,11
Parameters of the Atoms in Realgar

Atom	x	y	z
As(1)	0.118	0.024	−0.241
As(2)	0.425	−0.140	−0.142
As(3)	0.318	−0.127	0.181
As(4)	0.038	−0.161	−0.290
S(1)	0.346	0.008	−0.295
S(2)	0.213	0.024	0.120
S(3)	0.245	−0.225	−0.363
S(4)	0.115	−0.215	0.048

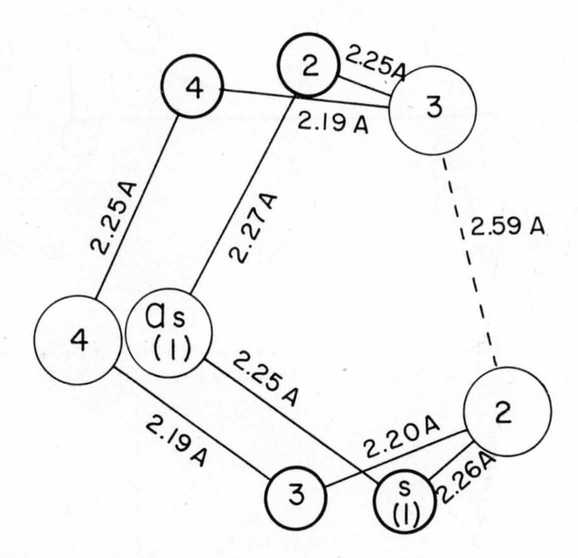

Fig. III,57. The bond dimensions of the As_4S_4 molecules in realgar.

The structure (Fig. III,58) is a packing of S_4N_4 molecules that have the shape indicated in Figure III,59. The sulfur atoms form a tetrahedron which is nearly regular and the nitrogen atoms together are at the corners of a square. All the S–N bonds = 1.60 A. The S–S distance is 2.71 A. for sulfur atoms linked through atoms of nitrogen; for those not so linked it is a little less, 2.58 A. The bond angles are N–S–N = 102° and S–N–S = 115°. Between molecules the closest atomic approaches are S–N = 3.3 A. and S–S = 3.7 A.

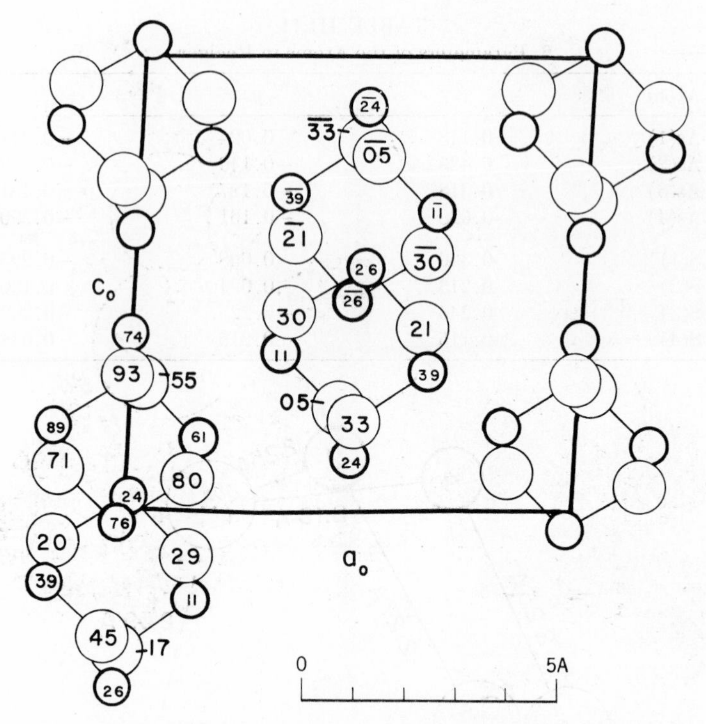

Fig. III,58a. A projection along b_0 of the monoclinic structure of S_4N_4. Sulfur atoms are the larger circles. Origin in lower left.

TABLE III,12
Parameters of the Atoms in S_4N_4

Atom	x	y	z
S(1)	0.984	0.932	0.296
S(2)	0.849	0.711	0.087
S(3)	0.148	0.800	0.067
S(4)	0.029	0.555	0.276
N(1)	0.995	0.763	0.980
N(2)	0.010	0.739	0.383
N(3)	0.167	0.610	0.160
N(4)	0.840	0.893	0.193

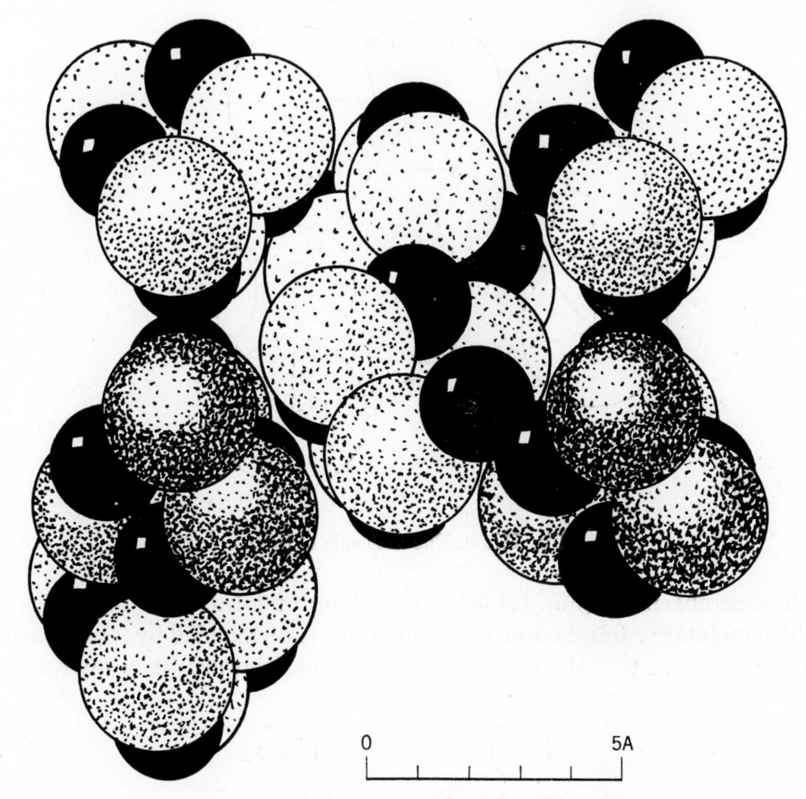

Fig. III,58b. A packing drawing of the monoclinic S_4N_4 structure viewed along the b_0 axis. The smaller, black atoms are the nitrogen.

III,h7. The recently discovered B_4Cl_4 is tetragonal. Its bimolecular unit has the edges:

$$a_0 = 8.09 \text{ A.}, \qquad c_0 = 5.45 \text{ A.}$$

Atoms are in the following special positions of D_{4h}^{15} ($P4/nmc$):

$$(8g) \quad 0uv;\ u0\bar{v};\ {}^1/_2,u+{}^1/_2,{}^1/_2-v;\ u+{}^1/_2,{}^1/_2,v+{}^1/_2;$$

$$0\bar{u}v;\ \bar{u}0\bar{v};\ {}^1/_2,{}^1/_2-u,{}^1/_2-v;\ {}^1/_2-u,{}^1/_2,v+{}^1/_2$$

For chlorine, $u = 0.2756$, $v = 0.2050$; for boron, $u' = 0.1043$, $v' = 0.3873$.

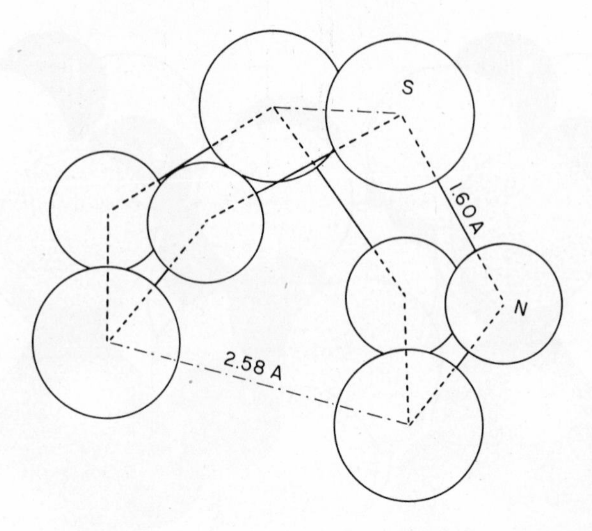

Fig. III,59. The molecule of S_4N_4.

The arrangement (Fig. III,60) is a packing of tetrahedral molecules in which an interior tetrahedron of boron atoms is surrounded by four chlorine atoms placed so that the B–Cl bonds are directed towards the center of the

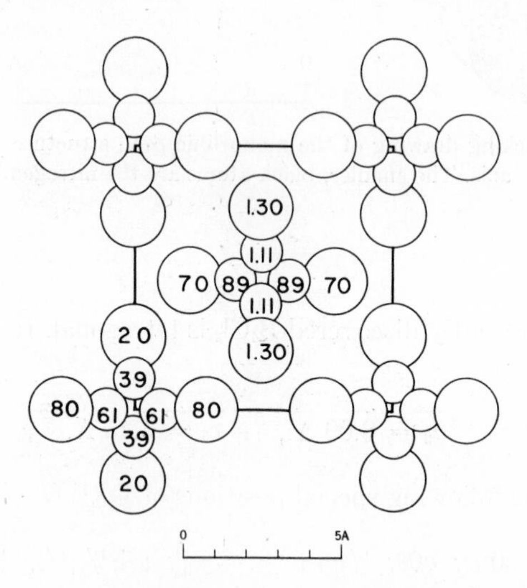

Fig. III,60a. A projection along c_0 of the tetragonal structure of B_4Cl_4. The larger circles are chlorine. Origin in the lower right.

tetrahedron. All B–Cl distances are 1.70 A. Four of the B–B separations equal 1.71 A., the other two 1.69 A. Between molecules the shortest Cl–Cl distance is 3.63 A.

III,h8. The compound B_8Cl_8 crystallizes in two forms, the simpler of which has the following structure. The symmetry is orthorhombic with four molecules in a cell of the dimensions:

$$a_0 = 13.64 \text{ A.}; \; b_0 = 7.85 \text{ A.}; \; c_0 = 12.91 \text{ A.}$$

Atoms are in general positions of V^4 ($P2_12_12_1$):

$$(4a) \quad xyz; \; {}^1/_2-x,\bar{y},z+{}^1/_2; \; x+{}^1/_2,{}^1/_2-y,\bar{z}; \; \bar{x},y+{}^1/_2,{}^1/_2-z$$

with the parameters of Table III,13.

The structure, as shown in Figure III,61, is composed of B_8Cl_8 molecules. In each molecule (Fig. III,62) the boron atoms are dodecahedrally distributed so that each is bound to four others, as well as to an atom of chlorine. The B–B bonds lie between 1.76 and 1.86 A.; B–Cl bonds range from 1.65 to 1.75 A.

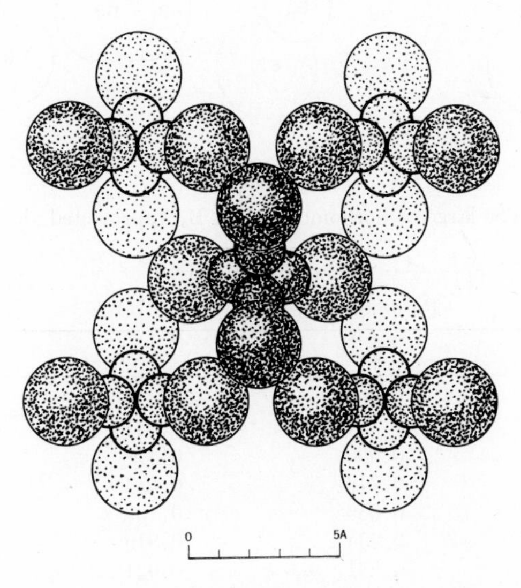

Fig. III,60b. A packing drawing of the tetragonal B_4Cl_4 structure viewed along the c_0 axis. The larger circles are the chlorine atoms.

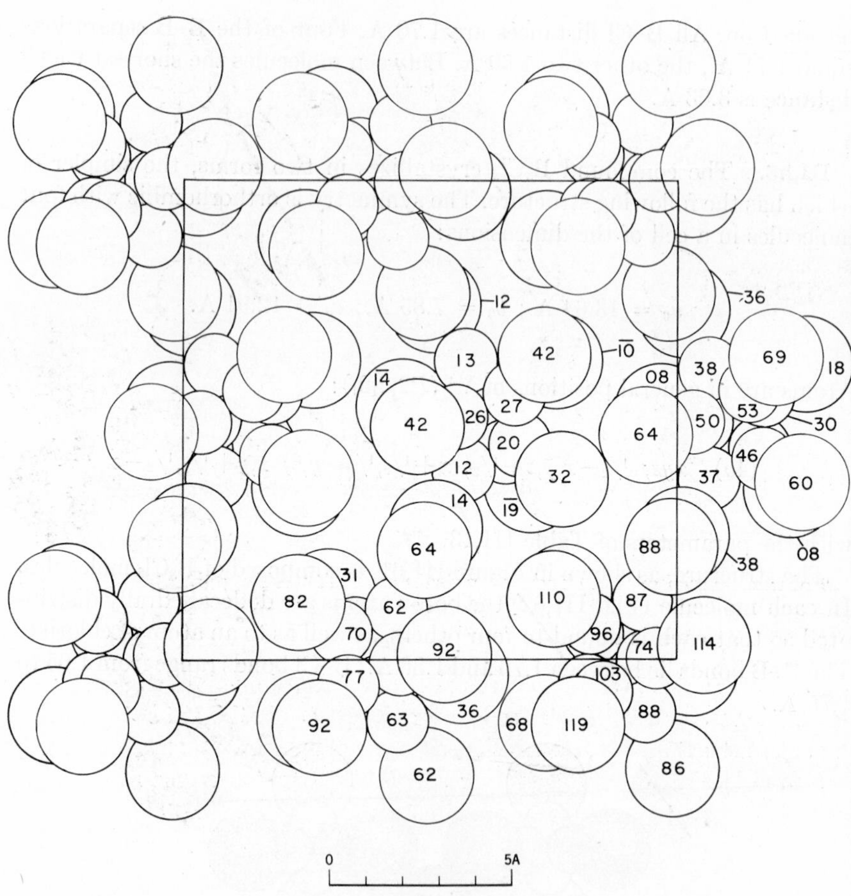

Fig. III,61a. The large orthorhombic cell of B_8Cl_8 projected along its b_0 axis.

TABLE III,13

Parameters of the Atoms in B_8Cl_8

Atom	x	y	z
Cl(1)	0.297	0.310	0.129
Cl(2)	0.491	0.641	0.228
Cl(3)	0.990	0.878	0.239
Cl(4)	0.776	0.584	0.138
Cl(5)	0.056	0.139	0.023
Cl(6)	0.234	0.819	0.120
Cl(7)	0.531	0.921	0.017
Cl(8)	0.747	0.100	0.128

(continued)

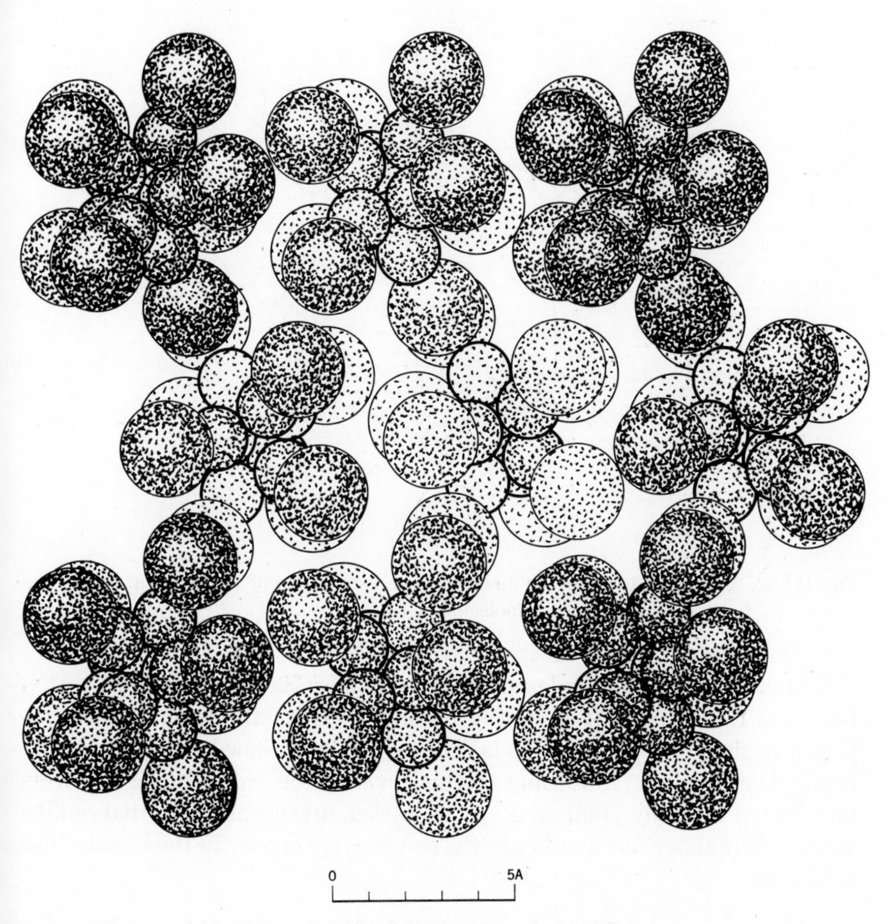

Fig. III,61b. A packing drawing of the B_8Cl_8 structure seen along its b_0 axis. The boron atoms are the smaller circles.

TABLE III,13 (*continued*)

Atom	x	y	z
B(1)	0.346	0.475	0.064
B(2)	0.435	0.625	0.108
B(3)	0.933	0.866	0.118
B(4)	0.835	0.735	0.075
B(5)	0.962	−0.001	0.001
B(6)	0.324	0.703	0.058
B(7)	0.956	0.739	0.008
B(8)	0.827	0.962	0.069

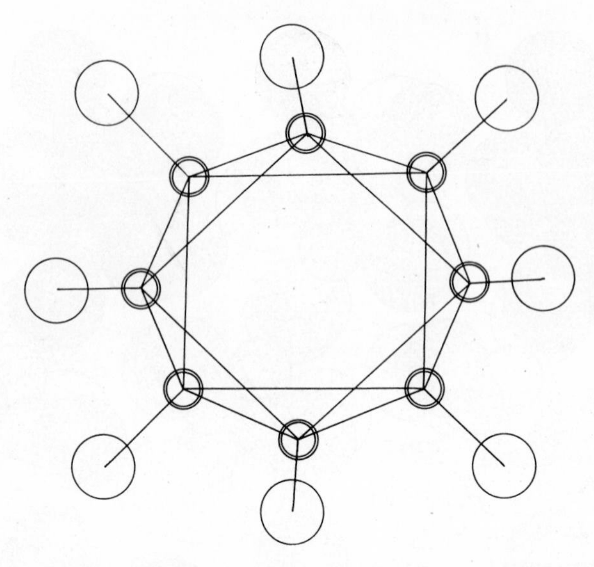

Fig. III,62. A drawing to show the dodecahedral distribution of the boron atoms in the molecule of B_8Cl_8.

III,h9. Discrete molecules are not found in crystals of *boron nitride*, BN, but as in the graphite it so closely resembles (**II,i2**) all atoms are held together in sheets by valency bonds. The originally assigned structure has recently been revised. According to this revision the boron–nitrogen sheets are shifted laterally compared to the earlier arrangement so that unlike atoms occur above one another in the consecutive layers. In the bimolecular cell, with

$$a_0 = 2.50399 \text{ A.}, \qquad c_0 = 6.6612 \text{ A.} \quad (35°\text{C.})$$

atoms are in the following special positions of D_{6h}^4 ($C6/mmc$):

$$\text{B:} \quad (2c) \quad \pm(^1/_3\ ^2/_3\ ^1/_4)$$
$$\text{N:} \quad (2d) \quad \pm(^1/_3\ ^2/_3\ ^3/_4)$$

In this structure (Fig. III,63), the B–N separation in the layer is 1.45 A. and between layers, 3.33 A. There is evidence that stacking faults are common.

Recently, two more forms of BN have been described. One of these is cubic with the zinc blende, ZnS, structure (**III,c1**). Its unit has:

$$a_0 = 3.615 \text{ A.} \quad (25°\text{C.})$$

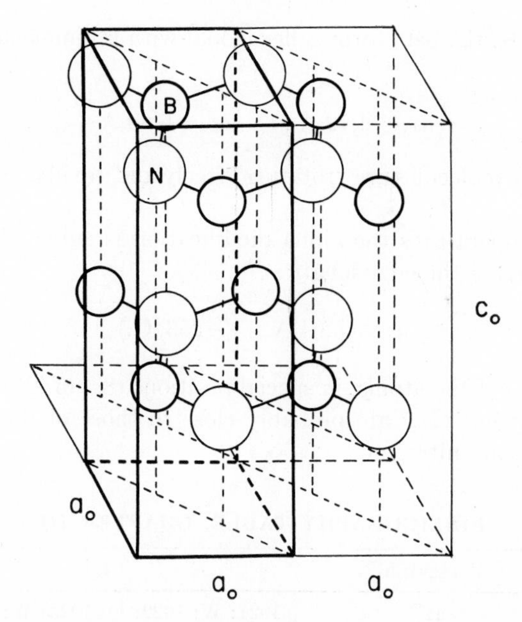

Fig. III,63. A perspective drawing of the contents of four unit cells of the new structure of BN.

This has been prepared from the above form by heating to 1800°C. under 85,000 atm. pressure.

A third, graphitic, form has been obtained by fusion. It is hexagonal with a unit of the dimensions:

$$a_0 = 2.504 \text{ A.}, \qquad c_0 = 10.01 \text{ A.}$$

Described as having the structure of beta-graphite (**II,i3**), its atoms would be in the positions:

B: 000; rh
N: 0 0 $^1/_3$; rh

This arrangement is illustrated in Figure II,13. As with the corresponding forms of graphite, this modification and the first consist of B–N sheets in stackings along the c_0 axis which repeat themselves as do the layers in the cubic and hexagonal close-packings.

III,h10. Solid *carbon monoxide*, CO, is dimorphous with structures strictly isomorphous with those of solid nitrogen (**II,j1** and **II,j2**).

Above 61.5°K. the beta form is hexagonal with a bimolecular cell of the dimensions:

$$a_0 = 4.12 \text{ A.}, \qquad c_0 = 6.80 \text{ A.}$$

Presumably the molecules are "rotating" freely and are in a hexagonal close-packing.

At lower temperatures the alpha modification is cubic with a tetramolecular unit having the edge length:

$$a_0 = 5.64 \text{ A.} \quad (-253°\text{C.})$$

The parameters of the atoms, in special positions (8*c*) of T^4 ($P2_13$), are not exactly known, but they are probably close to those of the atoms in the alpha form of solid nitrogen.

BIBLIOGRAPHY TABLE, CHAPTER III

Compound	Paragraph	Literature
AgBr	a1	1921: W; 1922: D; 1923: W; 1925: B&L; W; 1926: L; 1927: L; 1936: W&B; 1940: W; 1943: H; 1951: B; K&M
AgCd	b1	1923: B; 1927: N&F; 1928: A&W; 1929: W&A
AgCe	b1	1943: B&K
AgCl	a1	1921: W; 1922: D; 1923: W; 1925: B&L; B; W; 1940: W
AgF	a1	1926: O
AgI	c1, c2	1921: W; 1922: A; D; 1923: W; 1925: B&L; W; 1926: L; 1927: L; 1928: K,vD&B; 1929: L&R; 1931: B&M; 1934: K&vH; S; 1935: H; 1938: J; 1940: W; 1953: S; 1954: H&M; 1957: H; 1958: NBS; 1959: NBS
AgLa	b1	1943: B&K
AgMg	b1	1926: O&P
AgO	e6	1957: S,B&Z; 1958: S,B&Z; S&W; 1959: S&W; 1960: S,B&S; 1961: S,B&S
AgZn	b1	1923: B; 1925: W&P; 1926: W&P; 1929: R&C; W&A
AlAs	c1	1928: N&P
Al_2CO	c2	1961: A&J
AlN	c2	1924: O; 1935: S&S; 1955: J&P; 1956: J,P&M
AlNd	b1	1934: S&J
AlNi	b1	1923: B&E; 1929: W&A; 1937: B&T

(continued)

BIBLIOGRAPHY TABLE, CHAPTER III (*continued*)

Compound	Paragraph	Literature
AlP	c1	1928: P; 1960: A
AlSb	c1	1924: O&P; 1953: NBS; 1958: G&P; 1959: U&K
AmO	a1	1949: Z; 1960: MW,W,C,A,E&Z
AsS	h5	1935: B; 1952: I,M&S; 1954: T,Z&Z
AuCN	g8	1945: Z&S; 1959: NBS
AuCd	b1	1932: O
AuGa	d3	1950: P&S
AuI	g7	1956: W&W; 1959: J
AuMg	b1	1936: B&H
AuSn	d1	1927: P&O; 1931: S&W; 1932: B&J; 1933: J&B
AuZn	b1	1925: W&P; 1926: O&P
BAs	c1	1958: P,LP&P
B_4Cl_4	h7	1953: A&L; 1958: J&L; 1959: A&L; J&L
B_8Cl_8	h8	1958: J&L; 1959: A&L; J&L
BN	c1, h9	1923: T&T; 1926: H; J&W; 1937: B; 1939: B; 1940: B&Z; 1941: B,H&P; 1950: P; 1952: P; 1957: W; 1958: H,M&P
BP	c1	1957: P&I; R; 1958: P,LP&P
BaNH	a1	1934: H,F&E
BaO	a1	1921: G; 1922: G; 1926: G; 1933: B; 1942: H&W
BaS	a1	1923: H; 1956: G&F
BaSe	a1	1923: S; 1925: S; 1960: M,K&C
BaTe	a1	1927: H; 1928: H; 1929: G; 1960: M,K&C
BeCo	b1	1936: M
BeCu	b1	1928: D,H&M; 1935: M
BeO	c2	1922: G; MK; 1925: A; Z; 1926: C; Z; 1927: H; 1941: N; 1953: NBS; 1956: J,P&M
BePd	b1	1936: M
BeS	c1	1926: Z
BeSe	c1	1926: G; Z
BeTe	c1	1926: G; Z
BiSe	a1	1954: S
BiTe	a1	1954: S
CNBr	h3	1955: G&S
CNCl	h3	1956: H&C
CNI	h2	1939: K&Z

(*continued*)

BIBLIOGRAPHY TABLE, CHAPTER III (*continued*)

Compound	Paragraph	Literature
CO	h10	1930: V; 1934: V
CaCl	g4	1953: E&G; 1954: E&G
CaNH	a1	1933: F,B&H; 1934: H,F&E
CaO	a1	1920: D&H; 1921: G; 1922: G; 1923: D; 1927: H; O; 1928: R; 1929: N&P; 1942: H&W; 1953: NBS
CaS	a1	1923: D; H; 1927: O; R; 1956: G&F
CaSe	a1	1923: D; 1927: O
CaTe	a1	1927: O
CaTl	b1	1933: Z&B
CdCe	b1	1937: I&B
CdLa	b1	1937: I&B
CdO	a1	1920: D&H; 1922: S; 1926: G; 1927: W; 1929: B&A; F; N&P; P; 1931: K; 1933: G; M&L; 1938: F&M; 1953: NBS
CdPr	b1	1937: I&B
CdS	c1, c2	1920: B; 1925: U&Z; 1933: M&L; 1940: K; 1943: R&S; 1953: NBS; 1955: S; T&B; 1957: C
CdSb	c4	1948: A
CdSe	c2	1926: G; Z; 1955: G,K&FK
CdTe	c1	1925: Z; 1926: G; Z; 1955: S; 1960: L,N&Y; P,T,K&R
CeAs	a1	1937: I&B
CeBi	a1	1937: I&B
CeN	a1	1937: I&B
CeP	a1	1936: I&B
CeS	a1	1949: Z; 1955: I
CeSb	a1	1937: I&B
CeSe	a1	1955: I; 1959: G&B
CeTe	a1	1955: I
CoAs	d3	1934: F; 1950: P&S; 1957: V; H&C
CoB	d4	1933: B
CoO	a1	1922: H; 1925: G,B&L; 1926: B; N&R; 1928: N&S; 1929: H&K; N&P; P; 1950: T&R; G&S; 1951: R&T; 1953: G; 1959: NBS
CoP	d3	1934: F; 1939; B&H; 1950: P&S
CoS	d1	1925: A; 1927: dJ&W; 1932: C&R; 1938: L&W

(*continued*)

BIBLIOGRAPHY TABLE, CHAPTER III *(continued)*

Compound	Paragraph	Literature
CoSb	d1	1927: dJ&W; O; 1938: F&H
CoSe	d1	1927: dJ&W; O; 1955: B,G,H&P
CoTe	d1	1927: dJ&W; O
CrAs	d3	1938: N&A; 1950: P&S; 1960: Y
CrB	d5	1949: K; S; 1951: F
CrN	a1	1929: B; 1957: P&K
CrP	d3	1938: N&H; 1950: P&S; 1954: S
CrS_x	d1, e8	1927: dJ&W; 1937: H; 1957: Y,W,N&M
CrSb	d1	1927: dJ&W; O; 1953: W
CrSe	d1	1927: dJ&W; 1938: H&M
CrTe	d1	1927: O; 1946: G&B; 1960: H&C
CsBr	b1	1921: W; 1922: D; 1923: D; 1952: NBS 1957: H
CsCN	b1	1931: N&P
CsCl	a1, b1	1921: D&W; 1923: D; 1924: H,M&B; 1929: B,O&P; 1933: W&L; 1934: W; 1936: W&L; 1951: M,U&W; 1953: NBS; 1957: H
CsF	a1	1922: P&W; 1923: D
CsH	a1	1931: Z&H
CsI	b1	1921: D&W; 1922: C&D; 1923: C&D; D; 1951: R&H; 1953: NBS; 1955: S&K
$CsNH_2$	b1, b3	1959: J&M
Cs_2O_2	g18	1957: F
CsSH	b1	1934: W; 1939: T&K
CsSeH	b1	1939: T&K
CuBr	c1, c2	1922: D; W&P; 1925: B&L; 1942: V&S; 1952: H; K&S
CuCl	c1, c2	1922: D; W&P; 1925: B&L; 1942: V&S; 1953: NBS; 1956: L&P
CuF	c1	1933: E&W
CuH	c2	1926: M&B; 1955: G&A
CuI	c1, c2	1922: A; D; W&P; 1925: B&L; 1926: L; 1927: L; 1929: L&R; 1942: V&S; 1952: K&S; M,H&T; 1953: NBS
CuO	e5	1922: H; N; 1933: T,P&K; 1935: T,P&K
CuPd	b1	1924: H&S; 1925: J&L; 1927: J&L
CuS	e10	1927: G&M; 1929: R&K; 1931: A; 1932: O; 1953: NBS; 1954: B; 1958: D
CuSe	e10	1949: E; 1954: B; 1960: T&U

(continued)

BIBLIOGRAPHY TABLE, CHAPTER III (*continued*)

Compound	Paragraph	Literature
CuSn	**d1**	1928: W&P
CuZn	**b1**	1921: A; 1923: O&P; 1925: W&P
DyAs	**a1**	1960: B
DyN	**a1**	1956: K&W
DySb	**a1**	1960: B
DyTe	**a1**	1960: B
ErAs	**a1**	1960: B
ErN	**a1**	1956: K&W
ErSb	**a1**	1960: B
ErTe	**a1**	1960: B
EuN	**a1**	1956: K&W; E,B&E
EuO	**a1**	1956: E,B&E; 1959: D,F&G
EuS	**a1**	1938: N; 1939: K&S
EuSe	**a1**	1939: K&S; 1959: G&B
EuTe	**a1**	1939: K&S
FeAs	**d3**	1928: H; 1929: H; 1934: F; 1950: P&S; 1960: NBS
FeB	**d4**	1929: B&A; W; 1930: H&K; 1933: B
FeO	**a1**	1925: W&C; 1926: G; 1928: G; 1933: J&F; 1934: E&T; 1936: B&C; 1937: B; 1944: A&G; 1951: C&B; 1952: C&B; 1953: W&R; 1954: B; C; 1956: F&W; 1957: S; 1960: R
FeP	**d3**	1934: F; 1950: P&S
FeS_x	**d1**	1925: A; dJ; 1927: dJ&W; 1932: J&B; 1933: H; H&S; 1934: M&C; 1937: S&H; 1939: H; 1941: H; 1945: B; 1946: L; 1947: B; L; 1950; U,I&M; 1952: B; 1956: B; 1957: E&D
FeSb	**d1**	1927: dJ&W; O; 1928: H; 1929: H; O
$FeSe_x$	**d1**	1925: A; 1927: dJ&W; 1955: O&H; 1956: O&H
FeSn	**d1**	1947: N
FeTe	**d1**	1928: O
GaAs	**c1**	1926: G; 1958: G&P; W&S; 1959: G&F
GaN	**c2**	1938: J&H
GaP	**c1**	1926: G; 1958: G&P
GaS	**e9**	1955: H&F
GaSb	**c1**	1926: G; 1958: G&P
GaSe	**e9**	1953: S&D; 1955: S,D&K; 1956: T,A&P

(*continued*)

BIBLIOGRAPHY TABLE, CHAPTER III (*continued*)

Compound	Paragraph	Literature
GdAs	a1	1960: B
GdN	a1	1948: E; 1956: K&W
GdSb	a1	1960: B
GdSe	a1	1959: G&B
GeS	a9	1932: Z
GeSe	a9	1958: O; 1960: K,K&C
HCN	g9	1951: D&L
HF	g10	1939: B,B&S; G,H&S; 1954: A&L
H_2O_2	g14	1951: A,C&L
HfC	a1	1936: MK; 1953: G,M&P; 1954: C,D&J; 1959: N,K,B&B
Hg_2Br_2	g1	1925: H; 1926: H; 1927: V; V&H
Hg_2Cl_2	g1	1925: H; 1926: H; M&S; 1927: V; V&H
Hg_2F_2	g1	1933: E&W; 1956: G&D
Hg_2I_2	g1	1925: H; 1926: H; 1927: H&M; 1953: NBS
HgO	a6, a7	1927: Z; 1928: L; 1952: NBS; 1954: A; 1956: A; R; 1957: A&C; 1958: A&C; 1959: NBS; V&K
HgS	a6, c1	1923: M; 1924: K,B&K; L; 1925: B&V; vO; R; 1926: G; H; dJ&W; 1927: G&M; 1936: M&K; 1943: R&S; 1950: A; 1953: NBS
HgSe	c1	1926: G; H; dJ; Z; 1950: E; 1954: Z
HgTe	c1	1925: Z; 1926: H; dJ; Z; 1954: Z; 1960: L,N&Y
HoAs	a1	1960: B; 1961: B
HoBi	a1	1961: B
HoN	a1	1956: K&W
HoP	a1	1961: B
HoS	a1	1961: B
HoSb	a1	1960: B; 1961: B
HoSe	a1	1961: B
HoTe	a1	1960: B; 1961: B
ICl	h4	1956: B,vdH,V&W
InAs	c1	1941: I; 1958: G&P; W&S; 1961: S&O
InBi	e1	1948: M; 1956: B
InBr	g2	1950: S&M
InI	g2	1955: J&T
InN	c2	1938: J&H
InP	c1	1941: I; 1958: G&P

(*continued*)

BIBLIOGRAPHY TABLE, CHAPTER III *(continued)*

Compound	Paragraph	Literature
InS	a9	1954: S,D&G
InSb	c1	1926: G; 1941: I; 1951: L&P; 1953: NBS; 1956: B; 1957: S&R; 1958: G&P
InSe	e9	1954: S,D&G; 1958: S
InTe	d7	1955: S,D&K
IrGe	d3	1950: P&S
IrSb	d1	1958: K
IrTe	d1	1955: GM
KBr	a1	1913: B; 1914: G; 1915: G; 1921: D; W; 1923: D; 1924: H,M&B; 1925: B; S; 1926: E; O; 1927: L; V&H; 1928: O; 1953: NBS; 1956: T&S; 1957: T&S
KCN	a1, a5	1921: C; 1922: B; C; 1931: N&P; 1940: B&L; 1953: NBS; 1961: E&H
KCl	a1	1913: B; 1914: G; 1915: G; 1921: D; 1923: D; 1924: H,M&B; 1925: B; S; 1926: O; 1927: V&H; 1928: O; 1951: K; 1953: H; NBS; 1954: G; 1956: T&S
KF	a1	1919: H; 1921: D; W; 1922: P&W; 1923: D; 1929: B,O&P; 1953: NBS
KH	a1	1931: Z&H
KI	a1	1919: H; 1921: D; W; 1922: C&D; 1923: C&D; D; W; 1924: H,M&B; 1925: vO; 1927: L; 1953: H; NBS; 1956: T&S; 1957: T&S
K_2O_2	g17	1957: F
KOH	a1, a8, g2	1939: T&K; 1947: E&S; 1948: E; 1960: I,K&S
KSH	a1, a3	1934: W; 1939: T&K
KSb	d6	1961: B&L
KSeH	a1, a3	1939: T&K
LaAs	a1	1937: I&B; 1960: B
LaBi	a1	1937: I&B
LaN	a1	1937: I&B; 1952: Y&Z; 1956: K&W
LaP	a1	1936: I&B
LaS	a1	1955: I
LaSb	a1	1937: I&B; 1960: B
LaSe	a1	1955: I; 1959: G&B
LaTe	a1	1955: I; 1960: B

(continued)

BIBLIOGRAPHY TABLE, CHAPTER III *(continued)*

Compound	Paragraph	Literature
LiAg	b1	1930: P; 1931: P; 1933: Z&B
LiAs	d6	1959: C
LiBr	a1	1922: P&W; 1923: D; O; 1953: NBS
LiCN	g3	1942: L&B
LiCl	a1	1922: P&W; 1923: D; O; 1938: J,S&K
LiD	a1	1935: Z&H
LiF	a1	1916: D&S; 1919: H; 1923: D; 1925: F; 1926: O; 1937: M; 1940: H&J; 1952: N,S&Y; 1955: T; 1956: T&S
LiH	a1	1922: B&K; 1929: B&F; 1931: Z&H; 1932: B&K; 1935: Z&H
LiHg	b1	1933: Z&B
LiI	a1	1922: P&W; 1923: D; O; W&P; 1926: G
$LiNH_2$	g11	1951: J&O
Li_2O_2	g15	1938: A&G; F; 1952: C; 1953: F,vW&D; 1957: F
LiOH	e1	1932: E; 1933: E; 1959: D
LiSH	g11	1954: J&L
LiTl	b1	1933: Z&B
LuN	a1	1956: K&W
MgCe	b1	1942: N
MgHg	b1	1936: B&H
MgLa	b1	1942: N
MgO	a1	1919: H; 1920: D&H; 1921: G; G&P; S; W; 1922: G; H; 1927: B; 1929: H&K; N&P; P; 1935: B,B&G; 1956: T&S; 1958: R&AP
MgPr	b1	1933: R&I
MgS	a1	1923: H; 1927: B; 1956: G&F
MgSe	a1	1926: G; 1927: B
MgSr	b1	1942: N
MgTe	c2	1927: Z; 1951: K&W
MgTl	b1	1933: Z&B
MnAs	d1, d3	1927: dJ&W; 1928: O; 1934: F; 1950: P&S
MnBi	d1	1939: H&G
MnO	a1	1924: L; 1926: F; G; O; 1927: B; 1929: N&P; P; 1933: L&W; 1934: E&T; 1935: R; 1946: E; 1954: NBS; 1957: G&G
MnP	d3	1934: F; 1937: A&N; 1943: F; 1950: P&S

(continued)

BIBLIOGRAPHY TABLE, CHAPTER III *(continued)*

Compound	Paragraph	Literature
MnS	**a1, c1, c2**	1921: W; 1926: G; O; 1932: S; 1933: S; 1934: E&T; 1940: K; 1951: R&T; 1953: NBS; 1956: C,E&H
MnSb	**d1**	1927: dJ&W; O
MnSe	**a1, c1, c2**	1927: B; 1938: B; 1959: NBS
MnTe	**c2, d1**	1927: O; 1953: G; 1958: G&G
MoB	**d8**	1947: K
MoC	**f1**	1932: T; 1952: K&H; N&K
MoP	**f4**	1954: S; 1955: B,N&K
ND_4Br	**b1**	1938: S,T&K; 1951: L&P; 1953: L&P
ND_4Cl	**b1**	1942: V&H
NH_2OH	**g13**	1955: M&L
NH_4Br	**a1, b1, b3**	1921: B&L; V; 1924: H,M&B; 1926: E; 1927: L; V&D; 1934: K; 1936: W&S; 1942: S&T; 1953: NBS; 1959: A&S
NH_4CN	**b2**	1944: L&B
NH_4Cl	**a1, b1**	1921: B&L; V; 1922: W; 1924: H,M&B; 1929: W&A; 1933: L&U; 1942: D; V&H; 1948: T&L; 1952: G&H; L&P; 1953: NBS; 1956: S&V; 1959: A&S; K
NH_4F	**c2**	1926: G; 1927: Z
NH_4I	**a1, b1, b3**	1917: V; 1921: B&L; 1924: H,M&B; 1926: S&vS; 1927: L; 1934: K; 1942: S&T; 1953: NBS
NH_4SCN	**VI**	
NH_4SH	**b3**	1934: W
N_2H_5Br	**g6**	1952: S&T
N_2H_5Cl	**g5**	1952: S&T
N_2O_2	**h1**	1953: D,M&L
N_4S_4	**h6**	1936: B; 1943: H&V; 1944: L&D; 1952: C
NaBr	**a1**	1921: D; W; 1923: D; 1926: O; 1927: V&D; 1938: J,S&K; 1952: NBS
NaCN	**a1, a5**	1931: N&P; 1938: V&B; 1949: S; 1953: NBS
NaCl	**a1**	1913: B; 1914: B; G; 1915: G; R; 1925: B; C,B&DF; S; 1927: B&L; W&J; 1936: S&S; 1937: M; 1939: B,G,H&P; 1940: I&S; 1941: vB; 1953: NBS; 1954: B; E; 1956: T&S
NaF	**a1**	1921: D; 1922: P&W; 1923: D; 1927: B&L

(continued)

BIBLIOGRAPHY TABLE, CHAPTER III (*continued*)

Compound	Paragraph	Literature
NaH	**a1**	1931: Z&H; 1948: S,W,M&D
NaI	**a1**	1921: D; W; 1923: D; 1953: NBS
$NaNH_2$	**g12**	1955: J,W&O; 1956: J,W&O; Z&T
Na_2O_2	**g16**	1938: F; 1957: F; T,M&B
NaOH	**g2**	1946: E
NaSH	**a1, a3**	1934: W; 1939: T&K
NaSb	**d6**	1959: C
NaSeH	**a1, a3**	1939: T&K
NbC	**a1**	1925: B&E; 1940: U; 1946: K&U; 1947: N&K; 1954: B,R&W; 1958: E&K; 1960: K,S&F
NbN	**a1, c2, d1, f1**	1925: B&E; 1940: U; 1952: B&J; 1954: S; 1959: S&K; 1961: B&E
NbO	**a1, a2**	1940: K; 1941: B; 1957: A&M; 1958: A,S&G
NbP	**f2**	1954: S
NbS	**d1**	1954: S
NdAs	**a1**	1937: I&B
NdBi	**a1**	1956: I
NdN	**a1**	1937: I&B; 1956: K&W
NdP	**a1**	1937: I&B
NdS	**a1**	1955: I
NdSb	**a1**	1937: I&B
NdSe	**a1**	1955: I; 1959: G&B
NdTe	**a1**	1955: I
NiAs	**d1**	1923: A; 1925: A; dJ; 1927: dJ&W; 1933: F
NiB	**d5**	1952: B; 1959: R
NiGe	**d3**	1950: P&S
NiO	**a1**	1920: D&H; 1921: D; 1922: H; 1925: B; C, A&W; G,B&L; L&T; 1926: B; N&R; 1929: H&K; N&P; P; 1931: B,C&O; K; 1933: C&O; 1934: P; 1943: R; 1948: R; S&N; 1951: B; 1955: S
NiS	**c5, d1**	1925: A; 1926: O; 1927: dJ&W; W; 1931: K&M; 1935: L&B; 1947: L; 1956: L&B; 1959: L; 1960: NBS
NiSb	**d1**	1925: A; dJ; 1927: dJ&W; O
NiSe	**c5, d1**	1925: A; dJ&W; 1935: L&B; 1956: G&J; 1960: H&W
NiSn	**d1**	1928: O; 1947: N

(*continued*)

BIBLIOGRAPHY TABLE, CHAPTER III *(continued)*

Compound	Paragraph	Literature
NiTe	d1	1927: dJ&W; O; 1943: K&F; 1957: S&I
NpN	a1	1949: Z
NpO	a1	1949: Z
OsC	f4	1960: K&N
PH_4Br	b3	1955: S,B&F
PH_4I	b3	1922: D
PaO	a1	1952: Z
PbO	e1, e3	1924: D&F; L; 1926: L&N; 1927: H&P; 1932: D; 1941: M&P; M&T; 1943: B; 1945: B; 1953: NBS; 1961: K; L
PbS	a1	1922: D; 1924: K; L; 1925: B&B; R; 1926: B; 1935: vZ; 1951: W; 1953: NBS; 1958: R&AP
PbSe	a1	1925: vO; R; 1926: G; 1950: E; 1954: NBS; 1960: N&O; 1961: K,G&K
PbTe	a1	1925: R; 1926: G; 1957: R; 1958: F,G&G; 1960: N&O
PdGe	d3	1950: P&S
PdH	a1	1957: W,W&S
PdO	e4	1926: L&F; 1927: L; Z; 1939: F&K; 1941: M&P; 1953: NBS; W,L&P
PdS	e7	1932: B; 1937: G; 1956: G&R
PdSb	d1	1928: T
PdSe	e7	1957: S,B,B,G,H,L,V,W&W
PdSi	d3	1950: P&S
PdSn	d1, d3	1947: N; 1950: P&S
PdTe	d1	1929: T; 1955: GM; 1956: G&R
PrAs	a1	1937: I&B
PrBi	a1	1937: I&B
PrN	a1	1937: I&B
PrP	a1	1936: I&B
PrS	a1	1955: I
PrSb	a1	1937: I&B
PrSe	a1	1955: I; 1959: G&B
PrTe	a1	1955: I
PtB	d1	1959: A,S&A
PtBi	d1	1959: Z,S&Z
PtGe	d3	1950: P&S
PtO	e4	1941: M&P

(continued)

BIBLIOGRAPHY TABLE, CHAPTER III (*continued*)

Compound	Paragraph	Literature
PtS	e4	1932: B
PtSb	d1	1929: T
PtSi	d3	1950: P&S
PtSn	d1	1928: O; 1932: J&B; 1933: J&B
PuAs	a1	1957: G
PuB	a1	1960: MD&S
PuC	a1	1949: Z
PuN	a1	1949: Z
PuO	a1	1949: Z; 1958: B,G,M&R
PuP	a1	1957: G
PuS	a1	1949: Z
PuTe	a1	1957: G
RbBr	a1	1921: D; 1923: D; 1924: H,M&B; 1926: O
RbCN	a1	1931: N&P
RbCl	a1, b1	1921: W; 1923: D; 1924: H,M&B; 1925: vO; 1926: O; 1936: W&L; 1953: NBS; 1957: V&K
RbF	a1	1922: P&W; 1923: D; W&P; 1926: G
RbH	a1	1931: Z&H
RbI	a1, b1	1921: W; 1922: D; 1923: D; 1924: H,M&B; 1938: J; 1953: NBS; 1957: V&K
$RbNH_2$	a1	1959: J&M
Rb_2O_2	g18	1957: F
RbSH	a1, a3	1934: W; 1939: T&K
RbSeH	a1, a3	1939: T&K
RhB	d1	1959: A,A&S; A,S&A
RhBi	d1	1953: G&Z
RhGe	d3	1955: G
RhSb	d3	1950: P&S
RhSn	d1	1947: N
RhTe	d1	1955: G
RuC	f4	1960: K&N
RuP	d3	1960: R
ScAs	a1	1960: B
ScN	a1	1925: B&E
ScSb	a1	1960: B
ScTe	d1	1961: M,K,S&S

(*continued*)

BIBLIOGRAPHY TABLE, CHAPTER III (*continued*)

Compound	Paragraph	Literature
SiC	c1, c2, c3	1918: B&O; 1919: H; 1920: H&K; H; 1921: E; 1925: O; 1926: B; O; 1928: O; 1930: B; 1932: H&G; 1933: B&S; 1944: R; T; 1945: R; Z&M; 1947: R; Z&M; 1948: L; 1949: T; 1950: T&L; 1951: R&K; 1952: R&K; 1953: R&M; 1954: M; 1955; G; 1956: G; 1958: M,B&ES; 1959: A&M; M&A; T; 1960: T
SmAs	a1	1956: I
SmBi	a1	1956: I
SmN	a1	1956: E,B&E; I; K&W
SmO	a1, c1	1953: E&Z; 1956: E,B&E
SmP	a1	1956: I
SmS	a1	1956: I
SmSb	a1	1956: I
SmSe	a1	1959: G&B
SmTe	a1	1956: I
SnAs	a1	1928: G; 1934: W&E; 1935: H&H
SnO	e1	1924: L; 1926: L&N; 1932: W&M; 1933: S&S; 1941: M&P; 1953: NBS
SnS	a9	1935: H
SnSb	a1	1926: G; 1929: O; 1930: MJ&B; 1935: H&H
SnSe	a1, a9	1953: M,Y&O; 1956: O&U; 1960: N&O; 1961: K,G&K
SnTe	a1	1926: G; 1960: N&O; 1961: K,G&K
SrNH	a1	1934: H,F&E
SrO	a1	1921: G; 1922: G; 1926: G; 1928: W; 1933: B; 1954: NBS
SrS	a1	1923: H; 1927: R; 1956: G&F
SrSe	a1	1922: S; 1923: S; 1925: S
SrTe	a1	1926: G
SrTl	b1	1933: Z&B
TaB	d5	1949: K
TaC	a1	1924: vA; 1925: B&E; 1930: B&E; 1933: vS&S; 1934: B&B; 1936: MK; 1946: K&U; 1947: N&K; 1958: E&K; 1961: B
TaN	c2, f3	1924: vA; 1925: B&E; 1953: B&Z; 1954: B&Z; S
TaO	a1	1954: S
TaP	f2	1954: S; 1955: B,N&K

(*continued*)

BIBLIOGRAPHY TABLE, CHAPTER III (*continued*)

Compound	Paragraph	Literature
TbAs	**a1**	1960: B; 1961: B
TbBi	**a1**	1961: B
TbN	**a1**	1956: K&W
TbP	**a1**	1961: B
TbS	**a1**	1961: B
TbSb	**a1**	1960: B; 1961: B
TbSe	**a1**	1961: B
TbTe	**a1**	1960: B; 1961: B
ThAs	**a1**	1955: F
ThC	**a1**	1958: L&N
ThP	**a1**	1939: M
ThS	**a1**	1941: S&Z; 1949: Z
ThSb	**a1**	1941: Z; 1956: F
ThSe	**a1**	1952: DE,S&M
ThTe	**b1**	1954: DE&S; 1955: F
TiAs	**f1**	1954: L&T; T&L; 1955: B,N&K
TiB	**d4**	1952: P&G; 1954: D&K
TiC	**a1**	1924: vA; 1925: B&E; 1931: B; 1932: vS&S; 1934: B&B; 1936: H&S; 1940: D&R; 1941: U&K; 1946: K&U; 1947: M; N&K; 1953: M&S; 1958: E&K
Ti_2CS	**f1**	1960: K&R
TiN	**a1**	1924: vA; 1925: B&E; 1936: H&S; 1939: B; 1940: D&R; 1953: M&S; 1954: S
TiO_x	**a1, f3**	1928: B; 1959: A; 1960: S&L
TiP	**f1**	1954: S; 1955: B,N&K
TiS	**d1, d2**	1949: E; 1954: H&S; S; 1956: H&H; 1958: MT&W
TiSe	**d1**	1948: E; 1949: E; 1954: H&S; 1958: MT&W
TiTe	**d1**	1948: E; 1949: E; 1954: H&S; 1958: MT&W
TlBi	**b1**	1929: S
TlBr	**b1**	1924: vA; 1925: B&L; L; 1926: L; 1927: L; 1936: W&L; 1955: S&K
TlCN	**b1**	1934: S
TlCl	**b1**	1921: D&W; 1924: vA; 1925: B&L; L; 1933: M; 1936: W&L; 1937: M; 1953: H; NBS; 1955: S&K; 1959: M
TlF	**a4**	1935: K

(*continued*)

BIBLIOGRAPHY TABLE, CHAPTER III (*continued*)

Compound	Paragraph	Literature
TlI	b1, g2	1925: B&L; 1926: L; 1927: B; L; 1936: H; 1953: NBS
TlS	d7	1949: H&K; 1956: S&F
TlSb	b1	1927: B; 1928: P&W; 1929: S
TlSe	d7	1939: K,H,M&P
TmAs	a1	1960: B
TmN	a1	1956: K&W
TmSb	a1	1960: B
TmTe	a1	1960: B
UAs	a1	1952: F; I
UBi	a1	1952: F
UC	a1	1948: L,G&C; R,B,W&MD; 1958: L&N; R,W,B&T; 1959: A; V&K
UN	a1	1948: R,B,W&MD; 1952: F; 1955: P; 1958: R,W,B&T
UO	a1	1948: R,B,W&MD; 1958: R,W,B&T; R,B&W
UP	a1	1941: Z; 1952: F
US	a1	1949: Z
USb	a1	1952: F
USe	a1	1954: F
UTe	a1	1954: F
VAs	d3	1955: B&N
VB	d5	1952: B
VC	a1	1925: B&E; 1928: O&O; 1930: O&O; 1940: D&R; 1947: N&K; 1954: S
VN	a1	1925: B&E; 1940: D&R; E&B; 1946: E&B; 1947: E&O; 1949: H; 1960: S&W
VO_x	a1	1932: M,S&S; 1942: K&G; 1946: E&B; 1959: M,A,K,A,W,H&N; V,T&K
VP	d1	1954: S
VS	d1	1939: B&K; H&K
VSe	d1	1939: H&K
VTe_x	d1	1949: E; 1958: G,H&H
WB	d5, d8	1947: K; 1952: P&G
WC	f4	1925: B&E; 1926: W&P; 1928: B; 1931: O&T; 1947: M; N&K; 1954: P&R; 1961: L
WN	f4	1954: S; 1958: K&P
WP	d3	1954: S; 1955: B,N&K
YAs	a1	1960: B

(*continued*)

BIBLIOGRAPHY TABLE, CHAPTER III (*continued*)

Compound	Paragraph	Literature
YN	a1	1957: K,K&MG
YSb_x	a1	1958: D,F,G&L; 1960: B
YTe	a1	1960: B
YbAs	a1	1960: B
YbN	a1	1956: E,B&E; K&W
YbO	a1	1958: A&T
YbSb	a1	1960: B
YbSe	a1	1939: S&K
YbTe	a1	1939: S&K
ZnCe	b1	1937: I&B
ZnLa	b1	1937: I&B
ZnO	c2	1916: F; 1920: B; 1921: A; 1922: A; H; W; 1925: B&B; 1927: B; 1929: F; N&P; 1933: I,F&S; 1934: F&W; 1935: B; 1943: L&M; 1950: H,MG&W; 1952: R&A; 1953: A
ZnPr	b1	1937: I&B
ZnS	c1, c2, c3	1913: B&B; 1914: B; E; 1920: B; 1922: G; 1923: A; R; 1924: L; 1925: U&Z; 1928: dJ; 1929: F; 1933: L,P&Y; 1934: B; 1936: A&B; M&K; 1939: C&M; K; T&M; 1940: K; 1948: F&P; 1950: Z; 1951: NBS; 1952: M; 1953: NBS; 1955: S&S; 1959: E&MK; 1960: S&B
ZnSb	c4	1948: A; 1960: T
ZnSe	c1, c2	1923: D; 1926: G; Z; 1952: NBS; 1955: G,K&FK; 1959: G&F; 1960: C&C; P,T,K&R; 1961: K
ZnTe	c1, c2	1925: Z; 1926: Z; 1955: G,K&FK; 1960: P,T,K&R; 1961: C&C
ZrAs	f1	1956: T,W&L; 1958: T,W&L
ZrB	a1	1952: P&G; 1953: G&P
ZrC	a1	1924: vA; 1925: B&E; 1934: B&B; 1946: K&U; 1947: N&K; 1958: E&K; 1959: N,K,B&B
Zr_2CS	f1	1960: K&R
ZrN	a1	1924: vA; 1925: B&E; 1958: B
ZrO	a1	1957: S
ZrP	a1, f1	1954: S; 1955: B,N&K
ZrS	a1, e2	1954: H&S; 1957: H,H,M&S; 1958: MT&W
ZrTe	d1	1957: H&N; 1958: MT&W; 1959: H&N

BIBLIOGRAPHY, CHAPTER III

1913

Bragg, W. H., "The Reflection of X-Rays by Crystals," *Proc. Roy. Soc.* (*London*), **89A,** 246.

Bragg, W. H., and Bragg, W. L., "The Structure of the Diamond," *Nature,* **91,** 557; *Proc. Roy. Soc.* (*London*), **89A,** 277.

Bragg, W. L., "The Structures of Some Crystals as Indicated by their Diffraction of X-Rays," *Proc. Roy. Soc.* (*London*), **89A,** 248.

1914

Bragg, W. L., "The Analysis of Crystals by the X-Ray Spectrometer," *Proc. Roy. Soc.* (*London*), **89A,** 468.

Ewald, P. P., "The Intensity of Interference Spots with Zinc Blende and the Zinc Blende Grating," *Ann. Physik,* **44,** 257.

Glocker, R., "The Interference of X-Rays," *Physik. Z.,* **15,** 401.

1915

Glocker, R., "Crystal Structure and Interference of X-Rays," *Ann. Physik,* **47,** 377.

Rinne, F., "X-Ray Photographs of Crystals," *Ber. Sachs. Gesell. Wiss. Leipzig* (*Math.-Phys. Kl.*), **67,** 303.

1916

Debye, P., and Scherrer, P., "Interference of X-Rays Using Irregularly Oriented Substances," *Physik. Z.,* **17,** 277.

Fedorov, E. S., "The Structure of Crystals," *Bull. Acad. Sci. Petrograd,* p. 359.

1917

Vegard, L., "Results of Crystal Analysis IV," *Phil. Mag.,* **33,** 395.

1918

Burdick, C. L., and Owen, E. A., "The Structure of Carborundum Determined by X-Rays," *J. Am. Chem. Soc.,* **40,** 1749.

1919

Hull, A. W., "The Crystal Structure of Carborundum," *Phys. Rev.,* **13,** 292.

Hull, A. W., "The Positions of Atoms in Metals," *Proc. Am. Inst. Elec. Eng.,* **38,** 1171.

1920

Bragg, W. L., "The Crystalline Structure of Zinc Oxide," *Phil. Mag.,* **39,** 647.

Davey, W. P., and Hoffmann, E. O., "Crystal Analysis of Metallic Oxides," *Phys. Rev.,* **15,** 333.

Hauer, F., and Koller, P., "X-Ray Photographs of Carborundum," *Z. Krist.,* **55,** 260.

Hull, A. W., "The Crystal Structure of Carborundum," *Phys. Rev.,* **15,** 545.

1921

Aminoff, G., "On the Laue Photographs and Structure of Zincite," *Z. Krist.,* **56,** 495.

Andrews, M. R., "X-Ray Analysis of Three Series of Alloys," *Phys. Rev.,* **18,** 245.

Bartlett, G., and Langmuir, I., "Crystal Structures of the Ammonium Halides above and below the Transition Temperatures," *J. Am. Chem. Soc.*, **43**, 84.

Cooper, P. A., "X-Ray Structure of Potassium Cyanide," *Nature*, **107**, 745.

Davey, W. P., "Cubic Shapes of Certain Ions, as Confirmed by X-Ray Crystal Analysis," *Phys. Rev.*, **17**, 402.

Davey, W. P., "The Absolute Sizes of Certain Monovalent Ions," *Phys. Rev.*, **18**, 102.

Davey, W. P., and Wick, F. G., "The Crystal Structures of Two Rare Halogen Salts," *Phys. Rev.*, **17**, 403.

Espig, H., "X-Ray Investigations of Carborundum," *Abhandl. Saechs. Akad. Wiss. Leipzig, Math.-Phys. Kl.*, **38**, 53.

Gerlach, W., "Crystal Lattice Structure Investigations with X-Rays and a Simple X-Ray Tube," *Physik. Z.*, **22**, 557.

Gerlach, W., and Pauli, O., "The Crystal Lattice of Magnesium Oxide," *Z. Physik*, **7**, 116.

Schiebold, E., "The Crystal Structure of Periclase," *Z. Krist.*, **56**, 430.

Vegard, L., "The Constitution of Mixed Crystals and the Space Occupied by Atoms," *Z. Physik*, **5**, 17.

Wilsey, R. B., "The Crystal Structure of Silver Halides," *Phil. Mag.*, **42**, 262.

Wyckoff, R. W. G., "The Crystal Structure of Magnesium Oxide," *Am. J. Sci.*, **1**, 138.

Wyckoff, R. W. G., "The Crystal Structure of Alabandite," *Am. J. Sci.*, **2**, 239.

Wyckoff, R. W. G., "The Crystal Structures of the Alkali Halides," *J. Wash. Acad. Sci.*, **11**, 429.

1922

Aminoff, G., "On the Crystal Structure of Silver Iodide, Cuprous Iodide and Miersite," *Geol. Foren. Stockholm Forh.*, **44**, 444.

Aminoff, G., "On the Crystal Structure of Silver Iodide," *Z. Krist.*, **57**, 180.

Aminoff, G., "On the Powder Photograph of Zinc Oxide," *Z. Krist.*, **57**, 204.

Bijvoet, J. M., and Karssen, A., "The Crystal Structure of Lithium Hydride," *Proc., Acad. Sci. Amsterdam*, **25**, 27.

Bozorth, R. M., "The Crystal Structure of Potassium Cyanide," *J. Am. Chem. Soc.*, **44**, 317.

Clark, G. L., and Duane, W., "A Study of Secondary Valence by Means of X-Rays," *Phys. Rev.*, **20**, 85.

Cooper, P. A., "The X-Ray Structure of Potassium Cyanide," *Nature*, **110**, 544.

Davey, W. P., "The Absolute Sizes of Certain Monovalent and Bivalent Ions," *Phys. Rev.*, **19**, 248.

Davey, W. P., "Precision Measurements of Crystals," *Phys. Rev.*, **19**, 538.

Dickinson, R. G., "The Crystal Structure of Phosphonium Iodide," *J. Am. Chem. Soc.*, **44**, 1489.

Gerlach, W., "The Space Lattice Structure of the Alkaline Earth Oxides," *Z. Physik*, **9**, 184.

Gerlach, W., "The *K*-Alpha Doublet, Including a New Determination of the Lattice Constants of Some Crystals," *Physik. Z.*, **23**, 114.

Hedvall, J. A., "Changes in Properties Produced by Different Methods of Preparation of Some Ignited Oxides, as Studied by X-Ray Interference," *Arkiv Kemi Mineral. Geol.*, **8**, No. 11; *Z. Anorg. Chem.*, **120**, 327.

McKeehan, L. W., "The Crystal Structures of Beryllium and Beryllium Oxide," *Proc. Natl. Acad. Sci.*, **8**, 270.

Niggli, P., "The Crystal Structures of Several Oxides," *Z. Krist.*, **57**, 253.

Posnjak, E. W., and Wyckoff, R. W. G., "The Crystal Structures of the Alkali Halides II," *J. Wash. Acad. Sci.*, **12**, 248.

Scherrer, P., "The Space Lattice of Cadmium Oxide," *Z. Krist.*, **57**, 186.

Slattery, M. K., "The Crystal Structure of Strontium Selenide," *Phys. Rev.*, **20**, 84.

Weber, L., "The Structure of Zinc Oxide," *Z. Krist.*, **57**, 398.

Wyckoff, R. W. G., "The Crystallographic and Atomic Symmetries of Ammonium Chloride," *Am. J. Sci.*, **3**, 177.

Wyckoff, R. W. G., "On the Crystal Structure of Ammonium Chloride," *Am. J. Sci.*, **4**, 469.

Wyckoff, R. W. G., and Posnjak, E. W., "The Crystal Structures of the Cuprous Halides," *J. Am. Chem. Soc.*, **44**, 30.

1923

Aminoff, G., "Investigations upon the Crystal Structures of Wurtzite and Nickel Arsenide," *Z. Krist.*, **58**, 203.

Bain, E. C., "The Nature of Solid Solutions," *Chem. Met. Eng.*, **28**, 21, 576.

Becker, K., and Ebert, F., "X-Ray Spectroscopy of Metal Compounds," *Z. Physik*, **16**, 165.

Clark, G. L., and Duane, W., "A New Method of Crystal Analysis and the Reflection of Characteristic X-Rays," *J. Opt. Soc. Am.*, **7**, 455.

Clark, G. L., and Duane, W., "The Reflection by a Crystal of X-Rays Characteristic of Chemical Elements in It," *Proc. Natl. Acad. Sci.*, **9**, 126.

Davey, W. P., "Precision Measurements of Crystals of the Alkali Halides," *Phys. Rev.*, **21**, 143.

Davey, W. P., "Precision Measurements of the Crystal Structures of CaO, CaS and CaSe," *Phys. Rev.*, **21**, 213.

Davey, W. P., "Crystal Structure and Densities of Cu_2Se and ZnSe," *Phys. Rev.*, **21**, 380.

Dickinson, R. G., "Some Anomalous Spots on Laue Photographs," *Phys. Rev.*, **22**, 199.

Holgersson, S., "The Structure of the Sulfides of Magnesium, Calcium, Strontium and Barium," *Z. Anorg. Chem.*, **126**, 179.

Mauguin, C., "The Arrangement of the Atoms in Crystals of Cinnabar," *Compt. Rend.*, **176**, 1483.

Ott, H., "The Space Lattice of the Lithium Halides," *Physik. Z.*, **24**, 209.

Owen, E. A., and Preston, G. D., "X-Ray Analysis of Solid Solutions," *Proc. Phys. Soc. (London)*, **36**, 14.

Owen, E. A., and Preston, G. D., "X-Ray Analysis of Zinc–Copper Alloys," *Proc. Phys. Soc. (London)*, **36**, 49.

Rinne, F., "Observations and X-Ray Investigations on the Transformation and Decomposition of Crystal Structures," *Z. Krist.*, **59**, 230.

Slattery, M. K., "Precision Measurements of the Crystal Structure of Barium and Strontium Selenides," *Phys. Rev.*, **21**, 213.

Tiede, E., and Tomaschek, H., "X-Ray Analysis of Luminescent Boron Nitride," *Z. Elektrochem.*, **29**, 303.

Wilsey, R. B., "The Crystalline Structures of Silver Iodide," *Phil. Mag.*, **46**, 487.

Wyckoff, R. W. G., "On the Existence of an Anomalous Reflection of X-Rays in Laue Photographs of Crystals," *Am. J. Sci.*, **6**, 277.

Wyckoff, R. W. G., and Posnjak, E. W., "A Note on the Crystal Structure of Lithium Iodide and Rubidium Fluoride," *J. Wash. Acad. Sci.*, **13**, 393.

1924

Arkel, A. E. van, "The Structure of Mixed Crystals," *Physica*, **4**, 33.

Arkel, A. E. van, "Crystal Structure and Physical Properties," *Physica*, **4**, 286.

Dickinson, R. G., and Friauf, J. B., "The Crystal Structure of Tetragonal Lead Monoxide," *J. Am. Chem. Soc.*, **46**, 2457.

Havighurst, R. J., Mack, E., Jr. and Blake, F. C., "Precision Crystal Measurements on some Alkali and Ammonium Halides," *J. Am. Chem. Soc.*, **46**, 2368.

Holgersson, S., and Sedström, E., "Experimental Studies of the Crystal Structure of Various Alloys," *Ann. Physik*, **75**, 143.

Kolderup, N.-H., "The Crystal Structure of Lead, Galena, Lead Fluoride and Cadmium Fluoride," *Bergens Museums Aarbok, Naturvidensk. Raek.*, No. 2.

Kolkmeijer, N. H., Bijvoet, J. M., and Karssen, A., "The Crystal Structure of Mercuric Sulfide I and II," *Rec. Trav. Chim.*, **43**, 677, 894; *Proc. Acad. Sci. Amsterdam*, **27**, 390, 847; *Verslag Akad. Wetenschap. Amsterdam*, **33**, 327, 828.

Lehmann, W. M., "X-Ray Investigation of Natural and Synthetic Metacinnabarite," *Z. Krist.*, **60**, 379.

Levi, G. R., "Isomorphism of Stannous and Lead Oxides," *Nuovo Cimento*, **1**, 335.

Levi, G. R., "On the Crystalline Lattice of Manganous Oxide," *Gazz. Chim. Ital.*, **54**, 704; *Rend. Ist. Lombardo Sci. Lettere*, **57**, 619.

Ott, H., "The Lattice of Aluminum Nitride," *Z. Physik*, **22**, 201.

Owen, E. A., and Preston, G. D., "The Atomic Structure of Two Intermetallic Compounds," *Proc. Phys. Soc. (London)*, **36**, 341.

Simon, F., and Simson, C. v., "The Crystal Structure of Hydrogen Chloride," *Z. Physik*, **21**, 168.

1925

Alsén, N., "X-Ray Investigation of the Crystal Structures of Pyrrhotite, Breithauptite, Pentlandite, Millerite and Related Compounds," *Geol. Foren Stockholm Forh.*, **47**, 19.

Aminoff, G., "On Beryllium Oxide as a Mineral and its Crystal Structure," *Z. Krist.*, **62**, 113.

Barth, T., and Lunde, G., "Lattice Constants of the Cuprous and Silver Halides," *Norsk Geol. Tidsskr.*, **8**, 281.

Barth, T., and Lunde, G., "The Structure of Mixed Crystals," *Norsk Geol. Tidsskr.*, **8**, 293; *Z. Physik. Chem. (Leipzig)*, **122**, 293 (1926).

Becker, K., and Ebert, F., "The Crystal Structures of Various Binary Carbides and Nitrides," *Z. Physik*, **31**, 268.

Bragg, W. H., and Bragg, W. L., *X-Rays and Crystal Structure*, 5th Ed., G. Bell & Sons, Ltd., London.

Brentano, J., "A Focusing Method of Crystal Powder Analysis by X-Rays," *Proc. Phys. Soc. (London)*, **37**, 184.

Broomé, B. H., "An X-Ray Investigation of Mixed Crystals of the Systems Sodium Chloride–Silver Chloride and Potassium Chloride–Potassium Bromide," *Z. Anorg. Chem.*, **143**, 60.

Buckley, H. E., and Vernon, W. S., "The Crystal Structures of the Sulfides of Mercury," *Mineral. Mag.*, **20**, 382.

Clark, G. L., Asbury, W. C., and Wick, R. M., "An Application of X-Ray Crystallometry to the Structure of Nickel Catalysts," *J. Am. Chem. Soc.*, **47**, 2661.

Compton, A. H., Beets, H. N., and DeFoe, O. K., "The Grating Space of Calcite and Rock Salt," *Phys. Rev.*, **25**, 625.

Ferrari, A., "The Crystalline Lattices of Lithium and Magnesium Fluorides and Their Isomorphism," *Rend. Accad. Lincei*, **1**, 664.

Goldschmidt, V. M., Barth, T., and Lunde, G., "Isomorphy and Polymorphy of the Sesquioxides. The Lanthanide Contraction and its Consequences," *Skrifter Norske Videnskaps-Akad. Oslo I. Mat.-Naturv. Kl.*, **1925**, No. 7.

Havighurst, R. J., "Crystal Structure of the Mercurous Halides," *Am. J. Sci.*, **10**, 15.

Hylleraas, E., "The Arrangement of the Atoms in the Tetragonal Crystals of Hg_2Cl_2, Hg_2Br_2, Hg_2I_2 and the Calculation of the Optical Double Refraction of Hg_2Cl_2," *Physik. Z.*, **26**, 811.

Johansson, C. H., and Linde, J. O., "An X-Ray Determination of Atomic Arrangement in the Mixed Crystal Series Gold–Copper and Palladium–Copper," *Ann. Physik*, **78**, 439.

Jong, W. F. de, "Crystal Structure of Niccolite and Pyrrhotite," *Physica*, **5**, 194.

Jong, W. F. de, "Crystal Structure of Breithauptite, NiSb," *Physica*, **5**, 241.

Levi, G. R., and Tacchini, G., "The Non-Existence of Nickel Suboxide," *Gazz. Chim. Ital.*, **55**, 28.

Lunde, G., "The Crystal Structure of Thallous Chloride and Thallous Bromide," *Norsk Geol. Tidsskr.*, **8**, 217; *Z. Physik. Chem.*, **117**, 51.

Olshausen, S. v., "Crystal Structure Studies Using the Debye-Scherrer Method," *Z. Krist.*, **61**, 463.

Ott, H., "The Structure of Silicon Carbide," *Z. Krist.*, **61**, 515; *Naturwiss.*, **13**, 76.

Ott, H., "The Structure of Silicon Carbide," *Z. Krist.*, **62**, 201.

Ott, H., "The Structure of Silicon Carbide," *Naturwiss.*, **13**, 644.

Ramsdell, L. S., "The Crystal Structure of Some Metallic Sulfides," *Am. Mineralogist*, **10**, 281.

Sasahara, T., "X-Ray Analysis of the Solid Solutions of KCl and KBr," *Sci. Papers Inst. Phys. Chem. Res. (Tokyo)*, **2**, 277.

Siegbahn, M., *The Spectroscopy of X-Rays*, Oxford Univ. Press, London.

Slattery, M. K., "The Crystal Structure of Metallic Tellurium and Selenium and of Strontium and Barium Selenide," *Phys. Rev.*, **25**, 333.

Ulrich, F., and Zachariasen, W., "On the Crystal Structures of α- and β-CdS and of Wurtzite," *Z. Krist.*, **62**, 260.

Westgren, A., and Phragmén, G., "X-Ray Analysis of Copper–Zinc, Silver–Zinc and Gold–Zinc Alloys," *Phil. Mag.*, **50**, 311.

Wilsey, R. B., "X-Ray Analysis of Some Mixed Crystals of Silver Halides," *J. Franklin Inst.*, **200**, 739.

Wyckoff, R. W. G., and Crittenden, E. D., "The Preparation and Crystal Structure of Ferrous Oxide," *J. Am. Chem. Soc.*, **47**, 2876; *Z. Krist.*, **63**, 144 (1926).

Zachariasen, W. H., "The Crystal Structure of BeO," *Norsk Geol. Tidsskr.*, **8**, 189.

Zachariasen, W. H., "The Crystal Structure of the Tellurides of Zinc, Cadmium and Mercury," *Norsk Geol. Tidsskr.*, **8**, 302.

1926

Becker, K., "X-Ray Method of Determining Coefficient of Expansion at High Temperatures," *Z. Physik*, **40**, 37.

Bravo, F. M., "Determination of the Crystalline Structure of Nickel Oxide, Cobalt Oxide and Lead Sulfide," *Anal. Soc. Espanol. Fis. Quim.*, **24**, 116.

Claassen, A., "The Crystal Structure of Red Mercuric Iodide," thesis, Amsterdam.

Claassen, A., "The Crystal Structure of Beryllium Oxide," *Z. Physik. Chem.* (*Leipzig*), **124**, 139.

Erdal, A., "Contributions to the Analysis of Mixed Crystals and Alloys," *Z. Krist.*, **65**, 69.

Fontana, C., "The Structure of Manganese Oxide," *Gazz. Chim. Ital.*, **56**, 396.

Goldschmidt, V. M., "The Laws of Crystal Chemistry," *Skrifter Norske Videnskaps-Akad. Oslo I. Mat.-Naturv. Kl.*, **1926**, No. 2; *Naturwiss.*, **14**, 477.

Goldschmidt, V. M., "Researches on the Structure and Properties of Crystals," *Skrifter Norske Videnskaps-Akad. Oslo I. Mat.-Naturv. Kl.*, **1926**, No. 8.

Hartwig, W., "The Crystal Structure of Several Minerals of the Regular Mercuric Sulfide Series," *Sitzber. Preuss. Akad. Wiss. Berlin, Phys.-Math. Kl.*, **1926**, p. 79.

Hassel, O., "The Crystal Structure of Boron Nitride, BN," *Norsk Geol. Tidsskr.*, **9**, 266.

Havighurst, R. J., "Parameters in Crystal Structure. The Mercurous Halides," *J. Am. Chem. Soc.*, **48**, 2113.

Jaeger, F. M., and Westenbrink, H. G. K., "Remarks on the Crystal Form of Boron Nitride and Possible Ambiguity in the Analysis of Powder Spectrograms," *Proc. Acad. Sci. Amsterdam*, **29**, 1218; *Verslag Akad. Wetenschap. Amsterdam*, **35**, 857.

Jong, W. F. de, "The Structure of Tiemannite and Coloradoite," *Z. Krist.*, **63**, 466.

Jong, W. F. de, and Willems, H. W. V., "The Crystal Structure of Cinnabar," *Physica*, **6**, 129.

Levi, G. R., and Fontana, C., "Palladium Oxides," *Gazz. Chim. Ital.*, **56**, 388.

Levi, G. R., and Natta, G., "Isomorphism of the Oxides of Lead and Tin II," *Nuovo Cimento*, **3**, 114.

Lunde, G., "The Formation of Mixed Crystals by Precipitation," *Ber.*, **59B**, 2784.

Lunnon, R. G., "Atomic Dimensions," *Proc. Phys. Soc.* (*London*), **38**, 93.

Mark, H., and Steinbach, J., "The Space Lattice and Double Refraction of Calomel," *Z. Krist.*, **64**, 79.

Müller, H., and Bradley, A. J., "Copper Hydride and its Crystal Structure," *J. Chem. Soc.*, **1926**, 1669.

Natta, G., and Reina, A., "The Crystal Structures of Cobaltous Oxide and Hydroxide," *Rend. Accad. Lincei*, **4**, 48.

Ott, H., "The Structure of Silicon Carbide III and of the Amorphous Carbide," *Z. Krist.*, **63**, 1.

Ott, H., "The Structures of MnO, MnS, AgF, NiS, SnI_4, $SrCl_2$, BaF_2; Precision Measurements upon Various Alkali Halides," *Z. Krist.*, **63**, 222.

Owen, E. A., and Preston, G. D., "The Atomic Structure of AgMg and AuZn," *Phil. Mag.*, **2**, 1266.

Simon, F., and Simson, C. v., "A Transition Point of Ammonium Salts between $-30°$ and $-40°$," *Naturwiss.*, **14**, 880.

Westgren, A., and Phragmén, G., "X-Ray Analysis of the Systems Tungsten–Carbon and Molybdenum–Carbon," *Z. Anorg. Chem.*, **156**, 27.

Westgren, A., and Phragmén, G., "Structure analogies of Alloys," *Arkiv Mat. Astron. Fysik*, **19B**, No. 12.

Zachariasen, W. H., "The Crystal Structures of Beryllium Oxide and Beryllium Sulfide," *Z. Physik. Chem.* (*Leipzig*), **119**, 201.

Zachariasen, W. H., "The Crystal Structure of the Tellurides of Beryllium, Zinc, Cadmium and Mercury," *Z. Physik. Chem.* (*Leipzig*), **124**, 277.

Zachariasen, W. H., "Crystal Structures of the Selenides of Beryllium, Zinc, Cadmium and Mercury," *Z. Physik. Chem. (Leipzig)*, **124**, 436.

Zachariasen, W. H., "Some Evidence on the State of Ionization of Atoms in the Lattice of Beryllium Oxide Crystals," *Z. Physik*, **40**, 637.

1927

Barth, T., "The Lattice Dimensions of Zinc Oxide," *Norsk Geol. Tidsskr.*, **9**, 317.

Barth, T., "The System Thallium–Antimony," *Z. Physik. Chem. (Leipzig)*, **127**, 113.

Barth, T., "The Lattice Constant of Thallium Iodide," *Z. Physik. Chem. (Leipzig)*, **131**, 105.

Barth, T., and Lunde, G., "The Mineral Villiaumite," *Central. Mineral. Geol.*, **1927A**, 57.

Barth, T., and Lunde, G., "The Difference of the Lattice Constants of Rock Salt and of Chemically Pure Sodium Chloride," *Z. Physik. Chem. (Leipzig)*, **126**, 417.

Broch, E. "Precision Determinations of the Lattice Constants of the Compounds MgO, MgS, MgSe, MnO and MnSe," *Z. Physik. Chem. (Leipzig)*, **127**, 446.

Gossner, B., and Mussgnug, F., "The Crystal Structure of Cinnabar and Covellite," *Central. Mineral. Geol.*, **1927A**, 410.

Haase, M., "Optical Properties of the Highly Refractive Isostructural Compounds of Magnesium, Calcium, Strontium and Barium with Oxygen, Sulfur, Selenium and Tellurium," *Z. Krist.*, **65**, 509.

Halla, F., and Pawlek, F., "The Space Lattice of Yellow Lead Oxide," *Z. Physik. Chem. (Leipzig)*, **128**, 49.

Harrington, E. A., "X-Ray Diffraction Measurements on Some of the Pure Compounds Concerned in the Study of Portland Cement," *Am. J. Sci.*, **13**, 467.

Huggins, M. L., and Magill, P. L., "The Crystal Structures of Mercuric and Mercurous Iodides," *J. Am. Chem. Soc.*, **49**, 2357.

Johansson, C. H., and Linde, J. O., "Lattice Structure and Electrical Conductivity of the Mixed Crystal Systems Au–Cu, Pd–Cu and Pt–Cu," *Ann. Physik*, **82**, 449.

Jong, W. F. de, and Willems, H. W. V., "Compounds of the Lattice Type of Pyrrhotite (FeS)," *Physica*, **7**, 74.

Lunde, G., "The Constitution of Mixed Crystals," *Bull. Soc. Chim. France*, **41**, 304.

Lunde, G., "The Existence and Preparation of Certain Oxides of the Platinum Metals (with a Supplement Regarding Amorphous Oxides)," *Z. Anorg. Chem.*, **163**, 345.

Natta, G., and Freri, M., "X-Ray Analysis and the Crystalline Structure of Cadmium–Silver Alloys I, II and III," *Rend. Accad. Lincei*, **6**, 422, 505; **7**, 406 (1928).

Oftedal, I., "Some Crystal Structures of the Type NiAs," *Z. Physik. Chem. (Leipzig)*, **128**, 135.

Oftedal, I., "The Lattice Constants of CaO, CaS, CaSe, CaTe," *Z. Physik. Chem. (Leipzig)*, **128**, 154.

Preston, G. D., and Owen, E. A., "The Atomic Structure of AuSn," *Phil. Mag.*, **4**, 133.

Rumpf, E., "The Crystal Constants of the Calcium Sulfide– and the Strontium Sulfide–Samarium Mixed Phosphors," *Ann. Physik*, **84**, 313.

Vegard, L., "Lattice Variations in Mixed Crystals Prepared by Precipitation," *Z. Physik*, **43**, 299.

Vegard, L., and Dale H., "Investigations on Mixed Crystals and Alloys," *Skrifter Norske Videnskaps-Akad. Oslo I. Mat.-Naturv. Kl.*, **1927**, No. 14; *Z. Krist.*, **67**, 148 (1928).

Vegard, L., and Hauge, T., "Mixed Crystals and Their Production through the Contact of Solid Phases and through Precipitation," *Z. Physik*, **42**, 1.

Waller, I., and James, R. W., "On the Temperature Factors of X-Ray Reflexion for Sodium and Chlorine in the Rock Salt Crystal," *Proc. Roy. Soc. (London)*, **117A**, 214.

Walmsley, H. P., "The Structure of the Smoke Particles from a Cadmium Arc," *Proc. Phys. Soc. (London)*, **40**, 7.

Willems, H. W. V., "The Structure of Millerite," *Physica*, **7**, 203.

Zachariasen, W. H., "The Crystal Structure of Ammonium Fluoride," *Z. Physik. Chem. (Leipzig)*, **127**, 218.

Zachariasen, W. H., "The Crystal Structure of Palladium Oxide (PdO)," *Z. Physik. Chem. (Leipzig)*, **128**, 412.

Zachariasen, W. H., "The Crystal Structure of Magnesium Telluride," *Z. Physik. Chem. (Leipzig)*, **128**, 417.

Zachariasen, W. H., "The Crystal Structure of Mercuric Oxide," *Z. Physik. Chem. (Leipzig)*, **128**, 421.

1928

Åstrand, H., and Westgren, A., "X-Ray Analysis of the Silver–Cadmium Alloys," *Z. Anorg. Chem.*, **175**, 90.

Becker, K., "Crystal Structure and Linear Coefficient of Thermal Expansion of Tungsten Carbides," *Z. Physik*, **51**, 481.

Becker, K., "The Constitution of Tungsten Carbide," *Z. Elektrochem.*, **34**, 640.

Bräkken, H., "The Crystal Structure of Titanium Monoxide," *Z. Krist.*, **67**, 547.

Dahl, O., Holm, E., and Masing, G., "X-Ray Photographs of Beryllium–Copper Alloys," *Z. Metallk.*, **20**, 431; *Wiss. Veröff. Siemens-Konz.*, **8**, 154 (1929).

Goldschmidt, V. M., "Atomic Distances in Metals," *Z. Physik. Chem. (Leipzig)*, **133**, 397.

Groebler, H., "X-Ray Studies of the Structure of the Oxides of Iron," *Z. Physik*, **48**, 567.

Haase, M., "The Lattice Constant of Barium Telluride," *Z. Krist.*, **68**, 119.

Hägg, G., "X-Ray Studies of the Binary Systems of Iron with Phosphorus, Arsenic, Antimony and Bismuth," *Z. Krist.*, **68**, 470.

Jong, W. F. de, "Marmatite and Christophite," *Z. Krist.*, **66**, 515.

Kolkmeijer, N. H., Dobbenburgh, W. J. D. van, and Boekenoogen, H. A., "Density and Axial Relations of Hexagonal Silver Iodide Determined with X-Rays," *Verslag Akad. Wetenschap. Amsterdam*, **37**, 481; *Proc. Acad. Sci. Amsterdam*, **31**, 1014.

Levi, G. R., "Crystallographic Identity of the Two Forms of Mercuric Oxide," *Gazz. Chim. Ital.*, **58**, 417.

Natta, G., and Passerini, L., "Aluminum Arsenide," *Gazz. Chim. Ital.*, **58**, 458.

Natta, G., and Strada, M., "Oxides and Hydroxides of Cobalt," *Gazz. Chim. Ital.*, **58**, 419.

Oberlies, F., "Determination of the Lattice Constants of the Mixed Crystal System KCl–KBr," *Ann. Physik*, **87**, 238.

Oftedal, I., "X-Ray Studies of Manganese Arsenide, Iron Telluride, Nickel Stannide and Platinum Stannide," *Z. Physik. Chem. (Leipzig)*, **132**, 208.

Ott, H., "A New Modification of Silicon Carbide," *Festschrift "Arnold Sommerfeld,"* p. 208.

Ôya, S., and Ôsawa, A., "A Study of the Vanadium–Carbon System" *J. Study Metals*, **5**, 434.

Passerini, L., "Crystalline Structure of Some Phosphides of Bivalent and of Trivalent Metals," *Gazz. Chim. Ital.*, **58**, 655.

Persson, E., and Westgren, A., "X-Ray Analysis of the Thallium–Antimony Alloys," *Z. Physik. Chem. (Leipzig)*, **136**, 208.

Rumpf, E., "On the Lattice Constants of Calcium Oxide and Calcium Hydroxide," *Ann. Physik*, **87**, 595.

Thomassen, L., "The Preparation and Crystal Structure of the Mono- and Diantimonides of Palladium," *Z. Physik. Chem. (Leipzig)*, **135**, 383.

Westgren, A., and Phragmén, G., "X-Ray Analysis of the Copper–Tin Alloys," *Z. Anorg. Chem.*, **175**, 80.

Wilson, T. A., "The Crystal Structure of Strontium Oxide," *Phys. Rev.*, **31**, 1117.

1929

Bijvoet, J. M., and Frederikse, W. A., "The Scattering Power for X-Rays and the Electron Distribution of the H Ion," *Rec. Trav. Chim.*, **48**, 1041.

Bjurström, T., and Arnfelt, H., "X-Ray Analysis of the System Iron–Boron," *Z. Physik. Chem.*, **4B**, 469.

Blix, R., "X-Ray Analysis of the Chromium–Nitrogen System Including an Investigation of the Constitution of Ferrochrome Containing Nitrogen," *Z. Physik. Chem.*, **3B**, 229.

Brentano, J., and Adamson, J., "Precision Measurements of X-Ray Reflections from Crystal Powders. The Lattice Constants of Zinc Carbonate, Manganese Carbonate and Cadmium Oxide," *Phil. Mag.*, **7**, 507.

Broch, E., Oftedal, I., and Pabst, A., "New Determinations of the Lattice Constants of Potassium Fluoride, Cesium Chloride and Barium Fluoride," *Z. Physik. Chem.*, **3B**, 209.

Fuller, M. L., "Precision Measurements of X-Ray Reflexions from Crystal Powders," *Phil. Mag.*, **8**, 585.

Fuller, M. L., "The Crystal Structure of Wurtzite," *Phil. Mag.*, **8**, 658.

Fuller, M. L., "A Method of Determining the Axial Ratio of a Crystal from X-Ray Diffraction Data: The Axial Ratio and Lattice Constants of Zinc Oxide," *Science*, **70**, 196.

Goldschmidt, V. M., "The Lattice Constant of Barium Telluride," *Z. Krist.*, **69**, 411.

Hägg, G., "An X-Ray Study of the System Iron–Arsenic," *Z. Krist.*, **71**, 134.

Hägg, G., "X-Ray Studies on the Binary Systems of Iron with Nitrogen, Phosphorus, Arsenic, Antimony and Bismuth," *Nova Acta Reg. Soc. Sci. Upsaliensis IV*, **7**, No. 1.

Holgersson, S., and Karlsson, A., "X-Ray Examination of a System of Mixed Crystals with Monoxide Components," *Z. Anorg. Chem.*, **182**, 255.

Lunde, G., and Rosbaud, P., "Crystal Structure of Mixed Crystals of the Series CuI–AgI," *Z. Physik. Chem.*, **6B**, 115.

Natta, G., and Passerini, L., "Solid Solutions, Isomorphism and Symmorphism among Oxides of Bivalent Metals I. The Systems CaO–CdO, CaO–MnO, CaO–CoO, CaO–NiO, and CaO–MgO," *Gazz. Chim. Ital.*, **59**, 129.

Natta, G., and Passerini, L., "The Constitution of Rinmann Green, Thénard Blue and Other Colored Solid Derivatives of Cobalt Oxide," *Gazz. Chim. Ital.*, **59**, 620.

Oftedal, I., "Observations on the Lattice Dimensions of the System Fe_x–Sb_y," *Z. Physik. Chem.*, **4B**, 67.

Ôsawa, A., "An Intermetallic Compound Having a Simple Cubic Lattice," *Nature*, **124**, 14.

Passerini, L., "Solid Solutions, Isomorphism and Symmorphism among Oxides of Bivalent Metals II. The Systems CoO–NiO, CoO–MgO, CoO–MnO, CoO–CdO, NiO–MgO, NiO–MnO and NiO–CdO," *Gazz. Chim. Ital.*, **59**, 144.

Roberts, H. S., and Ksanda, C. J., "The Crystal Structure of Covellite," *Am. J. Sci.*, **17**, 489.

Roux, A., and Cournot, J., "X-Ray Study of the Internal Transformations in Silver–Zinc Alloys," *Compt. Rend.*, **188**, 1399.

Sekito, S., "Crystal Structure of Thallium," *J. Study Metals*, **6**, 372; *J. Inst. Metals*, **42**, 516; *Z. Krist.*, **74**, 189 (1930).

Thomassen, L., "The Crystal Structure of Some Binary Compounds of the Platinum Metals," *Z. Physik. Chem.*, **2B**, 349.

Thomassen, L., "The Crystal Structure of Some Binary Compounds of the Platinum Metals II.," *Z. Physik. Chem.*, **4B**, 277.

Westgren, A., and Almin, A., "Space Filling of the Atoms in Alloys," *Z. Physik. Chem.*, **5B**, 14.

Wever, F., "Iron–Beryllium and Iron-Boron Alloys; The Structure of Iron Boride," *Z. Tech. Physik*, **10**, 137.

Wyckoff, R. W. G., and Armstrong, A. H., "The X-Ray Diffracting Power of Chlorine and Ammonium in Ammonium Chloride," *Z. Krist.*, **72**, 319.

1930

Becker, K., and Ewest, H., "The Physical Properties of Tantalum Carbide," *Z. Tech. Physik*, **11**, 148.

Bräkken, H., "The Crystal Structure of Cubic SiC," *Z. Krist.*, **75**, 572.

Bräkken, H., "The Crystal Structure of Silver Cyanide," *Kgl. Norske Videnskab. Selskabs, Forh. II*, **1929**, No. 48, p. 169.

Hendricks, S. B., and Kosting, P. R., "The Crystal Structure of Fe_2P, Fe_2N, Fe_3N and FeB," *Z. Krist.*, **74**, 511.

Morris-Jones, W., and Bowen, E. G., "The Compound SnSb," *Nature*, **126**, 846.

Natta, G., "The Crystal Structure of Hydrogen Iodide and its Relation to That of Xenon," *Nature*, **126**, 97.

Ôsawa, A., and Ôya, M., "An Investigation of the Vanadium–Carbon System," *Sci., Rept. Tôhoku Imp. Univ.*, **19**, 95.

Pastorello, S., "X-Ray Analysis of the System Lithium–Silver," *Gazz. Chim. Ital.*, **60**, 493.

Vegard, L., "Crystal Structure and Luminous Power of Solid Carbon Monoxide," *Z. Physik*, **61**, 185.

1931

Alsén, N., "Crystal Structure of Covellite (CuS) and Copper Glance (Cu_2S)," *Geol. Foren. Stockholm Forh.*, **53**, 111.

Bennett, O. G., Cairns, R. W., and Ott, E., "Crystal Form of Nickel Oxides," *J. Am. Chem. Soc.*, **53**, 1179.

Bloch, R., and Möller, H., "The Modifications of Silver Iodide," *Z. Physik. Chem. (Leipzig)*, **152A**, 245.

Brantley, L. R., "The Size of the Unit Cell of Titanium Carbide," *Z. Krist.*, **77**, 505.

Kolkmeijer, N. H., and Moesveld, A. L. Th., "The Density and Structure of Millerite, Rhombohedral NiS," *Z. Krist.*, **80**, 91.

Ksanda, C. J., "Comparison Standards for the Powder Spectrum Method: NiO and CdO," *Am. J. Sci.*, **22,** 131.

Natta, G., "The Crystal Structure and Polymorphism of Hydrogen Halides," *Nature*, **127,** 235.

Natta, G., "The Dimensions of Atoms and Univalent Ions in the Lattices of Crystals," *Mem. Accad. Italia, Cl. Sci. Fis. Mat. Nat.*, **2,** *Chim.*, No. 3, 5.

Natta, G., and Passerini, L., "The Structure of Alkali Cyanides and Their Isomorphism with Halides," *Gazz. Chim. Ital.*, **61,** 191.

Ôsawa, A., and Takeda, S., "An X-Ray Investigation of Alloys of the Iron–Tungsten System and Their Carbides," *Kinzoku-no-kenkyu*, **8,** 181; *J. Inst. Metals*, **47,** 534.

Pastorello, S., "Thermal Analysis of the System Lithium–Silver," *Gazz. Chim. Ital.*, **61,** 47.

Stenbeck, S., and Westgren, A., "X-Ray Analysis of Gold–Tin Alloys," *Z. Physik. Chem.*, **14B,** 91.

Zintl, E., and Harder, A., "Alkali Hydrides," *Z. Physik. Chem.*, **14B,** 265.

1932

Bannister, F. A., "Determination of Minerals in Platinum Concentrates from the Transvaal by X-Ray Methods," *Mineral. Mag.*, **23,** 188.

Bivjoet, J. M., and Karssen, A., "Crystal Structure of Lithium Hydride," *Z. Physik. Chem.*, **15B,** 414. (See also Reply, Zintl, E., and Harder, A., *ibid.*, **15B,** 416.)

Bottema, J. A., and Jaeger, F. M., "On the Law of Additive Atomic Heats in Intermetallic Compounds IX. The Compounds of Tin and Gold, and of Gold and Antimony," *Proc. Acad. Sci. Amsterdam*, **35,** 916.

Caglioti, V., and Roberti, G., "X-Ray Investigation of a Subsulfide of Cobalt Employed as a Catalyst in the Hydrogenation of Phenol," *Gazz. Chim. Ital.*, **62,** 19.

Darbyshire, J. A., "An X-Ray Examination of the Oxides of Lead," *J. Chem. Soc.*, **1932,** 211.

Ernst, Th., "Crystal Structure of Lithium Hydroxide," *Naturwiss.*, **20,** 124.

Hengstenberg, J., and Garrido, J., "Electron Distribution in Carborundum," *Anal. Soc. Espanol, Fis. Quim.*, **30,** 409.

Jaeger, F. M., and Bottema, J. A., "The Exact Measurement of the Specific Heats of Solid Substances at High Temperatures VI. The Law of Neumann-Joule-Kopp-Regnault Concerning the Molecular Heat of Chemical Compounds as a Function of the Atomic Heats," *Proc. Acad. Sci. Amsterdam*, **35,** 352.

Juza, R., and Biltz, W., "The Systematic Doctrine of Affinity LVII. The Phase Diagram of Pyrite, Pyrrhotite, Troilite and Sulfur Vapor, Criticized in View of Sulfur Vapor Pressures, X-Ray Diagrams, Densities and Magnetic Measurements," *Z. Anorg. Chem.*, **205,** 273.

Mathewson, C. H., Spire, E., and Samans, C. H., "Division of the Iron–Vanadium–Oxygen System into Some of its Constituent Binary and Ternary Systems," *Trans. Am. Soc. Steel Treating*, **20,** 357.

Oftedal, I., "The Crystal Structure of Covellite," *Z. Krist.*, **83,** 9.

Ölander, A., "The Crystal Structure of AuCd," *Z. Krist.*, **83,** 145.

Ruhemann, B., and Simon, F., "Crystal Structures of Krypton, Xenon, Hydrogen Iodide and Hydrogen Bromide in Relation to the Temperature," *Z. Physik. Chem.*, **15B,** 389.

Schnaasse, H., "The Crystal Structure of Red Manganese Sulfide," *Naturwiss.*, **20,** 640.

Schwarz, M. v., and Summa, O., "New Determination of the Lattice Constant of Titanium Carbide," *Z. Elektrochem.*, **38**, 743.

Tutiya, H., "X-Ray Observation of Molybdenum Carbides Formed at Low Temperatures," *Bull. Inst. Phys. Chem. Res. (Tokyo)*, **11**, 1150.

Weiser, H. B., and Milligan, W. O., "X-Ray Studies on the Hydrous Oxides III. Stannous Oxide," *J. Phys. Chem.*, **36**, 3039.

Zachariasen, W. H., "The Crystal Structure of Germano Sulfide," *Phys. Rev.*, **40**, 917.

1933

Bjurström, T., "X-Ray Analysis of the Iron–Boron, Cobalt–Boron and Nickel–Boron Systems," *Arkiv Kemi Mineral. Geol.*, **11A**, No. 5.

Borrmann, G., and Seyfarth, H., "Precision Measurements of the Lattice Constants of Carborundum," *Z. Krist.*, **86**, 472.

Burgers, W. G., "X-Ray Investigation of the Behavior of BaO–SrO Mixtures on Ignition," *Z. Physik*, **80**, 352.

Cairns, R. W., and Ott, E., "X-Ray Studies of the System Nickel–Oxygen–Water I. Nickelous Oxide and Hydroxide," *J. Am. Chem. Soc.*, **55**, 527.

Ebert, F., and Woitinek, H., "Crystalline Structure of Fluorides II. HgF, HgF_2, CuF and CuF_2," *Z. Anorg. Chem.*, **210**, 269.

Ernst, T., "Preparation and Crystal Structure of Lithium Hydroxide," *Z. Physik. Chem.*, **20B**, 65.

Faber, W., "Niccolite," *Z. Krist.*, **84**, 408.

Franck, H. H., Bredig, M. A., and Hoffmann, G., "The Crystal Structure of Calcium–Nitrogen Compounds," *Naturwiss.*, **21**, 330.

Greenwood, G., "The Debye-Scherrer Photograph," *Indian J. Physics*, **8**, 269.

Hägg, G., "Vacant Positions in the Iron Lattice of Pyrrhotite," *Nature*, **131**, 167.

Hägg, G., and Sucksdorff, I., "The Crystal Structure of Troilite and Pyrrhotite," *Z. Physik. Chem.*, **22B**, 444.

Ivannikov, P. Y., Frost, A. V., and Shapiro, M. I., "The Effect of Ignition Temperature on the Catalytic Activity of Zinc Oxide," *Compt. Rend. Acad. Sci. (URSS)*, **1933**, 124.

Jaeger, F. M., and Bottema, J. A., "The Exact Determination of Specific Heats at Elevated Temperatures IV. Law of Neumann-Joule-Kopp-Regnault Concerning the Additive Property of Atomic Heats of Elements in their Chemical Combinations," *Rec. Trav. Chim.*, **52**, 89.

Jette, E. R., and Foote, F., "An X-Ray Study of the Wüstite (FeO) Solid Solutions," *J. Chem. Phys.*, **1**, 29.

Jette, E. R., and Foote, F., "A Study of the Homogeneity Limits of Wüstite by X-Ray Methods." *Trans. AIME*, **105**, Iron Steel Div., 276.

Lark-Horowitz, K., Purcell, E. M., and Yearnian, H. J., "Electron Diffraction from Vacuum Sublimed Layers," *Bull. Am. Phys. Soc.*, No. 5, p. 6.

Laschkarew, W. E., and Usyskin, J. D., "The Determination of the Positions of the Hydrogen Ions in the NH_4Cl Lattice by Electron Scattering," *Z. Physik*, **85**, 618.

LeBlanc, M., and Wehner, G., "Contribution to Knowledge about Oxides of Manganese," *Z. Physik. Chem. (Leipzig)*, **168A**, 59.

Moeller, K., "A Standard Substance for Accurate Debye-Scherrer Determination of Lattice Constants," *Naturwiss.*, **21**, 223.

Müller, W. J., and Löffler, G., "Concerning the Color of Precipitated CdS." *Z. Angew. Chem.*, **46**, 538.

Natta, G., "Structure and Polymorphism of Hydrohalic Acids," *Gazz. Chim. Ital.*, **63**, 425.

Natta, G., and Vecchia, O., "Structure and Polymorphism of Silver Cyanide," *Gazz. Chim. Ital.*, **63**, 439; *Rend. R. Instit. Lombardo,* **66**, 895.

Rossi, A., and Iandelli, A., "The Crystalline Structure of PrMg," *Rend. Accad. Lincei,* **18**, 156.

Schnaase, H., "Crystal Structure of Manganous Sulfide and its Mixed Crystals with Zinc Sulfide and Cadmium Sulfide," *Z. Physik. Chem.*, **20B**, 89.

Schwarz, M. v., and Summa, O., "The Crystal Structure of Tantalum Carbide," *Metallwirtschaft,* **12**, 298.

Straumanis, M., and Strenk, C., "Concerning SnO," *Z. Anorg. Allgem. Chem.*, **213**, 301.

Tunell, G., Posnjak, E., and Ksanda, C. J., "The Crystal Structure of Tenorite (Cupric Oxide)," *J. Washington Acad. Sci.*, **23**, 195.

Wagner, G., and Lippert, L., "Note on the Establishment of the Sodium Chloride Lattice in Cesium Chloride," *Z. Physik. Chem.*, **21B**, 471.

Zintl, E., and Brauer, G., "Metals and Alloys X. The Valence Electron Rule and the Atomic Radius of Non-Noble Metals in Alloys," *Z. Physik. Chem.*, **20B**, 245.

1934

Braekken, H., "Empty Places in the Crystal Structure of Iron-Containing ZnS," *Norske Videnskab. Forh.*, **7**, 119.

Burgers, W. G., and Basart, J. C. M., "Formation of High-Melting Metallic Carbides by Igniting a Carbon Filament in the Vapor of a Volatile Halogen Compound of the Metal," *Z. Anorg. Chem.*, **216**, 209.

Ellefson, B. S., and Taylor, N. W., "Crystal Structures and Expansion Anomalies of MnO, MnS, FeO, Fe_3O_4 between 100° K and 200° K," *J. Chem. Phys.*, **2**, 58.

Finch, G. J., and Wilman, H., "The Lattice Dimensions of ZnO," *J. Chem. Soc.*, **1934**, 751.

Fylking, K. E., "Phosphides and Arsenides with Modified NiAs Structure," *Arkiv Kemi Mineral. Geol.*, **11B**, 6.

Hartmann, H., Fröhlich, H. J., and Ebert, F., "A New Pernitride of Strontium and Calcium; Imides of the Alkaline Earth Metals," *Z. Anorg. Chem.*, **218**, 181.

Ketelaar, J. A. A., "Crystal Structure of the Low-Temperature Modification of Ammonium Bromide," *Nature,* **134**, 250.

Kolkmeijer, N. H., and Hengel, J. W. A. van, "Cubic and Hexagonal Silver Iodide," *Z. Krist.*, **88**, 317.

Michel, A., and Chaudron, G., "Transformations of Pyrrhotite and Ferrous Sulfide," *Compt. Rend.*, **198**, 1913.

Preston, B. A., "The Structure of Oxide Films on Ni," *Phil. Mag.*, **17**, 466.

Stillwell, C. W., and Jukkola, E. E., "The Crystal Structure of NdAl," *J. Am. Chem. Soc.*, **56**, 56.

Strada, M., "The Crystalline Structure of Thallium Cyanide," *Rend. Accad. Lincei,* **19**, 809.

Strock, L. W., "Crystal Structure of High Temperature Silver Iodide α-AgI," *Z. Physik. Chem.*, **25B**, 441.

Vegard, L., "Structure of the β-Form of Solid Carbon Monoxide," *Z. Physik,* **88**, 235.

West, C. D., "On the High-Temperature Modification of CsCl," *Z. Krist.*, **88**, 94.

West, C. D., "The Crystal Structures of Some Alkali Hydrosulfides and Monosulfides," *Z. Krist.*. **88**, 97.

West, C. D., "The Structure and Twinning of AgCN Crystals," *Z. Krist.*, **88,** 173.

Willott, W. H., and Evans, E. J. "X-Ray Investigation of the Arsenic–Tin System of Alloys " *Phil. Mag.* , **18,** 114.

1935

Buerger M. J., "The Unit Cell and Space Group of Realgar," *Am. Mineralogist,* **20,** 36.

Bunn, C. W., "The Lattice Dimensions of ZnO," *Proc. Phys. Soc.* (*London*), **47,** 835.

Büssem, W., Bluth, M., and Grochtmann, G., "X-Ray Expansion Measurements of Crystalline Masses," *Ber. Deut. Keram. Ges.*, **16,** 381.

Hägg, G., and Hybinette, A. G., "X-Ray Studies on the Systems Sn–Sb and Sn–As," *Phil. Mag.*, **20,** 913.

Helmholz, L., "The Crystal Structure of Hexagonal AgI," *J. Chem. Phys.*, **3,** 740.

Hofmann, W., "The Structures of Complex Sulfides. I. The Structure of SnS and Teallite, $PbSnS_2$," *Z. Krist.*, **92A,** 161.

Ketelaar, J. A. A., "The Crystal Structure of TlF," *Z. Krist.*, **92A,** 30.

Levi, G. R., and Baroni, A., "Structure and Alteration of Structure of NiS and NiSe," *Z. Krist.*, **92A,** 210.

Misch, L., "Structures of Intermetallic Compounds of Beryllium with Cu, Ni, and Fe," *Z. Physik. Chem.*, **29B,** 42.

Ruheman, B., "Temperature Variation of the Lattice Constant of MnO," *Phys. Z. Sowjetunion,* **7,** 590.

Stackelberg, M. v., and Spiesz, K. F., "The Structure of Aluminum Carbonitride," *Z. Physik. Chem.*, **175A,** 127.

Tunell, G., Posnjak, E., and Ksanda, C. J., "Geometrical and Optical Properties and Crystal Structure of Tenorite," *Z. Krist.*, **90A,** 120.

Zeipel, E. v., "The Lattice Constant of Galena Determined with a New X-Ray Spectrometer," *Arkiv Mat. Astron. Fysik,* **25A,** 1.

Zintl, E., and Harder, A., "Lattice Dimensions of LiH and LiD," *Z. Physik. Chem.*, **28B,** 478.

1936

Ackermann, P., and Mayer, J. E., "Determination of Molecular Structure by Electron Diffraction," *J. Chem. Phys.*, **4,** 377.

Aminoff, G., and Broome, B., "Oxidation of Single Crystals of Zinc Sulphide Studied by Electron Diffraction," *Nature,* **137,** 995.

Benard, J., and Chaudron, G., "Contribution to the Study of the Decomposition of FeO," *Compt. Rend.*, **202,** 1336.

Brauer, G., and Haucke, W., "Crystal Structure of the Intermetallic Phases MgAu and MgHg," *Z. Physik. Chem.*, **33B,** 304.

Buerger, M. J., "Crystals of the Realgar Type: The Symmetry, Unit Cell and Space Group of NS," *Am. Mineralogist,* **21,** 575.

Finch, G. J., and Fordham, S., "The Effect of Crystal Size on Lattice Dimensions," *Proc. Phys. Soc.* (*London*), **48,** 85.

Grether, W., "Determination of Atomic Separations in the Thallium and Tellurium Halides by Electron Diffraction," *Ann. Physik,* **26,** 1.

Helmholz, L., "The Crystal Structure of the Low Temperature Modification of TlI," *Z. Krist.*, **95A,** 129.

Hofmann, W., and Schrader, A., "TiC in Gray Iron," *Arch. Eisenhutt.*, **10,** 65.

Iandelli, A., and Botti, E., "On the Crystal Structure of Compounds of the Rare Earths with Metalloids of the V Group. Phosphides of La, Ce and Pr," *Rend. Atti. R. Accad. Lincei*, **24,** 459.

Lark-Horowitz, J. K., and Miller, E. P., "X-Ray Scattering in Molten Salts," *Phys. Rev.*, **49,** 418.

McKenna, P. M., "TaC in its Relation to Other Hard Refractory Compounds," *J. Ind. Eng. Chem.*, **28,** 767.

Misch, L., "Crystal Structure Investigations of Various Beryllium Alloys," *Metallwirtschaft*, **15,** 163.

Moltzau, R., and Kolthoff, I. M., "Mixed Crystal Formation of ZnS Post-Precipitated with HgS. The Aging of HgS and of ZnS," *J. Phys. Chem.*, **40,** 637.

Straumanis, M., and Jevins, A., "The Lattice Constants of NaCl and Rock Salt," *Z. Physik*, **102,** 353.

Wagner, C., and Beyer, J., "On the Nature of Disorder Phenomena in AgBr," *Z. Physik. Chem.*, **32B,** 113.

Wagner, G., and Lippert, L., "Polymorphic Transformations in Simple Ion Lattices. Investigations in the Transformation NaCl–CsCl Lattice," *Z. Physik. Chem.*, **33B,** 297; "The Transformation of the CsCl into the NaCl Lattice by Heat," *ibid.*, **31B,** 263.

Weigle, J., and Saini, H., "The Structure of NH_4Br at Low Temperatures," *Helv. Phys. Acta*, **9,** 515.

1937

Årstad, O., and Nowotny, H., "X-Ray Investigations in the System Manganese–Phosphorous," *Z. Physik. Chem.*, **38B,** 356.

Bannister, F. A., "The Discovery of Braggite," *Z. Krist.*, **96A,** 201.

Bénard J., "On the Parameter of FeO," *Compt. Rend.*, **205,** 912.

Bradley, J., and Taylor, A., "An X-Ray Analysis of the Nickel–Aluminum System," *Proc. Roy. Soc. (London)*, **159A,** 56.

Brager, A., "X-Ray Examination of the Structure of BN," *Acta Physicochem. URSS*, **7,** 699.

Gaskell, T. F., "The Structure of Braggite and PdS," *Z. Krist.*, **96A,** 203.

Haraldsen, H., "Phase Relations in the System Chromium–Sulfur," *Z. Anorg. Allgem. Chem.*, **234,** 372.

Hendricks, S. B., and Mosley, V. M., "Interatomic Distances in the Alkali Halides," *Phys. Rev.*, **51,** 1000.

Iandelli, A., and Botti, R., "On the Crystal Structure of Compounds of the Rare Earths with Metalloids of the V Group. Nitrides of La, Ce and Pr," *Rend. Atti R. Accad. Lincei*, **25,** 129; "Compounds 1:1 with Bi," *ibid.*, p. 233; "Arsenides and Antimonides of La, Ce and Pr," *ibid.*, p. 498; "Compounds of Nd," *ibid.*, p. 638; "On the Crystal Structures of Some Intermetallic Compounds of the Rare Earths," *Gazz. Chim. Ital.*, **67,** 638.

Maxwell, L. R., Hendricks, S. B., and Mosley, V. M., "Interatomic Distances of the Alkali Halide Molecules by Electron Diffraction," *Phys. Rev.*, **52,** 968.

Moeller, K., "Precision Determinations of Lattice Constants According to the Method of Debye-Scherrer," *Z. Krist.*, **97A,** 170.

Sidhu, S. S., and Hicks, V., "The Space Lattice and 'Super-Lattice' of Pyrrhotite," *Phys. Rev.*, **52,** 667.

1938

Aguzzi, A., and Genoni, F., "The Dehydration of $Li_2O_2 \cdot H_2O_2 \cdot 3H_2O$," *Gazz. Chim. Ital.*, **68**, 816.

Baroni, A., "Polymorphism and Isomorphism of the Sulfides and Selenides of Ni, Co, Cd and Hg," *Atti Congr. Intern. Chim., 10°, Rome* (pub. 1939), **2**, 586; "On the Polymorphism of MnSe," *Z. Krist.*, **99A**, 336.

Faivre, R., and Michel, A., "Variations of the Crystalline Parameter of CdO through Insertion of Cd Atoms into the Lattice," *Compt. Rend.*, **207**, 159.

Feher, F., "The Structure of Alkali Peroxides," *Z. Angew. Chem.*, **51**, 497.

Fürst, M., and Halla, F., "X-Ray Investigations in the Systems Mn–Bi, Co–Sb and Ni–Sb," *Z. Physik. Chem.*, **40B**, 285.

Haraldsen, H., and Mehmed, F., "Magnetochemical Investigations, XXX. Phase Ratios and Magnetic Properties in the System Cr–Se," *Z. Anorg. Allgem. Chem.*, **239**, 369.

Jacobs, R. B., "Polymorphic Transitions in Metallic Halides," *Phys. Rev.*, **54**, 468; "X-Ray Diffraction of Substances under High Pressures," *Phys. Rev.*, **54**, 325.

Jevins, A., Straumanis, M., and Karlsons, K., "Precision Determination of Lattice Constants of Hygroscopic Compounds," *Z. Physik. Chem.*, **40B**, 146.

Juza, R., and Hahn, H., "On the Crystal Structures of Cu_3N, GaN and InN, Metal Amides and Nitrides, V.," *Z. Anorg. Allgem. Chem.*, **239**, 282.

Klemm, W., "Investigations of the Chalcogenides of V and their Relations to the Chalcogenides of the Other Transition Elements," *Z. Angew. Chem.*, **51**, 756.

Lundquist, D., and Westgren, A., "X-Ray Investigation of the System Cobalt–Sulfur," *Z. Anorg. Allgem. Chem.*, **239**, 85.

Nowacki, W., "The Crystal Structure of EuS," *Z. Krist.*, **99A**, 339.

Nowotny, H., and Årstad, O., "X-Ray Investigations in the System Cr–CrAs," *Z. Physik. Chem.*, **38B**, 461.

Nowotny, H., and Henglein, E., "Investigations in the System Chromium–Phosphorus," *Z. Anorg. Allgem. Chem.*, **239**, 14.

Saur, E., and Stasiv, O., "The Effects of Small Amounts of Ionic Impurities on the Lattice Constants of the Alkali Halide Crystals," *Nachr. Ges. Wiss. Goettingen, Fachgruppe II*, **3**, 77.

Smits, A., Tollenaar, D., and Kröger, K. A., "Nature of the Low Temperature Transformation of ND_4Br," *Z. Physik. Chem.*, **41B**, 215.

Verweel, H. J., and Bijvoet, J. M., "The Crystal Structure of NaCN," *Z. Krist.*, **100A**, 201.

1939

Bauer, S. H., Beach, J. Y., and Simons, J. H., "The Molecular Structure of Hydrogen Fluoride," *J. Am. Chem. Soc.*, **61**, 19.

Biltz, W., and Heimbrecht, H., "On the Phosphide of Cobalt," *Z. Anorg. Allgem. Chem.*, **241**, 349.

Biltz, W., and Köcher, A., "Contribution to Systematic Affinity Theory, LXXXVIII. On the System V–S," *Z. Anorg. Allgem. Chem.*, **241**, 324.

Boochs, H., "Exact Determination of Lattice Constants by Electron Diffraction with Crystallites of Various Sizes," *Ann. Physik*, **35**, 333.

Brager, A., "An X-Ray Examination of TiN, I. Single Crystal Investigation," *Acta Physicochimica URSS*, **10**, 593; "III. Investigation by the Powder Method," *ibid.*, **11**, 617; "A Correction to X-Ray Examination of the Structure of BN," *ibid.*, **10**, 902.

Brill, R., Grimm, H. G., Hermann, C., and Peters, C., "Application of X-Ray Fourier Analysis to Questions of Chemical Combination," *Ann. Physik*, **34**, 393.

Chudoba, K. F., and Mackowsky, M. T., "The Isomorphism of Iron and Zinc in Zinc Blende," *Zentr. Min.*, **1939**, 12.

Fordham, S., and Khalsa, R. G., "Single Crystal Pd Films and Their Interaction with Gases," *J. Chem. Soc.*, **1939**, 406.

Germer, L. H., "Electron Diffraction Studies of Thin Films, I. Structure of Very Thin Films," *Phys. Rev.*, **56**, 58.

Günther, P., Holm, K., and Strunz, H., "The Structure of Solid HF," *Z. Physik. Chem.*, **43B**, 229.

Haraldsen, H., "The Transformations of FeS," *Z. Elektrochem.*, **45**, 370; "Transformations in the Troilite–Solid Solution Range," *Tidsskr. Kjemi, Bergvesen*, **19**, 144.

Hocart, R., and Guillaud, C., "On the Alloy MnBi," *Compt. Rend.*, **209**, 443.

Hoschek, E., and Klemm, W., "Vanadium Selenides," *Z. Anorg. Allgem. Chem.*, **242**, 49.

Ketelaar, J. A. A., t'Hart, W. H., Moerel, M., and Polder, D., "The Crystal Structure of Thallous, Thallic or Thallosic Selenide," *Z. Krist.*, **101A**, 396.

Ketelaar, J. A. A., and Zwartsenberg, J. W., "The Crystal Structure of the Cyanogen Halides, I. The Crystal Structure of (CN),I," *Rec. Trav. Chim.*, **58**, 449.

Klemm, W., and Senff, H., "Measurements on Di- and Quadrivalent Compounds of the Rare Earths, VIII. Chalcogenides of Divalent Eu," *Z. Anorg. Allgem. Chem.*, **241**, 259.

Kröger, F. A., "Formation of Solid Solutions in the System ZnS–MnS," *Z. Krist.*, **100A**, 543.

Maxwell, L. R., and Mosley, V. M., "Internuclear Distances in the Gas Molecules Se_2, HgCl, Cu_2Cl_2, Cu_2Br_2 and Cu_2I_2 by Electron Diffraction," *Phys. Rev.*, **55**, 238.

Meisel, K., "The Crystal Structure of Th Phosphides," *Z. Anorg. Allgem. Chem.*, **240**, 300.

Schmitz-Dumont, O., "The Polymeric Forms of AgCN," *Ber.*, **72B**, 298.

Senff, H., and Klemm, W., "Measurements on Di- and Quadrivalent Compounds of the Rare Earths, IX, Yb Chalcogenides," *Z. Anorg. Allgem. Chem.*, **242**, 92.

Teichert, W., and Klemm, W., "The Hydrosulfides and Hydroselenides of the Alkali Metals," *Z. Anorg. Allgem. Chem.*, **243**, 86; "The Structure of the High Temperature Form of KOH," *ibid.*, **243**, 138.

Thiessen, P. A., and Moliere, K., "On the Influence of Absorption on the Refraction of Electron Rays, I. Measurements of the Inner Potential on the Polar Tetrahedral Faces of Zinc Blende," *Ann. Physik*, **34**, 449.

1940

Bijvoet, J. M., and Lely, J. A., "Orthorhombic Modification of KCN. The Dependence of the Transition Point of NaCN on Admixtures," *Rec. Trav. Chim.*, **59**, 908.

Breger, A. K., "Dependence of the Unit Cube Edge of the Compounds TiC, TiN and TiO on the Radius of the Metalloid Atom," *Acta Physicochim. URSS,* **13,** 723.

Breger, A. K., and Zhdanov, G., "Nature of the Chemical Bond in Graphite and BN," *Compt. Rend. Acad. Sci. URSS,* **28,** 629.

Dawihl, W., and Rix., W., "The Lattice Constants of the Carbide and Nitride of Ti and V," *Z. Anorg. Allgem. Chem.,* **244,** 191.

Epelbaum, V., and Breger, A. K., "X-Ray Investigation of VN, II. Precise Determination of the Unit Cube Edge of VN," *Acta Physicochim. URSS,* **13,** 600.

Hutchison, C. A., and Johnston, H. L., "Determination of Crystal Density by the Temperature-of-Flotation Method. Density and Lattice Constant of LiF," *J. Am. Chem. Soc.,* **62,** 3165.

Ievins, A., and Straumanis, M., "Lattice Constant of Calcite Determined by the Rotating-Crystal Method," *Z. Physik,* **116,** 194.

Inuzuka, H., "The Crystal Structure of Silicon Monoxide," *Mazda Kenkiju Ziho,* **15,** 305.

Kröger, F. A., "Solid Solutions in the Ternary System ZnS–CdS–MnS," *Z. Krist.,* **102A,** 132.

Kubaschewski, O., "Reduction of Cb_2O_5 to Cb_2O with Hydrogen," *Z. Elektrochem.,* **46,** 284.

Umanskii, Y. S., "X-Ray Investigation of some Cb Compounds," *J. Phys. Chem. USSR,* **14,** 332.

Wilman, H., "The Orientation of Silver Halides," *Proc. Phys. Soc.* (*London*), **52,** 323.

1941

Bergen, H. van, "Precision Measurements of Lattice Constants with a Compensation Method, II.," *Ann. Physik,* **39,** 553.

Brauer, G., "The Oxides of Cb," *Z. Anorg. Allgem. Chem.,* **248,** 1.

Brill, R., Hermann, C., and Peters, C., "Remarks on the Paper by Breger and Zhdanov on the Nature of the Chemical Bond in Graphite + BN," *Nature,* **29,** 784.

Haraldsen, H., "Iron Sulfide Mixed Crystals," *Z. Anorg. Allgem. Chem.,* **246,** 169.

Iandelli, A., "The Structure of the Compounds InP, InAs and InSb," *Gazz. Chim. Ital.,* **71,** 58.

Kordes, E., "Ion Radii and the Periodic Table, II. Calculation of Ion Radii with the Help of Atom Physical Quantities," *Z. Physik. Chem.,* **48B,** 91.

Meldau, R., and Teichmüller, M., "The Morphology of Very Fine PbO Sublimates," *Z. Elektrochem.,* **47,** 191.

Moore, W. J., Jr., and Pauling, L., "The Crystal Structure of the Tetragonal Monoxides of Pb, Sn, Pd and Pt," *J. Am. Chem. Soc.,* **63,** 1392.

Nitka, H., "X-Ray High Temperature Diagrams of BeO," *Nature,* **29,** 336.

Strotzer, E. F., and Zumbusch, W., "Contribution to the Systematic Study of Affinity, XCVIII, Thorium Sulfide," *Z. Anorg. Allgem. Chem.,* **247,** 415.

Umanskii, Y. S., and Khedekel, S. S., "X-Ray Study of TiC," *J. Phys. Chem. USSR,* **15,** 983.

Zumbusch, M., "Structural Analogy of U and Th Phosphides," *Z. Anorg. Allgem. Chem.,* **245,** 402.

1942

Dinichert, P., "The Transformation of NH_4Cl Observed by the Diffraction of X-Rays," *Helv. Phys. Acta,* **15,** 462.

Huber, H., and Wagener, S., "Crystallographic Structure of Alkaline Earth Oxide Mixtures. Study of Oxide Cathodes by X-Rays and Electrons," *Z. Tech. Physik,* **23,** 1.

Klemm, W., and Grimm, L., "Lower Vanadium Oxides," *Z. Anorg. Allgem. Chem.,* **250,** 42.

Laves, F., and Wallbaum, H. J., "Some New Examples of the NiAs Type and Their Crystallographic Meaning," *Z. Angew. Mineral.,* **4,** 17.

Lely, J. A., and Bijvoet, J. M., "The Crystal Structure of LiCN," *Rec. Trav. Chim.,* **61,** 244.

Nowotny, H., "The Crystal Structures of Ni_5Ce, etc.," *Z. Metallk.,* **34,** 247.

Smits, A., and Tollenaar, D., "Nature of the Low Temperature Transformations of ND_4I," *Z. Physik. Chem.,* **52B,** 222.

Vegard, L., and Hillesund, S., "Structure of a Few D Compounds and Comparison with That of the Corresponding H Compounds," *Avhandl. Norske Videnskaps-Akad. Oslo, I. Mat.-Naturv. Kl.,* **1942,** No. 8, 24.

Vegard, L., and Skofteland, G., "X-Ray Investigation of the Series of Binary Solid Solutions Formed from the Substances CuCl, CuBr and CuI," *Arch. Mat. Natuurvidenskab.,* **45,** 163.

1943

Ageew, N. V., and Makarov, E. S., "Physicochemical Study of the Phases with a NiAs Structure in the Systems Fe–Sb–Ni and Fe–Sb–Co," *Bull. Acad. Sci. URSS, Classe Sci. Chim.,* **1943,** 161; *J. Gen. Chem. USSR,* **13,** 242.

Böttcher, C. J. F., "Method for Calculating the Polarizabilities and Radii of Molecules and Ions," *Rec. Trav. Chim.,* **62,** 325.

Bommer, H., and Krose, E., "Densities of the La Alloys with Cu, Ag, Au," *Z. Anorg. Allgem. Chem.,* **252,** 62.

Byström, A., "Roentgenographic Studies of Rhombic PbO," *Arkiv Kemi, Mineral. Geol.,* **17B,** No. 8, p. 1.

Fylking, K. E., "X-Ray Analysis of Twinned MnP Crystals," *Arkiv Kemi, Mineral. Geol.,* **17A,** No. 7, p. 9.

Hassel, O., and Viervoll, H., "The Structural Formulas of S_4N_4 and Related Compounds," *Tidsskr. Kjemi, Bergvesen Met.,* **3,** 7.

Hess, B., "Lattice Widening and the Latent Image. X-Ray Investigation of the Developability of Photographic Films," *Physik. Z.,* **44,** 245.

Klemm, W., and Fratini, N., "The System NiTe–$NiTe_2$," *Z. Anorg. Allgem. Chem.,* **251,** 222.

Lu, Chai-Si, and Malmberg, E. W., "ZnO Smoke as Reference Standard in Electron Wavelength Calibration," *Rev. Sci. Instr.,* **14,** 271.

Rittner, E. S., and Schulman, J. H., "The Coprecipitation of CdS and HgS," *J. Phys. Chem.,* **47,** 537.

Rooskby, H. P., "Structure of NiO," *Nature,* **152,** 304.

1944

Arkharov, V. I., and Graevskii, K. M., "Precision X-Ray Investigation of Fe, Co and Ni Tarnish. I.," *J. Tech. Phys. (USSR),* **14,** 132.

Lely, J. A., and Bijvoet, J. M., "The Crystal Structure and Anisotropic Vibration of Ammonium Cyanide," *Rec. Trav. Chim.*, **63**, 39.

Lu, S. Chia-Si, and Donohue, J., "An Electron Diffraction Investigation of S_4N_4, As_4S_4 (Realgar) and As_4S_6 (Orpiment)," *J. Am. Chem. Soc.*, **66**, 818.

Ramsdell, L. S., "The Crystal Structure of SiC, Type IV," *Am. Mineralogist*, **29**, 431.

Thibault, N. W., "Morphological and Structural Crystallography and Optical Properties of SiC, Parts I and II," *Am. Mineralogist*, **29**, 249, 327.

1945

Byström, A., "Monoclinic Pyrrhotites," *Arkiv Kemi, Mineral. Geol.*, **19B**, No. 8, p. 8.

Byström, A., "The Decomposition Products of Lead Peroxide and Oxidation Products of Lead Oxide," *Arkiv Kemi, Mineral. Geol.*, **A20**, No. 11, p. 31.

Ramsdell, L. S., "The Crystal Structure of SiC, Type VI," *Am. Mineralogist*, **30**, 519.

Zhdanov, G. S., and Minervina, Z. V., "Analysis of the Crystal Structure of SiC,V (51-Layered Packing)," *Compt. Rend. Acad. Sci. URSS*, **48**, 182.

Zhdanov, G. S., and Shugam, E. A., "The Crystal Structure of Cyanides. III. Structure of Gold Cyanide," *Acta Physicochem. URSS*, **20**, 253.

1946

Epelbaum, V. A., and Breger, A. K., "X-Ray Examination of Vanadium Nitride. III. System VN–VO," *Acta Physicochim. URSS*, **21**, 764; *J. Phys. Chem. USSR*, **20**, 459.

Epprecht, W., "The Manganese Minerals of Gonzen and Their Paragenesis," *Schweiz. Mineral. Petrog. Mitt.*, **26**, 19.

Ernst, T., "Crystal Structure of α-Sodium Hydroxide," *Nachr. Akad. Wiss. Goettingen, Math.-Physik. Kl., Math.-Physik.-Chem. Abt.*, **1946**, 76.

Guillaud, C., and Barbezat, S., "Ferromagnetic Properties of the Definite Compound CrTe," *Compt. Rend.*, **222**, 386.

Koval'skii, A. E., and Umanskii, Y. S., "X-Ray Investigation of Pseudobinary Systems. I. TaC–TiC; CbC–TiC; TaC–ZrC; CbC–ZrC," *J. Phys. Chem. USSR*, **20**, 769.

Koval'skii, A. E., and Umanskii, Y. S., "Interaction of the Monocarbides of Tungsten, Tantalum and Columbium (X-Ray Analysis) II.," *J. Phys. Chem.* (*USSR*), **20**, 773.

Lindroth, G. T., "The Crystal Structure of Pyrrhotite in Iron and Sulfide Ores of Sweden," *Tek. Tidskr.*, **76**, 383.

1947

Buerger, M. J., "The Cell and Symmetry of Pyrrhotite," *Am. Mineralogist*, **32**, 411.

Epelbaum, V. A., and Ormont, B. F., "Some Properties of Real Crystals of Vanadium Nitride," *J. Phys. Chem.* (*USSR*), **21**, 3.

Ernst, T., and Schober, R., "Crystal Structure of α-Potassium Hydroxide," *Nachr. Akad. Wiss. Goettingen, Math.-Physik. Kl., Math.-Physik.-Chem. Abt.*, **1947**, 49.

Kiessling, R., "Crystal Structures of Molybdenum and Tungsten Borides," *Acta Chem. Scand.*, **1**, 893.

Lipin, S. V., "The Nature of Pyrrhotite and Troilite," *Mem. Soc. Russe Mineral.*, **75**, 273; *Chem. Zentr.*, **1947**, I, 1167.

Lundqvist, D., "X-Ray Studies of the Binary System Ni–S," *Arkiv Kemi, Mineral. Geol.*, **24A**, No. 21, 12 pp.

Metcalfe, A. G., "The Mutual Solid Solubility of Tungsten Carbide and Titanium Carbide," *J. Inst. Metals*, **73**, 591.

Nial, O., "X-Ray Studies on Binary Alloys of Tin with Transition Metals. I. Binary Systems of Tin with Iron, Cobalt, Nickel, Manganese, Rhenium and Chromium," *Svensk Kem. Tidskr.*, **59**, 165; "II. Binary Systems of Tin with Platinum Metals," *ibid.*, **59**, 172.

Nowotny, H., and Kieffer, R., "X-Ray Investigation of Carbide Systems," *Metallforschung*, **2**, 257.

Ramsdell, L. S., "Studies on Silicon Carbide," *Am. Mineralogist*, **32**, 64.

Zhdanov, G. S., and Minervina, Z. V., "Crystal Structure of SiC, VI and Geometrical Theory of Silicon Carbide Structures, *J. Exptl. Theoret. Phys.* (*USSR*), **17**, 3.

1948

Almin, K. E., "The Crystal Structure of CdSb and ZnSb," *Acta Chem. Scand.*, **2**, 400.

Ehrlich, P., "Structure and Synthesis of Titanium Selenides and Tellurides," *Angew. Chem.*, **60A**, 68.

Endter, F., "Crystal Structure of Gadolinium Nitride," *Z. Anorg. Chem.*, **257**, 127.

Ernst, T. "The Alkali Hydroxides," *Z. Angew. Chem.*, **60A**, 77.

Frondel, C., and Palache, C., "Three New Polymorphs of Zinc Sulfide," *Science*, **107**, 602.

Litz, L. M., Garrett, A. B., and Croxton, F. C., "Preparation and Structure of the Carbides of Uranium," *J. Am. Chem. Soc.*, **70**, 1718.

Lundquist, D., "Crystal Structure of Silicon Carbide and its Content of Impurities," *Acta Chem. Scand.*, **2**, 177.

Makarov, E. S., "Crystal Structure of InBi," *Dokl. Akad. Nauk SSSR*, **59**, 899.

Rooksby, H. P., "Structure of Nickel Oxide at Subnormal and Elevated Temperatures," *Acta Cryst.*, **1**, 226.

Rundle, R. E., Baenziger, N. C., Wilson, A. S., and McDonald, R. A., "The Structures of Carbides, Nitrides and Oxides of Uranium," *J. Am. Chem. Soc.*, **70**, 99.

Shimomura, Y., and Nishiyama, Z., "The Crystal Structure of Nickel Oxide," *Mem. Inst. Sci. Ind. Res., Osaka Univ.*, **6**, 30.

Shull, C. G., Wollan, E. O., Morton, G. A., and Davidson, W. L., "Neutron-Diffraction Studies of NaH and NaD," *Phys. Rev.*, **73**, 842.

Trillat, J. J., and Laloeuf, A., "The Structure of Ammonium Chloride," *Compt. Rend.*, **227**, 67.

1949

Earley, J. W., "Studies of Natural and Artificial Selenides. I. Klockmannite, CuSe," *Am. Mineralogist*, **34**, 435.

Ehrlich, P., "Structure and Composition of the Chalcogenides of the Transition Elements," *Z. Anorg. Chem.*, **260**, 19.

Hahn, H., "Metal Amides and Metal Nitrides. XVIII. The System Vanadium–Nitrogen," *Z. Anorg. Chem.*, **258**, 58.

Hahn, H., and Klingler, W., "X-Ray Studies on the Systems Tl–S, Tl–Se and Tl–Te," *Z. Anorg. Chem.*, **260**, 110.

Kiessling, R., "The Binary System Chromium–Boron. I. Phase Analysis and Structure of the ζ and θ Phases." *Acta Chem. Scand.*, **3**, 595.

Kiessling, R., "The Borides of Tantalum," *Acta Chem. Scand.*, **3**, 603.

Siegel, L. A., "Molecular Rotation in Sodium Cyanide and Sodium Nitrate," *J. Chem. Phys.*, **17**, 1146.

Sindeband, S. J., "Properties of Chromium Boride and Sintered Chromium Boride," *J. Metals,* **1,** No. 2, *Trans.*, 198.
Thibault, N. W., "Alpha-Silicon Carbide, Type 51*R*," *Am. Mineralogist,* **33,** 588.
Zachariasen, W. H., "Crystal-Chemical Studies of the 5*f* Series of Elements. X. Sulfides and Oxysulfides," *Acta Cryst.*, **2,** 291.
Zachariasen, W. H., "Crystal-Chemical Studies of the 5*f* Series of Elements. XII. New Compounds Representing Known Structure Types," *Acta Cryst.*, **2,** 388.
Zvonkova, Z. V., and Zhdanov, G. S., "X-Ray Study of Ammonium Thiocyanate," *Zh. Fiz. Khim.*, **23,** 1495.

1950

Aurivillius, K. L., "On the Structure of Cinnabar," *Acta Chem. Scand.*, **4,** 1413.
Earley, J. W., "Description and Synthesis of Selenide Minerals," *Am. Mineralogist,* **35,** 337.
Greenwald, S., and Smart, J. S., "Deformations in the Crystal Structures of Antiferromagnetic Compounds," *Nature,* **166,** 523.
Heller, R. B., McGannon, J., and Weber, A. H., "Precision Determination of the Lattice Constants of Zinc Oxide," *J. Appl. Phys.*, **21,** 1283.
Pease, R. S., "Crystal Structure of Boron Nitride," *Nature,* **165,** 722.
Pfisterer, H., and Schubert, K., "New Phases of the MnP type," *Z. Metallk.*, **41,** 358.
Stephenson, N. C., and Mellor, D. P., "The Crystal Structure of Indium Monobromide," *Australian J. Sci. Res.*, **A3,** 581.
Taylor, A., and Laidler, D. S., "The Formation and Crystal Structure of Silicon Carbide," *Brit. J. Appl. Phys.*, **1,** 174.
Tombs, N. C., and Rooksby, H. P., "Structure of Monoxides of Some Transition Elements at Low Temperatures," *Nature,* **165,** 442.
Ueta, R., Ichinokawa, T., and Mitsui, T., "X-Ray Study of Synthetic Pyrrhotite, FeS_x," *Busseiron Kenkyu,* No. 33, 55.
Zemann, J., "Brunckite, Cryptocrystalline Sphalerite," *Tschermaks Mineral. Petrog. Mitt.*, **1,** 417.

1951

Abrahams, S. C., Collin, R. L., and Lipscomb, W. N., "The Crystal Structure of Hydrogen Peroxide," *Acta Cryst.*, **4,** 15.
Berry, C. R., "Data on Silver Bromide," *Phys. Rev.*, **82,** 422.
Bogatskii, D. P., "State Diagram of the System Nickel–Oxygen and Physico-Chemical Nature of the Solid Phases in that System," *Zh. Obshch. Khim.*, **21,** 3.
Cirilli, V., and Brisi, C., "The Limits of Composition of Wüstite," *Ann. Chim.* (*Rome*), **41,** 508.
ulmage, W. J., and Lipscomb, W. N., "Crystal Structures of Hydrogen Cyanide," *Acta Cryst.*, **4,** 330.
Frueh, A. J., Jr., "Confirmation of the Structure of Chromium Boride," *Acta Cryst.*, **4,** 66.
Juza, R., and Opp, K., "Metallic Amides and Metallic Nitrides. XXIV. The Crystal Structure of Lithium Amide," *Z. Anorg. Allgem. Chem.*, **266,** 313.
Kato, N., "Measurement of Lattice Constant by Electron Diffraction; Lattice Constant of Potassium Chloride Referred to Gold," *J. Phys. Soc. Japan,* **6,** 502.
Keith, H. D., and Mitchell, J. W., "Lattice Defects in Silver Bromide at Room Temperature," *Phil. Mag.*, **42,** 1331.

Klemm, W., and Wahl, K., "Magnesium Telluride," *Z. Anorg. Allgem. Chem.*, **266,** 289.

Levy, H. A., and Peterson, S. W., "The Nature of the Second-Order Transition in ND_4Br," *Phys. Rev.*, **83,** 1270.

Liu, T. S., and Peretti, E. A., "Lattice Parameter of InSb," *J. Metals, Trans. Sec.*, **3,** No. 9, 791.

Menary, J. W., Ubbelohde, A. R., and Woodward, I., "The Thermal Transition in Cesium Chloride in Relation to Crystal Structure," *Proc. Roy. Soc.* (*London*), **A208,** 158.

Ramsdell, L. S., and Kohn, J. A., "Disagreement between Crystal Symmetry and X-Ray Diffraction Data as Shown by a Type of Silicon Carbide, 10 *H*," *Acta Cryst.*, **4,** 111.

Rooksby, H. P., and Tombs, N. C., "Changes of Crystal Structure in Antiferromagnetic Compounds," *Nature*, **167,** 364.

Rymer, T. B., and Hambling, P. G., "The Lattice Constant of Cesium Iodide," *Acta Cryst.*, **4,** 565.

Wasserstein, B., "Precision Lattice Measurements of Galena," *Am. Mineralogist*, **36,** 102.

1952

Bertaut, F., "The Structure of Pyrrhotite," *Compt. Rend.*, **234,** 1295.

Blum, P., "The Structure of Nickel Boride," *J. Phys. Radium*, **13,** 430.

Blumenthal, H., "Vanadium Monoboride," *J. Am. Chem. Soc.*, **74,** 2942.

Brauer, G., and Jander, J., "The Nitrides of Columbium," *Z. Anorg. Chem.*, **270,** 160.

Cirilli, V., and Burdese, A., "Calcium Oxide–Wüstite System," *Proc. Intern. Symposium Reactivity of Solids, Gothenburg*, 1952, Pt. 2, p. 867.

Clark, D., "Structure of Sulfur Nitride," *J. Chem. Soc.*, **1952,** 1615.

Cohen, A. J., "Observations on Several Compounds of Lithium and Oxygen. I.," *J. Am. Chem. Soc.*, **74,** 3762.

D'Eye, R. W. M., Sellman, P. G., and Murray, J. R., "The Thorium–Selenium System," *J. Chem. Soc.*, **1952,** 2555.

Ferro, R., "Alloys of Uranium with Bismuth," *Atti Accad. Nazl. Lincei, Rend. Classe Sci. Fis., Mat. Nat.*, **13,** 401.

Gasilova, E. B., Beletskii, M. S., and Sokhor, M. I., "Determination of the New SiC, VII Structure," *Dokl. Akad. Nauk SSSR*, **82,** 57.

Gasilova, E. B., and Sokhor, M. I., "Crystal Structure of SiC, VIII," *Dokl. Akad. Nauk SSSR*, **82,** 249.

Goldschmidt, G. H., and Hurst, D. G., "The Structure of Ammonium Chloride by Neutron Diffraction," *Phys. Rev.*, **86,** 797.

Hoshino, S., "Crystal Structure and Phase Transition of Some Metallic Halides. II. Phase Transition and the Crystal Structures of Cuprous Bromide," *J. Phys. Soc. Japan*, **7,** 560.

Iandelli, A., "On Arsenides of Uranium," *Atti Accad. Nazl. Lincei, Rend. Classe Sci. Fis., Mat. Nat.*, **13,** 138.

Ito, T., Morimoto, N., and Sadanaga, R., "The Crystal Structure of Realgar (AsS)," *Acta Cryst.*, **5,** 775.

Krug, J., and Sieg, L., "The Structures of the High Temperature Modifications of CuBr & CuI," *Z. Naturforsch.*, **7A,** 369.

Kuo, K., and Hägg, G., "A New Molybdenum Carbide," *Nature*, **170,** 245.

Levy, H. A., and Peterson, S. W., "Neutron Diffraction Study of the Crystal Structure of Ammonium Chloride," *Phys. Rev.*, **86,** 766.

Miyake, S., Hoshino, S., and Takenaka, T., "The Phase Transition in Cuprous Iodide," *J. Phys. Soc. Japan*, **7**, 19.

Müller, H., "The One-Dimensional Transformation Zinc Blende–Würtzite and the Accompanying Anomalies," *Neues Jahrb. Mineral. Abhandl.*, **84**, 43.

Novoselova, A. V., Simanov, Y. P., and Yarembash, E. I., "Thermal and X-Ray Analysis of the Lithium–Beryllium Fluoride System," *Zh. Fiz. Khim.*, **26**, 1244.

Nowotny, H., and Kieffer, R., "A Note on the Existence of Cubic Molybdenum Carbide," *Z. Anorg. Allgem. Chem.*, **267**, 261.

Pease, R. S., "An X-Ray Study of Boron Nitride," *Acta Cryst.*, **5**, 356.

Post, B., and Glaser, F. W., "Borides of Some Transition Metals," *J. Chem. Phys.*, **20**, 1050.

Ramsdell, L. S., and Kohn, J. A., "Developments in Silicon Carbide Research," *Acta Cryst.*, **5**, 215.

Rymer, T. B., and Archard, G. D., "Lattice Constants of Zinc Oxide," *Research*, **5**, 292.

Sakurai, K., and Tomiie, Y., "The Crystal Structure of Hydrazinium Bromide," *Acta Cryst.*, **5**, 289.

Sakurai, K., and Tomiie, Y., "Crystal Structure of Hydrazinium Chloride," *Acta Cryst.*, **5**, 293.

Young, R. A., and Ziegler, W. T., "Crystal Structure of Lanthanum Nitride," *J. Am. Chem. Soc.*, **74**, 5251.

Zachariasen, W. H., "Identification and Crystal Structure of Protoactinium Metal and of Protoactinium Monoxide," *Acta Cryst.*, **5**, 19.

1953

Archard, G. D., "Anomalous Lattice Constants of Zinc Oxide," *Acta Cryst.*, **6**, 657.

Atoji, M., and Lipscomb, W. N., "Molecular Structure of B_4Cl_4," *J. Chem. Phys.*, **21**, 172.

Atoji, M., and Lipscomb, W. N., "The Crystal and Molecular Structure of B_4Cl_4," *Acta Cryst.*, **6**, 547.

Brauer, G., and Zapp, K. H., "Crystal Structure of Tantalum Nitride, TaN," *Naturwiss.*, **40**, 604.

Brisi, C., and Burdese, A., "The Crystal Structure of Wüstite," *Ann. Chim. (Rome)*, **43**, 69.

Dulmage, W. J., Meyers, E. A., and Lipscomb, W. N., "The Crystal and Molecular Structure of N_2O_2," *Acta Cryst.*, **6**, 760.

Ehrlich, P., and Gentsch, L., "Calcium Monochloride," *Naturwiss.*, **40**, 460.

Ellinger, F. H., and Zachariasen, W. H., "Crystal Structure of Samarium Metal and of Samarium Monoxide," *J. Am. Chem. Soc.*, **75**, 5650.

Feher, F., Wilicki, I. von, and Dost, G., "Hydrogen Peroxide and its Derivatives. VII. The Crystal Structure of Lithium Peroxide, Li_2O_2," *Chem. Ber.*, **86**, 1429.

Glagoleva, V. P., and Zhdanov, G. S., "Structure of Superconductors. III. X-Ray Investigation of the Structure and Solubility of Components in BiRh," *Zh. Eksperim. Teoret. Fiz.*, **25**, 248.

Glaser, F. W., Moskowitz, D., and Post, B., "Some Binary Hafnium Compounds," *J. Metals*, **5**, *Trans. AIME*, **197**, 1119.

Glaser, F. W., and Post, B., "The System: Zirconium Boron," *J. Metals*, **5**, *Trans. AIME*, **197**, 1117.

Greenwald, S., "The Antiferromagnetic Structure Deformations in CoO and MnTe," *Acta Cryst.*, **6**, 396.

Hambling, P. G., "The Lattice Constants and Expansion Coefficients of Some Halides," *Acta Cryst.*, **6**, 98.

Hoch, M., and Johnston, H. L., "Formation, Stability and Crystal Structure of Solid Silicon Monoxide," *J. Am. Chem. Soc.*, **75**, 5224.

Jacobs, G., "The Structure of Silicon Monoxide," *Compt. Rend.*, **236**, 1369.

Levy, H. A., and Peterson, S. W., "Neutron Diffraction Determination of the Crystal Structure of Ammonium Bromide in Four Phases," *J. Am. Chem. Soc.*, **75**, 1536.

Levy, H. A., and Peterson, S. W., "Neutron Diffraction Study of Sodium Chloride Type Modification of Ammonium d_4 Bromide and Iodide," *J. Chem. Phys.*, **21**, 366.

Matukura, Y., Yamamoto, T., and Okazaki, A., "A Preliminary Study of the Electrical Properties of SnSe Single Crystals," *Mem. Fac. Sci. Kyushu Univ., Ser. B*, **I**, No. 3, 98.

Münster, A., and Sagel, K., "X-Ray Studies of Surface Layers of Titanium Nitride and Titanium Carbide," *Z. Elektrochem.*, **57**, 571.

Ramsdell, L. S., and Mitchell, R. S., "A New Hexagonal Polymorph of Silicon Carbide, 19 *H*," *Am. Mineralogist*, **38**, 56.

Schubert, K., and Dörre, E., "Crystal Structures of GaSe," *Naturwiss.*, **40**, 604.

Schubert, K., and Fricke, H., "Crystallochemistry of β-Subgroup Metals. II. Trigonal Distorted NaCl Structures," *Z. Metallk.*, **44**, 457.

Sieg, L., "The Structure of Silver Iodide," *Naturwiss.*, **40**, 439.

Wasser, J., Levy, H. A., and Peterson, S. W., "The Structure of PdO," *Acta Cryst.*, **6**, 661.

Willis, B. T. M., "Crystal Structure and Antiferromagnetism of CrSb," *Acta Cryst.*, **6**, 425.

Willis, B. T. M., and Rooksby, H. P., "Change of the Structure of Ferrous Oxide at Low Temperature," *Acta Cryst.*, **6**, 827.

1954

Atoji, M., and Lipscomb, W. N., "The Crystal Structure of Hydrogen Fluoride," *Acta Cryst.*, **7**, 173.

Aurivillius, K., "Mercury (II) Oxide Chlorides and Mercury (II) Oxide," *Acta Chem. Scand.*, **8**, 523.

Batuecas, T., "Determination of Atomic Masses by the Pykno-X-Ray Method," *Nature*, **173**, 345.

Bénard, J., "The Limiting Parameters of the Phase FeO," *Acta Cryst.*, **7**, 214.

Berry, L. G., "The Crystal Structure of Covellite, CuS, and Klockmannite, CuSe," *Am. Mineralogist*, **39**, 504.

Brauer, G., Renner, H., and Wernet, J., "Carbides of Columbium," *Z. Anorg. Allgem. Chem.*, **277**, 249.

Brauer, G., and Zapp, K. H., "The Nitrides oɪ Tantalum," *Z. Anorg. Allgem. Chem.*, **277**, 129.

Collongues, R., "A Martensitic Transformation of the Ferrous Oxide Phase," *Acta Cryst.*, **7**, 213.

Curtis, C. E., Doney, L. M., and Johnson, J. R., "Properties of Hafnium Oxide, Hafnium Silicate, Calcium Hafnate and Hafnium Carbide," *J. Am. Ceram. Soc.*, **37**, 458.

Decker, B. F., and Kasper, J. S., "The Crystal Structure of Titanium Boride," *Acta Cryst.*, **7**, 77.

D'Eye, R. W. M., and Sellman, P. G., "Thorium–Tellurium System," *J. Chem. Soc.*, **1954**, 3760.

Ehrlich, P., and Gentsch, L., "Structure of Calcium Monochloride," *Naturwiss.*, **41**, 211.

Ericsson, G., "Synthetic Sodium Chloride Crystals; Refractive Index and Lattice Constant," *Arkiv Fysik*, **7**, 415.

Ferro, R., "Several Selenium and Tellurium Compounds of Uranium," *Z. Anorg. Allgem. Chem.*, **275**, 320.

Glover, R. E., III, "Lattice Extension of Potassium Chloride Monocrystals between 20° and 600°," *Z. Physik*, **138**, 222.

Hägg, G., and Schönberg, N., "X-Ray Studies of Sulfides of Titanium, Zirconium, Columbium and Tantalum," *Arkiv Kemi*, **7**, 371.

Hoshino, S., and Miyake, S., "Crystal Structure of α-Silver Iodide," *Sci. Ind. Phot.*, **25**, 154.

Juza, R., and Laurer, P., "Lithium Hydrogen Sulfide," *Z. Anorg. Allgem. Chem.*, **275**, 79.

Lukaszewicz, K., and Trzebiatowski, W., "Crystal Structure of TiAs," *Bull. Acad. Polon. Sci. Classe III*, **2**, 277.

Mitchell, R. S., "A Group of SiC Structures," *J. Chem. Phys.*, **22**, 1977.

Pfau, H., and Rix, W., "Crystal Form of Tungsten Carbide, and Carbon Distribution within the Lattice," *Z. Metallk.*, **45**, 116.

Schönberg, N., "An X-Ray Study of the Tantalum–Nitrogen System," *Acta Chem. Scand.*, **8**, 199.

Schönberg, N., "The Molybdenum–Nitrogen and the Tungsten–Nitrogen Systems," *Acta Chem. Scand.*, **8**, 204.

Schönberg, N., "Some Features of the Niobium–Nitrogen and Niobium–Nitrogen–Oxygen Systems," *Acta Chem. Scand.*, **8**, 208.

Schönberg, N., "An X-Ray Investigation of Transition Metal Phosphides," *Acta Chem. Scand.*, **8**, 226.

Schönberg, N., "An X-Ray Investigation of the Tantalum–Oxygen System," *Acta Chem. Scand.*, **8**, 240.

Schönberg, N., "Composition of the Phases in the Vanadium–Carbon System," *Acta Chem. Scand.*, **8**, 624.

Schönberg, N., "The Tungsten Carbide and Nickel Arsenide Structures," *Acta Met.*, **2**, 427.

Schubert, K., Dörre, E., and Günzel, E., "Crystal Chemical Phenomena on Phases of B Elements," *Naturwiss.*, **41**, 448.

Semiletov, S. A., "Electron Diffraction Investigation of the Structure of Sublimed Films of the Composition Bi–Se and Bi–Te," *Tr. Inst. Kristallogr., Akad. Nauk SSSR*, **10**, 76.

Tobelko, K. I., Zvonkova, Z. V., and Zhdanov, G. S., "Structure of Realgar and the Atomic Radius of Arsenic," *Dokl. Akad. Nauk SSSR*, **96**, 749.

Trzebiatowski, W., and Lukaszewicz, K., "Structure of Titanium Arsenide," *Roczniki Chem.*, **28**, 150.

Zorll, U., "Lattice Constants of Tellurium, Mercury Telluride and Mercury Selenide," *Z. Physik*, **138**, 167.

1955

Asprey, L. B., Ellinger, F. H., Fried, S., and Zachariasen, W. H., "Evidence for Quadrivalent Curium; X-Ray Data on Curium Oxides," *J. Am. Chem. Soc.*, **77**, 1707.

Bachmayer, K., and Nowotny, H., "Investigation in the System Vanadium–Arsenic," *Monatsh.*, **86**, 741.

Bachmayer, K., Nowotny, H., and Kohl, A., "The Structure of TiAs," *Monatsh.*, **86**, 39.

Bøhm, F., Grønvold, F., Haraldsen, H., and Prydz, H., "X-Ray and Magnetic Study of the System Cobalt–Selenium," *Acta Chem. Scand.*, **9**, 1510.

Ferro, R., "The Crystal Structure of Thorium Arsenides," *Acta Cryst.*, **8**, 360.

Ferro, R., "Thorium Monotelluride and Thorium Oxytelluride," *Atti Accad. Lincei, Rend., Classe Sci. Fis. Mat. Nat.*, **18**, 641.

Gasilova, E. B., "New Structures of SiC. A System for the Classification of the SiC Structures," *Dokl. Akad. Nauk SSSR,* **101**, 671.

Geller, S., "The Rhodium–Germanium System. I. The Crystal Structures of Rh_2Ge, Rh_5Ge_3 and RhGe," *Acta Cryst.*, **8**, 15.

Geller, S., "The Crystal Structures of RhTe and $RhTe_2$," *J. Am. Chem. Soc.*, **77**, 2641.

Geller, S., and Schawlow, A. L., "Crystal Structure and Quadripole Coupling of Cyanogen Bromide, BrCN," *J. Chem. Phys.*, **23**, 779.

Geller, S., and Thurmond, C. D., "On the Question of the Existence of a Crystalline SiO," *J. Am. Chem. Soc.*, **77**, 5285.

Goedkoop, J. A., and Andersen, A. F., "The Crystal Structure of Copper Hydride," *Acta Cryst.*, **8**, 118.

Goryunova, N. A., Kotovich, V. A., and Frank-Kamenetskii, V. A., "X-Ray Investigation of the Isomorphism of Gallium and Zinc Compounds," *Dokl. Akad. Nauk SSSR,* **103**, 659.

Goryunova, N. A., Kotovich, V. A., and Frank-Kamenetskii, V. A., "Mixed Crystals of Hexagonal Cadmium Selenide with ZnSe, InAs and In_2Se_3," *Zh. Tekh. Fiz.*, **25**, 2419.

Groenefeld Meijer, W. O. J., "Synthesis, Structures and Properties of Platinum Metal Tellurides," *Am. Mineralogist,* **40**, 646.

Hahn, H., and Frank, G., "Crystal Structure of GaS," *Z. Anorg. Allgem. Chem.*, **278**, 340.

Iandelli, A., "Monochalcogenides of Lanthanum, Cerium, Praseodymium and Neodymium," *Gazz. Chim. Ital.*, **85**, 881.

Jeffrey, G. A., and Parry, G. S., "Crystal Structure of Aluminum Nitride," *J. Chem. Phys.*, **23**, 406.

Jones, R. E., and Templeton, D. H., "The Crystal Structure of Indium(I) Iodide," *Acta Cryst.*, **8**, 847.

Juza, R., Weber, H. H., and Opp, K., "Crystal Structure of Sodium Amide," *Naturwiss.*, **42**, 125.

Meyers, E. A., and Lipscomb, W. N., "The Crystal Structure of Hydroxylamine," *Acta Cryst.*, **8**, 583.

Okazaki, A., and Hirakawa, K., "The X-Ray Study of $FeSe_x$. I," *Busseiron Kenkyu,* No. 87, p. 36.

Okazaki, A., and Hirakawa, K., "The X-Ray Study of $FeSe_x$. II," *Busseiron Kenkyu,* No. 90, p. 59.

Pério, P., "Crystallographic Study of the Uranium–Oxygen System," Comm. Energie At. (France), Rappt., No. 363, 107 pp.

Scatturin, V., Bellon, P. L., and Frasson, E., "The Crystal Structure of Phosphonium Bromide, PH_4Br," *Atti Ist. Veneto Sci. Lettere Arti, Classe Sci. Mat. Nat.*, **114**, 67 (1955–56).

Schubert, K., Dörre, E., and Kluge, M., "Crystal Chemistry of the B-Group Metals. III. Crystal Structure of GaSe and InTe," *Z. Metallk.*, **46**, 216.

Semiletov, S. A., "Electron Diffraction Investigation of the System Cadmium–Tellurium," *Tr. Inst. Kristallogr., Akad. Nauk SSSR*, **1955**, No. 11, 121.

Shimomura, Y., "Two Modifications of Nickel Oxide. I. The Determination of the Crystal Structure," *Bull. Naniwa Univ.*, **A3**, 175.

Short, M. A., and Steward, E. G., "X-Ray Powder Diffraction Data for Hexagonal Zinc Sulfide," *Acta Cryst.*, **8**, 733.

Smakula, A., and Kalnajs, J., "Precision Determination of Lattice Constants with a Geiger-Counter X-Ray Diffractometer," *Phys. Rev.*, **99**, 1737.

Smith, F. G., "Lattice Dimensions of Cadmium Sulfide," *Am. Mineralogist*, **40**, 696.

Thewlis, J., "Unit Cell Dimensions of Lithium Fluoride made from Li^6 and Li^7," *Acta Cryst.*, **8**, 36.

Traill, R. J., and Boyle, R. W., "Hawleyite, Isometric Cadmium Sulfide, a New Mineral," *Am. Mineralogist*, **40**, 555.

1956

Aurivillius, K., "The Crystal Structure of Mercury(II) Oxide," *Acta Cryst.*, **9**, 685.

Bertaut, E.-F., "Structure of Stoichiometric FeS," *Bull. Soc. Franc. Mineral. Crist.*, **79**, 276.

Binnie, W. P., "The Structural Crystallography of Indium Bismuthide," *Acta Cryst.*, **9**, 686.

Boswijk, K. H., Heide, J. van der, Vos, A., and Wiebenga, E. H., "The Crystal Structure of α-Iodine Chloride," *Acta Cryst.*, **9**, 274.

Corliss, L. M., Elliott, N., and Hastings, J. M., "Magnetic Structures of the Polymorphic Forms of Manganous Sulfide," *Phys. Rev.*, **104**, 924.

Eick, H. A., Baenziger, N. C., and Eyring, L., "Lower Oxides of Samarium and Europium. The Preparation and Crystal Structure of $SmO_{0.4-4.6}$, SmO and EuO," *J. Am. Chem. Soc.*, **78**, 5147.

Eick, H. A., Baenziger, N. C., and Eyring, L., "The Preparation, Crystal Structure and Some Properties of SmN, EuN and YbN," *J. Am. Chem. Soc.*, **78**, 5987.

Ferro, R., "The Crystal Structures of Thorium Antimonides," *Acta Cryst.*, **9**, 817.

Foster, P. K., and Welch, A. J. E., "Metal Oxide Solid Solutions. I. Lattice Constant and Phase Relations in Ferrous Oxide (Wüstite) and in Solid Solutions of Ferrous Oxide and Manganous Oxide," *Trans. Faraday Soc.*, **52**, 1626.

Gasilova, E. B., "Studies of SiC Crystals by the Polychromatic X-Ray Method," *Tr. Inst. Kristallogr., Akad. Nauk SSSR*, **1956**, No. 12, 41.

Grdenić, D., and Djordjević, C., "The Mercury–Mercury Bond Length in the Mercurous Ion. II. The Crystal Structure of Mercurous Fluoride," *J. Chem. Soc.*, **1956**, 1316.

Grønvold, F., and Jacobsen, E., "X-Ray and Magnetic Study of Nickel Selenides in the Range NiSe and $NiSe_2$," *Acta Chem. Scand.*, **10**, 1440.

Grønvold, F., and Røst, E., "On the Sulfides, Selenides and Tellurides of Palladium," *Acta Chem. Scand.*, **10**, 1620.

Güntert, O. J., and Faessler, A., "Lattice Constants of the Alkaline Earth Sulfides MgS, CaS, SrS and BaS," *Z. Krist.*, **107**, 357.

Hahn, H., and Harder, B., "The Crystal Structures of the Titanium Sulfides," *Z. Anorg. Allgem. Chem.*, **288**, 241.

Heiart, R. B., and Carpenter, G. B., "Crystal Structure of Cyanogen Chloride," *Acta Cryst.*, **9**, 889.

Iandelli, A., "Some Samarium Compounds of NaCl-Type," *Z. Anorg. Allgem. Chem.*, **288,** 81.

Jeffrey, G. A., Parry, G. S., and Mozzi, R. L., "Study of the Wurtzite-Type Binary Compounds. I. Structures of Aluminum Nitride and Beryllium Oxide," *J. Chem. Phys.*, **25,** 1024.

Juza, R., Rabenau, A., and Pascher, G., "Solid Solutions in the Systems ZnS/MnS, ZnSe/MnSe, ZnTe/MnTe," *Z. Anorg. Allgem. Chem.*, **285,** 61.

Juza, R., Weber, H. H., and Opp, K., "Crystal Structure of Sodium Amide," *Z. Anorg. Allgem. Chem.*, **284,** 73.

Klemm, W., and Winkelmann, G., "Nitrides of the Rare Earth Metals," *Z. Anorg. Allgem. Chem.*, **288,** 87.

Laffitte, M., and Bénard, J., "Limits of the Stability of Hexagonal Nickel Sulfide," *Compt. Rend.*, **242,** 518.

Lorenz, M. R., and Prener, J. S., "Preliminary Study of a High-Temperature Phase in Cuprous Chloride," *Acta Cryst.*, **9,** 538.

Okazaki, A., and Hirakawa, K., "Structural Study of Iron Selenides, $FeSe_x$. I. Ordered Arrangements of Defects of Fe Atoms," *J. Phys. Soc. Japan*, **11,** 930.

Okazaki, A., and Ueda, I., "The Crystal Structure of Stannous Selenide, SnSe," *J. Phys. Soc. Japan*, **11,** 470; *Mem. Fac. Sci. Kyushu Univ., Ser. B*, **2,** 46.

Roth, W. L., "The Structure of Mercuric Oxide," *Acta Cryst.*, **9,** 277.

Scatturin, V., and Frasson, E., "Thallium Sulfides from Ammonium Polysulfide: Formation and Structure of the Thallium Monosulfide, TlS," *Ric. Sci.*, **26,** 3382.

Stasova, M. M., and Vainshtein, B. K., "More Precise Determination of the Structure of the NH_4 Group in Ammonium Chloride," *Tr. Inst. Kristallogr., Akad. Nauk SSSR*, **1956,** No. 12, 18.

Tatarinova, L. I., Auleitner, Y. K., and Pinsker, Z. G., "An Electron Diffraction Study of GaSe," *Kristallografiya*, **1,** 537.

Trzebiatowski, W., Weglowski, S., and Lukaszewicz, K., "Crystal Structure of Zirconium Arsenides," *Roczniki Chem.*, **30,** 353.

Tucker, C. W., Jr., and Sénio, P., "Rapid Lattice-Parameter Measurements of Moderate Precision on Single Crystals," *U.S. At. Energy Comm. Rept.* AECU-3428, 5 pp.

Weiss, A., and Weiss, A., "Gold(I) Iodide," *Z. Naturforsch.*, **11b,** 604.

Zalkin, A., and Templeton, D. H., "The Crystal Structure of Sodium Amide," *J. Phys. Chem.*, **60,** 821.

1957

Andersson, G., and Magnéli, A., "Crystal Structure of Niobium Monoxide," *Acta Chem. Scand.*, **11,** 1065.

Aurivillius, K., and Carlsson, I. B., "The Crystal Structure of Hexagonal HgO and Hg_2O_2NaI," *Acta Chem. Scand.*, **11,** 1069.

Černý, P., "Second Occurrence of Hawleyite—β-CdS," *Casopis Mineral. Geol.*, **2,** 13.

Eliseev, E. N., and Denisov, A. P., "X-Ray Studies of Pyrrhotite," *Vestn. Leningr. Univ.*, **12,** No. 18, *Ser. Geol. i Geogr.* No. 3, 68.

Föppl, H., "Crystal Structures of the Alkali Peroxides," *Z. Anorg. Allgem. Chem.*, **291,** 12.

Goldsmith, J. R., and Graf, D. L., "The System CaO–MnO–CO_2; Solid-Solution and Decomposition Relations," *Geochim. Cosmochim. Acta*, **11,** 310.

Gorum, A. E., "Crystal Structures of PuAs, PuTe, PuP and PuOSe," *Acta Cryst.*, **10,** 144.

Hahn, H., Harder, B., Mutschke, U., and Ness, P., "The Crystal Structures of Some Compounds and Phases of the Zirconium–Sulfur System," *Z. Anorg. Allgem. Chem.*, **292**, 82.

Hahn, H., and Ness, P., "The Structure of the Phases Appearing in the System Zirconium–Tellurium," *Naturwiss.*, **44**, 534.

Heyding, R. D., and Calvert, L. D., "Arsenides of Transition Metals; The Arsenides of Iron and Cobalt," *Can. J. Chem.*, **35**, 449.

Hoshino, S., "Crystal Structure and Phase Transition of Some Metallic Halides. IV. On the Anomalous Structure of α-AgI," *J. Phys. Soc. Japan*, **12**, 315.

Hovi, V., "X-Ray Studies on CsCl, CsBr and CsCl–CsBr Solid Solutions," *Ann. Univ. Turku., Ser. A I*, **26**, 8 pp.

Jellinek, F., "The Structures of the Chromium Sulfides," *Acta Cryst.*, **10**, 620; *Congr. Intern. Chim. Pure Appl. 16e, Paris 1957, Mem. Sect. Chim. Minerale* (pub. 1958), p. 187.

Kempter, C. P., Krikorian, N. H., and McGuire, J. C., "The Crystal Structure of Yttrium Nitride," *J. Phys. Chem.*, **61**, 1237.

Pinsker, Z. G., and Kaverin, S. V., "Electron Diffraction Study of Nitrides and Carbides of Transition Metals," *Kristallografiya*, **2**, 386.

Pinsker, Z. G., Kaverin, S. V., and Troitskaya, N. V., "Electron Diffraction Study of Molybdenum Nitrides," *Kristallografiya*, **2**, 179.

Popper, P., and Ingles, T. A., "Boron Phosphide, a III–V Compound of Zinc-Blende Structure," *Nature*, **179**, 1075.

Reimer, L., "Structure of PbTe Evaporated Films," *Naturwiss.*, **44**, 416.

Rundqvist, S., "Crystal Structure of Boron Phosphide, BP," *Congr. Intern. Chim. Pure Appl. 16e, Paris 1957, Mem. Sect. Chim. Minerale* (pub. 1958), p. 539.

Sal'dau, E. P., "Changes in the Lattice Dimensions of Iozite by Oxidation to Magnetite and Maghemite," *Zap. Vses. Mineralog. Obshchestva*, **86**, 324.

Samsonov, G. V., "Intermediate Stages in the Formation of Carbides of Titanium, Zirconium, Vanadium, Niobium and Tantalum," *Ukr. Khim. Zh.*, **23**, 287.

Scatturin, V., Bellon, P., and Zannetti, R., "Crystal Structure of Silver Oxide AgO," *Ric. Sci.*, **27**, 2163.

Schneider, A., and Imhagen, K. H., "Temperature Dependence of the Lattice Constants in the Mixed Crystal Series NiTe–$NiTe_2$," *Naturwiss.*, **44**, 324.

Schubert, K., Breiner, H., Burkhardt, W., Günzel, E., Haufler, R., Lukas, H. L., Vetter, H., Wegst, J., and Wilkens, M., "Structural Data on Metallic Phases. II.," *Naturwiss.*, **44**, 229.

Semiletov, S. A., and Rozsibal, M., "Electron Diffraction Study of InSb Films," *Kristallografiya*, **2**, 287.

Tallman, R., Margrave, J. L., and Bailey, S. W., "Crystal Structure of Sodium Peroxide," *J. Am. Chem. Soc.*, **79**, 2979.

Teatum, E. T., and Smith, N. O., "Lattice Constants of Potassium Bromide–Potassium Iodide Solid Solutions," *J. Phys. Chem.*, **61**, 697.

Ventriglia, U., "Structural Investigations on Cobalt Arsenides," *Periodico Mineral. (Rome)*, **26**, 345.

Vereshchagin, L. F., and Kabalkina, S. S., "The Crystal Structure of Rubidium Halides at High Pressure," *Dokl. Akad. Nauk SSSR*, **113**, 797.

Wentorf, R. H., Jr., "The Cubic Form of Boron Nitride," *J. Chem. Phys.*, **26**, 956; *Congr. Intern. Chim. Pure Appl. 16e, Paris 1957, Mem. Sect. Chim. Minerale* (pub. 1958), p. 535.

Worsham, J. E., Jr., Wilkinson, M. K., and Shull, C. G., "Neutron Diffraction Observations on the Palladium–Hydrogen and Palladium–Deuterium Systems," *J. Phys. Chem. Solids*, **3**, 303.

Yuzuri, M., Watanabe, H., Nagasaki, S., and Maeda, S., "Magnetic Properties of Chromium Sulfides," *J. Phys. Soc. Japan*, **12**, 385.

1958

Alyamovskii, S. I., Shveikin, G. P., and Gel'd, P. V., "Lower Oxides of Niobium," *Zh. Neorgan. Khim.*, **3**, 2437.

Archard, J. C., and Tsoucaris, G., "Radiocrystallographic Study of YbO," *Compt. Rend.*, **246**, 285.

Aurivillius, K., and Carlsson, I. B., "Structure of Hexagonal Mercury(II) Oxide," *Acta Chem. Scand.*, **12**, 1297.

Baker, T. W., "The Coefficient of Thermal Expansion of Zirconium Nitride," *Acta Cryst.*, **11**, 300.

Ball, J. G., Greenfield, P., Mardon, P. G., and Robertson, J. A. L., "Crystal Structure of Plutonium, δ and ϵ Phases," At. Energy Res. Estab. (Gt. Brit.), M/R 2416, 11 pp.

Djurle, S., "An X-Ray Study on the System Cu–S," *Acta Chem. Scand.*, **12**, 1415.

Domange, L., Flahaut, J., Guittard, M., and Loriers, J., "Ytterbium Sulfides," *Compt. Rend.*, **247**, 1614.

Elliott, R. O., and Kempter, C. P., "Thermal Expansion of Some Transition Metal Carbides," *J. Phys. Chem.*, **62**, 630.

Feltynowski, A., Glass, I., and Grelewicz, L., "Fine Structure of Photoconducting PbTe Films," *Exptl. Tech. Physik*, **6**, 17.

Giesecke, G., and Pfister, H., "Determination of Precise Lattice Constants of $A^{III}B^{V}$-Compounds," *Acta Cryst.*, **11**, 369.

Grazhdankina, N. P., and Gurfel, D. I., "X-Ray Investigation of the Thermal Expansion of the Antiferromagnetic Compound MnTe," *Zh. Eksperim. i Teoret. Fiz.*, **35**, 907.

Grønvold, F., Hagberg, O., and Haraldsen, H., "Vanadium Tellurides," *Acta Chem. Scand.*, **12**, 971.

Hérold, A., Marzluf, B., and Pério, P., "Preparation and Structure of Boron Nitride," *Compt. Rend.*, **246**, 1866.

Jacobson, R. A., and Lipscomb, W. N., "The B_8Cl_8 Structure: A New Boron Polyhedron in Small Molecules," *J. Am. Chem. Soc.*, **80**, 5571.

Khitrova, V. I., and Pinsker, Z. G., "Electron Diffraction Investigation of Tungsten Nitride," *Kristallografiya*, **3**, 545.

Kuz'min, R. H., "X-Ray Diffraction Study of the Structure of IrSb," *Kristallografiya*, **3**, 366.

Laube, E., and Nowotny, H., "The System: UC–ThC," *Monatsh.*, **89**, 312.

McTaggart, F. K., and Wadsley, A. D., "The Sulfides, Selenides and Tellurides of Titanium, Zirconium, Hafnium and Thorium," *Australian J. Chem.*, **11**, 445.

Mitchell, R. S., Barakat, N., and El Shazly, E. M., "A Study of a Silicon-Carbide Crystal Containing a New Polytype, 27*H*," *Z. Krist.*, **111**, 63.

Okazaki, A., "The Crystal Structure of Germanium Selenide GeSe," *J. Phys. Soc. Japan*, **13**, 1151.

Perri, J. A., LaPlaca, S., and Post, B., "New Group III–Group V Compounds: BP and BAs," *Acta Cryst.*, **11**, 310.

Riano, E., and Amorós Portolés, J. L., "Thermal Expansion of NaCl-Type Compounds. I. Thermal Expansion of Galena between −150° and 150°," *Bol. Real. Soc. Espan. Hist. Nat., Secc. Geol.*, **56**, 345.

Riano, E., and Amorós Portolés, J. L., "Thermal Expansion of NaCl-Type Compounds. II. Thermal Expansion of MgO between −150° and 200°," *Bol. Real. Soc. Espan. Hist. Nat., Secc. Geol.*, **56**, 391.

Rundle, R. E., Baenziger, N. C., and Wilson, A. S., "X-Ray Study of the Uranium–Oxygen System," U.S. At. Energy Comm. Rept. TID-5290, Book 1, 131.

Rundle, R. E., Wilson, A. S., Baenziger, N. C., and Tevebaugh, A. D., "X-Ray Analysis of the Uranium–Carbon System," U.S. At. Energy Comm. Rept. TID-5290, Book 1, 67.

Scatturin, V., Bellon, P. L., and Zannetti, R., "Planar Coordination of the Group IB Elements: Crystal Structure of Ag(II) Oxide," *J. Inorg. Nuclear Chem.*, **8**, 462.

Semiletov, S. A., "Electron Diffraction Study of the Structure of InSe," *Kristallografiya*, **3**, 288.

Stehlík, B., and Weidenthaler, P., "The Crystal Structure of Silver Peroxide," *Chem. Listy*, **52**, 402.

Trzebiatowski, W., Weglowski, S., and Lukaszewicz, K., "The Crystal Structure of Zirconium Arsenides, ZrAs and $ZrAs_2$," *Roczniki Chem.*, **32**, 189.

Woolley, J. C., and Smith, B. A., "Solid Solution in Zinc Blende Type $A_2^{III}B_3^{VI}$ Compounds," *Proc. Phys. Soc. (London)*, **72**, 867.

1959

Adamsky, R. F., and Merz, K. M., "Synthesis and Crystallography of the Wurtzite Form of Silicon Carbide," *Z. Krist.*, **111**, 350.

Andersson, S., "Crystal Structure of So-Called δ-Titanium Oxide and its Structural Relation to the ω-Phase of Some Binary Alloy Systems of Titanium," *Acta Chem. Scand.*, **13**, 415.

Anselmo, V. C., and Smith, N. O., "Lattice Constants of Ammonium Chloride–Ammonium Bromide Solid Solutions," *J. Phys. Chem.*, **63**, 1344.

Aronsson, B., Aselius, J., and Stenberg, E., "Borides and Silicides of the Platinum Metals," *Nature*, **183**, 1318.

Aronsson, B., Stenberg, E., and Aselius, J., "Borides of Rhenium and the Platinum Metals. The Crystal Structures of Re_7B_3, ReB_3, Rh_7B_3, $RhB_{\sim 1.1}$, $IrB_{\sim 1.1}$ and PtB," U.S. Dept. Comm., Office Tech. Serv. PB Rept. 145, 866, 9 pp., *Acta Chem. Scand.*, **14**, 733 (1960).

Atoji, M., and Lipscomb, W. N., "Molecular and Crystal Structure of B_8Cl_8. I. Preliminary X-Ray Diffraction Study," *J. Chem. Phys.*, **31**, 601.

Austin, A. E., "Carbon Positions in Uranium Carbides," *Acta Cryst.*, **12**, 159.

Bogatskii, D. P., and Mineeva, I. A., "X-Ray Study of Nickel Oxides and Their Solid Solutions," *Fiz. Tverd. Tela, Akad. Nauk SSSR, Sb. Statei*, **2**, 301.

Cromer, D. T., "The Crystal Structure of LiAs," *Acta Cryst.*, **12**, 36.

Cromer, D. T., "The Crystal Structure of NaSb," *Acta Cryst.*, **12**, 41.

Dachs, H., "Determination of the Position of Hydrogen in LiOH by Neutron Diffraction," *Z. Krist.*, **112**, 60.

Domange, L., Flahaut, J., and Guittard, M., "The Sulfides and Oxysulfide of Europium," *Compt. Rend.*, **249**, 697.

Evans, H. T., Jr., and McKnight, E. T., "New Wurtzite Polytypes from Joplin, Mo.," *Am. Mineralogist,* **44,** 1210.

Goryunova, N. A., and Fedorova, N. N., "Solid Solution in the ZnSe–GaAs System," *Soviet Phys. Solid State,* **1,** 307; *Fiz. Tverd. Tela,* **1,** 344.

Guittard, M., and Benacerraf, A., "The Selenides, MSe, of the Lanthanides from Lanthanum to Gadolinium," *Compt. Rend.,* **248,** 2589.

Hahn, H., and Ness, P., "On the System Zirconium–Tellurium," *Z. Anorg. Allgem. Chem.,* **302,** 136.

Jacobson, R. A., and Lipscomb, W. N., "Molecular and Crystal Structure of B_8Cl_8. II. Solution of the Three-Dimensional Structure," *J. Chem. Phys.,* **31,** 605.

Jagodzinski, H., "The Crystal Structure of AuI," *Z. Krist.,* **112,** 80.

Juza, R., and Mehne, A., "Metal Amides and Nitrides. XXXVI. Crystal Structure of the Alkali Metal Amides," *Z. Anorg. Allgem. Chem.,* **299,** 33.

Kuwabara, S., "Accurate Determination of H Positions in NH_4Cl by Electron Diffraction," *J. Phys. Soc. Japan,* **14,** 1205.

Laffitte, M., "Crystal Chemistry of Nickel Monosulfide," *Bull. Soc. Chim. France,* **1959,** 1211.

Laffitte, M., "Crystal Structure and Thermodynamic Properties of Hexagonal Nickel Sulfide," *Rev. Nickel,* **25,** 79, 109.

Magnéli, A., Andersson, S., Kihlborg, L., Åsbrink, S., Westman, S., Holmberg, B., and Nordmark, C., "The Crystal Chemistry of Titanium, Vanadium and Molybdenum Oxides at Elevated Temperatures," U. S. At. Energy Comm. NP-8054, 141 pp.

Merz, K. M., and Adamsky, R. F., "Synthesis of the Wurtzite Form of Silicon Carbide," *J. Am. Chem. Soc.,* **81,** 250.

Meyerhoff, K., "Measurements of Precise Lattice Constants of TlCl by Electron Diffraction," *Acta Cryst.,* **12,** 330.

Nowotny, H., Kieffer, R., Benesovsky, F., and Brukl, C., "Partial Systems Titanium Carbide–Hafnium Carbide and Zirconium Carbide–Hafnium Carbide," *Monatsh.,* **90,** 86.

Rundqvist, S., "X-Ray Investigation of the Nickel–Boron System. Crystal Structures of Orthorhombic and Monoclinic Ni_4B_3," *Acta Chem. Scand.,* **13,** 1193.

Stehlík, B., and Weidenthaler, P., "Crystal Structure of Silver(II) Oxide," *Collection Czech. Chem. Commun.,* **24,** 1416.

Storms, E. K., and Krikorian, N. H., "The Variations of Lattice Parameter with Carbon Content of Niobium Carbide," *J. Phys. Chem.,* **63,** 1747.

Tomita, T., "Structure Analysis of Silicon Carbide of 174 Layers," *Sci. Rept. Saitama Univ. Ser. A,* **3,** No. 2, 115.

Ust'yanov, V. I., and Khaikina, R. M., "Electron Diffraction Investigation of Thin Layers of Aluminum Antimonide," *Fiz. Tverd. Tela, Akad. Nauk SSSR Sb. Statei,* **1,** 144.

Vogel, R. E., and Kempter, C. P., "Mathematical Technique for the Precision Determination of Lattice Constants," U.S. At. Energy Comm. LA-2317, 30 pp.

Vol'f, E., Tolkachev, S. S., and Kozhina, I. I., "X-Ray Study of the Lower Oxides of Titanium and Vanadium," *Vestn. Leningr. Univ.,* **14,** No. 10, *Ser. Fiz. Khim.,* No. 2, 87.

Zhuravlev, N. N., Stepanova, A. A., and Zyuzin, N. I., "Superconductivity of BiPt," *Zh. Eksperim. i Teor. Fiz.,* **37,** 880.

1960

Addamiano, A., "X-Ray Data for the Phosphides of Aluminum, Gallium and Indium," *Acta Cryst.*, **13,** 505.

Brixner, L. H., "Structure and Electric Properties of Some New Rare Earth Arsenides, Antimonides and Tellurides," *J. Inorg. Nuclear Chem.*, **15,** 199.

Chistyakov, Y. D., and Cruceaunu, E., "Preparation and Study of Certain Properties of Zinc Selenide," *Acad. Rep. Populare Romine, Studii Cercetari Met.*, **5,** 517.

Folberth, O. G., and Pfister, H., "The Crystal Structure of $ZnSnAs_2$," *Acta Cryst.*, **13,** 199.

Gel'd, P. V., Alyamovskiĭ, S. I., and Matveenko, I. I., "Structural Peculiarities of Vanadium Oxide," *Fiz. Metal. i Metalloved., Akad. Nauk SSSR*, **9,** 315.

Hiller, J. E., and Wegener, W., "The System Nickel–Selenium," *Neues Jahrb. Mineral., Abhandl.*, **94,** 1147.

Hirone, T., and Chiba, S., "Magnetic Anisotropy of Single-Crystal CrTe," *J. Phys. Soc. Japan*, **15,** 1991.

Ibers, J. A., Kumamoto, J., and Snyder, R. G., "Structure of Potassium Hydroxide: An X-Ray and Infrared Study," *J. Chem. Phys.*, **33,** 1164.

Kannewurf, C. R., Kelly, A., and Cashman, R. J., "Comparison of Three Structure Determinations for Germanium Selenide, GeSe," *Acta Cryst.*, **13,** 449.

Kempter, C. P., and Nadler, M. R., "Preparation and Crystal Structures of RuC and OsC," *J. Chem. Phys.*, **33,** 1580.

Kempter, C. P., Storms, E. K., and Fries, R. J., "Lattice Dimensions of NbC as a Function of Stoichiometry," *J. Chem. Phys.*, **33,** 1873.

Kudielka, H., and Rohde, H., "Structural Investigations of Carbosulfides of Titanium and Zirconium," *Z. Krist.*, **114,** 447.

Lawson, W. D., Nielsen, S., and Young, A. S., "Preparation and Properties of HgTe and Mixed Crystals of HgTe–CdTe," *Solid State Phys. Electron. Telecommun. Proc. Intern. Conf. Brussels 1958*, **2,** Pt. 2, 830 (pub. 1960).

McDonald, B. J., and Stuart, W. I., "The Crystal Structure of Some Plutonium Borides," *Acta Cryst.*, **13,** 447.

McWhan, D. B., Wallmann, J. C., Cunningham, B. B., Asprey, L. B., Ellinger, F. H., and Zachariasen, W. H., "Preparation and Crystal Structure of Americium Metal," *J. Inorg. Nuclear Chem.*, **15,** 185.

Miller, E., Komarek, K., and Cadoff, I., "Preparation and Properties of Barium, Barium Telluride and Barium Selenide," *Trans. AIME*, **218,** 978.

Nishiyama, A., and Okada, T., "Crystal Structure of the Compounds $(SnSe)_{1-x-y}$-$(SnTe)_x(PbTe)_y$," *Mem. Fac. Sci. Kyushu Univer. Ser. B*, **3,** 3.

Pashinkin, A. S., Tishchenko, G. N., Korneeva, I. V., and Ryzhenko, B. N., "Polymorphism of Some Selenides and Tellurides of Zinc and Cadmium," *Kristallografiya*, **5,** 261.

Roth, W. L., "Defects in the Crystal and Magnetic Structures of Ferrous Oxide," *Acta Cryst.*, **13,** 140.

Rundqvist, S., "Phosphides of the Platinum Metals," *Nature*, **185,** 31.

Sándor, E., and Wooster, W. A., "Diffuse Streaks in the Diffraction Pattern of Vanadium Single Crystals," *Acta Cryst.*, **13,** 339.

Scatturin, V., Bellon, P. L., and Salkind, A. J., "The Structure of Silver Oxide AgO Determined by Means of Neutron Diffraction," *Ric. Sci.*, **30,** 1034.

Skinner, B. J., and Barton, P. B., Jr., "The Substitution of Oxygen for Sulfur in Wurtzite and Sphalerite," *Am. Mineralogist,* **45,** 612.

Straumanis, M. E., and Li, H. W., "Lattice Constants, Linear Coefficients of Expansion, Densities, Cavity Arrangement and Structure of the Titanium(II) Oxide Phase," *Z. Anorg. Allgem. Chem.,* **305,** 143.

Taylor, C. A., and Underwood, F. A., "A Twinning Interpretation of "Superlattice" Reflections in X-Ray Photographs of Synthetic Klockmannite, CuSe," *Acta Cryst.,* **13,** 361.

Toman, K., "On the Structure of ZnSb," *Phys. Chem. Solids,* **16,** 160.

Tomita, T., "Crystal Structure of Silicon Carbide of 174 Layers," *J. Chem. Soc. Japan,* **15,** 99.

Yuzuri, M., "Magnetic Properties of Cr_2As and Cu_2Sb," *J. Phys. Soc. Japan,* **15,** 2007.

1961

Amma, E. L., and Jeffrey, G. A., "Study of the Wurtzite-Type Compounds. V. Structure of Aluminum Oxycarbide, Al_2CO; A Short-Range Wurtzite-Type Superstructure," *J. Chem. Phys.,* **34,** 252.

Bowman, A. L., "The Variation of Lattice Parameter with Carbon Content of Tantalum Carbide," *J. Phys. Chem.,* **65,** 1596.

Brauer, G., and Esselborn, R., "Nitride Phases of Niobium," *Z. Anorg. Allgem. Chem.,* **309,** 151.

Bruzzone, G., "Structures and Magnetic Properties of MX Compounds from Holmium and Metalloids of the Fifth and Sixth Groups," *Atti. Accad. Lincei, Classe Sci. Fis. Mat. Nat.,* **30,** 208.

Busmann, E., and Lohmeyer, S., "The Behaviour of the Alkali Metals toward Metalloids. VIII. The Crystal Structure of KSb," *Z. Anorg. Allgem. Chem.,* **312,** 53.

Chistyakov, Y. D., and Cruceaunu, E., "Crystal Structure of Zinc Telluride," *Rev. Phys. Acad. Rep. Populaire Roumaine,* **6,** 211.

Elliott, N., and Hastings, J., "Neutron Diffraction Investigation of KCN," *Acta Cryst.,* **14,** 1018.

Flahaut, J., Domange, L., and Patrie, M., "Face-Centered Cubic Solid Solutions formed by Yttrium Sulfide with Cubic Sulfides of the NaCl Type," *Compt. Rend.,* **253,** 1454.

Higashi, I., "Determination of Lattice Constants of UC and UC_2," *Rika Gaku Kenkyusho Hokoku,* **37,** 271.

Kay, M. I., "A Neutron Diffraction Study of Orthorhombic PbO," *Acta Cryst.,* **14,** 80.

Klemm, D. D., "The Solid Solution Series Sphalerite–Stilleite," *Neues Jahrb. Mineral.,* **1961,** 253.

Krebs, H., Grün, K., and Kallen, D., "Structures and Properties of Metalloids. XIV. Mixed Crystal Systems between Semiconducting Chalcogenides of the Fourth Group," *Z. Anorg. Allgem. Chem.,* **312,** 307.

Leciejewicz, J., "Neutron Diffraction Study of Orthorhombic Lead Monoxide," *Acta Cryst.,* **14,** 66.

Leciejewicz, J., "A Note on the Structure of Tungsten Carbide," *Acta Cryst.,* **14,** 200.

Men'kov, A. A., Komissarova, L. N., Simanov, Y. P., and Spitsyn, V. I., "Scandium Chalcogenides," *Dokl. Akad. Nauk SSSR,* **141,** 364.

Olcese, G. L., "Structure and Magnetic Properties of MX Compounds from Terbium and Metalloids of the Groups V and VI," *Atti Accad. Lincei, Rend., Classe Sci. Fis. Mat. Nat.,* **30,** 195.

Scatturin, V., Bellon, P. L., and Salkind, A. J., "Structure of Ag Oxide Determined by Means of Neutron Diffraction," *J. Electrochem. Soc.*, **108**, 819.

Sirota, N. N., and Olekhnovich, N. M., "Distribution and Electron Density in Indium Arsenide," *Dokl. Akad. Nauk SSSR*, **136**, 660.

Toth, L., Nowotny, H., Benesovsky, F., and Rudy, E., "The System Uranium–Boron–Carbon," *Monatsh.*, **92**, 794.

Chapter IV

STRUCTURES OF THE COMPOUNDS RX_2

There is no obvious basis, at once simple and comprehensive, for classifying the structures of the several hundred RX_2 compounds whose atomic arrangements are now known. Nevertheless a majority belong to one or another of a few types, often related to one another by exhibiting similar coordinations, by having large anions whose packing is important in determining the atomic arrangement that prevails, or by being built up of sheets or strings of chemically like atoms. These types will be used to group together for description as many as possible of the analyzed RX_2 crystals.

Compounds that are predominantly ionic and contain rather large cations, such as CaF_2, follow the general principles that apply to the ionic RX crystals discussed in Chapter III: they pack so that the chief contacts are between atoms of opposite sign and so that each atom is surrounded by the maximum number of atoms of opposite sign. Such a rule is, however, inadequate as applied to many crystals with weakly electropositive metallic atoms; in such cases the contacts between chemically like atoms may be as numerous and as important as those between unlike. When dealing with moderately ionic structures such as that of CdI_2, this results in sheetlike arrangements that can be considered as approaches to spherical close-packings of the larger atoms with the smaller atoms distributed in various ways among the voids of these packings. In compounds with only weakly ionic characteristics, the chemically like atoms may be tied together in strings as well as in two-dimensional sheets. Metallic borides provide examples of such chainlike structures. Almost all RX_2 structures are like the foregoing in showing no associations of atoms indicative of definite molecules. There are, however, a few, such as the boron halides, in which molecules are very apparent. These have been grouped together.

Wherever feasible, considerations such as those discussed above have guided the arrangement of the structures to be treated in this chapter. Sometimes, however, it will be necessary to refer to the bibliographic table, or to the index, to find a compound in which one may be interested.

Fluorite-Like Structures

IV,a1. Crystals RX_2 in which R is especially big are likely to have the *fluorite*, CaF_2, arrangement. In this grouping (Fig. IV,1) each R atom is at

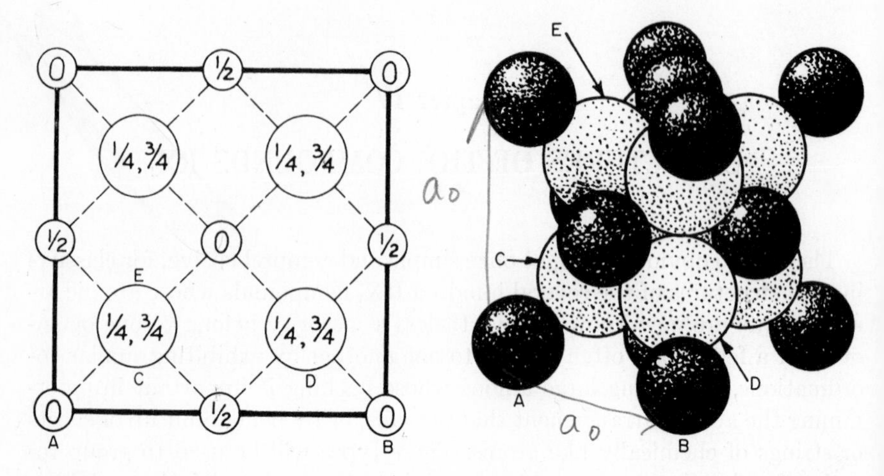

Fig. IV,1a (left). The positions of the atoms within the unit cell of fluorite, CaF_2, projected on a cube face. Lettered circles refer to the corresponding spheres of Figure IV,1b.
Fig. IV,1b (right). A perspective packing drawing showing the distribution of the atoms of CaF_2 within the unit cube. Atoms have been given their expected ionic sizes.

the center of eight X atoms situated at the corners of a surrounding cube; and each X atom has about it a tetrahedron of R atoms. The symmetry is cubic with the atoms of its four molecules per unit in the following positions of O_h^5 (*Fm3m*):

$$\text{R:} \quad (4a) \quad 000; \text{ F.C.}$$

$$\text{X:} \quad (8c) \quad \pm(\tfrac{1}{4}\,\tfrac{1}{4}\,\tfrac{1}{4}); \text{ F.C.}$$

A simple calculation shows that for spherical atoms contact between an R atom and its eight coordinated X atoms would occur only when the radius of R is greater than 0.73 times that of X, that is, when $r(\text{R})/r(\text{X}) \geqq 0.73$. This relation is fulfilled for compounds with the fluorite arrangement.

As will be seen from Table IV,1, the many fluorite-like crystals are of four kinds:

1. Halides, all but two of them fluorides, of the larger divalent cations.

2. Oxides, sulfides, etc. of univalent ions, mostly alkalis. This structure, with atom R the negative rather than the positive ion, is sometimes called the *anti-fluorite* arrangement; its radius ratio is commonly greater than unity.

3. Oxides of large quadrivalent cations.

4. Intermetallic compounds.

TABLE IV,1
Crystals with the Cubic Fluorite Structure (**IV,a1**)

Crystal	a_0, A.
AcH_x	5.670
AcOF	5.943
AgAsMg	6.240
AgAsZn	5.912[a]
AmO_2	5.376
$AuAl_2$	6.00
$AuGa_2$	6.063
$AuIn_2$	6.502
$AuSb_2$	6.656
$BaCl_2$	7.34
BaF_2	6.2001 (25°C.)
Be_2B	4.670
Be_2C	4.33
CaF_2	5.46295 (28°C.)
$CaCdNaYF_8$	5.432[b]
CdF_2	5.3880
CeH_2	5.590[c]
CeO_2	5.4110 (26°C.)[d]
CeOF	5.66–5.73 (5.714)
CmO_2	5.372
CoMnSb	5.900[a]
$CoSi_2$	5.356
CuBiMg	6.256
CuCdSb	6.262
CuMgSb	6.152
CuMnSb	6.066
DyH_2	5.201
ErH_2	5.123
EuF_2	5.796
$GeMg_2$	6.378
GdH_2	5.303
HfO_2	5.115
HgF_2	5.54
HoH_2	5.165
HoOF	5.523
$IrSn_2$	6.338
Ir_2P	5.535

(continued)

TABLE IV,1 (*continued*)

Crystal	a_0, A.
α-$KCeF_4$	5.906[e]
α-$KLaF_4$	5.944[e]
K_2O	6.436
K_2S	7.391
K_2Se	7.676
K_2Te	8.152
α-K_2ThF_6	6.006[e]
α-K_2UF_6	5.946[e]
LaH_2	5.667[f]
LaOF	5.756
LiMgN	4.970[g]
LiZnN	4.877[g]
Li_2NH	5.047
Li_2O	4.619
Li_2S	5.708
Li_2Se	6.005
Li_2Te	6.504
LuH_2	5.033
$NaYF_4$	5.459[b]
NaZnAs	5.912[a]
Na_2O	5.55
Na_2S	6.526
Na_2Se	6.809
Na_2Te	7.314
α-Na_2ThF_6	5.687[e]
α-Na_2UF_6	5.576[e]
NbH_2	4.563
NdH_2	5.470
NdOF	5.595
NiMgBi	6.166[a]
NiMgSb	6.048[a]
$NiSi_2$	5.395
NpO_2	5.4341
PaO_2	5.505
β-PbF_2	5.92732 (18°C.)
$PbMg_2$	6.836
α-PoO_2	5.687 (5.626 at −190°C.)
PrH_2	5.517

(*continued*)

TABLE IV,1 (*continued*)

Crystal	a_0, A.
PrO_2	5.4694
PrOF	5.644
$PtAl_2$	5.910
$PtGa_2$	5.911
$PtIn_2$	6.353
$PtSn_2$	6.425
PuO_2	5.3960
PuOF	5.71
RaF_2	6.368
Rb_2O	6.74
Rb_2S	7.65
Rh_2P	5.505
ScH_2	4.78315 (25°C.)
$SiMg_2$	6.39
SmH_2	5.376
SmOF	5.519
$SnMg_2$	6.765
$SrCl_2$	6.9767 (26°C.)
SrF_2	5.7996 (26°C.)
TbH_2	5.246
TbO_2	5.220
ThO_2	5.5997[h]
$(Th,U,Pb)O_2$	5.539–5.578
TmH_2	5.090
UN_2	5.31
UO_2	5.4682 (26°C.)[i]
YH_2	5.199
β-YOF	5.363
ZrO_2	5.07

[a] In these crystals the last two elements are either statistically distributed through $^1/_4\,^1/_4\,^1/_4$; $^3/_4\,^3/_4\,^3/_4$; F.C., or one is in $^1/_4\,^1/_4\,^1/_4$; F.C., the other in $^3/_4\,^3/_4\,^3/_4$; F.C.

[b] The metal atoms are statistically distributed through 000; F.C.

[c] $CeD_{2.48}$ has this structure with $a_0 = 5.530$ A.

[d] This structure is retained with a reduction of Ce to 20% Ce_2O_3.

[e] Unannealed samples, metal atoms statistically distributed.

[f] LaH_{2+x} exists with an excess of hydrogen; with extra D or H in $^1/_2$ 00; F.C.

[g] The two kinds of metal atoms are thought to be statistically distributed among the fluorine positions of CaF_2.

[h] Hydrated ThO_2 preparations of composition HThO(OH) have been described with $a_0 = 5.63$ A.

[i] UO_2 with excess oxygen is common; for $UO_{2.25}$, $a_0 = 5.4297$ A.

Rare-earth oxyfluorides and several intermetallic compounds of composition RXX′ have a structure like that of fluorite. Presumably their X atoms are in (4*c*) $^1/_4\,^1/_4\,^1/_4$; F.C. and their X′ atoms in (4*d*) $^3/_4\,^3/_4\,^3/_4$; F.C. of T_d^2 ($F\bar{4}3m$); or they may together fill in irregular fashion the positions of (8*c*), above.

Many crystals of more complex chemical composition, listed in later chapters, have their ions of various degrees of complexity distributed as are the atoms in fluorite.

IV,a2. The mineral *baddeleyite*, ZrO_2, has a structure that originally was described as a rather simple distortion of the fluorite arrangement. Its symmetry is monoclinic with a cell of the dimensions:

$$a_0 = 5.1454 \text{ A.};\ b_0 = 5.2075 \text{ A.};\ c_0 = 5.3107 \text{ A.};\ \beta = 99°14'$$

The atomic arrangement has been a subject for some discussion and a recent redetermination conflicts with the earlier results in placing all atoms in general positions of C_{2h}^5 ($P2_1/c$):

$$(4e)\quad \pm(xyz;\ \bar{x},y+{}^1/_2,{}^1/_2-z)$$

with the following parameters:

Atom	x	y	z
Zr	0.2758	0.0404	0.2089
O(1)	0.069	0.342	0.345
O(2)	0.451	0.758	0.479

In this arrangement (Fig. IV,2) the zirconium atoms exhibit a sevenfold coordination with Zr–O between 2.04 and 2.26 A. The shape of the coordination polyhedron is indicated in Figure IV,3. The two kinds of oxygen atoms are differently coordinated, O(1) having three zirconium neighbors at 2.04 to 2.15 A. and O(2) having four zirconiums at 2.16–2.26 A. The shortest O–O is 2.52 A.

The stable form of ZrO_2, synthetically produced as well as naturally occurring as baddeleyite, has this structure, and so does *hafnium dioxide*, HfO_2, for which the cell dimensions are

$$a_0 = 5.1156 \text{ A.};\ b_0 = 5.1722 \text{ A.};\ c_0 = 5.2948 \text{ A.};\ \beta = 99°11'$$

Three other forms of ZrO_2 have been described. One of these, the modification stable above ca. 1400°C., is cubic with the fluorite arrangement (**IV,a1**). This can be preserved by quenching to room temperature only if some impurity such as MgO is present.

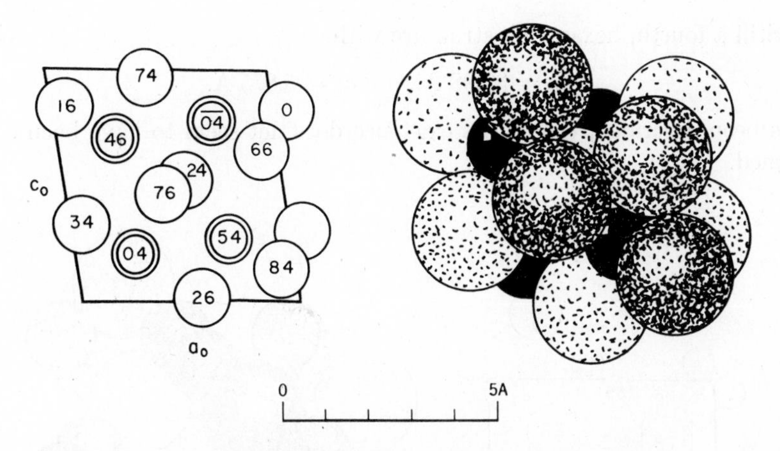

Fig. IV,2a (left). The ZrO_2 arrangement projected along the b_0 axis. The zirconium atoms are doubly ringed. Origin in lower left.

Fig. IV,2b (right). A packing drawing of the ZrO_2 arrangement viewed along the b_0 axis. The small black circles are zirconium.

A tetragonal form containing four molecules in a prism of the dimensions:

$$a_0 = 5.07 \text{ A.}, \qquad c_0 = 5.16 \text{ A.}$$

has been obtained by decomposing zirconium salts below about 600°C.; this too is said to be a distorted fluorite grouping, though a thorough study has not been made.

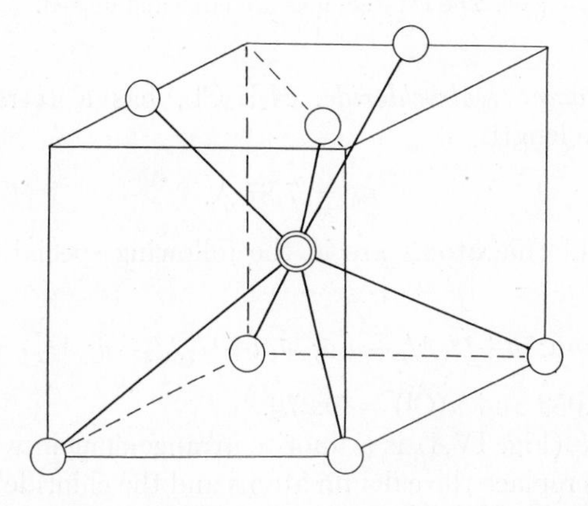

Fig. IV,3. A drawing to give an approximate picture of the sevenfold coordination of oxygen about zirconium in ZrO_2. The central zirconium atom is doubly ringed.

Still a fourth, hexagonal, structure with

$$a_0 = 3.598 \text{ A.}, \qquad c_0 = 5.875 \text{ A.}$$

has been described, though its existence does not seem to have been confirmed.

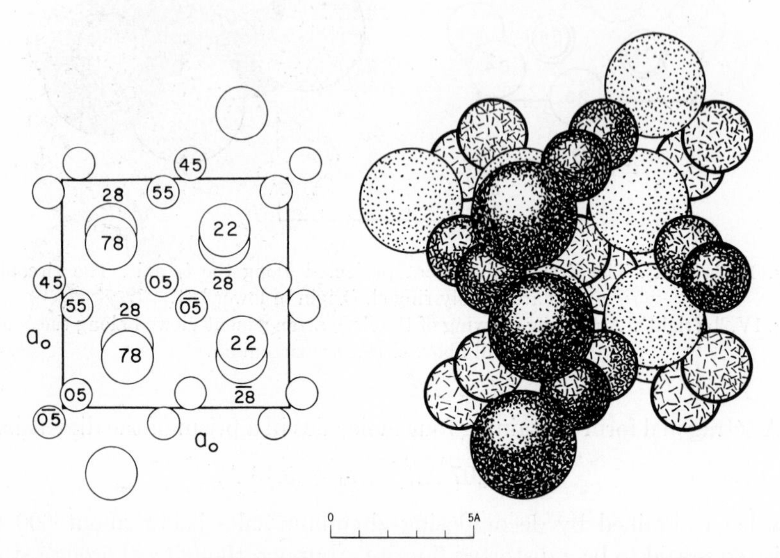

Fig. IV,4a (left). A projection along a cubic axis of the structure of $N_2H_6Cl_2$. The large circles are chlorine.

Fig. IV,4b (right). A packing drawing of the structure of $N_2H_6Cl_2$ viewed along a cubic axis. The chloride ions are large and dotted.

IV,a3. *Hydrazine hydrochloride*, $N_2H_6Cl_2$, has a tetramolecular unit cube of the edge length

$$a_0 = 7.87 \text{ A.}$$

Nitrogen and chlorine atoms are in the following special positions of T_h^6 ($Pa3$):

$$(8c) \quad \pm(uuu;\ u+{}^1/_2,{}^1/_2-u,\bar{u};\ \bar{u},u+{}^1/_2,{}^1/_2-u;\ {}^1/_2-u,\bar{u},u+{}^1/_2)$$

with $u(\text{N}) = 0.052$ and $u(\text{Cl}) = 0.279$.

This structure (Fig. IV,4) is a fluorite arrangement in which the centers of the N_2H_6 ions replace the calcium atoms and the chloride ions are slightly displaced, presumably because of the dumbbell shape of the cations, from the fluorine positions of fluorite. The N–N distance within the N_2H_6 ion is

1.42 A. Each nitrogen atom has four chlorine neighbors at distances of 3.10 A. The nearest approach of chlorine atoms to one another is 3.96 A.

The structure of hydrazine itself, N_2H_4, is described in paragraph **IV,i3.**

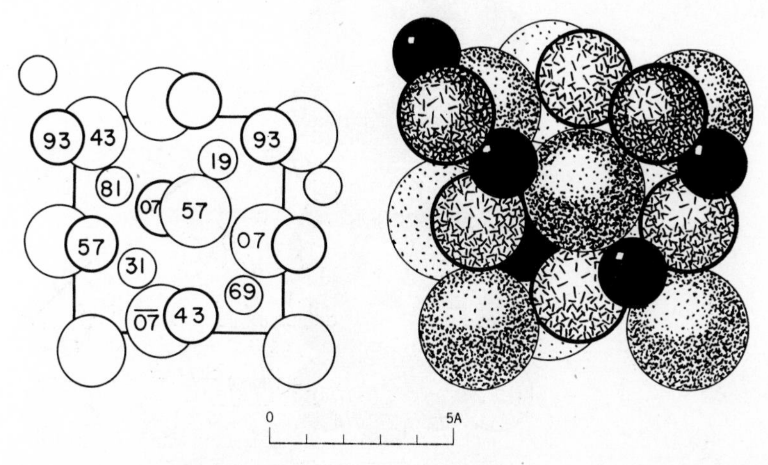

Fig. IV,5a (left). A projection along a cubic axis of the structure of ZrOS. The zirconium are the smallest and the sulfur the largest circles.

Fig. IV,5b (right). A packing drawing of the ZrOS structure viewed along a cubic axis. The zirconium atoms are black, the oxygens line-shaded.

IV,a4. The *oxysulfide* of *zirconium*, ZrOS, has a structure that is an extreme distortion of the fluorite grouping. It is cubic with a tetramolecular unit having:

$$a_0 = 5.696 \text{ A.}$$

All atoms are in special positions of T^4 ($P2_13$):

$$(4a) \quad uuu;\ u+{}^1/_2,{}^1/_2-u,\bar{u};\ {}^1/_2-u,\bar{u},u+{}^1/_2;\ \bar{u},u+{}^1/_2,{}^1/_2-u$$

with u(Zr) = 0.307, u(S) = 0.572 and u(O) = 0.928. In this arrangement (Fig. IV,5) each zirconium atom has four sulfur neighbors at Zr–S = 2.61 A. and 2.63 A. as well as three oxygen neighbors at 2.13 A. The shortest S–O and S–S separations are 2.96 A. and 3.59 A., respectively.

IV,a5. *Europium oxyfluoride*, EuOF, and the other oxyfluorides of Table IV,2 crystallize with hexagonal (rhombohedral) symmetry. The dimensions of the bimolecular unit rhombohedra, and of the three times larger hexagonal cells are given in the table.

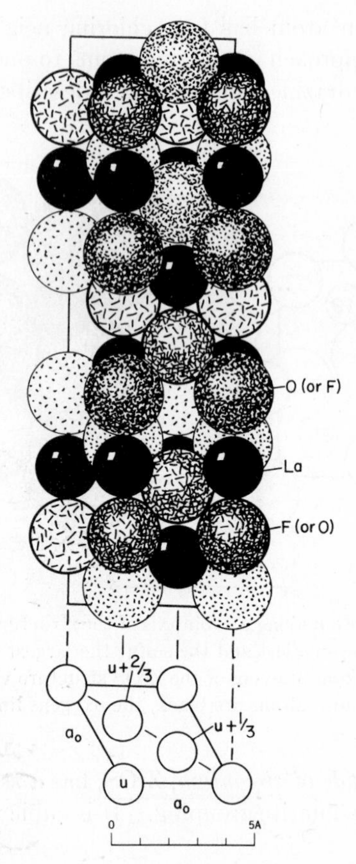

Fig. IV,6. Two projections of the hexagonal unit of EuOF.

TABLE IV,2
Cell Dimensions of Hexagonal EuOF and Similar Crystals (**IV,a5**)

	Rhombohedral axes		Hexagonal axes	
Compound	a_0, A.	α	a_0', A.	c_0', A.
EuOF	6.827	33°3′	3.884	19.344
GdOF	6.800	33°3′	3.868	19.268
β-LaOF	7.132	33°0′	4.051	20.212
β-NdOF	6.953	33°2′	3.953	19.702
PrOF	7.016	33°2′	3.989	19.881
SmOF	6.865	33°4′	3.907	19.451
TbOF	6.758	33°1′	3.841	19.150
β-YOF	6.697	33°12′	3.826	18.966

All atoms have been placed in the following special positions of D_{3d}^5 ($R\bar{3}m$):

(2*c*)	$\pm(uuu)$	for the rhombohedral axes
(6*c*)	$\pm(00u)$; rh	for the hexagonal axes

For LaOF, which is the best-studied member of the group, u(La) = 0.242, u(F) = 0.122 (or 0.370), and u(O) = 0.370 (or 0.122). It has been shown that the same parameters apply to YOF.

The structure, illustrated in Figure IV,6, is a relatively small distortion of that of fluorite (**IV,a1**). The measure of this distortion may be judged from the fact that, for the cubic structure expressed in rhombohedral terms, α would be 33°13′ and the parameters would be u(La) = 0.250, u(F) = 0.125, and u(O) = 0.375.

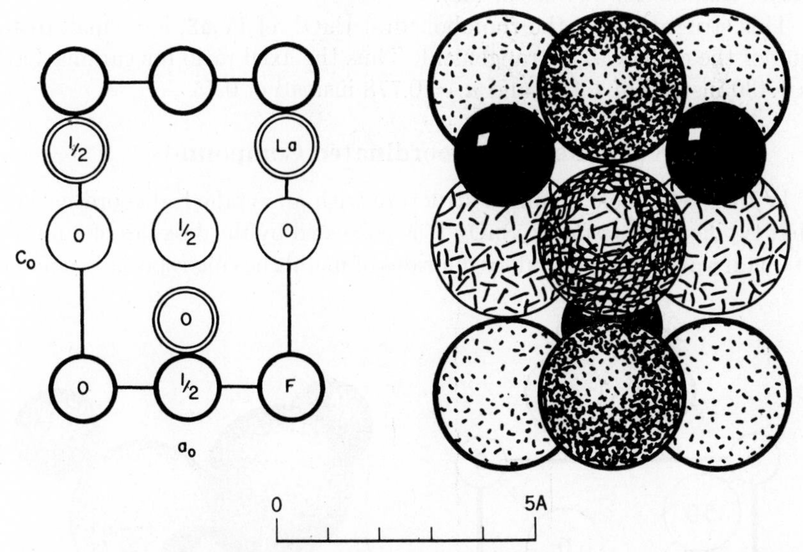

Fig. IV,7a (left). The tetragonal structure of γ-LaOF projected along the b_0 axis. Origin in lower left.

Fig. IV,7b (right). A packing drawing of the γ-LaOF structure viewed along the b_0 axis. The lanthanum atoms are black, the oxygens are line-shaded.

IV,a6. Most of the oxyhalides of trivalent metals that crystallize with tetragonal symmetry have been assigned the PbFCl arrangement (**IV,d3**) The three oxyfluorides, gamma-LaOF, gamma-YOF, and PuOF, are among the exceptions. Their bimolecular cells have the edges:

For gamma LaOF: a_0 = 4.091 A., c_0 = 5.852 A.
For gamma YOF: a_0 = 3.938 A., c_0 = 5.47 A.
For PuOF: a_0 = 4.05 A., c_0 = 5.72 A.

Though these dimensions are close to what would be expected from a PbFCl arrangement, and the space group is the same as for PbFCl (D_{4h}^7—*P*4/*nmm*), a different structure (Fig. IV,7) has been assigned. As with PbFCl, fluorine atoms are in

$$(2a) \quad 000;\ {}^1/_2\,{}^1/_2\,0$$

and the metal atoms are in

$$(2c) \quad 0\,{}^1/_2\,u;\ {}^1/_2\,0\,\bar{u}$$

But the value of u is somewhat different, 0.778, and the oxygen atoms are in

$$(2b) \quad 00\,{}^1/_2;\ {}^1/_2\,{}^1/_2\,{}^1/_2$$

rather than in another set of (2*c*).

This structure, like the rhombohedral EuOF of **IV,a5,** is a small distortion of the cubic CaF_2 arrangement. Thus the axial ratio for gamma-LaOF is 1.430 instead of 1.414, and u = 0.778 instead of 0.75.

Octahedrally Coordinated Compounds

IV,b1. The most common structure with an octahedral coordination is that typified by *cassiterite*, SnO_2. It is possessed by the dioxides of a number of quadrivalent metals and by fluorides of metals having especially small di-

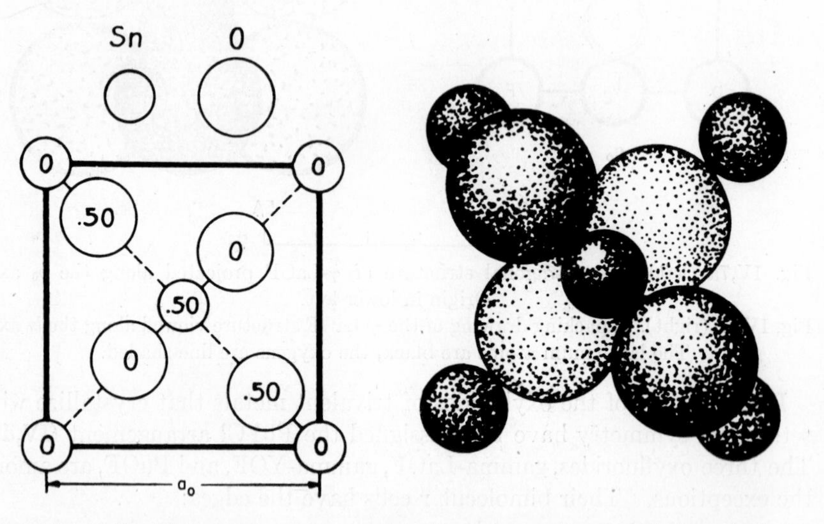

Fig. IV,8a (left). The atomic arrangement in the tetragonal unit of SnO_2 projected on the basal, *c*, face. The small circles represent tin atoms.

Fig. IV,8b (right). A drawing to show the way the atoms of SnO_2 pack together if they are given their expected ionic sizes. The large spheres are the oxygen ions.

TABLE IV,3
Crystals with the Tetragonal SnO_2 Structure (**IV,b1**)

Crystal	a_0, A.	c_0, A.	u
CoF_2	4.6951	3.1796	0.306
FeF_2	4.6966	3.3091	0.300
MgF_2	4.623	3.052 (27°C.)	0.303
MnF_2	4.8734	3.3099	0.305
NiF_2	4.6506	3.0836	0.302
PdF_2	4.931	3.367	—
ZnF_2	4.7034	3.1335 (25°C.)	0.303
CrO_2	4.41	2.91	—
$(Cr_{0.19}, Mo_{0.81})O_2$	4.760	2.848	—
$(Cr_{0.33}, Mo_{0.67})O_2$	4.696	2.886	—
$CrO_{2.14}$	4.423	2.917	—
GeO_2	4.395	2.859	0.307
IrO_2	4.49	3.14	—
β-MnO_2	4.396	2.871	0.302
MoO_2	4.86	2.79	—
NbO_2	4.77	2.96	—
OsO_2	4.51	3.19	—
PbO_2	4.946	3.379	—
RuO_2	4.51	3.11	—
SnO_2	4.73727	3.186383 (20–23°C.)	0.307
TaO_2	4.709	3.065	—
TeO_2	4.79	3.77	—
TiO_2 (rutile)	4.59373	2.95812 (25°C.)	0.3053
WO_2	4.86	2.77	—

valent ions. The symmetry is tetragonal with a flat unit containing two molecules. Atoms are in the following special positions of D_{4h}^{14} ($P4/mnm$):

$$\text{R:}\quad (2a)\quad 000;\ {}^1/_2\,{}^1/_2\,{}^1/_2$$

$$\text{X:}\quad (4f)\quad \pm(uu0;\ u+{}^1/_2,{}^1/_2-u,{}^1/_2)$$

As can be seen from Table IV,3, all the crystals with this structure (Fig. IV,8) have about the same axial ratio. Wherever determined, the parameter u has been not far from 0.30. The six X atoms about each R atom are of two sorts and the octahedron is not exactly regular, four being at a slightly different distance from the other two. In general, these observed R–X separations do not differ by more than about 0.10 A. and they agree well with the sums of the ionic radii. This structure brings anions close together, however, and in fact it makes one anion–anion contact considerably less than

TABLE IV,4

Crystals RMX_4 with the Simple Tetragonal SnO_2 Arrangement (**IV,b1**)

Crystal	a_0, A.	c_0, A.
$AlSbO_4$	4.510	2.961
$CrNbO_4$	4.635	3.005
$CrSbO_4$	4.577	3.042
$CrTaO_4$	4.626	3.009
$FeNbO_4$	4.68	3.05
$FeSbO_4$	4.623	3.011
$FeTaO_4$	4.672	3.042
$GaSbO_4$	4.59	3.03
$RhNbO_4$	4.686	3.014
$RhSbO_4$	4.601	3.100
$RhTaO_4$	4.684	3.020
$RhVO_4$	4.607	2.923

the ionic radial sum. Thus in the typical case of SnO_2, one O–O = 2.54 A., whereas the other close approaches of oxygen atoms are 2.90 A., which is near the uncorrected ionic radial sum.

A number of compounds of the type RMX_4 have substantially this atomic arrangement. Their cell dimensions are listed in Table IV,4. It has been said that their R and M atoms should be considered as statistically distributed among the positions of (2*a*) with the oxygen atoms in (4*f*). In the case of $AlSbO_4$, u(O) was assigned the value 0.305.

IV,b2. A structure which is only a slight distortion of the cassiterite grouping has been given the orthorhombic crystals of anhydrous *calcium chloride*, $CaCl_2$. The *bromide*, $CaBr_2$, has the same arrangement. Their bimolecular cells have the dimensions:

$$CaCl_2: \quad a_0 = 6.24 \text{ A.}; \; b_0 = 6.43 \text{ A.}; \; c_0 = 4.20 \text{ A.}$$

$$CaBr_2: \quad a_0 = 6.55 \text{ A.}; \; b_0 = 6.88 \text{ A.}; \; c_0 = 4.34 \text{ A.}$$

Atoms are in the following positions of V_h^{12} (*Pnnm*):

$$\text{Ca:} \quad (2a) \quad 000; \; \tfrac{1}{2}\tfrac{1}{2}\tfrac{1}{2}$$

$$\text{Cl (or Br):} \quad (4g) \quad \pm(uv0; \; u+\tfrac{1}{2},\tfrac{1}{2}-v,\tfrac{1}{2})$$

For $CaCl_2$, $u = 0.275$, $v = 0.325$. The bromine parameters have not been determined in $CaBr_2$. The nearest approach of the calcium and chlorine atoms, 2.70–2.76 A., and of the chlorine atoms to one another, 3.60 A., are in accord with the idea of ions.

This structure (Fig. IV,9) would be exactly the SnO_2 arrangement if $a_0 = b_0$ and $u = v$. The distortion it expresses corresponds to a small displacement of the halogen atoms from positions in the more symmetrical structure.

Another compound has now been found with this structure. *Chromous chloride*, $CrCl_2$, has a unit of the dimensions:

$$a_0 = 5.974 \text{ A.}; \; b_0 = 6.624 \text{ A.}; \; c_0 = 3.488 \text{ A.}$$

The parameters are: $u = 0.278$, $v = 0.362$. These give rise to the significant interatomic distances Cr–4Cl = 2.37 A., Cr–2Cl = 2.91 A.

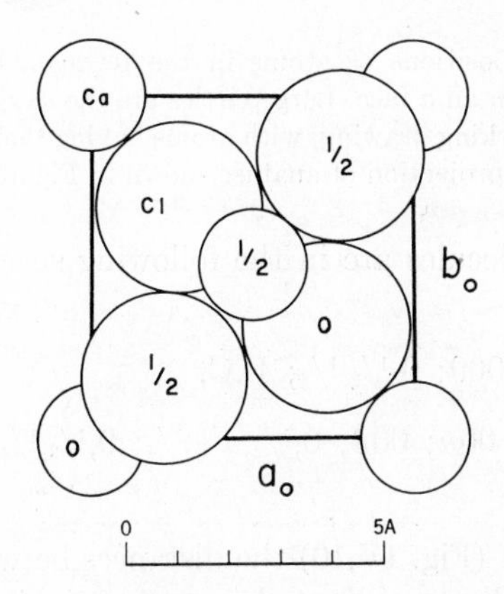

Fig. IV,9. A projection along the c_0 axis of the orthorhombic structure of $CaCl_2$. Comparison with Figure IV,8a brings out the close similarity to the SnO_2 arrangement. Origin in lower right.

IV,b3. Titanium dioxide, TiO_2, has two modifications in addition to the cassiterite form that occurs naturally as the mineral rutile. The simpler, found as the mineral *anatase*, is tetragonal with the elongated cell:

$$a_0 = 3.785 \text{ A.}, \qquad c_0 = 9.514 \text{ A.}$$

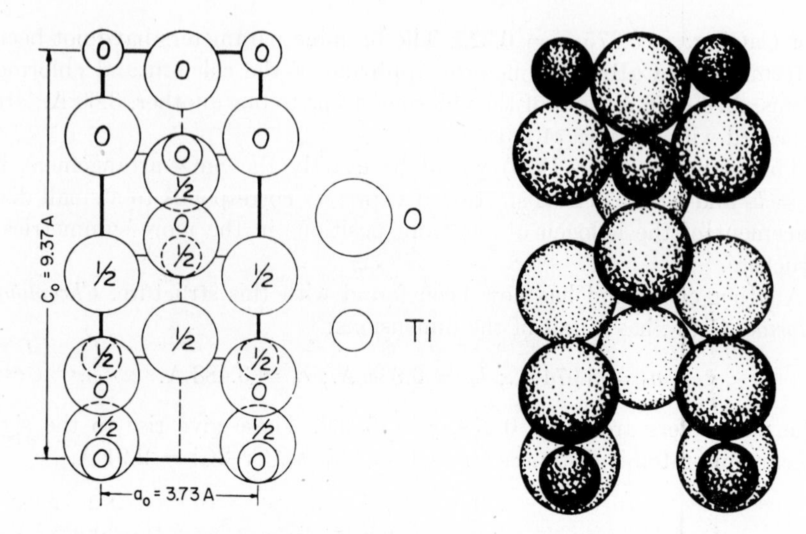

Fig. IV,10a (left). The positions of atoms in the tetragonal unit of anatase, TiO_2, projected on an *a* face. Large circles are the oxygen atoms.
Fig. IV,10b (right). A packing drawing, with atoms having their ionic sizes, corresponding to the projection of anatase shown in Figure IV,10a.

Atoms of its four molecules are in the following special positions of D_{4h}^{19} $(I4/amd)$:

$$\text{Ti: (4}a\text{)} \quad 000;\ 0\,{}^1/_2\,{}^1/_4;\ \text{B.C.}$$

$$\text{O: (8}e\text{)} \quad 00u;\ 00\bar{u};\ 0,{}^1/_2,u+{}^1/_4;\ 0,{}^1/_2,{}^1/_4-u;\ \text{B.C.}$$

with $u = 0.2066$.

In this arrangement (Fig. IV,10) the distances between a titanium atom and its six octahedrally coordinated oxygen neighbors are nearly equal (1.91–1.95 A.) to one another and to those in rutile, but the oxygen octahedron is not regular. In rutile there is one, in anatase there are two especially short oxygen-to-oxygen separations (2.43 A.).

IV,b4. The third form of TiO_2, the orthorhombic mineral *brookite*, has a more complicated crystal structure. Its eight-molecule unit has the edge lengths:

$$a_0 = 9.184 \text{ A.};\ b_0 = 5.447 \text{ A.};\ c_0 = 5.145 \text{ A.}$$

All atoms are in general positions of V_h^{15} $(Pbca)$:

$$(8c) \quad \pm(xyz;\ x+{}^1/_2,{}^1/_2-y,\bar{z};\ \bar{x},y+{}^1/_2,{}^1/_2-z;\ {}^1/_2-x,\bar{y},z+{}^1/_2)$$

with the recently determined values:

Atom	x	y	z
Ti	0.1290	0.0972	−0.1371
O(1)	0.0101	0.1486	0.1824
O(2)	0.2304	0.1130	−0.4629

These are more accurate than but not very different from the original values established many years ago.

Interatomic distances of this structure (Fig. IV,11) are similar to those in the other modifications, the six oxygen atoms about each titanium atom being at distances which range from ca. 1.87 A. to ca. 2.04 A. In brookite too there is close anion–anion contact, the oxygen–oxygen separations varying upwards from 2.49 A.

The mineral *tellurite*, TeO_2, is the only other substance known to have this structure. If its axes are taken so as to give a unit with the dimensions:

$$a_0 = 11.75 \text{ A.}; \ b_0 = 5.50 \text{ A.}; \ c_0 = 5.59 \text{ A.}$$

all atoms will be in positions (8*c*) listed above. The determined parameters referred to these axes are:

Atom	x	y	z
Te	0.118	0.027	−0.116
O(1)	−0.022	0.240	0.235
O(2)	0.174	0.164	−0.465

IV,b5. Besides the previously investigated rutile-like and brookite-like structures for *tellurium dioxide*, TeO_2, another tetragonal form has now been described. It is not, as might have been expected, like anatase but has a differently shaped tetramolecular cell with the dimensions:

$$a_0 = 4.805 \text{ A.}, \qquad c_0 = 7.609 \text{ A.}$$

Atoms have been found to be in the following positions of D_4^4 ($P4_12_12$) (or in the corresponding positions of the enantiomorphous D_4^8):

Te: (4*a*) $uu0; \bar{u}\bar{u}\,{}^1/_2; {}^1/_2-u,u+{}^1/_2,{}^1/_4; u+{}^1/_2,{}^1/_2-u,{}^3/_4$

O: (8*b*) $xyz; \bar{x},\bar{y},z+{}^1/_2; {}^1/_2-y,x+{}^1/_2,z+{}^1/_4; y+{}^1/_2,{}^1/_2-x,z+{}^3/_4;$
$yx\bar{z}; \bar{y},\bar{x},{}^1/_2-z; {}^1/_2-x,y+{}^1/_2,{}^1/_4-z; x+{}^1/_2,{}^1/_2-y,{}^3/_4-z$

with $u = 0.030$, $x = 0.177$, $y = 0.227$, and $z = 0.217$.

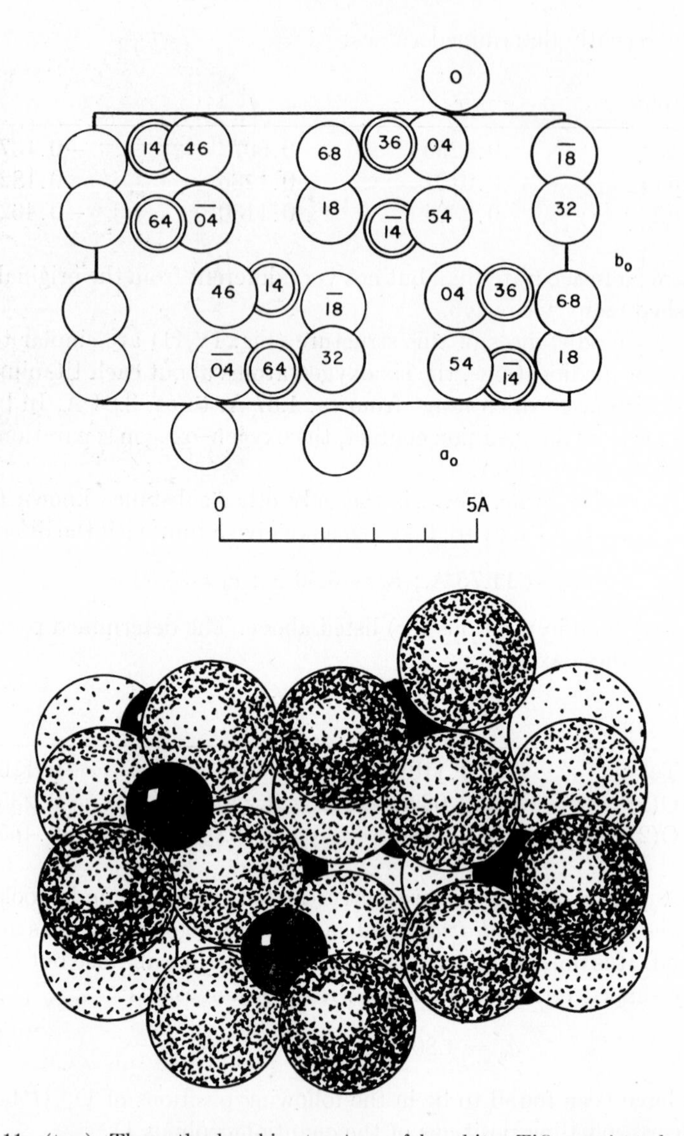

Fig. IV,11a (top). The orthorhombic structure of brookite, TiO_2, projected along its c_0 axis. The titanium atoms are doubly ringed. Origin in lower right.

Fig. IV,11b (bottom). A packing drawing of the brookite, TiO_2, arrangement viewed along its c_0 axis. The titanium atoms are black.

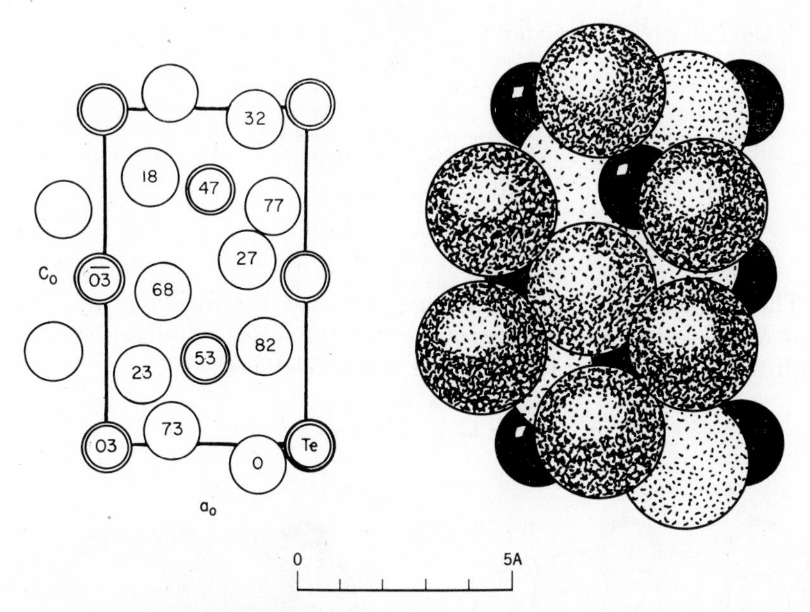

Fig. IV,12a (left). The tetragonal TeO_2 structure projected along its b_0 axis. Origin in lower left.

Fig. IV,12b (right). A packing drawing of the TeO_2 structure viewed along its b_0 axis. The tellurium atoms are black.

The resulting arrangement, as shown in Figure IV,12, can be thought of as a distortion of the rutile structure doubled in the c_0 direction. In the deformed octahedron of oxygen atoms enveloping a tellurium atom there are four Te–O = 2.03 A. and two Te–O = 2.67 A.

It is stated that synthetically prepared TeO_2 has invariably had this structure.

IV,b6. It has recently been shown that the rutile-like form of vanadium dioxide, VO_2, is a different distortion of this higher symmetry arrangement (**IV,b1**). The true symmetry for VO_2 is monoclinic and there are four molecules in a cell of the dimensions:

$$a_0 = 5.743 \text{ A.}; \; b_0 = 4.517 \text{ A.}; \; c_0 = 5.375 \text{ A.}; \; \beta = 122° \, 36'$$

All atoms are in the general positions of C_{2h}^5 ($P2_1/c$):

$$(4e) \quad \pm(xyz; \; x,{}^1/_2-y,z+{}^1/_2)$$

with the following parameters:

Atom	x	y	z
V	0.242	0.975	0.025
O(1)	0.10	0.21	0.20
O(2)	0.39	0.69	0.29

The originally proposed tetragonal cell had the dimensions: $a_0' =$ 4.54 A., $c_0' = 2.88$ A. The edges of the correct monoclinic cell are connected with this pseudocell by the following vector relations:

$$a_0 = 2c_0';\ b_0 = a_0';\ c_0 = a_0' - c_0'$$

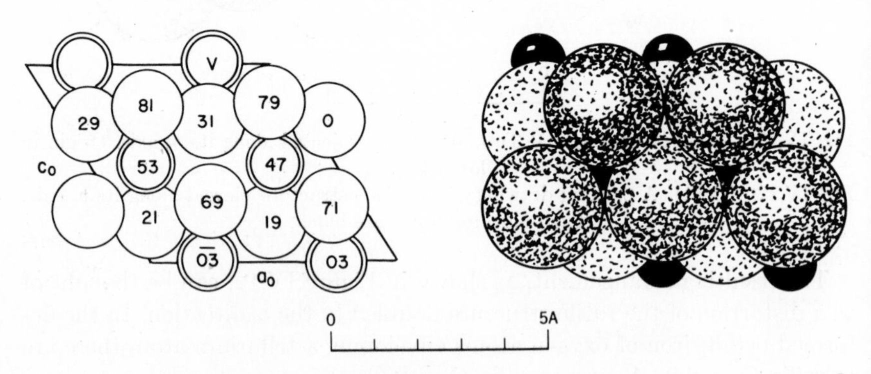

Fig. IV,13a (left). The monoclinic VO_2 arrangement projected along its b_0 axis. Origin in lower left.

Fig. IV,13b (right). A packing drawing of the VO_2 structure viewed along its b_0 axis. The vanadium atoms are black.

In this monoclinic structure (Fig. IV,13), metallic atoms have very nearly the same octahedral coordination of oxygen atoms that prevails in SnO_2 with V–O between 1.76 and 2.05 A. The O–O separations within an oxygen octahedron lie between 2.50 and 2.90 A.

The naturally occurring paramontroseite (**IV,d1**) is a different form of VO_2.

The structure described a number of years ago for *molybdenum dioxide*, MoO_2, is very near to that of VO_2. Though the space group was reported as C_2^2 ($P2$), it is now pointed out that the VO_2 arrangement based on C_{2h}^5 fits the data equally well. Parameters thus applicable to MoO_2 are about the same as those for VO_2:

Atom	x	y	z
Mo	0.232	0.000	0.017
O(1)	0.11	0.21	0.24
O(2)	0.39	0.70	0.30

The only important difference from the VO_2 structure is a small displacement of the metal atoms from the centers of their octahedra. This brings the molybdenum atoms rather close together with Mo–Mo = 2.48 A. Within an octahedron Mo–O varies between 1.9 and 2.1 A.

The cell dimensions of MoO_2 and of other compounds with this arrangement are:

$$MoO_2: \quad a_0 = 5.584 \text{ A.}; \ b_0 = 4.842 \text{ A.}; \ c_0 = 5.608 \text{ A.}; \ \beta = 120°59'$$

$$WO_2: \quad a_0 = 5.565 \text{ A.}; \ b_0 = 4.892 \text{ A.}; \ c_0 = 5.650 \text{ A.}; \ \beta = 120°42'$$

$$ReO_2: \quad a_0 = 5.562 \text{ A.}; \ b_0 = 4.838 \text{ A.}; \ c_0 = 5.561 \text{ A.}; \ \beta = 120°52'$$

$$TcO_2: \quad a_0 = 5.53 \text{ A.}; \ b_0 = 4.79 \text{ A.}; \ c_0 = 5.53 \text{ A.}; \ \beta = 120°$$

The relation to rutile is brought out by the fact that, described in terms of a pseudocell having the axial orientation of VO_2, the rutile axes would be:

$$a_0' = 5.918 \text{ A.}; \ b_0' = 4.593 \text{ A.}; \ c_0' = 5.464 \text{ A.}; \ \beta = 122°48'$$

IV,b7. The alpha form of *lead dioxide*, PbO_2, produced by electrolysis of neutral solutions, has been assigned the tetramolecular orthorhombic cell:

$$a_0 = 4.947 \text{ A.}; \ b_0 = 5.951 \text{ A.}; \ c_0 = 5.497 \text{ A.}$$

Atoms have been placed in the following positions of V_h^{14} (*Pbcn*):

Pb: (4*c*) $\pm(0\,u\,{}^1/_4;\ {}^1/_2,u+{}^1/_2,{}^1/_4)$,

O: (8*d*) $\pm(xyz;\ {}^1/_2-x,{}^1/_2-y,z+{}^1/_2;\ x+{}^1/_2,{}^1/_2-y,\bar{z};\ \bar{x},y,{}^1/_2-z)$

with $u = 0.178$, $x = 0.276$, $y = 0.410$, and $z = 0.425$.

In this structure (Fig. IV,14), Pb–O = 2.16–2.22 A. It is an arrangement that closely resembles that of columbite, $Nb_2(Fe,Mn)O_6$ (Chapter IX) with lead atoms replacing all metallic atoms in the mineral.

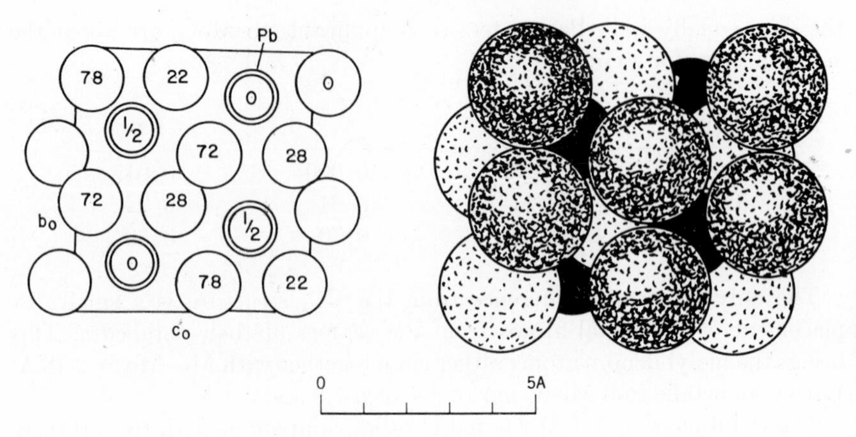

Fig. IV,14a (left). The orthorhombic structure of α-PbO_2 projected along its a_0 axis. Origin in lower left.

Fig. IV,14b (right). A packing drawing of the α-PbO_2 arrangement viewed along its a_0 axis. The lead atoms are black.

The modification of *rhenium dioxide,* ReO_2, stable in the region 300–1050°C. has this structure, though with a cell having a shorter c_0 axis:

$$a_0 = 4.8094 \text{ A.}; \; b_0 = 5.6433 \text{ A.}; \; c_0 = 4.6007 \text{ A.}$$

The influence of the oxygen atoms is too small to permit a precise determination of their positions, but it is said that the available powder data are satisfied by the following choice of parameters: $u(\text{Re}) = 0.110$, and for oxygen $x = 0.25$, $y = 0.36$, $z = 0.125$. Except for the very different value of z, these parameters are not far from those chosen for PbO_2. The structure viewed along the c_0 axis instead of along the a_0 axis, as in Figure IV,14, is shown in Figure IV,15. The ReO_6 octahedra share edges to yield the kind of strings through the crystal that occur in brookite (**IV,b4**); the strings themselves are tied together by sharing octahedral corners. Interatomic distances are Re–O = 1.95–2.10 A. and O–O = 2.6–3.1 A.

IV,b8. *Niobium dioxide,* NbO_2, was originally given the rutile structure (IV,b1). The principal x-ray reflections point to this structure, but additional weak reflections indicate a more complicated grouping. According to a recent determination the symmetry is tetragonal with a large unit

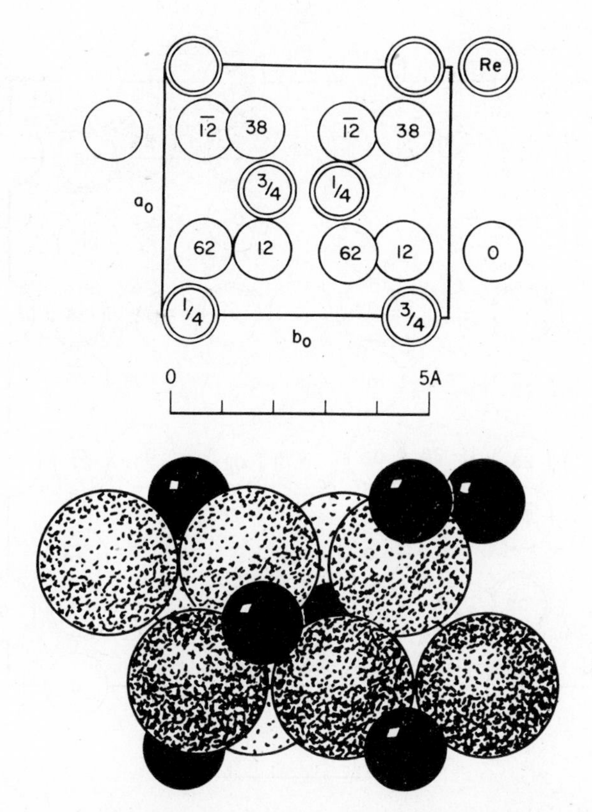

Fig. IV,15a (top). The α-PbO_2 structure, as illustrated by ReO_2, viewed along the c_0 axis instead of the a_0 axis. Origin in lower right.
Fig. IV,15b (bottom). A packing drawing of ReO_2, with the α-PbO_2 structure, viewed along its c_0 axis. The Re atoms are black.

containing 32 molecules and having the edges:

$$a_0 = 2\sqrt{2}\, a_r = 13.71 \text{ A.,} \qquad c_0 = 2c_r = 5.985 \text{ A.}$$

where a_r and c_r are the edges of the small rutile-like sub-unit.

The space group is C_{4h}^6 ($I4_1/a$) with all atoms in the general positions:

$$(16f) \quad xyz;\ \bar{x}\bar{y}z;\ x,y+{}^1/_2,{}^1/_4-z;\ \bar{x},{}^1/_2-y,{}^1/_4-z;$$

$$\bar{y}x\bar{z};\ y\bar{x}\bar{z};\ \bar{y},x+{}^1/_2,z+{}^1/_4;\ y,{}^1/_2-x,z+{}^1/_4;\ \text{B.C.}$$

The selected parameters are those of Table IV,5.

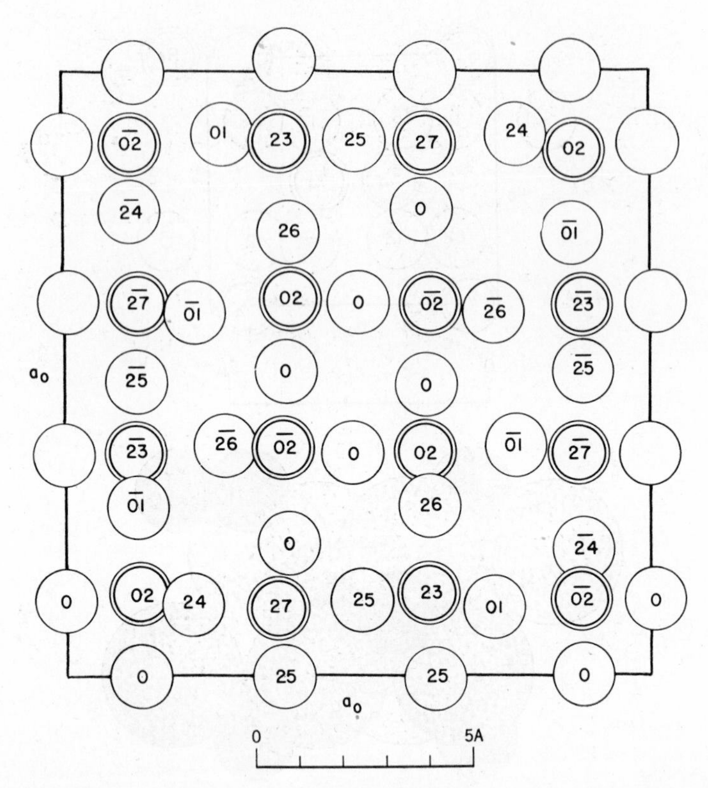

Fig. IV,16a. Half the contents of the large tetragonal unit of NbO_2 projected along its c_0 axis. The Nb atoms are doubly ringed.

TABLE IV,5
Parameters of the Atoms in NbO_2 (**IV,b9**)

Atom	x	y	z
Nb(1)	0.118	0.125	0.483
Nb(2)	0.133	0.125	0.016
O(1)	0.986	0.130	0.001
O(2)	0.971	0.125	0.501
O(3)	0.280	0.126	0.993
O(4)	0.266	0.118	0.503

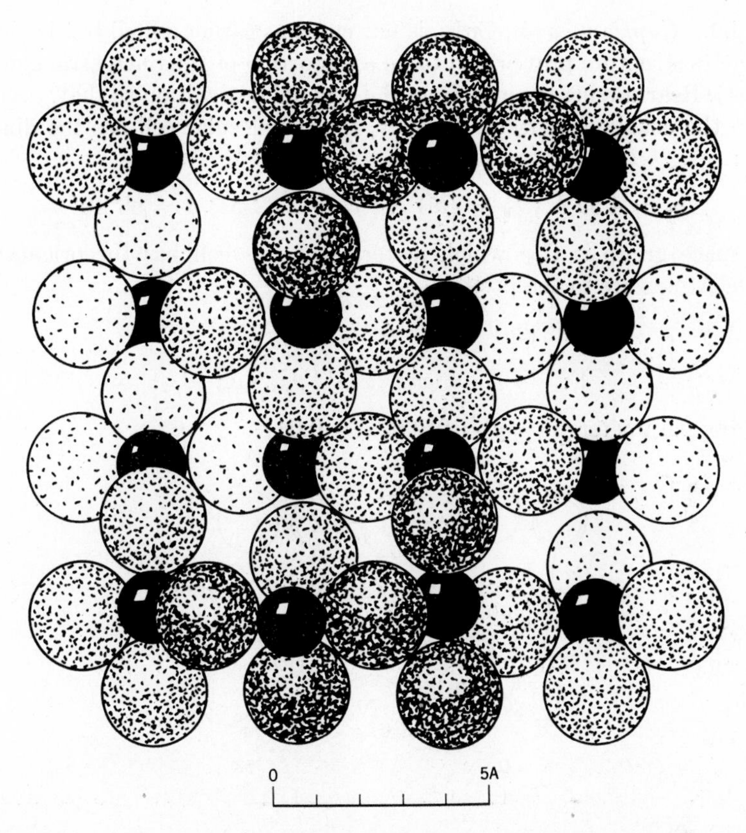

Fig. IV,16b. A packing drawing of half the atoms in the unit cell of NbO_2 viewed along the c_0 axis. The Nb atoms are black.

This structure can be pictured as having the same kind of NbO_6 octahedra sharing edges and corners that exist in the rutile arrangement. Along c_0 the Nb–Nb separations are alternately short and long (2.80 and 3.20 A.). The Nb–O distances range upwards from ca. 2.0 A.; the shortest O–O = 2.6 A. Figure IV,16 reproduces a lower half of the contents of a unit. The upper half cannot with clarity be included because it is almost identical in content and only slightly displaced in its projection along c_0. This is apparent from the way the coordinates of the table are paired in their x and y parameters.

IV,b9. *Cupric fluoride*, CuF_2, is not cubic as earlier stated but is monoclinic. Its structure proves also to be a distortion of the SnO_2 arrangement (**IV,b1**). Rearranging the axes from the original description so that, according to the usual convention, $\beta > 90°$, the bimolecular cell has the dimensions:

$$a_0 = 4.59 \text{ A.}; \; b_0 = 4.54 \text{ A.}; \; c_0 = 3.32 \text{ A.}; \; \beta = 96°40'$$

The space group is C_{2h}^5 which in terms of this cell has the orientation $P2_1/n$. Atoms are in the positions:

Cu: (2*a*) $000; \; {}^1/_2 \, {}^1/_2 \, {}^1/_2$

F: (4*e*) $\pm(xyz; \; x+{}^1/_2, {}^1/_2-y, z+{}^1/_2)$

with the parameters $x = y = 0.300$, $z = 0.044$.

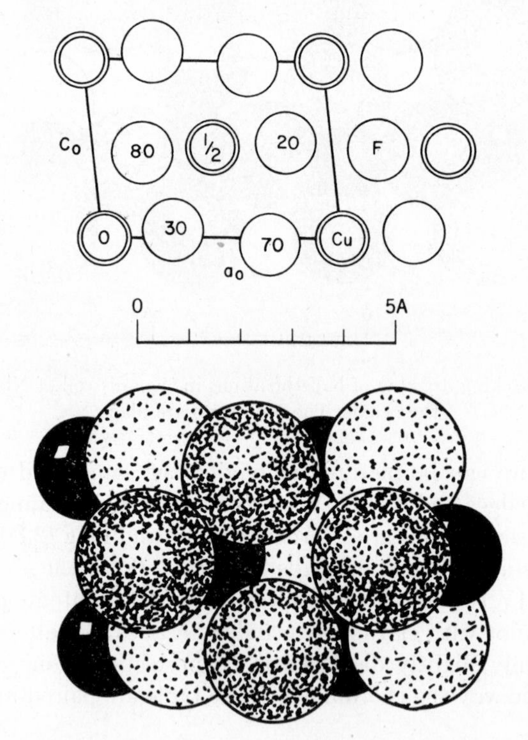

Fig. IV,17a (top). The monoclinic structure of CuF_2 projected along its b_0 axis. Origin in lower left.

Fig. IV,17b (bottom). A packing drawing of the CuF_2 arrangement viewed along its b_0 axis. The Cu atoms are black.

In this structure (Fig. IV,17) the metallic atoms have the same kind of distorted octahedral coordination as in SnO_2 with Cu–2F = 2.27 A. and Cu–4F = 1.93 A. The closest approaches of the fluorine atoms to one another are 2.61 and 2.73 A.

This arrangement appears to represent a departure from the usually fourfold, square coordination of the copper atom, but the four nearest fluorine atoms are in fact at the corners of a planar square with copper at its center.

Chromous fluoride, CrF_2, has the same arrangement, with

$$a_0 = 4.732 \text{ A.}; \; b_0 = 4.718 \text{ A.}; \; c_0 = 3.505 \text{ A.}; \; \beta = 96°30'$$

Parameters for the fluorine atoms are $x = y = 0.297$, $z = 0.044$. Within a CrF_6 octahedron, Cr–F = 1.98, 2.01, and 2.43 A.

IV,b10. *Iridium diselenide*, $IrSe_2$, is orthorhombic with an eight-molecule cell of the dimensions

$$a_0 = 20.94 \text{ A.}; \; b_0 = 5.93 \text{ A.}; \; c_0 = 3.74 \text{ A.}$$

All its atoms have been placed in the following special positions of V_h^{16} (*Pnam*):

$$(4c) \quad \pm(u\,v\,{}^1/_4;\; u+{}^1/_2,{}^1/_2-v,{}^1/_4)$$

with the parameters of Table IV,6.

The resulting arrangement, as shown in Figure IV,18, places each iridium atom at the center of an octahedron of six selenium atoms with Ir–Se = 2.42–2.54 A. Each selenium atom has one selenium and three iridium neighbors tetrahedrally distributed. In half these tetrahedra the Se(3)–Se(4) separation is short (2.57 A.), indicative of a close bonding; in the other half it is 3.27 A. This is a rather open structure that bears a certain relation to that of marcasite, FeS_2 (**IV,g6**).

TABLE IV,6
Parameters of the Atoms in $IrSe_2$ (**IV,b10**)

Atom	u	v
Ir(1)	−0.076	−0.427
Ir(2)	−0.304	−0.442
Se(1)	0.007	0.270
Se(2)	−0.121	−0.045
Se(3)	0.237	0.324
Se(4)	−0.363	−0.074

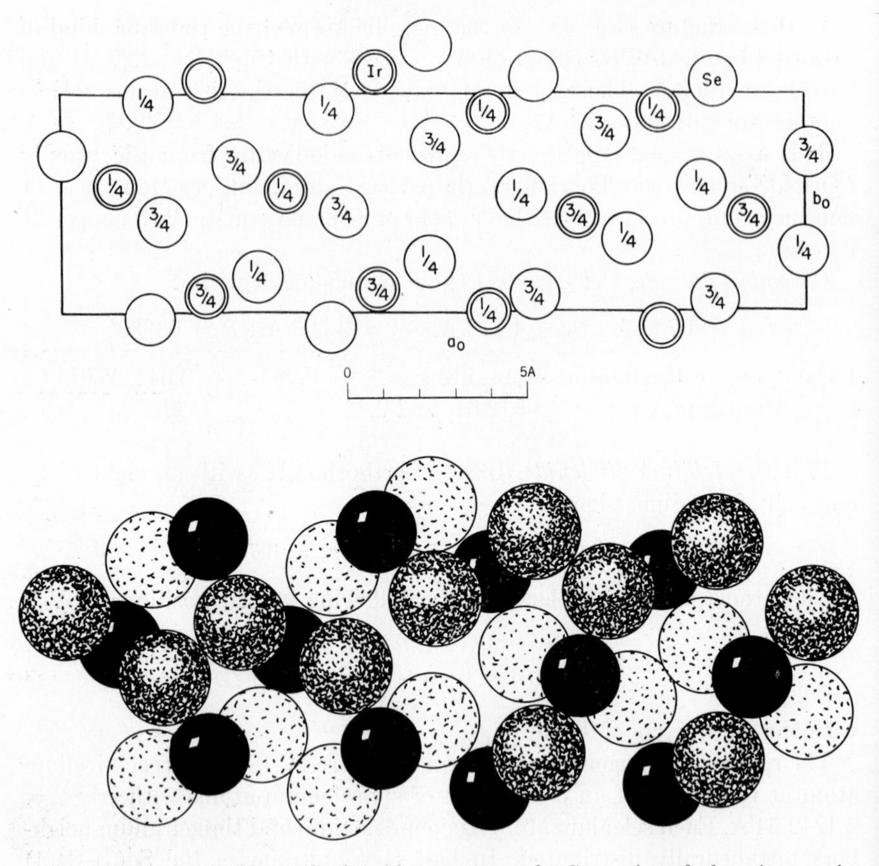

Fig. IV,18a (top). The orthorhombic structure of $IrSe_2$ projected along its c_0 axis. Origin in the lower right.

Fig. IV,18b (bottom). A packing drawing of the very open structure found for $IrSe_2$ viewed along the c_0 axis. Atoms have been given approximately their neutral radii, with Ir the black circles.

Cadmium Iodide-Like Structures

IV,c1. In the structures of the preceding paragraphs, crystals are held together by contacts between unlike atoms. Those now to be considered, though still involving an octahedral coordination of their cations, exhibit only anion–anion contacts in certain directions. They are distinctly layer-like both in atomic arrangement and in their physical properties. To the degree that they may be considered as ionic in character they can conveniently be viewed as more or less perfect packings of large anions, with the smaller cations lying in interstices of these packings.

The simplest and most familiar of these layer structures is the original CdI_2, or *cadmium hydroxide*, $Cd(OH)_2$, grouping. It is hexagonal with a single molecule per cell. The atoms are in the following special positions of D_{3d}^3 ($C\bar{3}m$):

R: (1*a*) 000

X: (2*d*) $^1/_3\,^2/_3\,u;\ ^2/_3\,^1/_3\,\bar{u}$ with u about $^1/_4$

More than 50 compounds (Table IV,7) have been found to have this arrangement (Fig. IV,19). They are for the most part hydroxides, bromides, and iodides of divalent metals, and sulfides, selenides, and tellurides of tetravalent metals.

With $u = {}^1/_4$ and $c_0/a_0 = 1.63$, the X atoms would be in a perfect hexagonal close-packing; in known crystals c_0/a_0 lies between 1.60 and 1.40.

In the typical example of CdI_2 the cadmium-to-iodine distance is 2.99 A. and the iodine-to-iodine separation is 4.21–4.24 A. These distances are about equal to the sums of the ionic radii, and from this standpoint the crystal can be viewed as a hexagonal close-packing of iodine ions with the small cadmium ions nested between alternate layers of iodine. In some

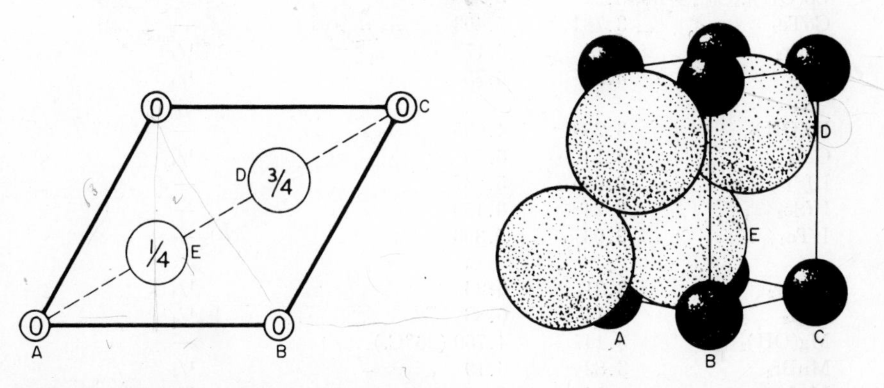

Fig. IV,19a (left). A basal projection of the atomic positions within the hexagonal unit prism of the $Cd(OH)_2$ arrangement. Letters refer to the correspondingly marked atoms of Figure IV,19b.

Fig. IV,19b (right). A perspective packing drawing of the atomic arrangement in $Cd(OH)_2$. In this figure the large and small spheres have been given the relative sizes of the I′ and Cd^{2+} ions.

other crystals with this structure, the X–X separations are not all alike and some are larger than the ionic sums. When better values of u are at hand it will be important to speculate on the significance of these differences.

Particularly detailed studies have been made by both x-ray and neutron diffraction to define the positions of the hydrogen atoms in $Ca(OH)_2$. From the x-ray measurements it was decided that the hydrogen atoms, like the oxygen atoms, are in ($2d$) with $u(H) = 0.395$. This results in an O–H separation of 0.79 A.

TABLE IV,7
Crystals with the Hexagonal $Cd(OH)_2$ Structure (**IV,c1**)

Crystal	a_0, A.	c_0, A.	u or Remarks
Ag_2F	2.989	5.710	ca. 0.3
BiTeBr	4.23	6.47	—
BiTeI	4.31	6.83	—
CaI_2	4.48	6.96	$^1/_4$
$Ca(OH)_2$	3.5844	4.8962	0.2330
CdI_2	4.24	6.84	—
$Cd(OH)_2$	3.48	4.67	—
$CoBr_2$	3.68	6.12	$^1/_4$
CoI_2	3.96	6.65	$^1/_4$
$Co(OH)_2$	3.173	4.640	0.22
$Co(OH)_{1.5}Br_{0.5}$	3.23	5.91	—
$Co(OH)_{1.5}Cl_{0.5}$	3.22	5.50	—
$CoTe_2$	3.784	5.403	—
$FeBr_2$	3.74	6.17	$^1/_4$
FeI_2	4.04	6.75	$^1/_4$
$Fe(OH)_2$	3.258	4.605	—
GeI_2	4.13	6.79	$^1/_4$
HfS_2	3.635	5.837	—
$HfSe_2$	3.748	6.159	—
$IrTe_2$	3.93	5.393	—
$MgBr_2$	3.81	6.26	$^1/_4$
MgI_2	4.14	6.88	$^1/_4$
$Mg(OH)_2$	3.147	4.769 (26°C.)	—
$MnBr_2$	3.82	6.19	$^1/_4$
MnI_2	4.16	6.82	$^1/_4$
$Mn(OH)_2$	3.34	4.68	—
$Ni(OH)_2$	3.117	4.595	—
$Ni(OH)_{1.5}Cl_{0.5}$	3.15	5.36	—

(continued)

Neutron diffraction measurements were made at room and at low temperatures. At 20°C.,

$$a_0 = 3.5918 \text{ A.}, \qquad c_0 = 4.9063 \text{ A.}$$

The parameters of oxygen and hydrogen were found as $u(\text{O}) = 0.2341$ and $u(\text{H}) = 0.4248$, leading to an O–H separation of 0.936 A. When measured at −140°C. the cell dimensions were found to be

$$a_0 = 3.5862 \text{ A.}, \qquad c_0 = 4.8801 \text{ A.}$$

TABLE IV,7 (*continued*)

Crystal	a_0, A.	c_0, A.	u or Remarks
$NiTe_2$	3.861	5.297	1/4
PbI_2	4.555	6.977 (25°C.)	0.265
$PdTe_2$	4.0365	5.1262	—
PtS_2	3.537	5.019	—
$PtSe_2$	3.724	5.062	—
$PtTe_2$	4.010	5.201	—
$RhTe_2$ (high)	3.92	5.41	0.25
$SiTe_2$	4.28	6.71	0.265
SnS_2	3.639	5.868	—
SnSSe	3.716	6.050	—
$SnSe_2$	3.811	6.137	—
α-TaS_2	3.35	5.86	—
ThI_2	4.13	7.02	—
$TiBr_2$	3.629	6.492	—
$TiCl_2$	3.561	5.875	0.25
TiI_2	4.110	6.820	—
TiS_2	3.412	5.695	—
$TiSe_2$	3.541	5.986	Unbroken solutions with TiSe
$TiTe_2$	3.757	6.513	Unbroken solutions with TiTe
TmI_2	4.520	6.967	—
VBr_2	3.768	6.180	—
VCl_2	3.601	5.835	—
VI_2	4.000	6.670	—
W_2C	2.98	4.71	By electron diffraction
YbI_2	4.503	6.972	—
ZnI_2(I)	4.25	6.54	0.25
ZrS_2	3.662	5.813	—
$ZrSe_2$	3.771	6.138–6.149	c_0 increases with Se content
$ZrTe_2$	3.950	6.630	Unbroken solutions with ZrTe

At this low temperature, u(O) was determined as 0.2346 and u(H) as 0.4280.

Since neutron diffraction should give the position of the proton nucleus and the x-ray diffraction a mean position between the oxygen and the hydrogen nuclei of the bonding electrons, the foregoing difference between the values of u(H) as established by the two methods of diffraction is to be expected.

Lead diiodide, PbI_2, which is one of the crystals showing this arrangement, resembles CdI_2 in having several forms that are layer structures differing from one another in the way these layers are repeated along the c_0 axis. These forms are described in paragraph **IV,c5.**

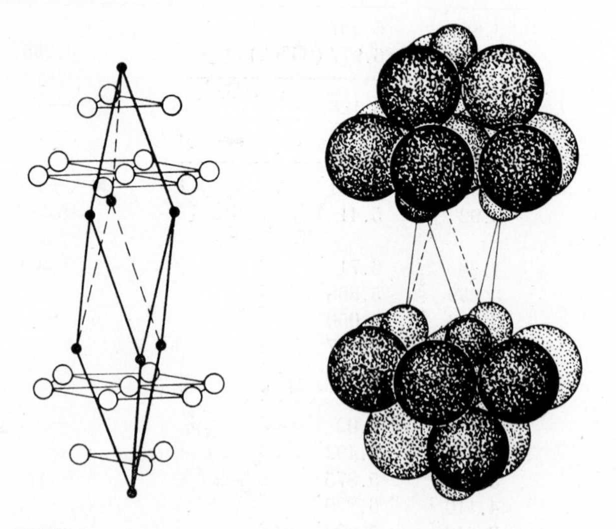

Fig. IV,20a (left). A perspective drawing of the unit rhombohedron of the $CdCl_2$ structure with some of the surrounding chlorine atoms shown as open circles.
Fig. IV,20b (right). A perspective drawing showing the packing of the cadmium and chlorine atoms of $CdCl_2$ considered as ions. The chloride ions are the larger spheres.

IV,c2. The *cadmium chloride*, $CdCl_2$, arrangement differs from the preceding $Cd(OH)_2$ structure in having its anions in a cubic rather than a hexagonal close-packing. Its unit cell, a rhombohedron containing a single molecule, has atoms in the following special positions of D_{3d}^5 ($R\bar{3}m$):

$$\text{R:} \quad (1a) \quad 000$$

$$\text{X:} \quad (2c) \quad uuu;\ \bar{u}\bar{u}\bar{u}$$

with $u = 0.25$ (Fig. IV,20a and b).

The corresponding cell containing three molecules has the dimensions:

$$a_0' = 3.85 \text{ A.}, \qquad c_0' = 17.46 \text{ A.}$$

Atoms are in the positions:

Cd: 000; rh

Cl: $00u$; $00\bar{u}$; rh

where u is the same as above. This cell (Fig. IV,20c) has a base of about the same size and a height three times greater than the CdI_2 unit (**IV,c1**).

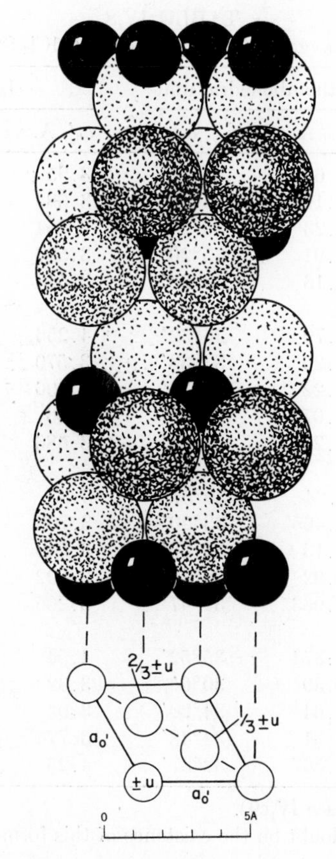

Fig. IV,20c. Two projections of the $CdCl_2$ arrangement in terms of its hexagonal unit. The atoms of the upper projection, looking normal to the $a_0'c_0'$ plane, have been shaded with the cadmium atoms black.

In both the chloride and the iodide, the cadmium atoms are at the centers of octahedra which can be thought of as joined together in sheets perpendicular to the threefold crystal axes by having X atoms in common. In the iodide these sheets are stacked one above another, and in $CdCl_2$ they are staggered according to the demands of a cubic close-packing of the chlorine atoms. In the latter crystal Cd–Cl = 2.74 A. and Cl–Cl = 3.68 A. are about equal to the sums of the radii of these atoms as ions.

Twenty substances have been reported to have this $CdCl_2$ structure. The dimensions of their rhombohedral and hexagonal cells are listed in Table IV,8. As stated there, the parameter u when determined has had a value close to 0.25.

TABLE IV,8
Crystals with the Hexagonal (Rhombohedral) $CdCl_2$ Structure (**IV,c2**)

Crystal	Rhombohedral unit		Hexagonal unit		
	a_0, A.	α	a_0', A.	c_0', A.	u
$CdBr_2$	6.63	34°42′	3.95	18.67	—
$CdBr_{0.6}(OH)_{1.4}$	6.198	33°32′	3.58	17.55	—
$CdCl_2$	6.23	36°2′	3.854	17.457	0.25
$CdCl_{0.75}(OH)_{1.25}$	5.91	35°20′	3.587	16.605	—
$CoCl_2$	6.16	33°26′	3.544	17.430	0.25
Cs_2O	6.79	36°32′	4.256	18.99	0.256
$FeCl_2$	6.20	33°33′	3.579	17.536	—
$MgCl_2$	6.22	33°36′	3.596	17.589	—
Mg(OH)Cl	6.078	32°6′	3.36	17.3	—
$MnCl_2$	6.20	34°35′	3.686	17.470	—
NbS_2[a]	6.24	30°57′	3.33	17.80	0.25
$NiBr_2$	6.465	33°20′	3.708	18.300	0.255
$NiCl_2$	6.13	33°36′	3.543	17.335	—
NiI_2	6.92	32°40′	3.892	19.634	0.250
Ni(OH)Cl	5.963	31°47′	3.265	16.99	—
PbI_2(II)	7.374	35°52′	4.54	20.7	0.26
γ-TaS_2	6.39	30°6′	3.32	18.29	—
$ZnBr_2$	6.64	34°20′	3.92	18.73	—
$ZnCl_2$[b]	6.31	34°48′	3.774	17.765	—
ZnI_2	7.567	32°37′	4.25	21.5	—

[a] This form is doubtful (see **IV,c9**).
[b] Recent work has cast doubt on the existence of this form of $ZnCl_2$ (see **IV,e2**).

IV,c3. Since the first studies were made years ago it has been reported that some crystalline preparations of *cadmium iodide* show diffraction lines that cannot be accounted for by the simple grouping described in **IV,c1.** Instead they point to a bimolecular cell with a base of the same dimensions as that of the simple structure but with twice the height:

$$a_0 = 4.24 \text{ A.}, \qquad c_0 = 13.67 \text{ A.}$$

The more complicated diffraction patterns they give indicate an atomic arrangement described by the following positions of C_{6v}^4 ($C6mc$):

Cd: (2*b*) $\frac{1}{3}\,\frac{2}{3}\,u;\ \frac{2}{3},\frac{1}{3},u+\frac{1}{2}$ with $u = 0$

I(1): (2*b*), with $u = \frac{3}{8}$

I(2): (2*a*) $00v;\ 0,0,v+\frac{1}{2}$ with $v = \frac{5}{8}$

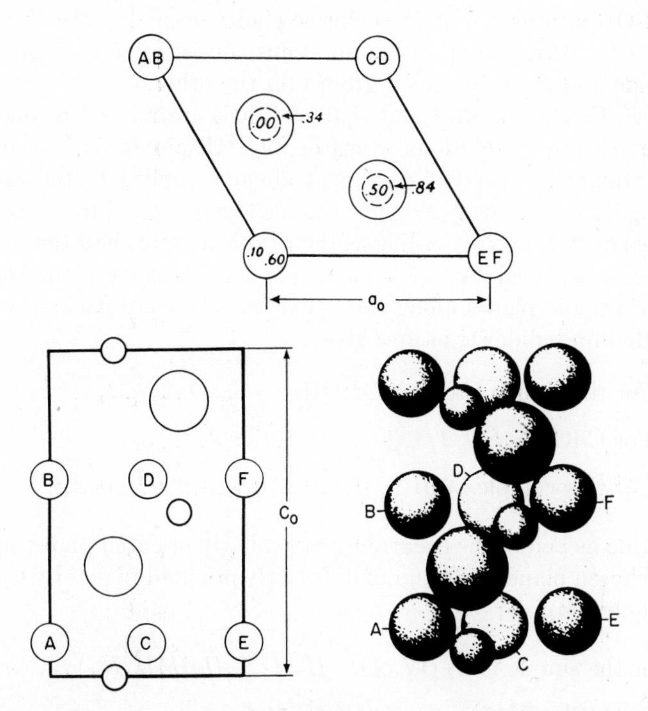

Fig. IV,21a (top). A basal projection of the atomic arrangement in Cd(OH)Cl. The smallest circles are the cadmium atoms, the largest are the chlorine atoms.

Fig. IV,21b (bottom left). A diagonal side face (11·0) projection of the atomic arrangement in Cd(OH)Cl.

Fig. IV,21c (bottom right). A drawing to show how the atoms of Cd(OH)Cl recorded in the projection of Figure IV,21b pack if they are assigned their usual ionic sizes. The letters identify equivalent atoms in each of the three drawings of this structure.

Such a grouping (Fig. IV,21) has the same sheets of linked CdX_6 octahedra as the preceding structures; but the stacking is different from that of either.

Two other substances have this structure with the cell dimensions:

$$Cd(OH)Cl: \quad a_0 = 3.66 \text{ A.}, \quad c_0 = 10.27 \text{ A.}$$

$$Ca(OH)Cl: \quad a_0 = 3.86 \text{ A.}, \quad c_0 = 9.90 \text{ A.}$$

For Cd(OH)Cl the cadmium and chlorine atoms are described as being in (2*b*) with u(Cd) = 0 and u(Cl) = 0.337, while the OH is in (2*a*) with v = 0.60. These parameters are not very different from those given CdI_2; their departure from $^3/_8$ and $^5/_8$ can be considered as a consequence of the different sizes of the chloride and hydroxyl ions. In this mixed structure, planes of OH alternate with the chlorine planes normal to the c_0 axis; hence the octahedra surrounding cadmium atoms consist of three chlorine atoms on one side and three hydroxyl groups on the other.

The two CdI_2 structures and that of $CdCl_2$ are related to one another somewhat as are the various forms of SiC (**III,c3**). This is brought out clearly through the type of treatment already applied to the carbide. In describing its several arrangements, atomic layers normal to the c_0 axis were designated as *0*, *1*, or *2* according as their typical atoms had the coordinates $00u$ or were displaced to $^1/_3\,^2/_3\,z$ or $^2/_3\,^1/_3\,z'$. A similar listing of the sequence of atomic planes along the c_0 axis for the close-packed ions of the three cadmium halide structures gives:

For the simple CdI_2, or $Cd(OH)_2$: *1,2,1,2,1,2, . . .*

For $CdCl_2$: *1,0,2,1,0,2,1,0,2,1,0,2, . . .*

For this complex CdI_2: *0,1,0,2,0,1,0,2,0,1,0,2, . . .*

Designating as before the occurrence of similarly oriented planes above and below a chosen plane by *H* and of differently oriented planes by *C*, the foregoing sequences give:

For the simple CdI_2 (**IV,c1**): *H,H,H,H,H,H,H, . . .*

For $CdCl_2$ (**IV,c2**): *C,C,C,C,C,C,C, . . .*

For the complex CdI_2 (**IV,c3**): *H,C,H,C,H,C,H,C, . . .*

The complex arrangement is thus the simplest possible mixture of cubic and hexagonal close-packings of the X atoms. Their sequence is the same as that of the carbon (or silicon) atoms in modification III of SiC.

The metallic hydroxy halides appear to provide complicated systems of solid solutions and compounds which have phases involving the foregoing packings. As already stated, $Cd(OH)Cl$ has the complex CdI_2 arrangement. The corresponding $Ni(OH)Cl$ and the basic $Cd(OH)_{1.25}Cl_{0.75}$ have the $CdCl_2$ structure. The more basic cadmium salt $Cd(OH)_{1.75}Cl_{0.25}$ has the simple $Cd(OH)_2$ arrangement; for it, $a_0 = 3.53$ A. and $c_0 = 5.03$ A.

The mixed packing of this paragraph has also been found for the beta modification of *tantalum disulfide*, TaS_2. Its unit has the dimensions:

$$a_0 = 3.32 \text{ A.}, \qquad c_0 = 12.30 \text{ A.}$$

A number of additional stacking modifications of *cadmium iodide* have recently been described. Designating them, as with the several forms of SiC in terms of the number of layers per cell [whereby the $Cd(OH)_2$ stacking is $2H$ and the complex CdI_2 described above is $4H$], other observed forms are:

$6H$: $a_0 = 4.24$ A., $c_0 = 20.505$ A., 3 molecules/cell (Type 1)

$6H$: $a_0 = 4.24$ A., $c_0 = 20.505$ A., 3 molecules/cell (Type 2)

$8H$: $a_0 = 4.24$ A., $c_0 = 27.34$ A., 4 molecules/cell

$10H$: $a_0 = 4.24$ A., $c_0 = 34.17$ A., 5 molecules/cell

$12H(1)$: $a_0 = 4.24$ A., $c_0 = 41.01$ A., 6 molecules/cell (Type 1)

$12H(2)$: $a_0 = 4.24$ A., $c_0 = 41.01$ A., 6 molecules/cell (Type 2)

$12H(3)$: $a_0 = 4.24$ A., $c_0 = 41.01$ A., 6 molecules/cell (Type 3)

$14H$: $a_0 = 4.24$ A., $c_0 = 47.84$ A., 7 molecules/cell.

It appears that the iodine stackings in these forms may follow the sequences:

$6H(1)$: $(HCHCHH)_n = (HC)_2H_2 \ldots$

$6H(2)$: $(HCCHCC)_n$

$8H$: $(HHHCHCHH)_n = (HC)_2H_4 \ldots$

$10H$: $(HCHCHCHCHH)_n = (HC)_4H_2 \ldots$

$12H(1)$: $(CCHCHCHCHHCH)_n$

$12H(2)$: $(CHHCHHHCHHCH)_n = (CH_2)_2H(CH_2)CH \ldots$

$12H(3)$: $(HCHCHCHCHHHH)_n = (HC)_4H_4 \ldots$

$14H$: $(HCHCHCHCHCHCHH)_n = (HC)_6H_2 \ldots$

IV,c4. The compounds $NiBr_2$ and $CdBr_2$ appear to form irregularly mixed close-packings. Preparations crystallized from solution or ground from the melt have given extraordinarily simple diffraction patterns which can be interpreted as due to an irregular repetition of the cubic and hexagonal close-packings. The patterns correspond to very small hexagonal cells containing only one third of a molecule RX_2:

$$CdBr_2: \quad a_0' = 2.30 \text{ A.}, \quad c_0' = 6.23 \text{ A.}$$

$$NiBr_2: \quad a_0' = 2.11 \text{ A.}, \quad c_0' = 6.08 \text{ A.}$$

The reflections from these fractional cells are those common to the $Cd(OH)_2$ and $CdCl_2$ structures.

A similar incomplete structure is said to be formed when divalent metallic hydroxides, such as $Cd(OH)_2$, are rapidly precipitated with alkali. If these structures are thrown down from pure solutions they are unstable and quickly pass over into a fully crystalline form, but they are stabilized by the presence of sugars and other impurities. For the most part their only x-ray diffractions have the indices $(hk{\cdot}0)$, though weak basal reflections may sometimes be seen. The values of a_0 calculated from those reflections that appear are about 0.1 A. shorter than a_0 for the completed CdI_2-like arrangement. For example, for crystalline $Cd(OH)_2$, $a_0 = 3.49$ A. while with the incomplete form it is 3.36 A.

IV,c5. Several forms of *lead diiodide*, PbI_2, have been described as having hexagonal layered structures differing from one another in the way these layers are repeated. The simplest has the $Cd(OH)_2$ type of structure (see **IV,c1**). Its monomolecular unit has the dimensions:

$$a_0 = 4.557 \text{ A.}, \quad c_0 = 6.979 \text{ A.}$$

The next more complicated form, with two molecules in the unit, has the dimensions:

$$a_0 = 4.557 \text{ A.}, \quad c_0 = 13.958 \text{ A.}$$

It has been given the same structure as the bimolecular form of CdI_2 (**IV,c3**) with atoms in the positions of C_{6v}^4 ($P6_3mc$):

Pb: (2*b*) $1/3\ 2/3\ u;\ 2/3,1/3,u+1/2$ with $u = 0$

I(1): (2*b*) with $u' = 0.367$

I(2): (2*a*) $00v;\ 0,0,v+1/2$ with $v = 0.633$

There is next a trimolecular form with the dimensions:

$$a_0 = 4.557 \text{ A.,} \qquad c_0 = 20.937 \text{ A.}$$

Its atoms have been assigned the following positions in C_{3v}^1 ($P3m1$):

$$(1a)\ \ 00u; \qquad (1b)\ \ {}^1/_3\,{}^2/_3\,v; \qquad (1c)\ \ {}^2/_3\,{}^1/_3\,w$$

Two lead atoms have been placed in ($1a$) with $u = 0$ and $^1/_3$, and the third in ($1c$) with $w = {}^2/_3$. Of the iodine atoms, one was put in ($1a$) with $u = {}^2/_3 - 0.088 = 0.579$, two in $\pm({}^1/_3\,{}^2/_3\,v)$ with $v = 0.088$, two more in $\pm({}^1/_3,{}^2/_3,{}^2/_3 + 0.088)$, and one more in $^1/_3,{}^2/_3,{}^1/_3 + 0.088$.

Two forms based on a rhombohedral rather than an hexagonal lattice have also been described. The simpler of these has one molecule in the unit rhombohedron, which has the dimensions:

$$a_0 = 7.458 \text{ A.,} \qquad \alpha = 35°34'$$

The corresponding trimolecular cell has the same dimensions as that of the trimolecular hexagonal form already described:

$$a_0' = 4.557 \text{ A.,} \qquad c_0' = 20.937 \text{ A.}$$

The atoms have been placed in the following positions of D_{3d}^5 ($R\bar{3}m$):

Pb: ($3a$) 000; rh

I: ($6c$) $\pm(00u)$; rh with $u = 0.245$

This is, of course, the $CdCl_2$ arrangement (**IV,c2**).

The other rhombohedral form has two molecules in the unit rhombohedron:

$$a_0 = 14.204 \text{ A.,} \qquad \alpha = 18°28'$$

Its hexagonal cell, containing six molecules, has the edges:

$$a_0' = 4.557 \text{ A.,} \qquad c_0' = 41.874 \text{ A.}$$

Atoms appear to be in the following positions* of D_{3d}^5 ($R\bar{3}m$):

Pb(1): ($3a$) 000; rh

Pb(2): ($3b$) $0\,0\,{}^1/_2$; rh

* The descriptions given here are not identical with those in the original paper but they seem to be what was intended.

$$\text{I(1)}: \quad (6c) \quad \pm(00u);\ \text{rh} \qquad \text{with } u = 0.123$$

$$\text{I(2)}: \quad (6c) \quad \text{with } u' = 0.289$$

If we wish to describe these complex forms in terms of the stackings of layers along the hexagonal axis, as was done for SiC and for CdI_2 (in **IV,c3**) we have:

4*H*: $a_0 = 4.557$ A., $c_0 = 13.958$ A. $(CHCH)_n$ for the iodine atoms

6*H*: $a_0 = 4.557$ A., $c_0 = 20.937$ A. $(HHCHCH)_n$ for the iodine atoms

6*R*: $a_0 = 4.557$ A., $c_0 = 20.937$ A. $(CCCCCC)_n$ for the iodine atoms

12*R*: $a_0 = 4.557$ A., $c_0 = 41.874$ A. $[CH(HCC)H(HCC)HHC]_n$ for the iodine atoms

IV,c6. The compound *cadmium bromoiodide*, CdBrI, appears as a separate phase midway between $CdBr_2$ and CdI_2. It has rhombohedral symmetry with a unit cell more complicated than that for $CdCl_2$. It contains two molecules and has the dimensions:

$$a_0 = 13.46 \text{ A.,} \qquad \alpha = 17°34'$$

The cadmium atoms are in 000; $\frac{1}{2}\frac{1}{2}\frac{1}{2}$ and the halogen atoms, on trigonal axes, have various values of u with the coordinates uuu. The structure, however, was so imperfect that it could not be determined whether the halogen atoms are distributed in an orderly fashion or haphazardly. If the latter, the (Br, I) atoms are in two sets of $\pm(uuu)$, with $u = 0.125$ and $u' = 0.29$. In this case the structure is identical with that described in the preceding paragraph (**IV,c5**) for the second rhombohedral form of PbI_2.

IV,c7. The delta modification of *tantalum disulfide*, TaS_2, provides a further type of the CdI_2 stacking. Its hexagonal cell, containing six molecules, has the edge lengths:

$$a_0 = 3.34 \text{ A.,} \qquad c_0 = 35.94 \text{ A.}$$

The corresponding bimolecular rhombohedral unit has:

$$a_0' = 12.14 \text{ A.,} \qquad \alpha = 15°50'$$

In terms of this cell all atoms are in the special positions (6*c*) of D_{3d}^5 ($R\bar{3}m$):

$$(6c) \quad \pm(00u);\ \text{rh}$$

with $u(\text{Ta}) = 1/12$, $u(\text{S},1) = 5/24$ and $u(\text{S},2) = 3/8$. This is a cell similar to that described in the preceding paragraph for CdBrI, but the stacking is different.

As indicated below, the other three known forms of TaS_2 have hexagonal units with bases of the same size as that of this delta TaS_2:

$$\text{For alpha } TaS_2: \quad a_0 = 3.35 \text{ A.}, \quad c_0 = 5.86 \text{ A.}$$

$$\text{For beta } TaS_2: \quad a_0 = 3.32 \text{ A.}, \quad c_0 = 12.30 \text{ A.}$$

$$\text{For gamma } TaS_2: \quad a_0 = 3.32 \text{ A.}, \quad c_0 = 18.29 \text{ A.}$$

Of these, the alpha TaS_2 has the $Cd(OH)_2$ arrangement (**IV,c1**) and the beta form that of the Cd(OH)Cl structure (**IV,c3**). The gamma TaS_2 has the rhombohedral $CdCl_2$ grouping (**IV,c2**).

IV,c8. *Thallous sulfide,* Tl_2S, would appear to have a structure that involves a complicated form of mixed packing. It has been given the surprisingly large rhombohedral unit containing nine molecules and having the dimensions:

$$a_0 = 13.61 \text{ A.}, \qquad \alpha = 82°.$$

The hexagonal cell containing 27 molecules has:

$$a_0' = 12.20 \text{ A.}, \qquad c_0' = 18.17 \text{ A.}$$

This a_0' is approximately three times that of the $Cd(OH)_2$-like bases. The arrangement within this cell has been said to be roughly one with all atoms in general positions of C_3^4 ($R3$). In the hexagonal cell these coordinates are

$$(9b) \quad xyz; \ \bar{y},x-y,z; \ y-x,\bar{x},z; \ \text{rh}$$

with the six sets of thallium and the three sets of sulfur parameters col-

TABLE IV,9
Parameters of the Atoms in Tl_2S (**IV,c8**)

Atom	x	y	z
Tl(1)	0.12	0.20	0
Tl(2)	0.12	0.20	0.33
Tl(3)	0.12	0.20	0.67
Tl(4)	0.23	0.09	0.152
Tl(5)	0.23	0.09	0.482
Tl(6)	0.23	0.09	0.822
S(1)	0.33	0.33	0.243
S(2)	0.67	0	0.573
S(3)	0	0.67	0.913

lected in Table IV,9. This layer structure, upon which more work should be done, represents a relatively slight distortion of the simple $Cd(OH)_2$ grouping.

IV,c9. *Molybdenite*, MoS_2, and probably WS_2, are examples of a six-fold coordinated layer structure in which the coordination is not octahedral. These crystals are hexagonal with the elongated bimolecular units:

$$MoS_2: \quad a_0 = 3.1604 \text{ A.}, c_0 = 12.295 \text{ A.} \quad (26°\text{C.})$$

$$WS_2: \quad a_0 = 3.18 \text{ A.}, \quad c_0 = 12.5 \text{ A.}$$

Atoms are in the following special positions of D_{6h}^4 ($C6/mmc$):

$$\text{Mo (or W)}: \quad (2c) \quad \pm(^1/_3\ ^2/_3\ ^1/_4)$$

$$\text{S}: \quad (4f) \quad \pm(^1/_3\ ^2/_3\ u;\ ^2/_3,^1/_3,u+^1/_2)$$

with $u = 0.629$ for MoS_2 and undoubtedly close to $^5/_8$ for WS_2 (Fig. IV,22). The sulfur atoms about the metallic atom are at the corners of right equilateral trigonal prisms which share vertical edges with one another to build up RS_2 layers normal to the c_0 axis. This prismatic distribution of sulfur atoms about the metallic atom has already been seen in the NiAs arrangement (**III,d1**). In MoS_2, however, the crystal is built up by repeating these complete layers one above another according to the alternating requirements of hexagonal close-packing; in NiAs, which is not a layer structure, arsenic atoms are shared vertically as well as horizontally by the coordinating prisms. In MoS_2 the S–S distance between double layers, 3.66 A., is approximately the radial sum for ionic sulfur. The other S–S separations are much less, the nearest sulfur atoms in a trigonal prism being only 2.98 A. apart. The Mo–S separation is 2.35 A.

Two other substances known to have this structure have cells of the dimensions:

$$WSe_2: \quad a_0 = 3.29 \text{ A.}, \quad c_0 = 12.97 \text{ A.}$$

$$MoTe_2: \quad a_0 = 3.5182 \text{ A.}, c_0 = 13.9736 \text{ A.}$$

The parameter u has been determined to be 0.621.

A recently synthesized form of *molybdenum disulfide*, MoS_2, is rhombohedral and bears the same kind of relation to hexagonal MoS_2 that applies, for instance, to the two kinds of graphite.

Its trimolecular hexagonal cell has the dimensions:

$$a_0' = 3.16 \text{ A.}, c_0' = 1.5 \times 12.30 \text{ A.} = 18.45 \text{ A.}$$

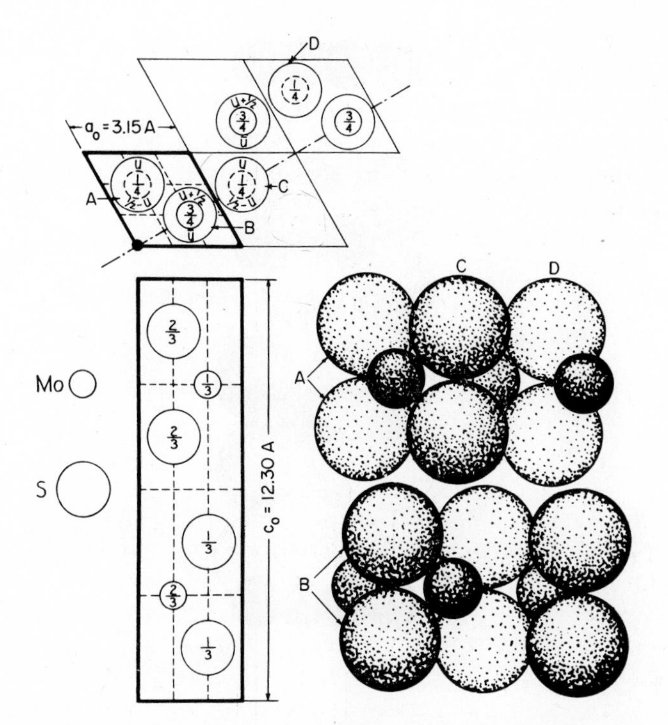

Fig. IV,22a (left). Projections of the MoS_2 structure. The top drawing is a basal projection with the hexagonal unit outlined by heavy lines. The projection of the atomic contents of one cell on a side face is shown below. Large circles are the sulfur atoms.

Fig. IV,22b (right). A packing drawing showing a portion of the MoS_2 structure viewed from a vertical plane through the diagonal dot-and-dash line of the top drawing of Figure IV,22a. The large spheres are the sulfur atoms. Letters refer to atoms similarly designated in the basal projection.

The space group is described as C_{3v}^5 ($R3m$) with atoms in the positions:

$$\text{Mo:}\quad (3a)\quad 00u;\ \text{rh}\qquad \text{with } u = 0$$

$$\text{S(1):}\quad (3a)\qquad \text{with } u' = {}^1/_4$$

$$\text{S(2):}\quad (3a)\qquad \text{with } u'' = {}^5/_{12}$$

The arrangement is pictured in Figure IV,23.

For *niobium disulfide*, NbS_2, the structure is now found to be like rhombohedral MoS_2 (rather than being of the previously assigned $CdCl_2$ type), with

$$a_0' = 3.33 \text{ A.,}\qquad c_0' = 17.91 \text{ A.}$$

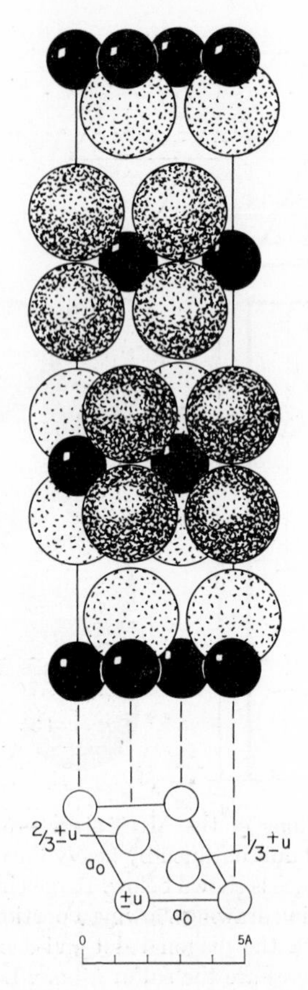

Fig. IV,23. Two projections of the rhombohedral MoS_2 structure in terms of its hexagonal unit. The black circles of the upper projection are molybdenum.

It has been given the same approximate parameters. Preparations with an excess of niobium in the range $Nb_{1+x}S_2$ with x = 0.12–0.25 also have this structure. There is, in addition, a hexagonal (non-rhombohedral) form of NbS_2 which has a_0 = 3.31 A., c_0 = 11.89 A. and a structure said to be unlike that of hexagonal MoS_2.

IV,c10. Crystals of the beta modification of *zinc hydroxychloride*, Zn(OH)Cl, are orthorhombic with a unit containing eight molecules and having the edge lengths:

$$a_0 = 5.86 \text{ A.}; \; b_0 = 6.58 \text{ A.}; \; c_0 = 11.33 \text{ A.}$$

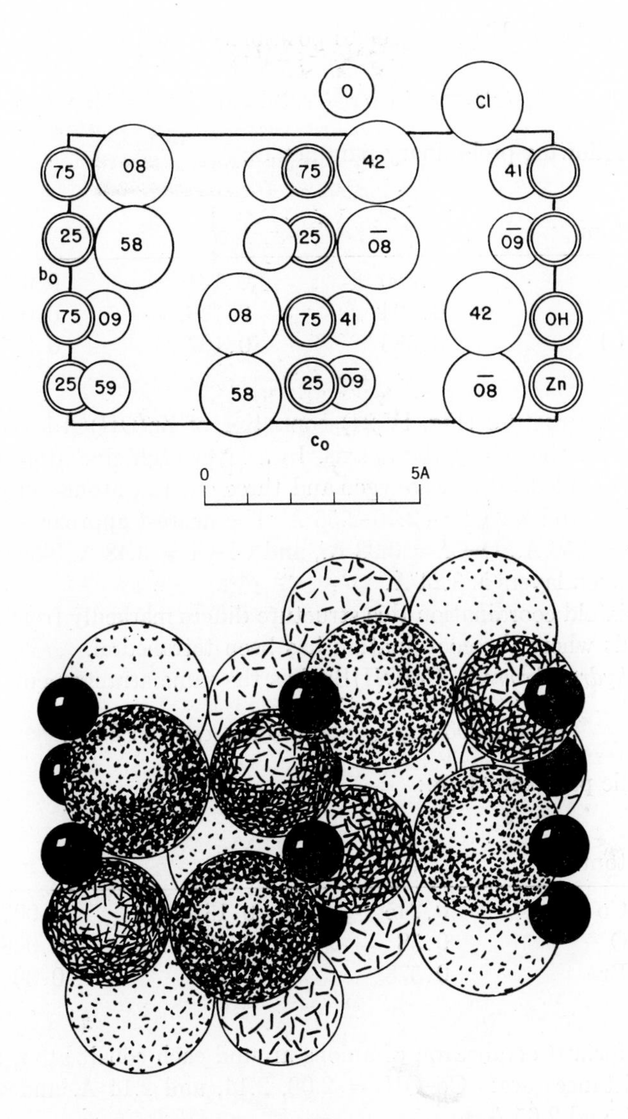

Fig. IV,24a (top). The orthorhombic structure of β-Zn(OH)Cl projected along its a_0 axis. Origin in lower left.

Fig. IV,24b (bottom). A packing drawing of the β-Zn(OH)Cl arrangement viewed along its a_0 axis. The zinc atoms are the black, the chlorine atoms the large dotted circles.

Atoms have been placed in general positions of V_h^{15} ($Pcab$):

$$(8c)\quad \pm(xyz;\ {}^1/_2-x,y+{}^1/_2,\bar{z};\ x+{}^1/_2,\bar{y},{}^1/_2-z;\ \bar{x},{}^1/_2-y,z+{}^1/_2)$$

with the following preliminary parameters:

Atom	x	y	z
Zn	0.250	0.131	0.000
O	0.594	0.126	0.071
Cl	0.583	0.107	0.357

This is a structure (Fig. IV,24) consisting of Zn(OH)Cl layers stacked one above another along the c_0 axis. In a layer each zinc atom is octahedrally surrounded by three oxygen and three chlorine atoms, with Zn–O = 2.01–2.17 A. and Zn–Cl = 2.46–2.56 A. The nearest approaches of anions are O–O = 2.56 A., O–Cl = 3.08 A., and Cl–Cl = 3.48 A. The separation O–Cl between layers is 3.24 A.

In its sixfold coordination this structure differs markedly from most zinc compounds where the coordination has been tetrahedral.

Cobalt hydroxybromide, Co(OH)Br, has the same arrangement, with

$$a_0 = 5.903 \text{ A.};\ b_0 = 6.700 \text{ A.};\ c_0 = 11.86 \text{ A.}$$

The atomic parameters are:

Atom	x	y	z
Co	0.245	0.137	0.002
O	0.58	0.103	0.068
Br	0.578	0.101	0.36

In the deformed octahedron of anions around each cobalt atom the interatomic distances are: Co–OH = 2.08, 2.14, and 2.14 A. and Co–Br = 2.49, 2.55, and 2.65 A.

It is possible that the following compounds also have this structure:

alpha Mn(OH)Cl: $a_0 = 6.07$ A.; $b_0 = 6.91$ A.; $c_0 = 11.46$ A.

beta Fe(OH)Cl: $a_0 = 5.93$ A.; $b_0 = 6.66$ A.; $c_0 = 11.33$ A.

Co(OH)Cl: $a_0 = 5.75$ A.; $b_0 = 6.60$ A.; $c_0 = 11.38$ A.

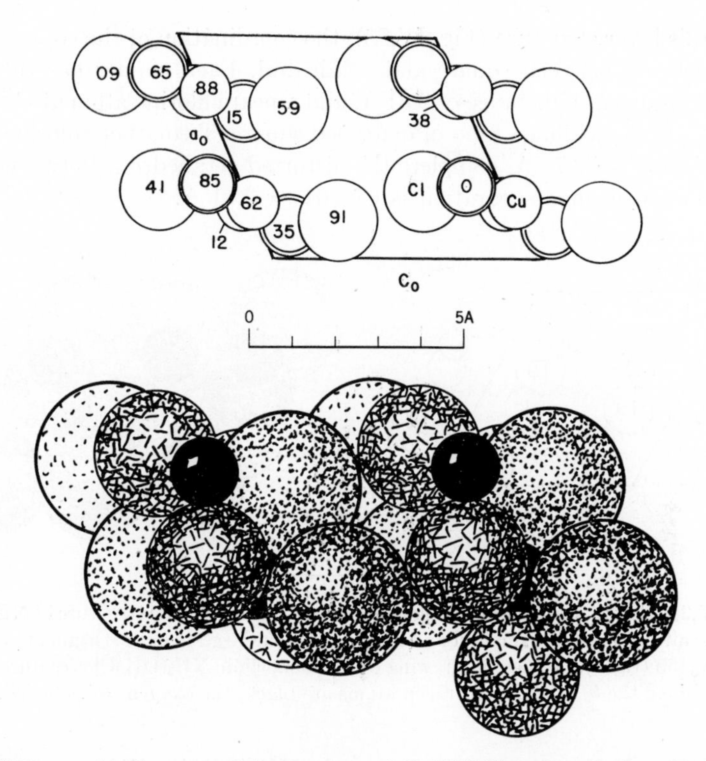

Fig. IV,25a (top). The monoclinic structure of Cu(OH)Cl projected along its b_0 axis. Origin in lower left.

Fig. IV,25b (bottom). A packing drawing of the Cu(OH)Cl structure viewed along its b_0 axis. The copper atoms are black, the oxygens line-shaded.

IV,c11. *Copper hydroxychloride*, Cu(OH)Cl, forms monoclinic crystals which have four molecules in a unit of the size:

$$a_0 = 5.555 \text{ A.}; \ b_0 = 6.671 \text{ A.}; \ c_0 = 6.127 \text{ A.}; \ \beta = 114°53'$$

The space group is C_{2h}^5 ($P2_1/a$) with all atoms in the general positions:

$$(4e) \quad \pm(xyz; \ x+{}^1/_2,{}^1/_2-y,z)$$

The parameters have been determined as:

Atom	x	y	z
Cu	0.2541	0.1173	−0.0319
Cl	0.3220	0.4101	−0.3120
O	0.1549	0.3493	0.1227

In this layer structure (Fig. IV,25), the coordination of the copper atoms is something between square and octahedral. Each copper has three hydroxyl groups at a distance of 2.01 A. and a near chlorine atom at Cu–Cl = 2.30 A., thus yielding a kind of deformed square. Two other chlorine atoms at a distance of 2.71 A. complete the distorted octahedron. Between layers the shortest atomic separation is Cl–OH = 3.21 A.

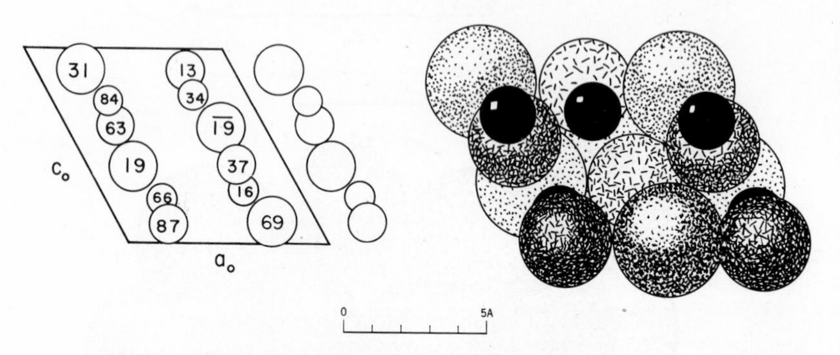

Fig. IV,26a (left). A projection along the b_0 axis of the monoclinic structure of $NH_3(OH)$-Cl. The nitrogens are the smallest and the chlorines the largest circles. Origin at lower left.
Fig. IV,26b (right). A packing drawing of the monoclinic $NH_3(OH)Cl$ structure viewed along the b_0 axis. The nitrogen atoms are black, the oxygens line-shaded.

IV,c12. *Hydroxyl ammonium chloride*, $NH_3(OH)Cl$, and the corresponding bromide, $NH_3(OH)Br$, are monoclinic with tetramolecular units of the dimensions:

$NH_3(OH)Cl$: a_0 = 6.95 A.; b_0 = 5.95 A.; c_0 = 7.69 A.; β = 120°48′

$NH_3(OH)Br$: a_0 = 7.29 A.; b_0 = 6.13 A.; c_0 = 8.04 A.; β = 120°48′

All atoms have been placed in general positions of C_{2h}^5 ($P2_1/c$):

$$(4e) \quad \pm(xyz;\ \bar{x},y+{}^1/_2,{}^1/_2-z)$$

with parameters for the chloride:

Atom	x	y	z
N:	0.283	0.664	0.222
OH:	0.253	0.870	0.106
Cl:	0.233	0.190	0.395

This is a layer structure (Fig. IV,26) in which N–O = 1.45 A. and in which each nitrogen atom has four chlorine neighbors at distances between 3.17 and 3.26 A. An oxygen atom has two close chlorine neighbors at 2.99 and 3.05 A., while the smallest Cl–Cl separation is 3.92 A.

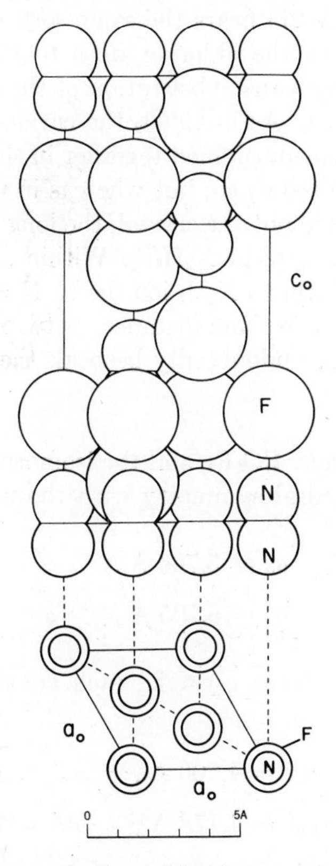

Fig. IV,27. Two projections of the hexagonal cell of $N_2H_6F_2$.

IV,c13. *Hydrazine hydrofluoride*, $N_2H_6F_2$, has a structure of lower symmetry than that of the chloride, $N_2H_6Cl_2$ (**IV,a3**). It is hexagonal with a unimolecular rhombohedral unit of the dimensions:

$$a_0 = 5.43 \text{ A.}, \qquad \alpha = 48°10'$$

The corresponding trimolecular hexagonal cell has the edges:

$$a_0' = 4.43 \text{ A.}, \qquad c_0' = 14.37 \text{ A.}$$

Atoms of nitrogen and of fluorine are in special positions of D_{3d}^5 ($R\bar{3}m$):

$$(6c) \quad \pm(00u);\ \text{rh}$$

with u(F) = 0.2435 and u(N) = 0.0495.

This structure (Fig. IV,27) bears the same sort of relation to the $CdCl_2$ arrangement (**IV,c2**) that the chloride does to CaF_2. Its rhombohedral symmetry is in accord with the substitution of the elongated N_2H_6 ion for the cadmium ion in $CdCl_2$. As in $CdCl_2$ the cations (in this case $N_2H_6^{2+}$) are at the centers of octahedra joined together in sheets stacked one above another in face-centered sequence; but whereas in $CdCl_2$ the anions are in substantial contact throughout the crystal, the long N_2H_6 ions prevent such uninterrupted anion contacts in $N_2H_6F_2$. Within an octahedral layer the closest F–F = 3.38 A. Within a N_2H_6 ion the N–N = 1.42 A. is the same as that found for $N_2H_6Cl_2$. The short distances between nitrogen and fluorine atoms, 2.62 and 2.80 A., undoubtedly bespeak the existence of hydrogen bonds between them.

IV,c14. *Uranyl fluoride*, UO_2F_2, and the isomorphous neptunyl fluoride, NpO_2F_2, have rhombohedral symmetry with the unimolecular cells:

$$UO_2F_2: \quad a_0 = 5.766 \text{ A.}, \qquad \alpha = 42°47'$$

$$NpO_2F_2: \quad a_0 = 5.795 \text{ A.}, \qquad \alpha = 42°16'$$

The hexagonal cells for these crystals, each containing three molecules, have the dimensions:

$$UO_2F_2: \quad a_0' = 4.206 \text{ A.}, \qquad c_0' = 15.692 \text{ A.}$$

$$NpO_2F_2: \quad a_0' = 4.178 \text{ A.}, \qquad c_0' = 15.80 \text{ A.}$$

The ideal arrangement chosen for UO_2F_2 has atoms in the following special positions of D_{3d}^5 ($R\bar{3}m$)

U: (3a) 000; rh

O: (6c) $\pm(00u)$; rh with u = 0.122

F: (6c) with u' = 0.294

The parameters have been selected to give expected interatomic distances. They result in a layer structure (Fig. IV,28) in which uranium has two oxygen neighbors at U–O = 1.91 A. and six fluorine neighbors at U–F =

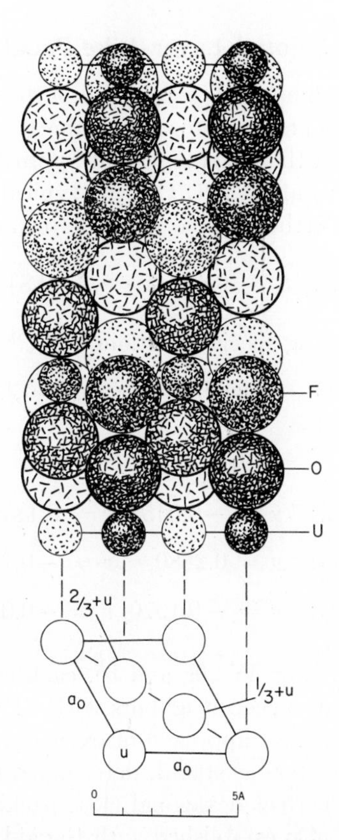

Fig. IV,28. Two projections of the rhombohedral UO_2F_2 structure in terms of its hexagonal axes. In the upper projection the uranium atoms are small and dotted and the oxygen atoms line-shaded. The similarity between this arrangement and that of $N_2H_6F_2$ becomes evident through a comparison of Figures IV,27 and IV,28.

2.50 A. Actual crystals have given a diffuseness of x-ray diffraction that has been interpreted as due to irregular displacements of successive layers.

Plutonyl fluoride, PuO_2F_2, is reported to have this structure with:

$$a_0 = 5.797 \text{ A.}, \qquad \alpha = 42°$$

Its trimolecular hexagonal cell has:

$$a_0' = 4.154 \text{ A.}, \qquad c_0' = 15.84 \text{ A.}$$

This arrangement differs from that of $N_2H_6F_2$ (**IV,c13**) only to the degree required by the difference in dimensions between the N_2H_6 and UO_2 cations.

Additional Layer-Like Packings

IV,d1. Besides the compounds considered in the preceding section there are numerous other RX_2 crystals for which the R coordination is more or less sixfold and in which the X arrangement approximates a close-packing.

The mineral *diaspore*, alpha AlO(OH), is typical of one of these structures. Its symmetry is orthorhombic with the tetramolecular unit:

$$a_0 = 4.396 \text{ A.}; \; b_0 = 9.426 \text{ A.}; \; c_0 = 2.844 \text{ A.} \; (25°\text{C.})$$

Atoms are in the following special positions of V_h^{16} (*Pbnm*):

$$(4c) \quad \pm(u\, v\, {}^1/_4;\; {}^1/_2-u, v+{}^1/_2, {}^1/_4)$$

The parameters are:

$$\text{Al:} \quad u = -0.0451, \; v = 0.1446$$

$$\text{O(1):} \quad u = 0.2880, \; v = -0.1989$$

$$\text{O(2):} \quad u = -0.1970, \; v = -0.0532$$

In this arrangement (Fig. IV,29a and b), each aluminum atom has six octahedrally distributed oxygen neighbors with Al–O separations between 1.85 and 1.98 A. The nearest approach of oxygen atoms to one another is 2.65 A. With the parameters as stated, determined by neutron diffraction, they are in an almost perfect hexagonal close-packing.

Hydrogen positions, also established with the aid of neutrons, are those of (4*c*) with $u = -0.4095$ and $v = -0.0876$. They are closest to the O(1) atoms, with O(1)–H = 0.990 A. The distance to the nearest O(2) atom is 1.694 A. and the direction O(1)–H makes an angle of 12° with the line O(1)–O(2). These hydrogen positions are shown in Figure IV,29c.

Several other substances with this structure have been completely studied. They are the following.

For the mineral *goethite*, alpha FeO(OH):

$$a_0 = 4.64 \text{ A.}; \; b_0 = 10.0 \text{ A.}; \; c_0 = 3.03 \text{ A.}$$

Its atomic parameters are:

$$\text{Fe:} \quad u = -0.03, \; v = 0.15$$

$$\text{O(1):} \quad u = 0.25, \; v = -0.294$$

$$\text{O(2):} \quad u = -0.25, \; v = -0.042.$$

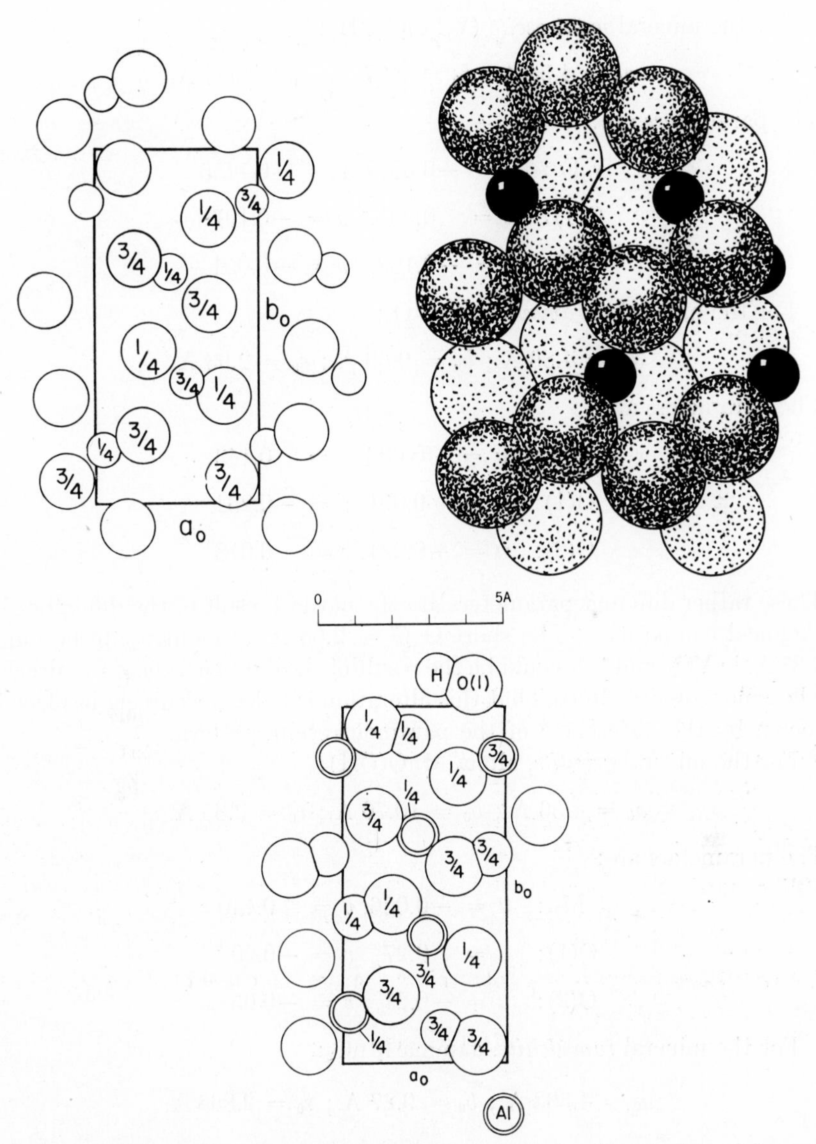

Fig. IV,29a (top, left). A projection along c_0 of the orthorhombic structure of diaspore, AlO(OH). The oxygens are the larger circles and no discrimination is made between O and OH. Origin in lower right.

Fig. IV,29b (top, right). A packing drawing of the orthorhombic structure for diaspore, AlO(OH), viewed along the c_0 axis. The black circles are aluminum.

Fig. IV,29c (bottom). The diaspore arrangement projected along its c_0 axis, as in Figure IV,29a, but with the hydrogen atoms added as small singly lined circles.

For the mineral *montroseite* (V,Fe)O(OH):

$$a_0 = 4.54 \text{ A.};\ b_0 = 9.97 \text{ A.};\ c_0 = 3.03 \text{ A.}$$

The parameters are:

$$\begin{aligned} &\text{V,Fe:} && u = -0.0517, && v = 0.1455 \\ &\text{O(1):} && u = 0.301, && v = -0.197 \\ &\text{O(2):} && u = -0.198, && v = -0.054 \end{aligned}$$

For the mineral *paramontroseite* VO_2:

$$a_0 = 4.89 \text{ A.};\ b_0 = 9.39 \text{ A.};\ c_0 = 2.93 \text{ A.}$$

The parameters are:

$$\begin{aligned} &\text{V:} && u = 0.088, && v = 0.143 \\ &\text{O(1):} && u = 0.091, && v = -0.254 \\ &\text{O(2):} && u = -0.231, && v = -0.018 \end{aligned}$$

These rather different parameters are the natural result of the difference in chemical composition. The short O–O = 2.63 A. in montroseite becomes 3.87 A. in VO_2, and this could be the result of the disappearance of hydrogen when montroseite alters. That this alteration is indeed occurring in place is shown by the diffuseness of the paramontroseite pattern.

For the mineral *groutite*, alpha MnO(OH):

$$a_0 = 4.56 \text{ A.};\ b_0 = 10.70 \text{ A.};\ c_0 = 2.85 \text{ A.}$$

The parameters are:

$$\begin{aligned} &\text{Mn:} && u = -0.036, && v = 0.140 \\ &\text{O(1):} && u = 0.27, && v = -0.20 \\ &\text{O(2):} && u = -0.21, && v = -0.05 \end{aligned}$$

For the mineral *ramsdellite*, gamma MnO_2:

$$a_0 = 4.533 \text{ A.};\ b_0 = 9.27 \text{ A.};\ c_0 = 2.866 \text{ A.}$$

Its parameters are:

$$\begin{aligned} &\text{Mn:} && u = -0.022, && v = 0.136 \\ &\text{O(1):} && u = 0.17, && v = -0.23 \\ &\text{O(2):} && u = -0.21, && v = -0.033 \end{aligned}$$

Some patterns of ramsdellite have shown additional broad lines which have been interpreted by saying that the crystals consist of irregularly repeated regions of pyrolusite (beta MnO_2) interspersed through the gamma MnO_2 structure.

IV,d2. A second modification of FeO(OH), the mineral *lepidochrosite*, is orthorhombic with a unit whose c_0 dimension is almost identical with that of the alpha form, goethite. Its other axes are different, however, and its structure is based on a different space group V_h^{17} (*Bbmm*). The four molecules in its unit:

$$a_0 = 12.4 \text{ A.}; \; b_0 = 3.87 \text{ A.}; \; c_0 = 3.06 \text{ A.}$$

are in the special positions:

$$(4c) \quad \pm(u\ {}^1/_4\ 0;\ u+{}^1/_2,{}^1/_4,{}^1/_2)$$

The atomic parameters are: u(Fe) = −0.178, u(O) = 0.21, u(OH) = 0.425. The resulting significant interatomic distances are Fe–O = 1.94–2.13 A., Fe–OH = 2.05 A., Fe–Fe = 2.88 A., O–O = O–OH = 2.80 A., and OH–OH = 2.70 A.

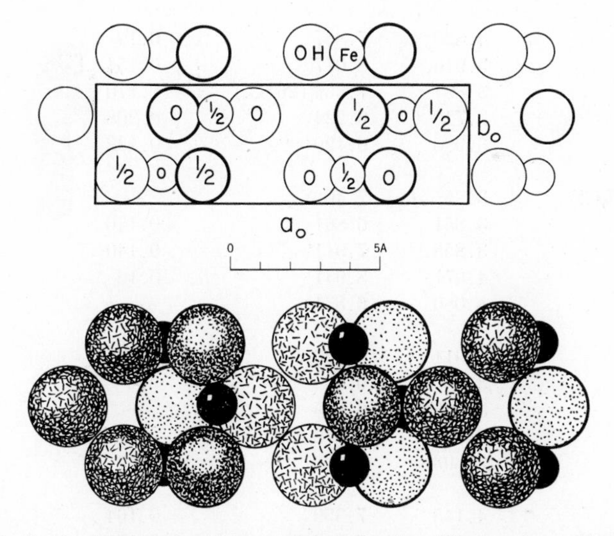

Fig. IV,30a (top). A projection along c_0 of the orthorhombic structure of lepidochrosite, FeO(OH). The small circles are iron, the hydroxyls are the light, larger circles. Origin in lower right.

Fig. IV,30b (bottom). A packing drawing of the orthorhombic structure of lepidochrosite, FeO(OH). The hydroxyls are line-shaded; iron atoms are small and black.

This structure (Fig. IV,30) is built up of well-defined layers parallel to the (100) face, each layer being made up of octahedra surrounding iron atoms and linked together by sharing corners. The octahedra are nearly regular and have four corners occupied by oxygen atoms and two by hydroxyl groups. The arrangement is related to those of FeOCl (**IV,d4**) and PbFCl (**IV,d3**).

Three other compounds are now known to have this structure. They are: For the mineral *boehmite*, gamma AlO(OH):

$$a_0 = 12.227 \text{ A.}; \; b_0 = 3.700 \text{ A.}; \; c_0 = 2.866 \text{ A. } (26°\text{C.})$$

The parameters are: $u(\text{Al}) = -0.166$, $u(\text{O}) = 0.213$, $u(\text{OH}) = 0.433$.

TABLE IV,10
Crystals with the Tetragonal PbFCl Arrangement (**IV,d3**)

Crystal	a_0, A.	c_0, A.	u(Metal)	u(Halogen)
AcOBr	4.29	7.42	—	—
AcOCl	4.25	7.08	—	—
AmOCl	4.00	6.78	0.18	—
BaHBr	4.564	7.428	0.175	0.67
BaHCl	4.408	7.202	0.215	0.65
BaHI	4.828	7.867	0.19	0.680
BiOBr	3.916	8.077	0.154	0.653
BiOCl	3.891	7.369 (26°C.)	0.170	0.645
BiOF	3.748	6.224	0.208	0.65
BiOI	3.985	9.129	0.132	0.668
BiO(OH,Cl)	3.85	7.40	—	—
CaHCl	3.851	6.861	0.146	0.695
CaHBr	3.858	7.911	0.140	0.67
CaHI	4.071	8.941	0.16	0.675
CeOCl	4.080	6.831	—	—
DyOCl	3.911	6.620	—	—
ErOCl	3.88	6.58	—	—
EuOCl	3.965	6.695	—	—
GdOCl	3.950	6.672	—	—
HoOCl	3.893	6.602	0.17	0.63
LaOBr	4.145	7.359	0.164	0.635
LaOCl	4.119	6.883	0.178	0.635
LaOI	4.144	9.126	0.135	0.660
NdOBr	4.017	7.619	0.16	0.64
NdOCl	4.018	6.782	0.18	0.64

(*continued*)

For gamma ScO(OH):

$$a_0 = 13.01 \text{ A.};\ b_0 = 4.01 \text{ A.};\ c_0 = 3.24 \text{ A.}$$

The parameters are: $u(\text{Sc}) = -0.182$, $u(0) = 0.218$, $u(\text{OH}) = 0.429$.
For *cupric hydroxide*, $Cu(OH)_2$:

$$a_0 = 10.59 \text{ A.};\ b_0 = 5.256 \text{ A.};\ c_0 = 2.949 \text{ A.}$$

The parameters are: $u(\text{Cu}) = -0.1825$, $u(\text{O},1) = 0.198$, $u(\text{O},2) = 0.435$.

IV,d3. The layered tetragonal structure for *lead fluochloride*, PbFCl, is also shown by about 50 substances (Table IV,10). Its unit, containing two

TABLE IV,10 (*continued*)

Crystal	a_0, A.	c_0, A.	u(Metal)	u(Halogen)
NpOS	3.825	6.654	0.200	0.638
PaOS	3.832	6.704	—	—
PbFBr	4.18	7.59	0.195	0.65
PbFCl	4.106	7.23	0.20	0.65
PrOCl	4.051	6.810	0.18	0.64
PuOBr	4.022	7.571	0.16	0.64
PuOCl	4.012	6.792	0.18	0.64
PuOI	4.042	9.169	0.13	0.67
PuOSe	4.151	8.369	—	—
SmOCl	3.982	6.721	0.17	0.63
SmOI	4.016	9.210	—	—
SrHBr	4.254	7.290	0.155	0.68
SrHCl	4.100	6.961	0.199	0.66
SrHI	4.371	8.450	0.20	0.700
TbOCl	3.927	6.645	—	—
ThOS	3.963	6.746	0.200	0.647
ThOSe	4.038	7.03	0.18	0.63
ThOTe	4.118	7.549	0.18	0.63
TmOI	3.895	9.184	0.125	0.680
UOCl	4.00	6.85	—	—
UOS	3.843	6.694	0.200	0.638
UOSe	3.908	6.996	—	—
UTe_2	4.006	7.471	—	—
YOCl	3.903	6.597	0.18	0.64
YbOI	3.878	9.179	—	—

molecules, has atoms in the following positions of D_{4h}^7 ($P4/nmm$):

F: (2*a*) 000; $\frac{1}{2}\frac{1}{2}0$

Cl: (2*c*) $0\,\frac{1}{2}\,u$; $\frac{1}{2}\,0\,\bar{u}$ with $u = 0.65$

Pb: (2*c*) with $u' = 0.20$.

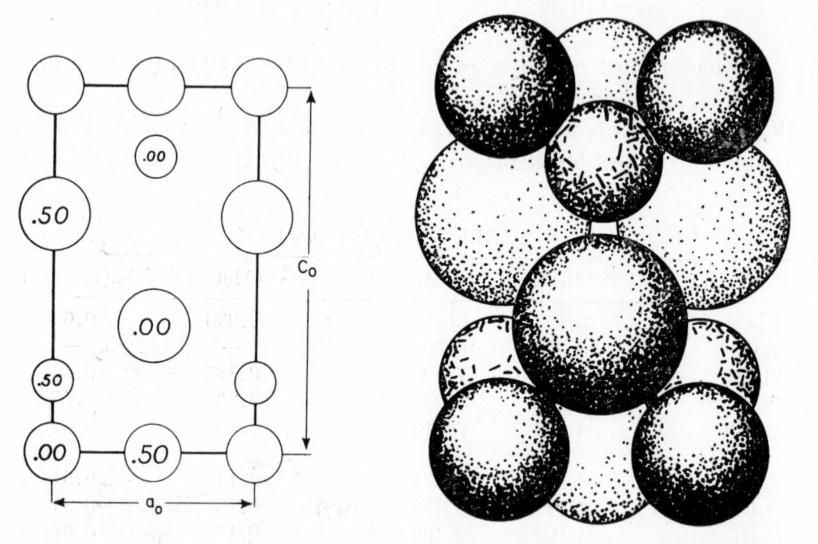

Fig. IV,31a (left). A projection along an a_0 axis of the atomic arrangement in the tetragonal unit of PbFCl. The smallest circles are the lead atoms, the largest are those of chlorine.

Fig. IV,31b (right). A drawing showing the way atoms of PbFCl pack together if they have their usual ionic sizes. The largest spheres are the chlorine ions, the line-shaded small spheres are those of Pb^{2+}.

With these parameters each lead atom has four fluorine atoms at the distance of 2.52 A. and five chlorine atoms at distances of 3.07 and 3.21 A. The halogen separations are Cl–Cl = 3.61 A., F–F = 2.89 A., and Cl–F = 3.24 A., all of which are close to the ionic sums.

In this arrangement (Fig. IV,31), layers of lead and fluorine atoms are related to one another much as they are in the fluorite modification of PbF_2 (**IV,a1**). They are bound together along the c_0 axis by a double layer of chlorine atoms.

IV,d4. The layer structure given the orthorhombic *ferric oxychloride*, FeOCl, is developed from the space group V_h^{13} ($Pmmn$). Its bimolecular cell has the dimensions:

$$a_0 = 3.75 \text{ A.}; \; b_0 = 3.3 \text{ A.}; \; c_0 = 7.65 \text{ A.}$$

Atoms are in the following special positions:

Fe: (2*b*) $0\,{}^1/_2\,u;\; {}^1/_2\,0\,\bar{u}$ with $u = 0.097$

Cl: (2*a*) $00v;\; {}^1/_2\,{}^1/_2\,\bar{v}$ with $v = 0.305$

O: (2*a*) with $v' = -0.083$.

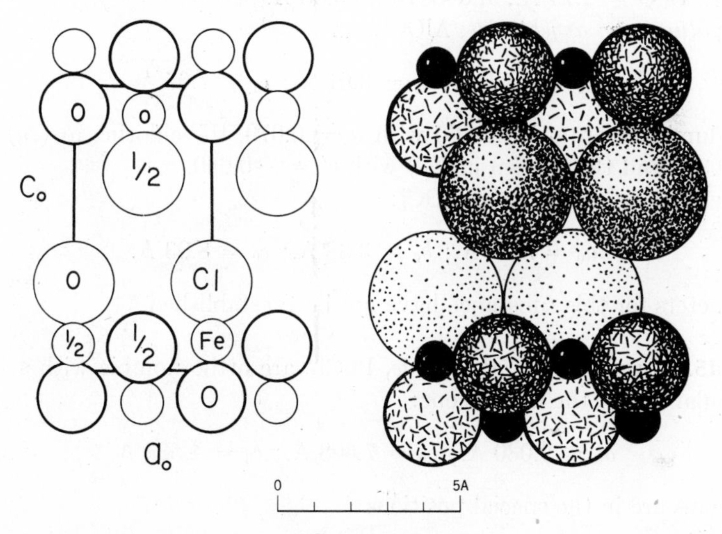

Fig. IV,32a (left). A projection along b_0 of the orthorhombic structure of FeOCl. Origin in lower left.

Fig. IV,32b (right). A packing drawing of the orthorhombic structure of FeOCl viewed along the b_0 axis. The oxygen atoms are line-shaded, the chlorine atoms are dotted.

As is obvious from Figure IV,32, this arrangement is an orthorhombic distortion of the tetragonal PbFCl structure of **IV,d3.** The significant interatomic distances are Fe–Cl = 2.29 A., Fe–O = 1.88 and 2.15 A., O–O = 2.80 A., Cl–Cl = 3.90 A., and O–Cl = 2.98 and 3.02 A.

Four other compounds are known to have this structure. They are:

For *indium oxychloride*, InOCl:

$$a_0 = 4.065 \text{ A.}; \; b_0 = 3.523 \text{ A.}; \; c_0 = 8.080 \text{ A.}$$

The indium atom is in (2*b*) with u = 0.121, the chlorine in (2*a*) with v = 0.345 and the oxygen in (2*a*) with v' = −0.03. The significant interatomic separations are In–O = 2.16 A., In–Cl = 2.51 A., Cl–Cl = 3.52 and 3.69 A., O–O = 2.74 A., and O–Cl = 3.05 A.

For *indium oxybromide*, InOBr:

$$a_0 = 4.049 \text{ A.}; \; b_0 = 3.611 \text{ A.}; \; c_0 = 8.649 \text{ A.}$$

The indium atom is in (2*b*) with u = 0.113, the bromine in (2*a*) with v = 0.336, and the oxygen in (2*a*) with v' = −0.02. The important interatomic distances are In–O = 2.15 and 2.18 A., In–Br = 2.64 A., Br–Br = 3.61 and 3.94 A., O–O = 2.74 A., and O–Br = 3.08 A.

For *aluminum oxychloride*, AlOCl:

$$a_0 = 3.62 \text{ A.}; \; b_0 = 3.61 \text{ A.}; \; c_0 = 7.67 \text{ A.}$$

The aluminum atom is in (2*b*) with u = 0.070, the chlorine in (2*a*) with v = 0.300 and the oxygen in (2*a*) with v' = −0.060.

For *titanium oxychloride*, TiOCl:

$$a_0 = 3.79 \text{ A.}; \; b_0 = 3.38 \text{ A.}; \; c_0 = 8.03 \text{ A.}$$

Parameters for this substance have not been established.

IV,d5. Crystals of *lead chloride*, $PbCl_2$, are orthorhombic with a tetramolecular cell of the dimensions:

$$a_0 = 9.030 \text{ A.}; \; b_0 = 7.608 \text{ A.}; \; c_0 = 4.525 \text{ A.}$$

All atoms are in the special positions

$$(4c) \quad \pm(u\,v\,{}^1/_4;\; {}^1/_2-u,v+{}^1/_2,{}^1/_4)$$

of V_h^{16} (*Pbnm*), with the parameters u(Pb) = 0.0956, v(Pb) = 0.2617, u(Cl,1) = 0.0742, v(Cl,1) = 0.8610, u(Cl,2) = 0.8370, and v(Cl,2) = 0.4768.

This structure (Fig. IV, 33a,b) can be thought of as a considerably distorted close-packing of halogen atoms with the lead atoms accommodated in the same plane with them; it bears definite relationships to those of PbFCl (**IV,d3**) and of NH_4CdCl_3 (Chapter VII). The cation coordination is hard to define. Each lead atom has about it two chlorine atoms at 2.67 A., one each at 3.05, 3.08, and 2.88 A., and two each at 3.13 and 3.29 A. Several of the Cl–Cl separations are shorter than the sum of the atomic radii, varying between ca. 3.20 and 3.47 A.

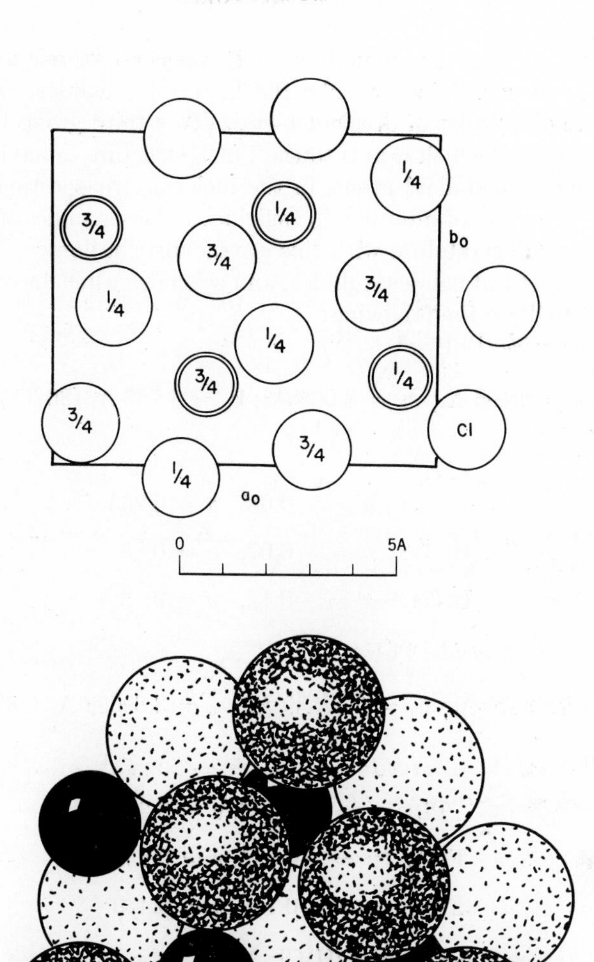

Fig. IV,33a (top). The orthorhombic structure of $PbCl_2$ projected along its c_0 axis. Origin in lower right.

Fig. IV,33b (bottom). A packing drawing of the $PbCl_2$ arrangement viewed along the c_0 axis. The lead atoms are black.

The numerous other compounds with this general atomic arrangement fall into four groups. Some are, like $PbCl_2$, metallic halides, or disulfides, and some are dihydrides of divalent metals. To a third group belong such compounds as Co_2P which is anti to the $PbCl_2$ structure in having twice as many metallic as metalloid atoms in the molecule. In addition, there are now known a number of intermetallic silicides and germanides of the transition metals which crystallize with this atomic distribution.

Other isostructural halides, sulfides, and selenides which have been completely analyzed are the following:

For *lead bromide*, $PbBr_2$:

$$a_0 = 9.466 \text{ A.};\ b_0 = 8.068 \text{ A.};\ c_0 = 4.767 \text{ A. } (26°\text{C.})$$

The parameters are:

$$\begin{aligned} \text{Pb:}\quad u &= 0.087,\ v = 0.265 \\ \text{Br(1):}\quad u &= 0.07,\ \ v = 0.86 \\ \text{Br(2):}\quad u &= -0.17,\ v = 0.48 \end{aligned}$$

For alpha *lead fluoride*, PbF_2:

$$a_0 = 7.63574 \text{ A.};\ b_0 = 6.42689 \text{ A.};\ c_0 = 3.89098 \text{ A. } (18°\text{C.})$$

The parameters are:

$$\begin{aligned} \text{Pb:}\quad u &= 0.103,\ v = 0.244 \\ \text{F(1):}\quad u &= 0.085,\ v = 0.858 \\ \text{F(2):}\quad u &= -0.186,\ v = 0.449 \end{aligned}$$

For *lead hydroxychloride*, Pb(OH)Cl:

$$a_0 = 9.7 \text{ A.},\ b_0 = 7.1 \text{ A.};\ c_0 = 4.05 \text{ A.}$$

The parameters are:

$$\begin{aligned} \text{Pb:}\quad u &= 0.088,\ v = 0.204 \\ \text{OH:}\quad u &= 0.125,\ v = 0.831 \\ \text{Cl:}\quad u &= -0.185,\ v = 0.469 \end{aligned}$$

For *lead hydroxyiodide*, Pb(OH)I:

$$a_0 = 10.41 \text{ A.};\ b_0 = 7.80 \text{ A.};\ c_0 = 4.19 \text{ A.}$$

The parameters are:

$$\text{Pb:}\quad u = 0.081,\ v = 0.181$$
$$\text{OH:}\quad u = 0.056,\ v = 0.855$$
$$\text{I:}\quad u = -0.180,\ v = 0.455$$

For *thorium disulfide*, ThS_2:

$$a_0 = 8.617\ \text{A.};\ b_0 = 7.263\ \text{A.};\ c_0 = 4.267\ \text{A.}$$

The parameters are:

$$\text{Th:}\quad u = 0.125,\ v = 0.250$$
$$\text{S(1):}\quad u = 0.068,\ v = 0.850$$
$$\text{S(2):}\quad u = -0.180,\ v = 0.465$$

For *thorium diselenide*, $ThSe_2$:

$$a_0 = 9.064\ \text{A.};\ b_0 = 7.610\ \text{A.};\ c_0 = 4.420\ \text{A.}$$

The parameters are:

$$\text{Th:}\quad u = 0.125,\ v = 0.250$$
$$\text{Se(1):}\quad u = 0.07,\ v = 0.88$$
$$\text{Se(2):}\quad u = -0.180,\ v = 0.47$$

The following two anti-$PbCl_2$ phosphides have cells of the same shape as the above, and substantially the same parameters:

For *dicobalt phosphide*, Co_2P:

$$a_0 = 6.608\ \text{A.};\ b_0 = 5.646\ \text{A.};\ c_0 = 3.513\ \text{A.}$$

The parameters are:

$$\text{P:}\quad u = 0.1249,\ v = 0.2461$$
$$\text{Co(1):}\quad u = 0.0647,\ v = 0.8560$$
$$\text{Co(2):}\quad u = -0.1657,\ v = 0.4685$$

For *diruthenium phosphide*, Ru_2P:

$$a_0 = 6.896\ \text{A.};\ b_0 = 5.902\ \text{A.};\ c_0 = 3.859\ \text{A.}$$

The parameters are:

$$\text{P:}\quad u = 0.1135,\ v = 0.2455$$

$$\text{Ru(1):}\quad u = 0.0736,\ v = 0.8585$$

$$\text{Ru(2):}\quad u = -0.1586,\ v = 0.4780$$

A third phosphide, *dirhenium phosphide*, Re_2P, has a differently shaped unit and parameters that depart considerably from the foregoing values. For it,

$$a_0 = 10.040 \text{ A.};\ b_0 = 5.540 \text{ A.};\ c_0 = 2.939 \text{ A.}$$

The parameters have been determined as:

$$\text{P:}\quad u = 0.110,\quad v = 0.395$$

$$\text{Re(1):}\quad u = 0.0655,\quad v = 0.8295$$

$$\text{Re(2):}\quad u = -0.2850,\ v = 0.3520$$

Other compounds of this type whose cell data indicate that they have the $PbCl_2$ structure are:

$$BaBr_2\text{:}\quad a_0 = 9.838 \text{ A.};\ b_0 = 8.247 \text{ A.},\ c_0 = 4.948 \text{ A.}$$

$$BaCl_2\text{:}\quad a_0 = 9.333 \text{ A.};\ b_0 = 7.823 \text{ A.},\ c_0 = 4.705 \text{ A.}$$

$$BaI_2\text{:}\quad a_0 = 10.566 \text{ A.};\ b_0 = 8.862 \text{ A.},\ c_0 = 5.268 \text{ A.}$$

$$EuCl_2\text{:}\quad a_0 = 8.914 \text{ A.};\ b_0 = 7.499 \text{ A.},\ c_0 = 4.493 \text{ A.}$$

$$SmCl_2\text{:}\quad a_0 = 8.973 \text{ A.};\ b_0 = 7.532 \text{ A.},\ c_0 = 4.497 \text{ A.}$$

$$SbTeI\text{:}\quad a_0 = 10.8 \text{ A.};\ b_0 = 9.18 \text{ A.},\ c_0 = 4.23 \text{ A.}$$

$$\text{Beta } US_2\text{:}\quad a_0 = 8.46 \text{ A.};\ b_0 = 7.11 \text{ A.},\ c_0 = 4.12 \text{ A.}$$

$$\text{Beta } USe_2\text{:}\quad a_0 = 8.98 \text{ A.};\ b_0 = 7.46 \text{ A.},\ c_0 = 4.26 \text{ A.}$$

Parameters have not been determined for these substances.

The tetramolecular orthorhombic unit of *strontium bromide*, $SrBr_2$, points to an atomic arrangement of the $PbCl_2$ type though with a_0 and b_0 axes reversed in relative lengths:

$$a_0 = 9.20 \text{ A.};\ b_0 = 11.42 \text{ A.};\ c_0 = 4.3 \text{ A.}$$

Changing the signs of the y parameters, i.e., the direction of the b_0 axis referred to the original description, one obtains the following parameters for

$SrBr_2$. They are very similar to those found for $PbCl_2$:

$$\text{Sr:}\quad u = 0.111,\ v = -0.189$$

$$\text{Br(1):}\quad u = 0.119,\ v = 0.103$$

$$\text{Br(2):}\quad u = -0.158,\ v = 0.614$$

In this arrangement (Fig. IV,33c,d) the metallic coordination is, as with $PbCl_2$, somewhat indefinite, each strontium atom having one bromine neighbor at 3.16 A., two at 3.21 A., one at 3.32 A., another at 3.44 A., and the next at 4.80 A.

The alkaline earth hydrides and the two rare-earth deuterides EuD_2 and YbD_2 have cells similar in shape to that of $PbCl_2$. The positions found for the metallic atoms, similar to those for lead in the chloride, place them in a nearly perfect hexagonal close-packing. It has been natural to expect the hydrogens in interstices of this packing. With this in mind the following hydrogen parameters were suggested for SrH_2. The cell dimensions for this hydride are:

$$SrH_2:\quad a_0 = 7.343\text{ A.};\ b_0 = 6.364\text{ A.};\ c_0 = 3.875\text{ A.}$$

The parameters as determined for strontium and proposed for hydrogen are:

$$\text{Sr:}\quad u = 0.010,\ v = 0.240$$

$$\text{H(1):}\quad u = 0.430,\ v = 0.240$$

$$\text{H(2):}\quad u = 0.742,\ v = 0.004$$

The hydrogen parameters are very different from the halogen parameters in the $PbCl_2$-like crystals.

Unit cells have also been measured for the following:

$$BaH_2:\quad a_0 = 7.829\text{ A.};\ b_0 = 6.788\text{ A.};\ c_0 = 4.167\text{ A.}$$

$$CaH_2:\quad a_0 = 6.838\text{ A.};\ b_0 = 5.936\text{ A.};\ c_0 = 3.600\text{ A.}$$

$$EuD_2:\quad a_0 = 7.16\text{ A.};\ b_0 = 6.21\text{ A.};\ c_0 = 3.77\text{ A.}$$

$$YbD_2:\quad a_0 = 6.763\text{ A.};\ b_0 = 5.871\text{ A.};\ c_0 = 3.561\text{ A.}$$

A determination of the ytterbium parameters in the last compound has resulted in the same values found for strontium in SrH_2: $u(\text{Yb}) = 0.010$, $v(\text{Yb}) = 0.240$.

The compound Rh_2Ge is typical of the intermetallic compounds that have $PbCl_2$-like unit cells. Its unit with axes in the *Pbnm* orientation chosen

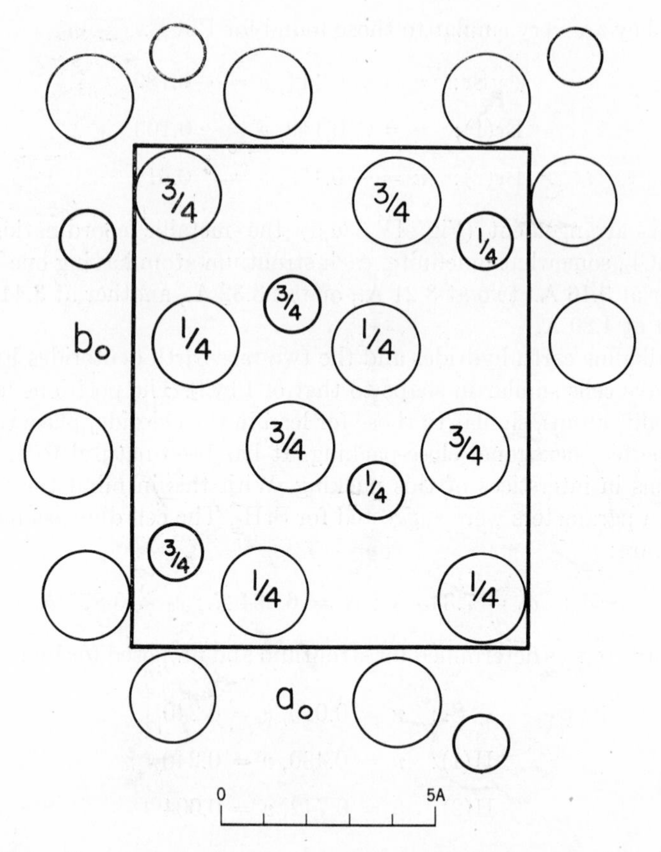

Fig. IV,33c. The structure of $SrBr_2$ projected, as with $PbCl_2$, along its c_0 axis. The changes in atomic environment that result from a reversal of the relative lengths of a_0 and b_0 are seen by comparing this drawing with that of Figure IV,33a.

for $PbCl_2$ has the dimensions:

$$a_0 = 7.57 \text{ A.};\ b_0 = 5.44 \text{ A.};\ c_0 = 4.00 \text{ A.}$$

The determined parameters have, however, been so different from those of $PbCl_2$ that the two structures must be considered as little more than formally related to one another. They are:

$$\text{Ge:}\quad u = 0.605,\ v = 0.713$$

$$\text{Rh(1):}\quad u = 0.207,\ v = 0.029$$

$$\text{Rh(2):}\quad u = 0.571,\ v = 0.163$$

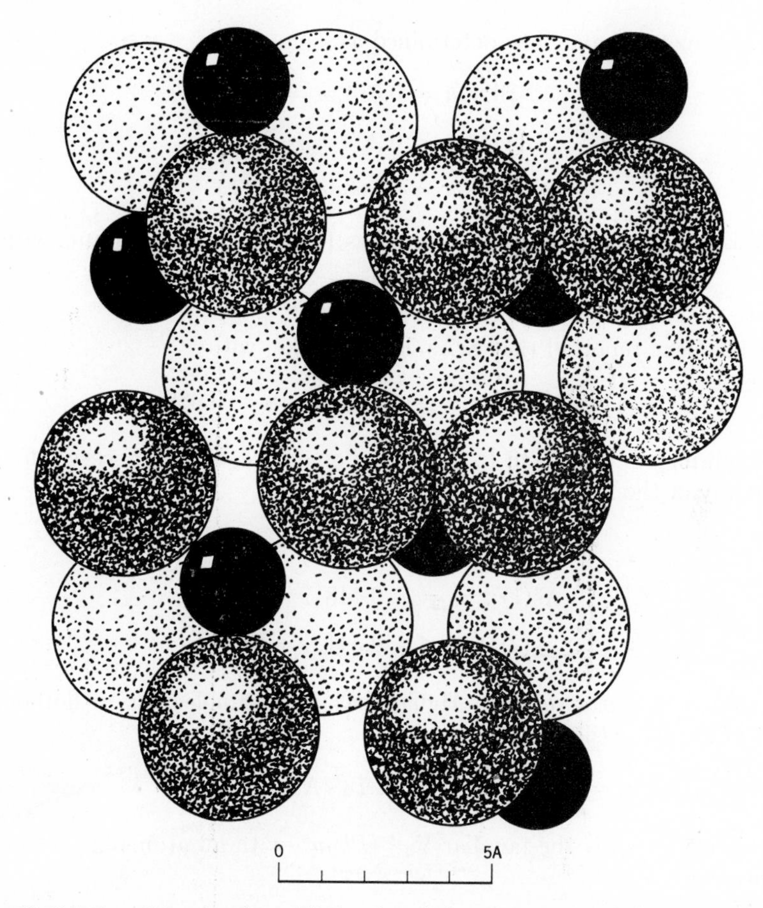

Fig. IV,33d. A packing drawing of $SrBr_2$ viewed along its c_0 axis. The strontium atoms are black.

Other intermetallic compounds known to have this structure are:

Co_2Si: $a_0 = 7.109$ A.; $b_0 = 4.918$ A.; $c_0 = 3.738$ A.

The parameters have been determined as:

Si: $u = 0.611$, $v = 0.702$

Co(1): $u = 0.218$, $v = 0.038$

Co(2): $u = 0.562$, $v = 0.174$

Ni_2Si: $a_0 = 7.03$ A.; $b_0 = 4.99$ A.; $c_0 = 3.72$ A.

The parameters have been determined as:

$$\text{Si:} \quad u = 0.614, v = 0.714$$

$$\text{Ni(1):} \quad u = 0.203, v = 0.042$$

$$\text{Ni(2):} \quad u = 0.563, v = 0.175$$

Probably Rh_2Si (studied mistakenly as Rh_2B) has this structure, with

$$a_0 = 7.44 \text{ A.}; \ b_0 = 5.42 \text{ A.}; \ c_0 = 3.98 \text{ A.}$$

It has been reported that $ZrAs_2$ is $PbCl_2$-like, with

$$a_0 = 9.027 \text{ A.}; \ b_0 = 6.801 \text{ A.}; \ c_0 = 3.689 \text{ A.}$$

The parameters that seem to have been given are, however, very different from any of the foregoing:

$$\text{Zr:} \quad u = -0.159, v = 0.225$$

$$\text{As(1):} \quad u = 0.148, v = 0.081$$

$$\text{As(2):} \quad u = 0.541, v = 0.122$$

IV,d6. *Mercuric chloride*, $HgCl_2$, has a tetramolecular orthorhombic unit of the dimensions:

$$a_0 = 4.325 \text{ A.}; \ b_0 = 12.735 \text{ A.}; \ c_0 = 5.963 \text{ A.}$$

The space group is the familiar V_h^{16} (*Pbnm*) with all atoms in the special positions:

$$(4c) \quad \pm (u\, v\, {}^1/_4;\ {}^1/_2-u, v+{}^1/_2, {}^1/_4)$$

The selected parameters are:

$$\text{Hg:} \quad u = 0.050, v = 0.126$$

$$\text{Cl(1):} \quad u = 0.406, v = 0.255$$

$$\text{Cl(2):} \quad u = 0.806, v = 0.496$$

This is an arrangement in which sheets of $HgCl_2$ molecules are stacked one above another along the c_0 axis in such a way that the nearest chlorine atoms are only 3.35 A. apart (Fig. IV, 34). These separations are considerably less than the sums of the ionic radii. Within a $HgCl_2$ molecule, the two Hg–Cl distances are 2.23 and 2.27 A.

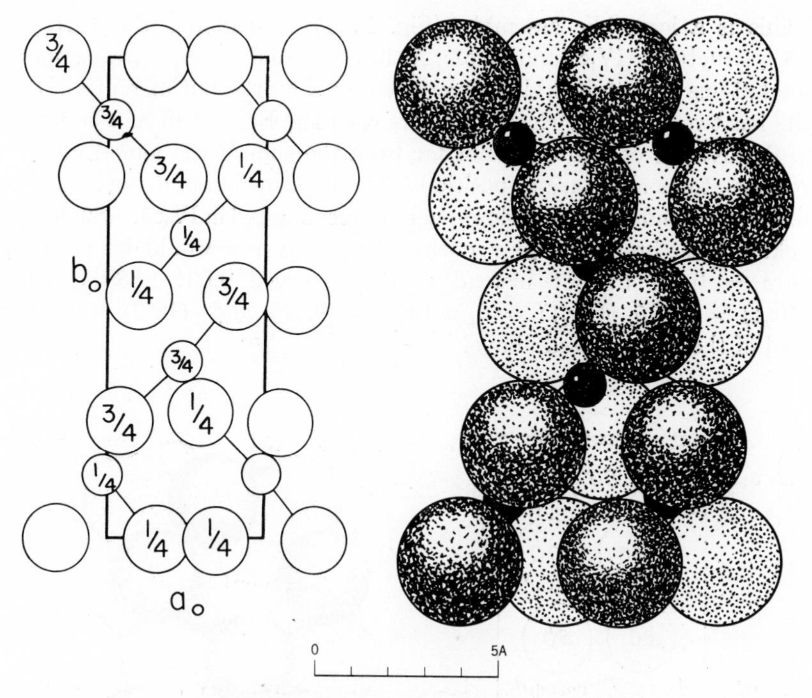

Fig. IV,34a (left). A projection along c_0 of the orthorhombic structure of $HgCl_2$. The larger circles are chlorine. Origin in lower right.

Fig. IV,34b (right). A packing drawing of the orthorhombic structure of $HgCl_2$ viewed along the c_0 axis. The dotted atoms are chlorine.

IV,d7. *Mercuric bromide*, $HgBr_2$, and the yellow form of *mercuric iodide*, HgI_2, have units of the same general shape as that of $HgCl_2$ (**IV,d6**), but the structure given them is an entirely different one. Their unit cells have the dimensions:

$$HgBr_2: \quad a_0 = 4.624 \text{ A.}; \ b_0 = 12.445 \text{ A.}; \ c_0 = 6.798 \text{ A.}$$

$$HgI_2: \quad a_0 = 4.674 \text{ A.}; \ b_0 = 13.76 \text{ A.}; \ c_0 = 7.32 \text{ A.}$$

The atoms in their four-molecule cells are all in the following special positions of C_{2v}^{12} ($Bm2b$):

$$(4a) \quad 0uv; \ 0,u+{}^1/_2,\bar{v}; \ {}^1/_2,u,v+{}^1/_2; \ {}^1/_2,u+{}^1/_2,{}^1/_2-v$$

Satisfactory parameters have not been published for the iodide, but for the bromide, $u(\text{Hg}) = 0$, $v(\text{Hg}) = 0.334$, $u(\text{Br},1) = 0.132$, $v(\text{Br},1) = 0.056$, $u(\text{Br},2) = 0.368$, and $v(\text{Br},2) = 0.389$.

This is a layer-like assembly (Fig. IV,35) resembling the CdI_2 and $CdCl_2$ structures (**IV,c1** and **IV,c2**). As in these other layered arrangements, metal atoms are octahedrally surrounded by X atoms, but mercury in these salts has two especially close neighbors with Hg–Br = 2.48 A. The nearest Br–Br distance of 3.74 A. is not far from the sum of the ionic radii. The bromine atoms alone are in positions that are a moderate distortion of the mixed cubic and hexagonal close-packing encountered in **IV,c3**. Such close-packing with bromine layers normal to the b_0 axis would be perfect if $a:b:c = 0.577:1.885:1$ instead of the observed $0.681:1.880:1$ and if $u(\text{Br},1) = 0.125$, $v(\text{Br},1) = 0$, $u(\text{Br},2) = 0.375$, and $v(\text{Br},2) = 0.33$ instead of the values listed above.

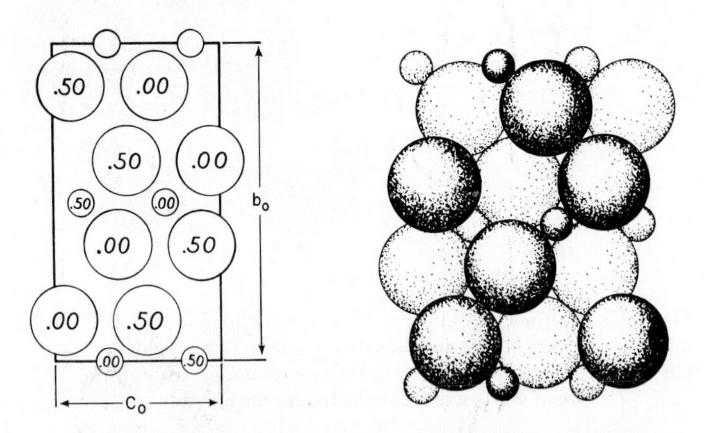

Fig. IV,35a (left). The orthorhombic structure of $HgBr_2$ projected along the a_0 axis. The large circles are the bromine atoms. Origin in lower left.

Fig. IV,35b (right). A drawing of $HgBr_2$ illustrating the way the atoms of Figure IV,35a pack together if they are assigned their ionic sizes. The large spheres are bromine ions.

IV,d8. The mixed halide phase, $Hg(Cl, Br)_2$, in which the two halogens are present in approximately equal atomic proportions, occurs in two forms having similar four-molecular orthorhombic cells. The beta modification has been studied in detail. Its cell has the dimensions:

$$a_0 = 4.10 \text{ A.};\ b_0 = 13.17 \text{ A.};\ c_0 = 6.78 \text{ A.}$$

All atoms are in the general positions of V^4 ($P2_12_12_1$):

$$(4a)\quad xyz;\ {}^1/_2-x,y+{}^1/_2,\bar{z};\ \bar{x},{}^1/_2-y,z+{}^1/_2;\ x+{}^1/_2,\bar{y},{}^1/_2-z$$

The chosen parameters are:

$$\text{Hg:} \quad x = 0.056; \; y = 0.372; \; z = 0.086$$

$$\text{(Cl,Br,1):} \quad x = 0.418; \; y = 0.500; \; z = 0.00$$

$$\text{(Cl,Br,2):} \quad x = 0.805; \; y = 0.728; \; z = 0.833$$

The resulting arrangement represents a considerable distortion of that found for $HgCl_2$ in **IV,d6.**

Tetrahedrally Coordinated Compounds

IV,e1. *Red mercuric iodide,* HgI_2, appears to be a strictly ionic compound with its iodide ions in an almost perfect cubic close-packing. It (Fig. IV,36) differs from the structures thus far discussed in having a

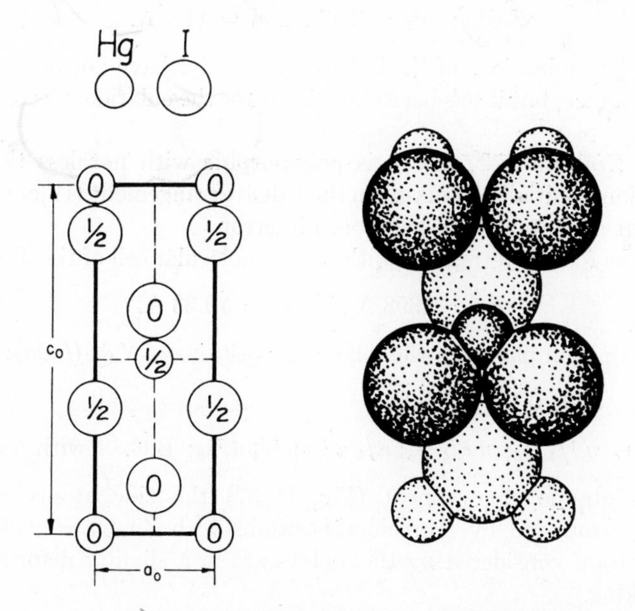

Fig. IV,36a (left). A projection along an a_0 axis of the contents of the tetragonal unit of red HgI_2. The small circles are mercury atoms.

Fig. IV,36b (right). A drawing to show how the atoms of Figure IV,36a pack together if they are given their usual ionic sizes.

metallic coordination of only four. Its symmetry is tetragonal with the bimolecular unit:

$$a_0 = 4.356 \text{ A.}, \qquad c_0 = 12.34 \text{ A.}$$

Atoms are in the following special positions of D_{4h}^{15} ($P4/nmc$):

Hg: (2*a*) 000; $^1/_2\,^1/_2\,^1/_2$

I: (4*d*) $0\,^1/_2\,u$; $^1/_2\,0\,\bar{u}$; $0,^1/_2,u+^1/_2$; $^1/_2,0,^1/_2-u$ with $u = 0.14$

Each mercury atom is tetrahedrally surrounded by four iodine atoms at a distance of 2.78 A.; the nearest approaches of iodine atoms to one another are 4.11 and 4.36 A. Cubic close-packing of the iodine atoms would be perfect if c/a were 2.828 instead of 2.84, and if $u = 0.125$ instead of the observed 0.14.

The gamma form of *zinc chloride*, $ZnCl_2$, and *zinc bromide*, $ZnBr_2$, as precipitated from aqueous solution at ca. 75°C., have this structure. They have the units:

$$ZnCl_2: \quad a_0 = 3.70 \text{ A.}, c_0 = 10.67 \text{ A.}$$

$$ZnBr_2: \quad a_0 = 3.87 \text{ A.}, c_0 = 11.4 \text{ A.}$$

Accurate determinations of the halogen parameter have not been made for either substance, but it has been stated that for the chloride, u is close to $^1/_8$.

IV,e2. *Zinc chloride*, $ZnCl_2$, is polymorphic with not less than three modifications. Initially it was given the $CdCl_2$ arrangement (**IV,c2**), but this assignment now appears to have been in error.

The *alpha* form is tetragonal with a tetramolecular cell of the dimensions:

$$a_0 = 5.398 \text{ A.}, \qquad c_0 = 10.33 \text{ A.}$$

Atoms have been placed in the following positions of V_d^{12} ($I\bar{4}2d$):

Zn: (4*a*) 000; $0\,^1/_2\,^1/_4$; B.C.

Cl: (8*d*) $u\,^1/_4\,^1/_8$; $\bar{u}\,^3/_4\,^1/_8$; $^3/_4\,u\,^7/_8$; $^1/_4\,\bar{u}\,^7/_8$; B.C. with u = ca. $^1/_4$

In this atomic arrangement (Fig. IV,37) the zinc atoms are tetrahedrally surrounded by four chlorine atoms with Zn–Cl = 2.30 A. The chlorine atoms considered by themselves are in a slightly deformed cubic close-packing.

The *beta* form of $ZnCl_2$ has been assigned the unusually large 12-molecular monoclinic but pseudo-orthorhombic cell:

$$a_0 = 6.54 \text{ A.}; \; b_0 = 11.31 \text{ A.}; \; c_0 = 12.33 \text{ A.}; \; \beta = 90°$$

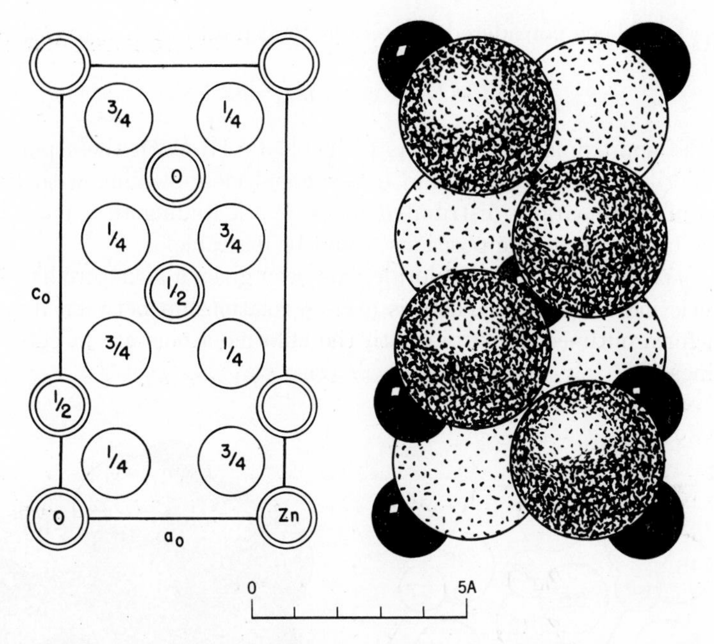

Fig. IV,37a (left). A projection down its b_0 axis of the tetragonal structure of α-$ZnCl_2$. Origin in lower left.

Fig. IV,37b (right). A packing drawing of the structure of α-$ZnCl_2$ viewed along its b_0 axis. The zinc atoms are black.

TABLE IV,11

Proposed Parameters of the Atoms in β-$ZnCl_2$ (**IV,e2**)

Atom	x	y	z
Zn(1)	$^1/_6$	$^1/_6$	$^1/_{16}$
Zn(2)	$^1/_6$	$^3/_6$	$^3/_{16}$
Zn(3)	$^4/_6$	$^4/_6$	$^3/_{16}$
Cl(1)	$^2/_6$	0	$^1/_8$
Cl(2)	$^2/_6$	$^2/_6$	$^1/_8$
Cl(3)	$^2/_6$	$^4/_6$	$^1/_8$
Cl(4)	$^5/_6$	$^1/_6$	$^1/_8$
Cl(5)	$^5/_6$	$^3/_6$	$^1/_8$
Cl(6)	$^5/_6$	$^5/_6$	$^1/_8$

Atoms have been considered to be in general positions of C_{2h}^5 ($P2_1/n$):

$$(4e) \quad \pm(xyz;\ x+{}^1/_2,{}^1/_2-y,z+{}^1/_2)$$

with the approximate parameters of Table IV,11. With these parameters the chlorine atoms would be in a hexagonal close-packing with the zinc atoms in tetrahedrally distributed holes. Work leading to a more precise determination of atomic positions would be desirable.

The *gamma* form of zinc chloride, has been given a structure like that of mercuric iodide (**IV,e1**). There is a close relationship between it and the alpha form outlined above; in both the chlorine atoms are in cubic close-packings.

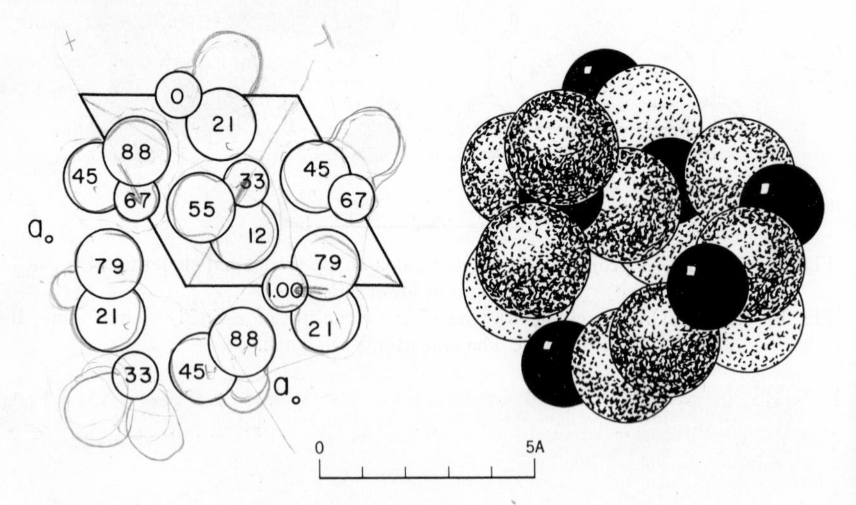

Fig. IV,38a (left). A basal projection of the hexagonal structure of low quartz, SiO_2. The smaller circles are silicon.

Fig. IV,38b (right). A packing drawing of the low quartz structure viewed down the c_0 axis. The larger atoms are oxygen.

IV,e3. Ordinary low-temperature *quartz*, SiO_2, has three molecules in a hexagonal unit of the dimensions:

$$a_0 = 4.91304 \text{ A.}, \ c_0 = 5.40463 \text{ A.} \ (25°\text{C.})$$

Its atoms are arranged (Fig. IV,38) according to the enantiomorphic pair of space groups D_3^4 and D_3^6 in the positions (of D_3^4—$P3_12$):

$$\text{Si:} \quad (3a) \quad \bar{u}\,\bar{u}\,{}^1/_3;\ u00;\ 0\,u\,{}^2/_3$$

O: (6c) xyz; $y-x,\bar{x},z+{}^1/_3$; $\bar{y},x-y,z+{}^2/_3$;

$x-y,\bar{y},\bar{z}$; $y,x,{}^2/_3-z$; $\bar{x},y-x,{}^1/_3-z$

A Fourier analysis of excellent data has led to the parameters u(Si) = 0.465, x = 0.415, y = 0.272, and z = 0.120. The tetrahedron of oxygen atoms about a silicon atom is almost regular, with Si–O = 1.61 A. Besides its two silicon neighbors, each oxygen has six adjacent oxygens at distances ranging between 2.60 and 2.67 A.

The absolute configuration has recently been determined by showing that for a levorotatory crystal the space group is D_3^4 ($P3_12$) rather than D_3^6 ($P3_22$). Thus the direction of light through quartz with this rotation follows the right-handed spirals that exist in its structure.

Two other compounds are known to have this low-quartz structure. Their cells are:

For soluble GeO_2: a_0 = 4.972 A., c_0 = 5.648 A.

For gamma BeF_2: a_0 = 4.74 A., c_0 = 5.15 A. (> 337°C.)

Parameters have not been determined.

IV,e4. *High-temperature*, or *beta*, *quartz*, SiO_2, is hexagonal with a unit of almost the same dimensions as that of low-quartz (**IV,e3**):

$$a_0 = 5.01 \text{ A.}, c_0 = 5.47 \text{ A.} \text{ (ca. 600°C.)}$$

Its space group is one or the other of the enantiomorphic pair D_6^4 or D_6^5, and the atoms of its three molecules are in the following special positions (choosing D_6^4—$P6_22$):

Si: (3c) ${}^1/_2\,{}^1/_2\,{}^1/_3$; ${}^1/_2\,0\,0$; $0\,{}^1/_2\,{}^2/_3$

O: (6j) $u\,\bar{u}\,{}^5/_6$; $\bar{u}\,u\,{}^5/_6$; $u,2u,{}^1/_2$; $\bar{u},2\bar{u},{}^1/_2$; $2u,u,{}^1/_6$; $2\bar{u},\bar{u},{}^1/_6$

with u = 0.197

This arrangement (Fig. IV,39) gives Si–O = 1.62 A. and the nearest O–O = 2.60 A. Adjacent SiO_4 tetrahedra in this and the other forms of silica are joined to one another by sharing a corner, i.e., by having a single oxygen atom in common.

The atomic arrangements in high- and low-temperature quartz are so closely alike that when a single crystal of low-quartz is cautiously heated above 575°C. it gradually and without violence changes into a single crystal of the high-temperature form, the symmetry at the same time passing from threefold to sixfold.

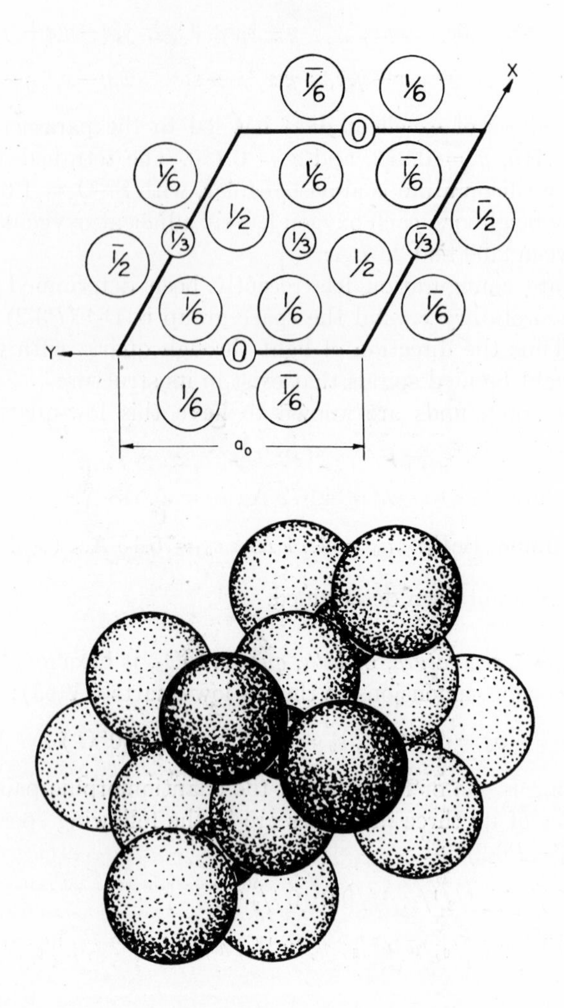

Fig. IV,39a (top). A projection on its base of atoms in and about the hexagonal unit of high, or beta, quartz. The large circles are the oxygen atoms.

Fig. IV,39b (bottom). A drawing of the beta-quartz structure showing how the atoms of Figure IV,39a pack together to provide an indefinitely extended system of linked tetrahedra.

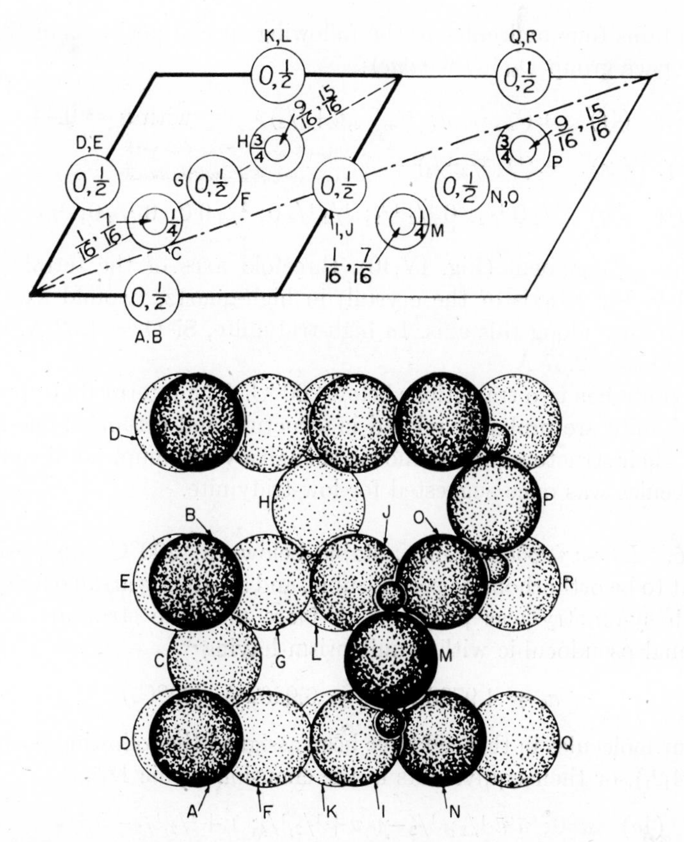

Fig. IV,40a (top). A projection on its base of the atomic contents of two of the hexagonal units of high, or beta, tridymite.

Fig. IV,40b (bottom). A packing drawing of the beta-tridymite structure viewed from a plane through the c axis and containing the dot-and-dash line of Figure IV,40a Letters refer to corresponding atoms in the two drawings. The small circles are silicon

IV,e5. The *high-temperature* form of *tridymite*, SiO_2, is hexagonal, but with a distribution of its silicon–oxygen tetrahedra different from that found in quartz (**IV,e3, e4**). Its larger unit prism has

$$a_0 = 5.03 \text{ A.}, \; c_0 = 8.22 \text{ A. above } 200°\text{C.}$$

and contains four molecules in the following special positions of the holohedral space group D_{6h}^4 ($P6/mmc$):

Si: (4f) $\pm(^1/_3\ ^2/_3\ u;\ ^2/_3,^1/_3,u+^1/_2)$ with $u = 0.44$

O(1): (2c) $\pm(^1/_3\ ^2/_3\ ^1/_4)$

O(2): (6g) $^1/_2\ 0\ ^1/_2;\ 0\ ^1/_2\ ^1/_2;\ ^1/_2\ ^1/_2\ 0;\ ^1/_2\ 0\ 0;\ 0\ ^1/_2\ 0;\ ^1/_2\ ^1/_2\ ^1/_2$

In this arrangement (Fig. IV,40), threefold axes of the tetrahedra are parallel to the c_0 axis of the crystal; in high-quartz, twofold axes of the tetrahedra are along this axis. In high-tridymite, Si–O = 1.52 A. and the nearest O–O is 2.57 A.

Tridymite has two transitions below 200°C. The intermediate form and low-tridymite are probably related in structure and orthorhombic in symmetry. Their structures are unknown, though a very complex cell containing 64 molecules was once suggested for low-tridymite.

IV,e6. *Low-cristobalite*, SiO_2, stable below ca. 200°C. was originally thought to be orthorhombic pseudocubic, and an atomic arrangement based on such symmetry was proposed. A more probable structure makes it tetragonal pseudocubic with a unit prism having

$$a_0 = 4.9733 \text{ A.},\ c_0 = 6.9262 \text{ A. (30°C.)}$$

The four molecules in this cell have been assigned the following positions of D_4^4 ($P4_12_1$), or their equivalents in the enantiomorphic D_4^8:

Si: (4a) $uu0;\ \bar{u}\bar{u}\ ^1/_2;\ ^1/_2-u,u+^1/_2,^1/_4;\ u+^1/_2,^1/_2-u,^3/_4$

O: (8b) $xyz;\ \bar{x},\bar{y},z+^1/_2;\ ^1/_2-y,x+^1/_2,z+^1/_4;\ y+^1/_2,^1/_2-x,z+^3/_4;$
$yx\bar{z};\ \bar{y},\bar{x},^1/_2-z;\ ^1/_2-x,y+^1/_2,^1/_4-z;\ x+^1/_2,^1/_2-y,^3/_4-z$

with u(Si) = 0.30, x = 0.245, y = 0.10, and z = 0.175.

This grouping (Fig. IV,41) is a relatively small distortion of that of high-cristobalite (**IV,e7**), its corresponding pseudocube that would contain eight molecules being diagonal to the above unit, with

$$a_0' = \sqrt{2}a_0 = 7.02 \text{ A.},\qquad c_0' = c_0 = 6.92 \text{ A.}$$

The atomic separations are practically the same as in the other forms of silica, with Si–O = 1.59 A. Each oxygen atom has six oxygen neighbors with O–O = 2.58–2.63 A., as well as its two adjacent silicon atoms.

Crystals of *beryllium fluoride*, BeF_2, have this arrangement, with

$$a_0 = 6.60 \text{ A.},\qquad c_0 = 6.74 \text{ A.}$$

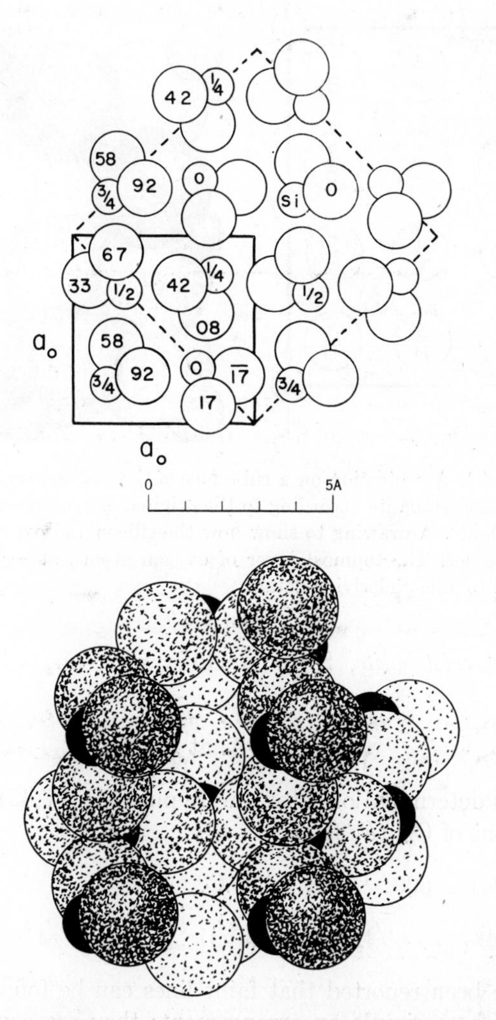

Fig. IV,41a (top). A projection along the c_0 axis of the tetragonal structure of the low-cristobalite form of SiO_2. The base of the larger pseudocube corresponding to the unit of high cristobalite (**IV,e7**) is outlined by the dashed lines. Origin in lower right.

Fig. IV,41b (bottom). A packing drawing of the tetragonal low-temperature form of cristobalite, SiO_2, viewed along the c_0 axis. The dotted atoms are oxygen.

Additional examples are furnished by BPO_4 and related compounds (see Chapter VIII).

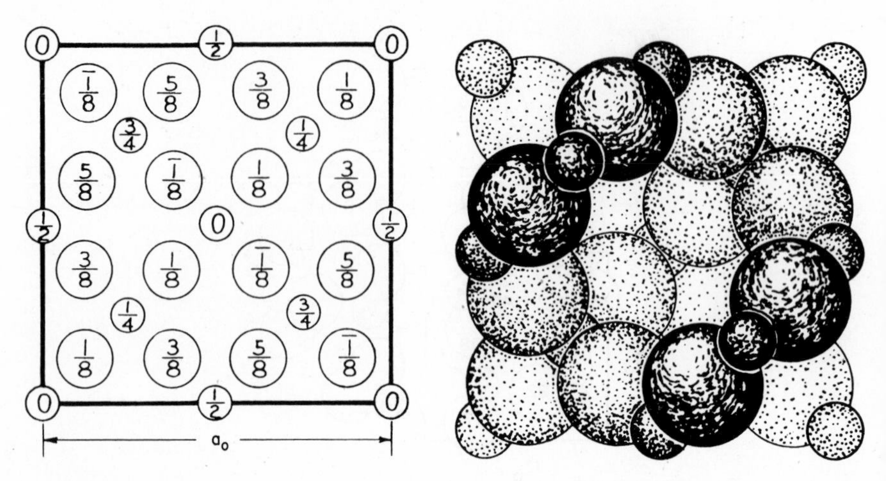

Fig. IV,42a (left). A projection on a cube face of the positions given the atoms in the unit of high, beta, cristobalite according to the original assignment of structure.

Fig. IV,42b (right). A drawing to show how the silicon and oxygen atoms of Figure IV,42a pack together. The topmost layer of oxygen atoms, at $-{}^1/_8 = {}^7/_8$, has been omitted so as not to hide underlying details.

IV,e7. *High-cristobalite*, SiO_2, is cubic with an eight-molecule unit having

$$a_0 = 7.16 \text{ A. at } 290°\text{C.}$$

In the original determination, atoms were put (Fig. IV,42) in the following special positions of O_h^7 ($Fd3m$):

Si: (8*a*) 000; ${}^1/_4\,{}^1/_4\,{}^1/_4$; F.C.

O: (16*c*) ${}^1/_8\,{}^1/_8\,{}^1/_8$; ${}^1/_8\,{}^3/_8\,{}^3/_8$; ${}^3/_8\,{}^1/_8\,{}^3/_8$; ${}^3/_8\,{}^3/_8\,{}^1/_8$; F.C.

It has since been reported that faint lines can be found which are not compatible with so simple an arrangement; they are considered to show that the atoms are somewhat removed from these highly symmetrical positions. Based on the tetrahedral space group T^4 ($P2_13$), the newer determination places the eight silicon and four of the oxygen atoms in three sets of

$$(4a) \quad uuu;\ u+{}^1/_2,{}^1/_2-u,\bar{u};\ {}^1/_2-u,\bar{u},u+{}^1/_2;\ \bar{u},u+{}^1/_2,{}^1/_2-u$$

with $u(\text{Si},1) = 0.255$, $u(\text{Si},2) = -0.008$, $u(\text{O}) = 0.125$. The other twelve oxygen atoms are in general positions:

$$(12b)\quad xyz;\ x+1/2,1/2-y,\bar{z};\ y+1/2,1/2-z,\bar{x};\ z+1/2,1/2-x,\bar{y};$$
$$yzx;\ \bar{x},y+1/2,1/2-z;\ \bar{y},z+1/2,1/2-x;\ \bar{z},x+1/2,1/2-y;$$
$$zxy;\ 1/2-x,\bar{y},z+1/2;\ 1/2-y,\bar{z},x+1/2;\ 1/2-z,\bar{x},y+1/2$$

with $x = y = 0.66$, $z = 0.06$. In this arrangement the nearest Si–O distances are between 1.58 and 1.69 A. (Fig. IV,43).

The compound *beryllium fluoride*, BeF_2, also has this type of high-temperature form, with

$$a_0 = 6.78 \text{ A.}$$

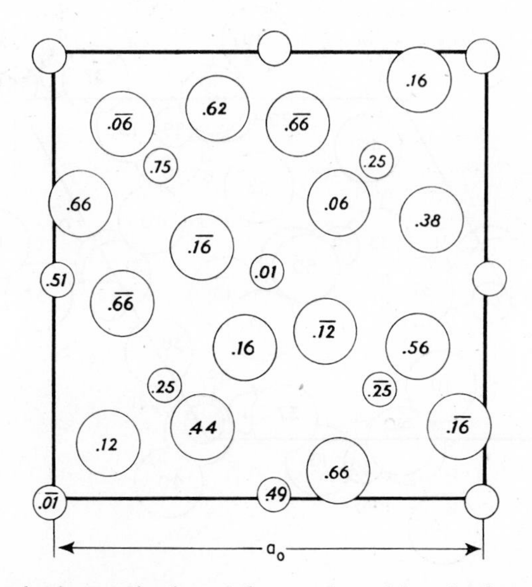

Fig. IV,43. A cube face projection of the atomic positions in the more recent structure given beta-cristobalite. The atomic positions are not sufficiently different from those of Figure IV,42a to alter appreciably the packing of Figure IV,42b.

IV,e8. *Coesite*, SiO_2, is one of the two new forms of silica for which structures have been determined. It is monoclinic with a unit that shows no departure from hexagonal. To emphasize this, the monoclinic unit has been chosen with c_0 rather than the b axis normal to the plane of the other two. This unit has the dimensions:

$$a_0 = b_0 = 7.17 \text{ A.};\ c_0 = 12.38 \text{ A.};\ \gamma = 120°$$

It has been concluded that the cell contains 16 molecules, though the measured density points to ca. 16.6 molecules. The space group is C_{2h}^6 and since the principal axis is c_0 its orientation symbol is $B2/b$. Atoms have been placed in the following positions:

$$(4a) \quad 000;\ 0\,{}^1/_2\,0;\ {}^1/_2\,0\,{}^1/_2;\ {}^1/_2\,{}^1/_2\,{}^1/_2$$

$$(4e) \quad \pm(0\,{}^1/_4\,u;\ {}^1/_2,{}^1/_4,u+{}^1/_2)$$

$$(8f) \quad \pm(xyz;\ x,y+{}^1/_2,\bar{z})$$

·d these positions around ${}^1/_2\,0\,{}^1/_2$, with the parameters of Table IV,12.

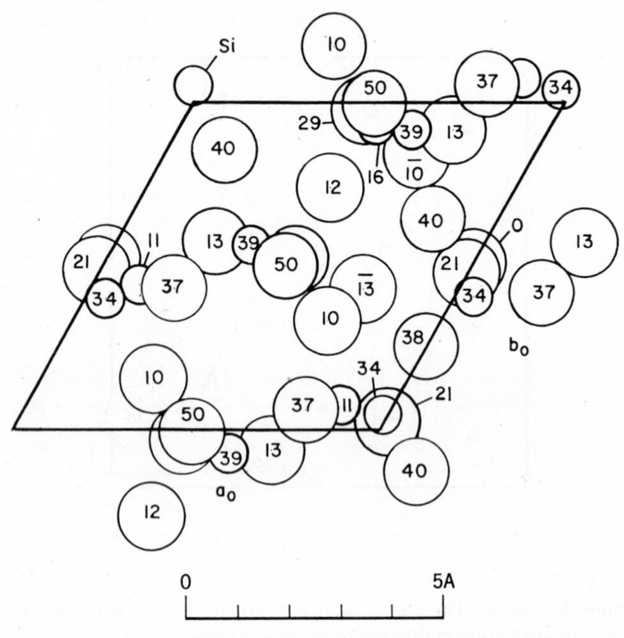

Fig. IV,44. A projection along the c_0 axis (which in this case is normal to the a_0b_0 plane), showing some of the atoms in the monoclinic structure of coesite. Origin in lower right.

In this structure silicon atoms are tetrahedrally surrounded by oxygens as in the other forms of silica, with Si–O between 1.60 and 1.63 A. The distances between oxygen atoms of a tetrahedron are 2.60–2.67 A. and each oxygen belongs to two tetrahedra. There is as yet no adequate explanation of the excess weight per cell.

This is an arrangement in which there is so much superposition of atoms in any projection one may choose to make that its graphical representation

TABLE IV,12
Positions and Parameters of the Atoms in Coesite, SiO_2 (**IV,e8**)

Atom	Position	x	y	z
Si(1)	(8*f*)	0.1403	0.0735	0.1084
Si(2)	(8*f*)	0.5063	0.5388	0.1576
O(1)	(4*a*)	0	0	0
O(2)	(4*e*)	0	$^1/_4$	0.3834
O(3)	(8*f*)	0.2694	0.9405	0.1256
O(4)	(8*f*)	0.3080	0.3293	0.1030
O(5)	(8*f*)	0.0123	0.4726	0.2122

is unsatisfactory. Some idea of the way the SiO_4 tetrahedra are linked together may, however, be obtained from Figure IV,44 which shows atoms of the lower half of the cell.

IV,e9. *Keatite*, SiO_2, is the other form of silica whose structure has recently been described. It has a tetragonal unit containing 12 molecules. The cell edges are

$$a_0 = 7.456 \text{ A.}, \qquad c_0 = 8.604 \text{ A.}$$

The space group is the same as that used for low cristobalite (**IV,e6**), namely D_4^4 ($P4_12$) (or the enantiomorphic D_4^8). Atoms are in the positions:

(4*a*) $uu0;\ \bar{u}\,\bar{u}\,{}^1/_2;\quad {}^1/_2-u,u+{}^1/_2,{}^1/_4;\quad u+{}^1/_2,{}^1/_2-u,{}^3/_4$

(8*b*) $xyz;\ \bar{x},\bar{y},z+{}^1/_2;\ {}^1/_2-y,x+{}^1/_2,z+{}^1/_4;\ y+{}^1/_2,{}^1/_2-x,z+{}^3/_4;$
$yx\bar{z};\ \bar{y},\bar{x},{}^1/_2-z;\ {}^1/_2-x,y+{}^1/_2,{}^1/_4-z;\ x+{}^1/_2,{}^1/_2-y,{}^3/_4-z$

with the parameters of Table IV,13.

TABLE IV,13
Positions and Parameters of the Atoms in Keatite, SiO_2 (**IV,e9**)

Atom	Position	x	y	z
Si(1)	(8*b*)	0.326	0.120	0.248
Si(2)	(4*a*)	0.410	0.410	0
O(1)	(8*b*)	0.445	0.132	0.400
O(2)	(8*b*)	0.117	0.123	0.296
O(3)	(8*b*)	0.344	0.297	0.143

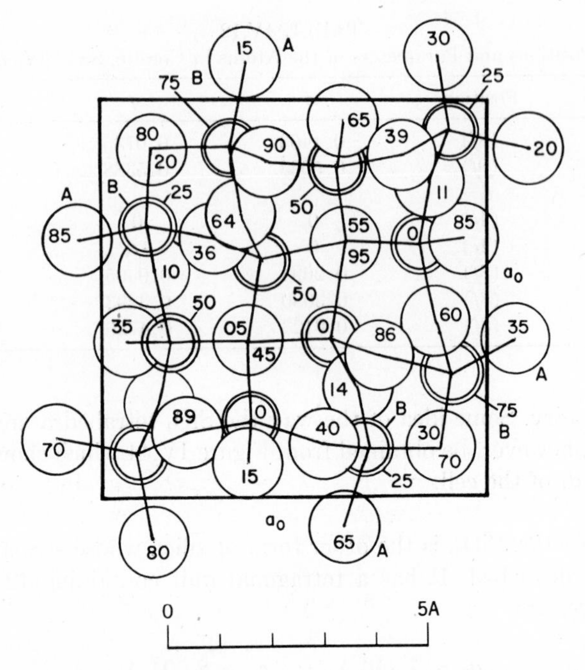

Fig. IV,45. A projection along its c_0 axis of the proposed tetragonal structure for keatite. Atoms involved in unexpectedly long atomic separations are marked A and B. The doubly ringed circles represent silicon.

The resulting structure projected along the c_0 axis is shown in Figure IV,45. Each silicon atom appears to be surrounded by a tetrahedron of oxygen atoms. Those around the silicon atoms in (4*a*) are approximately equidistant with Si–O = 1.57–1.61 A. Three of those around each silicon atom in (8*b*) are at about the same distance but the fourth is too far away (3.69 A.) to complete an SiO_4 tetrahedron. Involved in such excessive separations are the atoms labeled A and B in the figure. This is so improbable that there is something wrong either with the parameters as stated or in the structure itself.

IV,e10. The structure of *ice*, H_2O, is in a sense related to that of tridymite. The unit cells are of similar shapes and the oxygen atoms of ice have the same coordinates as the silicon atoms of high-tridymite (**IV,e5**).

The tetramolecular hexagonal unit of ordinary ice has the edges:

$$a_0 = 4.5227 \text{ A.}, \quad c_0 = 7.3671 \text{ A. at } 0°\text{C.}$$

$$a_0 = 4.5177 \text{ A.}, \; c_0 = 7.353 \text{ A.} \quad \text{at } -66°\text{C.}$$

$$a_0 = 4.493 \text{ A.}, \; c_0 = 7.337 \text{ A.} \quad \text{at } -110°\text{C.}$$

For heavy ice, D_2O, the cell dimensions are very little different:

$$a_0 = 4.5257 \text{ A.}, \; c_0 = 7.3687 \text{ A. at } 0°\text{C.}$$

$$a_0 = 4.5147 \text{ A.}, \; c_0 = 7.353 \text{ A.} \quad \text{at } -66°\text{C.}$$

$$a_0 = 4.495 \text{ A.}, \; c_0 = 7.335 \text{ A.} \quad \text{at } -130°\text{C.}$$

The oxygen atoms are in the positions:

$$(4f) \quad \pm(1/3 \; 2/3 \; u; \; 2/3, 1/3, u+1/2)$$

of D_{6h}^4 ($P6/mmc$), with $u = 1/16$ as determined at $-20°$C. (Fig. IV,46).

There have been extensive discussions of the distribution of the hydrogen atoms. A neutron diffraction study of D_2O supports the idea that the water molecules are statistically oriented within the structure in such a way that

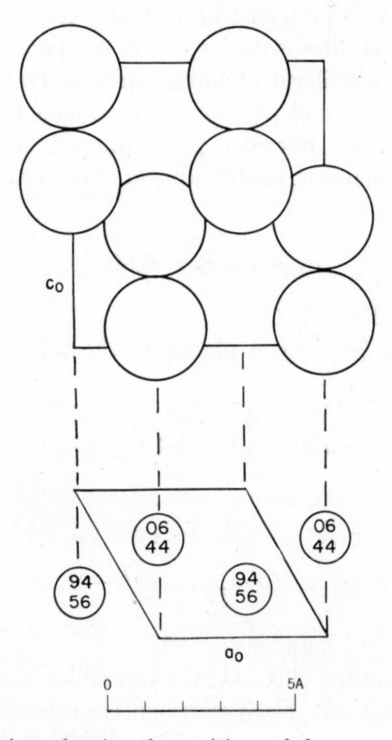

Fig. IV,46. Two projections showing the positions of the oxygen atoms in ordinary ice.

four "half-hydrogens" will be in (4f) as stated above and the others in

$$(12k) \quad \pm(u,2u,v;\ 2\bar{u},\bar{u},v;\ u\bar{u}v;\ \bar{u},2\bar{u},v+{}^1/_2;\ 2u,u,v+{}^1/_2;\ \bar{u},u,v+{}^1/_2)$$

When water vapor is slowly condensed at low pressures onto a cold surface *cubic ice* is formed. It has a cell of the edge length:

$$H_2O: \quad a_0 = 6.350 \text{ A. } (-130°\text{C.})$$

$$D_2O: \quad a_0 = 6.351 \text{ A. } (-130°\text{C.})$$

There are eight molecules per unit and this modification of ice has the same kind of relation to the high-cristobalite structure that ordinary ice bears to tridymite. Its oxygen atoms, like the silicon atoms of high-cristobalite, are in or near the positions (Fig. IV,42):

$$(8a) \quad 000;\ {}^1/_4\,{}^1/_4\,{}^1/_4;\ \text{F.C. of } O_h^7\ (Fd3m)$$

Electron-diffraction data favor for hydrogen the same kind of "half-hydrogen" model that is preferred for ordinary ice.

Years ago, x-ray studies were reported on two of the high-pressure modifications of ice, stabilized at liquid air temperatures. Both were described as orthorhombic, and positions were assigned to the oxygen atoms in each. Recent work has, however, failed to confirm these results.

Instead, one of these forms, *ice III*, appears to be tetragonal, pseudocubic with the cell:

$$a_0 = \text{ca. } c_0 = 6.80 \text{ A.}$$

containing 12 molecules of H_2O.

The oxygen atoms have been placed in the following positions of D_4^4 ($P4_12_12$):

$$(4a) \quad uu0;\ \bar{u}\,\bar{u}\,{}^1/_2;\ {}^1/_2-u,u+{}^1/_2,{}^1/_4;\ u+{}^1/_2,{}^1/_2-u,{}^3/_4$$

with $u = 0.392$

$$(8b) \quad xyz;\ \bar{x},\bar{y},z+{}^1/_2;\ {}^1/_2-y,x+{}^1/_2,z+{}^1/_4;\ y+{}^1/_2,{}^1/_2-x,z+{}^3/_4;$$

$$yx\bar{z};\ \bar{y},\bar{x},{}^1/_2-z;\ {}^1/_2-x,y+{}^1/_2,{}^1/_4-z;\ x+{}^1/_2,{}^1/_2-y,{}^3/_4-z$$

with $x = 0.095$, $y = 0.30$, and $z = 0.29$.

The resulting structure (Fig. IV,47) resembles that given the keatite modification of SiO_2 (**IV,e9**). Each oxygen atom is tetrahedrally surrounded by four others at distances between 2.73 and 2.90 A.

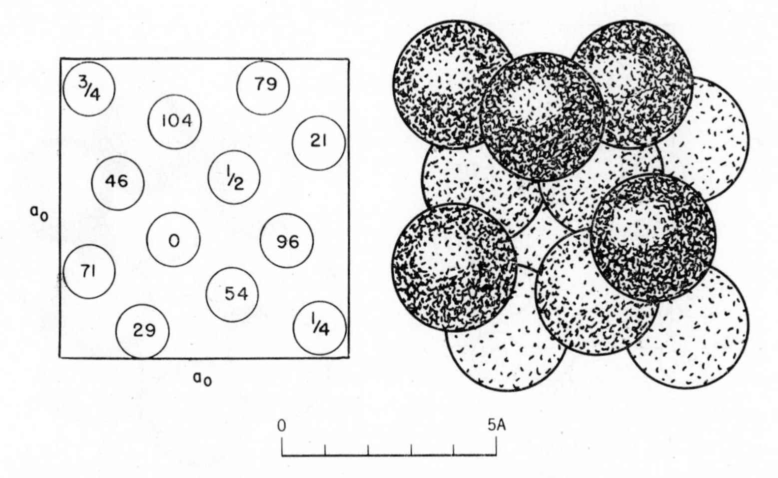

Fig. IV,47a (left). A projection along the c_0 axis showing the distribution of the oxygen atoms in the tetragonal, high-pressure modification III of ice.

Fig. IV,47b (right). A packing drawing of the oxygen atoms of ice III, viewed along the c_0 axis.

IV,e11. Crystals of *zinc hydroxide*, $Zn(OH)_2$, are orthorhombic with four molecules in a cell of the dimensions:

$$a_0 = 8.53 \text{ A.}; \; b_0 = 5.16 \text{ A.}; \; c_0 = 4.92 \text{ A.}$$

All atoms are in general positions of V^4 ($P2_12_12_1$):

$$(4a) \quad xyz; \; {}^1/_2-x,\bar{y},z+{}^1/_2; \; x+{}^1/_2,{}^1/_2-y,\bar{z}; \; \bar{x},y+{}^1/_2,{}^1/_2-z$$

with the parameters that follow:

Atom	x	y	z
Zn	0.125	0.100	0.175
O(1)	0.025	0.430	0.085
O(2)	0.325	0.125	0.370

This tetrahedral structure (Fig. IV,48) differs from the several forms of silica in that its metallic atom is so big that the surrounding X atoms are not in contact. The significant atomic separations are Zn–OH = 1.95 A., OH–OH = 2.83 A. in the same and 3.06–3.08 A. between adjacent tetrahedra.

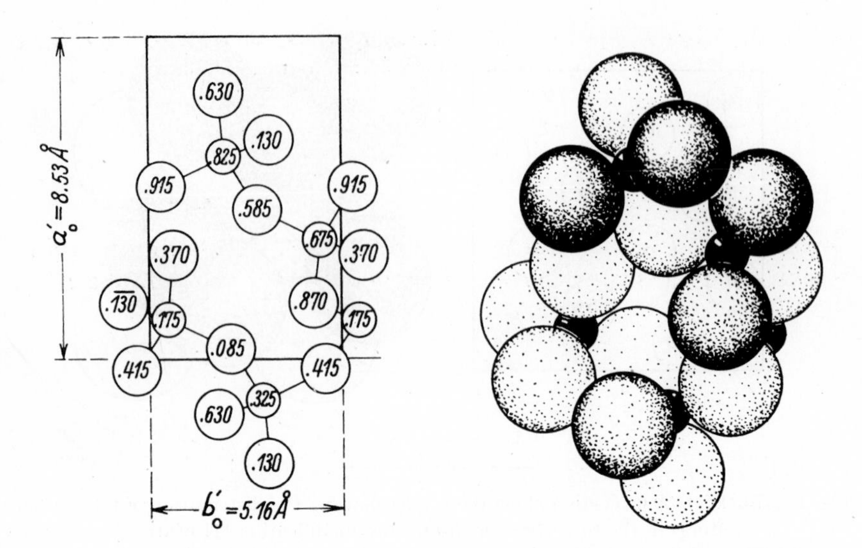

Fig. IV,48a (left). A projection of the orthorhombic structure of $Zn(OH)_2$ on its c face. The small circles are the zinc atoms.

Fig. IV,48b (right). A drawing to show how the zinc atoms and (OH) groups, considered as ions, pack together in $Zn(OH)_2$. The (OH) ions are the large spheres.

The beta modification of *beryllium hydroxide*, $Be(OH)_2$, has this structure, with

$$a_0 = 7.039 \text{ A.};\ b_0 = 4.620 \text{ A.};\ c_0 = 4.535 \text{ A.}$$

The established atomic parameters are

Atom	x	y	z
Be	0.125	0.047	0.220
OH(1)	0.015	0.345	0.090
OH(2)	0.285	0.140	0.440

In this arrangement Be–OH = 1.57–1.69 A. and the shortest OH–OH = 2.52 A.

IV,e12. Crystals of *germanium disulfide*, GeS_2, furnish an example of an especially complicated assembly of RX_4 tetrahedra linked together by sharing corners. The symmetry is orthorhombic, with a very large unit containing 24 molecules and having the dimensions:

$$a_0 = 11.66 \text{ A.};\ b_0 = 22.34 \text{ A.};\ c_0 = 6.86 \text{ A.}$$

An atomic arrangement (Fig. IV,49) based on the face-centered space group C_{2v}^{19} (*Fdd*) has been described. This places eight germanium atoms in special positions (8*a*) and all other atoms in general positions:

$$(8a) \quad 00u;\ {}^1/_4,{}^1/_4,u+{}^1/_4;\ \text{F.C.}$$

$$(16b) \quad xyz;\ \bar{x}\bar{y}z;\ {}^1/_4-x,y+{}^1/_4,z+{}^1/_4;\ x+{}^1/_4,{}^1/_4-y,z+{}^1/_4;\ \text{F.C.}$$

The assigned parameters are those of Table IV,14. The germanium–sulfur separations in the somewhat distorted tetrahedra vary from 2.07 to 2.26 A.; S–S distances are from 3.35 A. upwards.

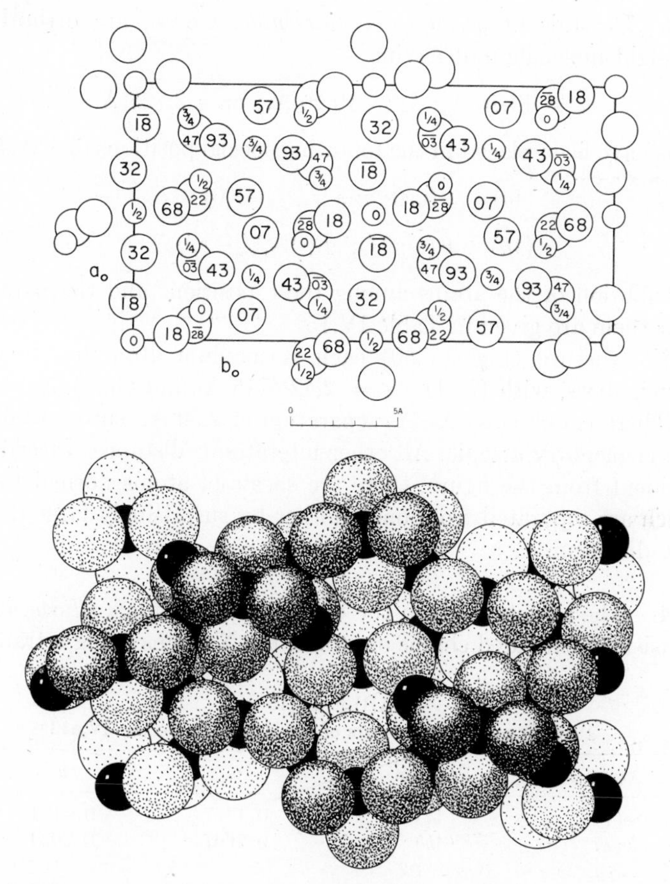

Fig. IV,49a (top). A projection along c_0 of the orthorhombic structure of GeS_2. Sulfur are the larger circles. Origin in lower left.

Fig. IV,49b (bottom). A packing drawing of the orthorhombic structure of GeS_2 viewed along the c_0 axis. Sulfur atoms are dotted.

TABLE IV,14
Positions and Parameters of the Atoms in GeS_2 (**IV,c12**)

Atom	Position	x	y	z
Ge(1)	(8*a*)	0	0	0
Ge(2)	(16*b*)	0.125	0.139	0
S(1)	(16*b*)	0.022	0.081	0.183
S(2)	(16*b*)	0.153	−0.014	−0.183
S(3)	(16*b*)	0.062	0.125	−0.278

IV,e13. Crystals of *germanium diarsenide*, $GeAs_2$, are orthorhombic with an eight-molecule unit having

$$a_0 = 14.76 \text{ A.}; \; b_0 = 10.16 \text{ A.}; \; c_0 = 3.728 \text{ A.}$$

All atoms have been placed in the following special positions of V_h^9 (*Pbam*):

$$(4g) \quad \pm(uv0;\; u+\tfrac{1}{2},\tfrac{1}{2}-v,0)$$

$$(4h) \quad \pm(u\,v\,\tfrac{1}{2};\; u+\tfrac{1}{2},\tfrac{1}{2}-v,\tfrac{1}{2})$$

The distribution of the atoms among these positions and the parameters found for them are given in Table IV,15.

In this structure (Fig. IV,50) each germanium atom has four close arsenic neighbors, with Ge(1)–As = 2.42–2.45 A. and Ge(2)–As = 2.45–2.51 A. There is one close As–As separation of 2.50 A., substantially that found in elementary arsenic. All other interatomic distances exceed 3.1 A. As is evident from the figure, there are sheets of atoms normal to the a_0 axis which are separated from one another by surprisingly large distances in this a_0 direction.

IV,e14. The structure given the conspicuously fibrous *silicon disulfide*, SiS_2, consists of extended chains of atoms, the X atoms being tetrahedrally

TABLE IV,15
Positions and Parameters of the Atoms in $GeAs_2$ (**IV,e13**)

Atom	Position	u	v
Ge(1)	(4*g*)	0.1377	0.4191
Ge(2)	(4*h*)	0.2667	0.2024
As(1)	(4*g*)	0.2953	0.3532
As(2)	(4*g*)	0.0383	0.2260
As(3)	(4*h*)	0.1117	0.1014
As(4)	(4*h*)	0.4021	0.0628

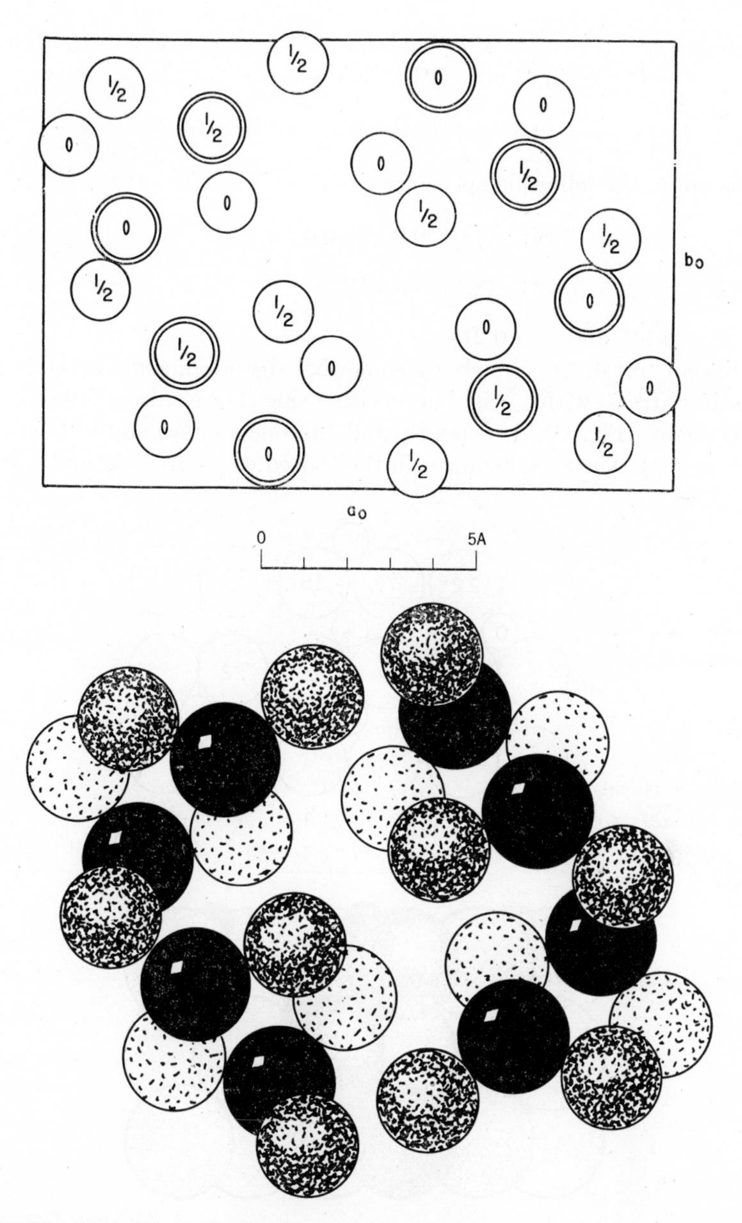

Fig. IV,50a (top). The orthorhombic structure of $GeAs_2$ projected along its c_0 axis. The germanium atoms are doubly ringed. Origin in lower right.

Fig. IV,50b (bottom) A packing drawing of the $GeAs_2$ structure viewed along its c_0 axis. The atoms have approximately their neutral radii. The germanium atoms are black.

distributed about central silicon atoms. The symmetry is orthorhombic with four molecules in a unit of the dimensions:

$$a_0 = 9.55 \text{ A.}; \; b_0 = 5.65 \text{ A.}; \; c_0 = 5.54 \text{ A.}$$

Atoms are in the following special positions of V_h^{26} (*Ibam*):

Si: (4*a*) $0\,0\,{}^1/_4;\; 0\,0\,{}^3/_4$; B.C.

S: (8*j*) $\pm(uv0;\; \bar{u}\,v\,{}^1/_2)$; B.C.

with $u = 0.117$ and $v = 0.217$.

Adjacent tetrahedra, which are somewhat distorted, share an edge (i.e., two sulfur atoms) with a neighbor on either side. The resulting linked chains of tetrahedra (Fig. IV,51) extend parallel to one another and to the c_0 axis of the crystal. The S–S distances in the shared edges are 3.24 and 3.58 A.,

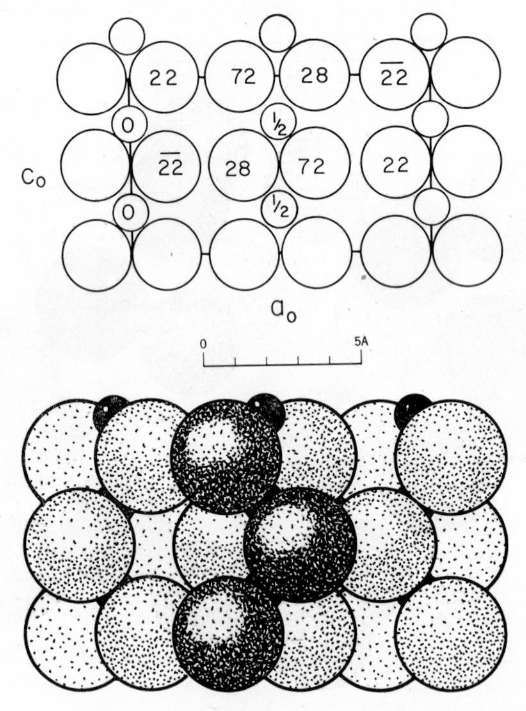

Fig. IV,51a (top). A projection along b_0 of the orthorhombic structure of SiS_2. The larger circles are sulfur. Origin at lower left.

Fig. IV,51b (bottom). A packing drawing of the orthorhombic SiS_2 arrangement viewed along the b_0 axis. The large atoms are sulfur.

the lengths of the other edges being 3.62 A. The Si–S separation is 2.14 A. The nearest approach of neighboring chains, through sulfur atoms, is 3.75 A.

Three other compounds are known with this structure.

For *beryllium chloride*, $BeCl_2$, the unit cell has the edges:

$$a_0 = 9.86 \text{ A.};\ b_0 = 5.36 \text{ A.};\ c_0 = 5.26 \text{ A.}$$

The chlorine parameters are $u = 0.109$, $v = 0.203$.

For *"fibrous silica,"* obtained by heating "SiO" at 1200–1400°C.:

$$a_0 = 8.36 \text{ A.};\ b_0 = 5.16 \text{ A.};\ c_0 = 4.72 \text{ A.}$$

The oxygen parameters have been given as $u = 0.110$, $v = 0.209$.

For silicon diselenide, $SiSe_2$, the unit has the edges:

$$a_0 = 9.76 \text{ A.};\ b_0 = 6.03 \text{ A.};\ c_0 = 5.76 \text{ A.}$$

Parameters for the selenium atoms have not been determined.

The compound $AlPS_4$ (Chapter VIII) has a very similar structure.

Compounds with Linear and Square Coordinations

IV,f1. *Cuprite*, Cu_2O, has a high-symmetry, cubic structure with less than a fourfold coordination of its metal atoms. There are two molecules in the unit cube:

$$a_0 = 4.2696 \text{ A. } (26°\text{C.}).$$

Atoms are in the special positions of T_h^2 ($Pn3$):

Cu: (4*b*) $\frac{1}{4}\frac{1}{4}\frac{1}{4};\ \frac{3}{4}\frac{3}{4}\frac{1}{4};\ \frac{3}{4}\frac{1}{4}\frac{3}{4};\ \frac{1}{4}\frac{3}{4}\frac{3}{4}$

O: (2*a*) $000;\ \frac{1}{2}\frac{1}{2}\frac{1}{2}$

In this structure (Fig. IV, 52) each metal atom has only two close oxygen neighbors, while each oxygen atom is surrounded by a tetrahedron of copper atoms. The Cu–O separation is 1.85 A. The oxygen atoms are far apart (O–O = 3.69 A.); the closer Cu–Cu = 3.01 A. is about the sum of the neutral radii.

Other substances with this structure have the cell edges:

Ag_2O: $a_0 = 4.72$ A.

Pb_2O: $a_0 = 5.38$ A.

$Cd(CN)_2$: $a_0 = 6.32$ A.

$Zn(CN)_2$: $a_0 = 5.905$ A. (25°C.)

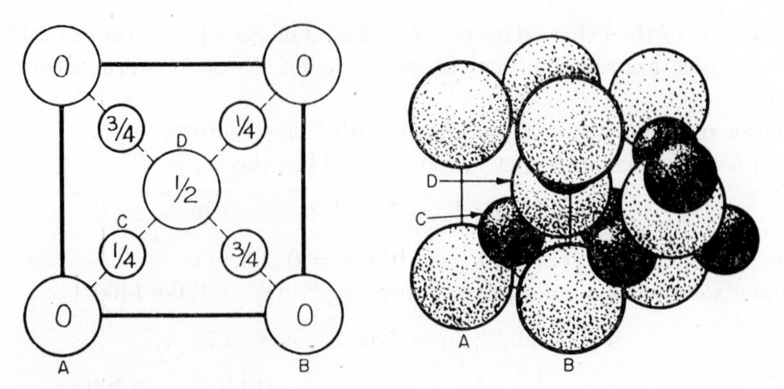

Fig. IV,52a (left). A projection on the cube face of atoms in the unit of Cu_2O. The small circles represent the copper atoms.

Fig. IV,52b (right). A perspective drawing showing the packing of the atoms in Cu_2O if they are given their ionic sizes. Letters identify corresponding atoms in this and in Figure IV,52a.

The structure of these cyanides is obviously "anti" to the others. It has been stated that the carbon and nitrogen atoms in them are in positions fixed by symmetry, that is they form nonrotating ions. Their atomic positions, based on the different space group T_d^1 ($P\bar{4}3m$), would be

$$\text{Zn or Cd:} \quad (1a) \quad 000, \text{ and } (1b) \quad \tfrac{1}{2}\tfrac{1}{2}\tfrac{1}{2}$$

$$\text{C and N, each:} \quad (4e) \quad uuu;\ u\bar{u}\bar{u};\ \bar{u}u\bar{u};\ \bar{u}\bar{u}u$$

These parameters have not been established.

IV,f2. Originally it was thought that the high-temperature forms of Ag_2S and Ag_2Se also had the cuprite structure (**IV,f1**) with units of the dimensions:

$$\text{For high } Ag_2S: \quad a_0 = 4.88 \text{ A. at } 250°\text{C.}$$

$$\text{For high } Ag_2Se: \quad a_0 = 4.983 \text{ A. at ca. } 250°\text{C.}$$

The telluride of silver and the high-temperature forms of Cu_2S and Cu_2Se also were described as cubic, but they were supposed to have the fluorite structure (**IV,a1**), their four-molecular cubes having been given the edge lengths:

$$\text{For high } Ag_2Te: \quad a_0 = 6.572 \text{ A. at ca. } 250°\text{C.}$$

$$\text{For high } Cu_2S: \quad a_0 = 5.564 \text{ A. at ca. } 170°\text{C.}$$

$$\text{For high } Cu_2Se: \quad a_0 = 5.840 \text{ A. at ca. } 170°\text{C.}$$

Later it was concluded that these simple atomic distributions were incorrect and that the metallic atoms in all these crystals should be considered as haphazardly distributed among a few of the points of more complicated special positions of the appropriate space groups. Much additional work is needed on these systems.

At temperatures below their inversions these compounds have been described sometimes as orthorhombic and sometimes as monoclinic. Pure *cuprous sulfide*, Cu_2S, above about 105°C. has been reported as hexagonal, though a cubic modification such as that referred to above does occur in preparations deficient in copper. This hexagonal, high, Cu_2S, has two molecules in a cell of the edge lengths:

$$a_0 = 3.961 \text{ A.}, \quad c_0 = 6.722 \text{ A.} \ (152°\text{C.})$$

Its atoms have been placed in the following positions of D_{6h}^4 ($P6/mcm$):

$$\text{Cu(1)}: \quad (2b) \quad \pm(0\,0\,{}^1/_4)$$

$$\text{Cu(2)}: \quad (2d) \quad \pm({}^1/_3\,{}^2/_3\,{}^3/_4)$$

$$\text{S}: \quad (2c) \quad \pm({}^1/_3\,{}^2/_3\,{}^1/_4)$$

It is now reported that this high Cu_2S changes over to the cubic form referred to above at sufficiently high temperatures (ca. 460°C.). At 465°C. it has been found to have the cubic edge, $a_0 = 5.725$ A.

The data on the room-temperature, orthorhombic, form of Cu_2S are conflicting. According to one study it has a very large unit which is pseudohexagonal and bears a "super-lattice" relation to the hexagonal high-Cu_2S. The cell edges have been stated as

$$a_0 = 11.90 \text{ A.}; \ b_0 = 27.28 \text{ A.}; \ c_0 = 13.41 \text{ A.}$$

These axial lengths are roughly three, four, and two times the lengths of the orthohexagonal axes of high-Cu_2S. The assigned space group was C_{2v}^{15} ($Ab2m$), but atomic positions were not determined.

According to another study, the data for low-Cu_2S are approximately satisfied through a pseudocell of the dimensions:

$$a_0 = 3.94 \text{ A.}; \ b_0 = 6.75 \text{ A.}; \ c_0 = 6.70 \text{ A.}$$

to which the different space group V_h^{19} ($Cmmm$) is applicable. Clearly there is more work to be done on this system.

There is a hexagonal high-temperature modification of *cuprous telluride*, Cu_2Te, which has a unit similar to that for high-Cu_2S:

$$a_0 = 4.237 \text{ A.}, \qquad c_0 = 7.274 \text{ A.}$$

Unlike the structure that has been given the high-Cu_2S, however, atoms have been said to be in the following positions of D_{6h}^1 ($P6/mmm$):

Cu: (4h) $\pm(^1/_3\ ^2/_3\ u;\ ^2/_3\ ^1/_3\ u)$ with $u = 0.160$

Te: (2e) $\pm(00v)$ with $v = 0.306$

Atomic positions have been proposed for the compound CuZnAs which is cubic and is considered to have the same type of structure as the cubic high-temperature forms of Ag_2Te and Cu_2Se, referred to above. Its tetramolecular unit has the edge:

$$\text{CuZnAs:}\quad a_0 = 5.872\ \text{A.}$$

It has been thought that the arsenic atoms are in 000; F.C., the zinc atoms in $^1/_4\ ^1/_4\ ^1/_4$; F.C., and that the copper atoms are irregularly distributed.

For high-Cu_2Se with

$$a_0 = 5.840\ \text{A.}\ (170°\text{C.})$$

it has been said that one copper and the selenium atoms have the foregoing

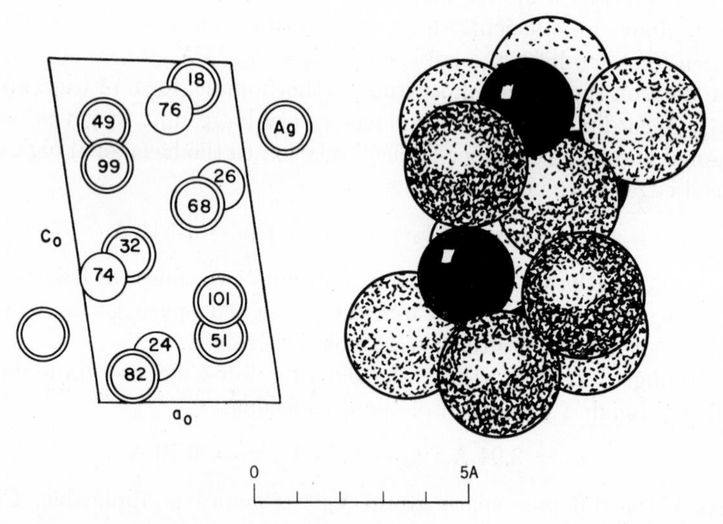

Fig. IV,53a (left). The monoclinic structure of the mineral acanthite, β-Ag_2S, projected along its b_0 axis. Origin in lower left.

Fig. IV,53b (right). A packing drawing of the β-Ag_2S arrangement viewed along its b_0 axis. The atoms have their neutral radii, the sulfur atoms being represented by the smaller, black circles.

ZnS type distribution and that the other half of the copper atoms are irregularly distributed.

Above 300°C., mixed $(Ag,Cu)_2S$ preparations have similar face-centered cubic cells. Between ca. 115° and 300°C. they are body-centered cubic like the Ag_2S mentioned at the beginning of this section (**IV,f2**).

A more satisfying study has recently been made of the low-temperature (beta) form of Ag_2S, the mineral *acanthite*. It is monoclinic with a tetramolecular unit of the dimensions:

$$a_0 = 4.23 \text{ A.}; \; b_0 = 6.91 \text{ A.}; \; c_0 = 7.87 \text{ A.}; \; \beta = 99°35'$$

Atoms are in the general positions of C_{2h}^5 ($P2_1/n$):

$$(4e) \quad \pm(xyz; \; x+1/2, 1/2-y, z+1/2)$$

with the following parameters:

Atom	x	y	z
Ag(1)	0.758	0.015	0.305
Ag(2)	0.285	0.320	0.435
S	0.359	0.239	0.134

It is to be noted that this arrangement (Fig. IV,53) results in substantially the same body-centered distribution of the sulfur atoms that was proposed for the high (alpha) Ag_2S. In this structure, Ag(1) has two sulfur neighbors at distances of 2.49 and 2.52 A.; Ag(2) has three with Ag(2)–S = 2.50, 2.61, and 2.69 A.

IV,f3. Crystals of the mixed sulfide mineral *stromeyerite*, AgCuS, are orthorhombic with the tetramolecular unit:

$$a_0 = 4.06 \text{ A.}; \; b_0 = 6.66 \text{ A.}; \; c_0 = 7.99 \text{ A.}$$

Atoms have been found to be in the following special positions of V_h^{17} (*Cmcm*):

Ag: $(4a)$ $000; \; 0\,0\,1/2$; B.C.

Cu: $(4c)$ $\pm(0\,u\,1/4; \; 1/2, u+1/2, 1/4)$ with $u = 0.46$

S: $(4c)$ with $u' = 0.80$

This structure (Fig. IV,54) has been described as involving chains of silver and sulfur atoms parallel to the c_0 axis tied together by atoms of copper.

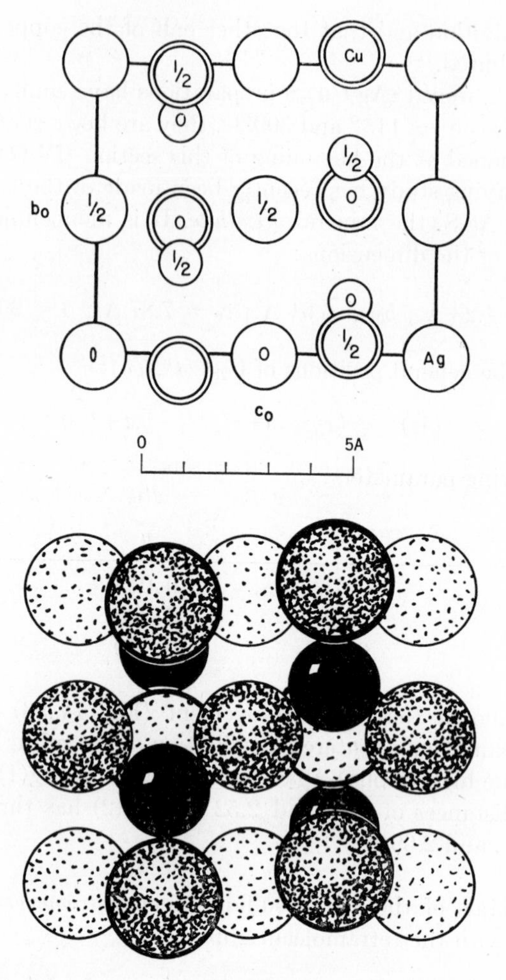

Fig. IV,54a (top). The orthorhombic structure of stromeyerite, AgCuS, projected along its a_0 axis. Origin in lower left.

Fig. IV,54b (bottom). A packing drawing of the rather open structure found for AgCuS viewed along its a_0 axis. Atoms have been given their neutral radii. The sulfur atoms are black, the dotted copper atoms are the more heavily ringed.

In the chains, the Ag–S distance is 2.40 A., the Ag–S–Ag angle equals 113°, and S–Ag–S = 180°. Each copper atom has two sulfur atoms at a distance of 2.29 A. and a third at 2.26 A.

IV,f4. The mineral *eucairite*, CuAgSe, appears to have a large orthorhombic cell containing ten molecules and having the edges:

$$a_0 = 4.105 \text{ A.}; \ b_0 = 20.35 \text{ A.}; \ c_0 = 6.31 \text{ A.}$$

Only a very few weak reflections require this long b_0 axis; if they are ignored it is one fifth as long, with $b_0' = 4.07$ A.

An approximate structure was deduced on the basis of this bimolecular cell. For it the space group was chosen as V_h^{13} ($Pmmn$), with atoms in the positions:

Ag: (2*a*) $\pm(^1/_4\ ^1/_4\ u)$ with $u = 0.449$

Se: (2*a*) with $u' = 0.873$

Cu: (2*b*) $\pm(^1/_4\ ^3/_4\ v)$ with $v = 0.105$

In this atomic arrangement each silver atom has one close selenium atom at 2.67 A. and four silver atoms at 2.96 A. The selenium atoms form flattened tetrahedra that share corners with one another and contain non-centered copper atoms with Cu–2Se = 2.06 A. and Cu–2Se = 2.50 A. The closest Cu–Ag = 2.98 A. It is not a structure in which atomic separations agree well with either neutral or ionic radii.

The exact way in which these subcells are stacked in the fivefold bigger unit is not known.

IV,f5. The structures given the two forms of $AuTe_2$ show a linear coordination of their gold atoms. The simpler of these modifications is the monoclinic, pseudo-orthorhombic mineral *calaverite*. Its bimolecular unit has the dimensions:

$$a_0 = 7.18 \text{ A.}; \ b_0 = 4.40 \text{ A.}; \ c_0 = 5.07 \text{ A.}; \ \beta = 90°$$

Atoms have been placed in the following special positions of C_{2h}^3 ($C2/m$):

Au: (2*a*) $000;\ ^1/_2\ ^1/_2\ 0$

Te: (4*i*) $u0v;\ \bar{u}0\bar{v};\ u+{}^1/_2,{}^1/_2,v;\ {}^1/_2-u,{}^1/_2,\bar{v}$

with $u = 0.69$ and $v = 0.29$. The arrangement that results (Fig. IV,55) is an assemblage of linear $AuTe_2$ molecules in which the Au–Te separation is 2.67 A. The nearest approach of tellurium atoms to one another is 3.77 A.

IV,f6. There are eight molecules in the orthorhombic unit of *krennerite*, $AuTe_2$. Its dimensions are

$$a_0 = 16.51 \text{ A.}; \ b_0 = 8.80 \text{ A.}; \ c_0 = 4.45 \text{ A.}$$

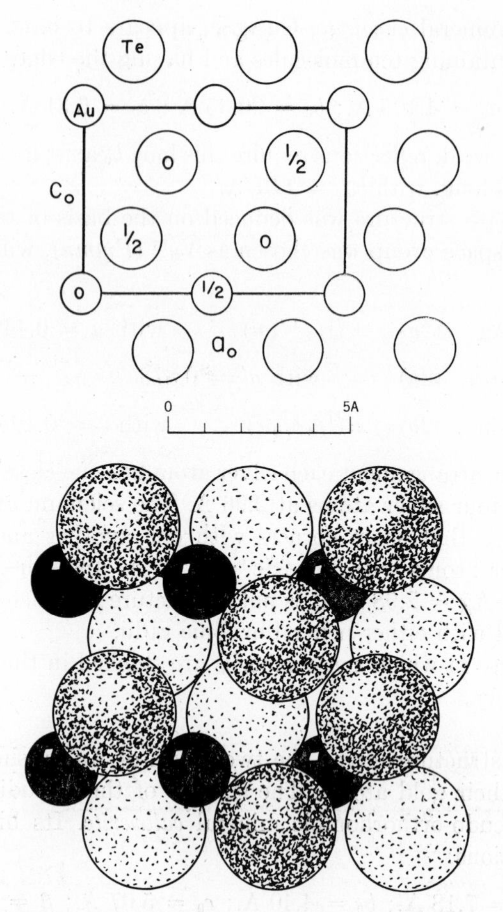

Fig. IV,55a (top). A projection along the b_0 axis of the pseudo-orthorhombic, monoclinic structure of calaverite, $AuTe_2$. Origin in lower left.

Fig. IV,55b (bottom). A packing drawing of the monoclinic structure of calaverite, $AuTe_2$, viewed along the b_0 axis. The dotted atoms are tellurium.

The assigned structure makes use of the following positions of the space group C_{2v}^4 (*Pma*):

$$(2a)\quad 00u;\ {}^1/_2\,0\,u$$

$$(2c)\quad {}^1/_4\,v\,w;\ {}^3/_4\,\bar{v}\,w$$

$$(4d)\quad xyz;\ \bar{x}\bar{y}z;\ {}^1/_2-x,y,z;\ x+{}^1/_2,\bar{y},z$$

Atoms are in the positions and have the parameters listed in Table IV,16.

In this structure, which is, like that of calaverite (**IV,f5**), an aggregate of linear $AuTe_2$ molecules, the nearest Au–Te = 2.56 A. The Te–Te separations of 3.02 A. are much shorter than those encountered in calaverite (Fig. IV,56).

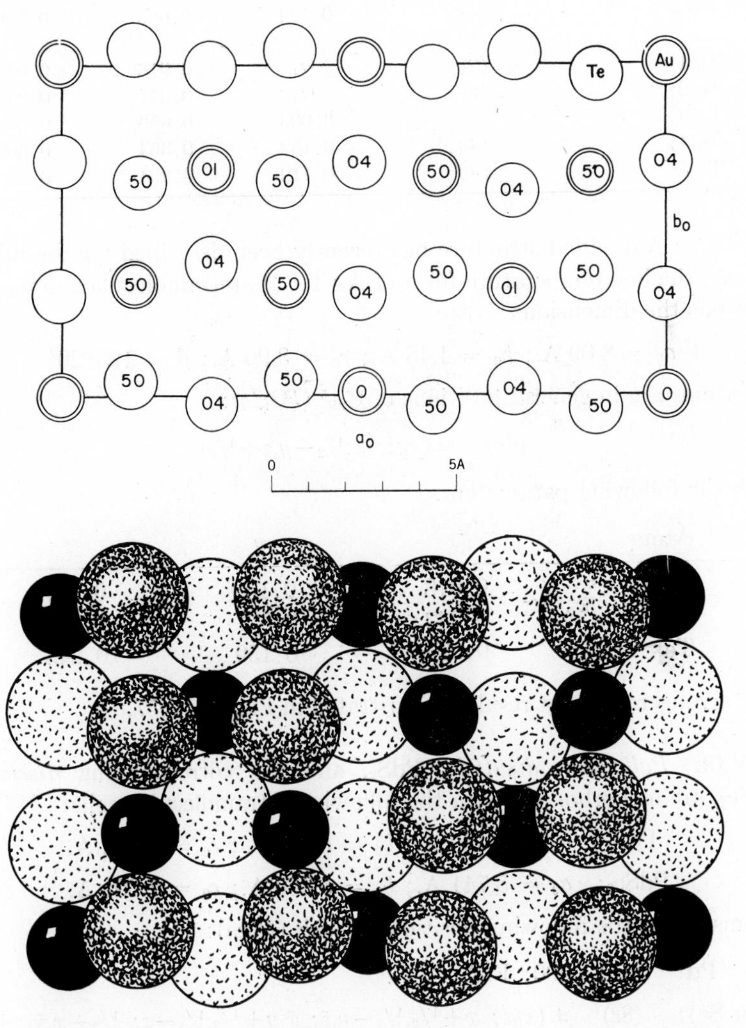

Fig. IV,56a (top). The orthorhombic structure of krennerite, $AuTe_2$, viewed along its c_0 axis. Origin in lower right.

Fig. IV,56b (bottom). A packing drawing of the krennerite structure viewed along its c_0 axis. The gold atoms are the black circles.

TABLE IV,16
Positions and Parameters of the Atoms in Krennerite (**IV,f6**)

Atom	Position	x	y	z
Au(or Ag)(1)	(2*a*)	0	0	0.00
Au(or Ag)(2)	(2*c*)	$^1/_4$	0.319	0.014
Au(or Ag)(3)	(4*d*)	0.124	0.666	0.500
Te(1)	(2*c*)	$^1/_4$	0.018	0.042
Te(2)	(2*c*)	$^1/_4$	0.617	0.042
Te(3)	(4*d*)	0.003	0.699	0.042
Te(4)	(4*d*)	0.132	0.364	0.500
Te(5)	(4*d*)	0.119	0.964	0.500

IV,f7. A detailed structure has recently been described for the mineral *hessite*, the low-temperature form of Ag_2Te. Its tetramolecular, monoclinic unit has the dimensions

$$a_0 = 8.09 \text{ A.};\ b_0 = 4.48 \text{ A.};\ c_0 = 8.96 \text{ A.};\ \beta = 123°20'$$

All atoms are in general positions of C_{2h}^5 ($P2_1/c$):

$$(4e)\quad \pm(xyz;\ x,{}^1/_2-y,z+{}^1/_2)$$

with the following parameters:

Atom	x	y	z
Ag(1)	0.018	0.152	0.371
Ag(2)	0.332	0.837	0.995
Te	0.272	0.159	0.243

The arrangement that results is illustrated in Figure IV,57.

IV,f8. *Palladium disulfide*, PdS_2, and the corresponding *diselenide*, $PdSe_2$, are orthorhombic with the following tetramolecular cells:

$$PdS_2:\ a_0 = 5.460 \text{ A.};\ b_0 = 5.541 \text{ A.};\ c_0 = 7.531 \text{ A.}$$

$$PdSe_2:\ a_0 = 5.741 \text{ A.};\ b_0 = 5.866 \text{ A.};\ c_0 = 7.691 \text{ A.}$$

Atoms are in the following positions of V_h^{15} ($Pbca$):

Pd: (4*a*) 000; F.C.

S (or Se): (8*c*) $\pm(xyz;\ x+{}^1/_2,{}^1/_2-y,\bar{z};\ \bar{x},y+{}^1/_2,{}^1/_2-z;\ {}^1/_2-x,\bar{y},z+{}^1/_2)$

For PdS_2, the sulfur parameters are $x = 0.107$, $y = 0.112$, and $z = 0.425$; for $PdSe_2$ the corresponding selenium parameters are $x = 0.112$, $y = 0.117$, and $z = 0.407$.

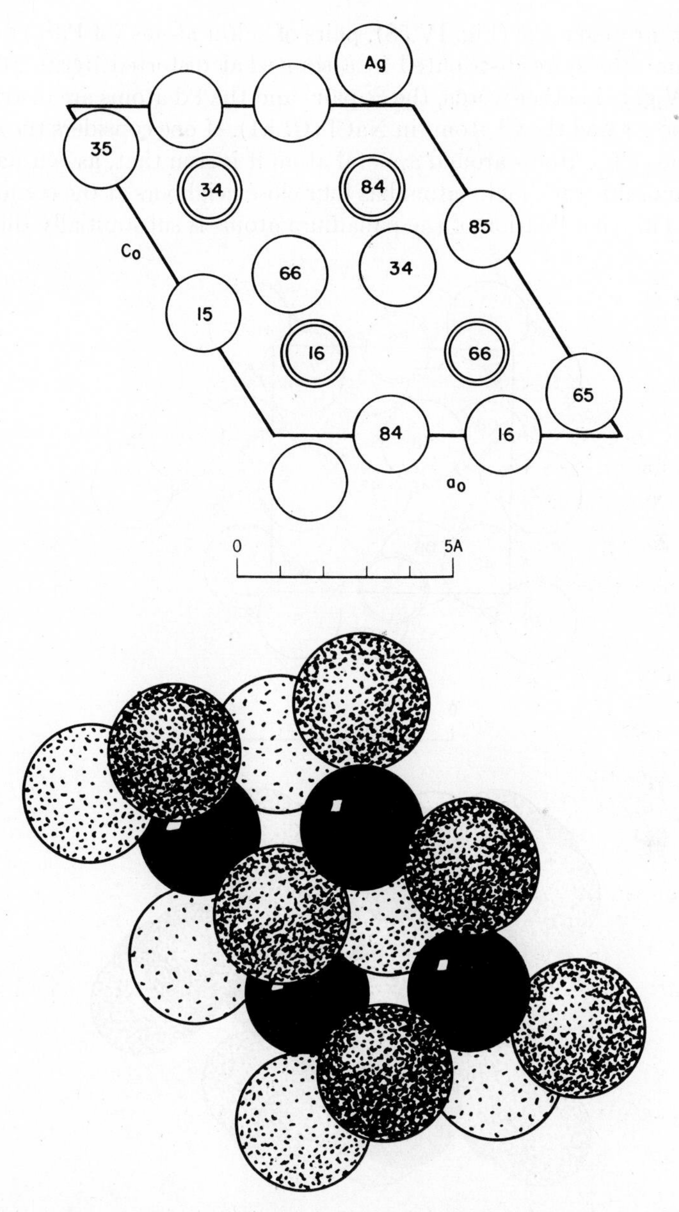

Fig. IV,57a (top). The monoclinic structure of hessite, Ag_2Te, projected along its b_0 axis. Origin in the lower left.

Fig. IV,57b (bottom). A packing drawing of the Ag_2Te arrangement viewed along its b_0 axis. Atoms have approximately their neutral radii. The tellurium atoms are black.

In this arrangement (Fig. IV,58), pairs of sulfur atoms (of PdS_2) and the palladium atoms are distributed in a somewhat distorted pyrite arrangement (**IV,g1**); in other words, the S_2 pairs and the Pd atoms are distributed as are the Na and the Cl atoms in NaCl (**III,a1**). If one considers the grouping of the sulfur atoms around a metal atom it is seen that, as is usual with palladium salts, each metal atom has four close neighbors at the corners of a square. This coordination of the palladium atoms is substantially the same

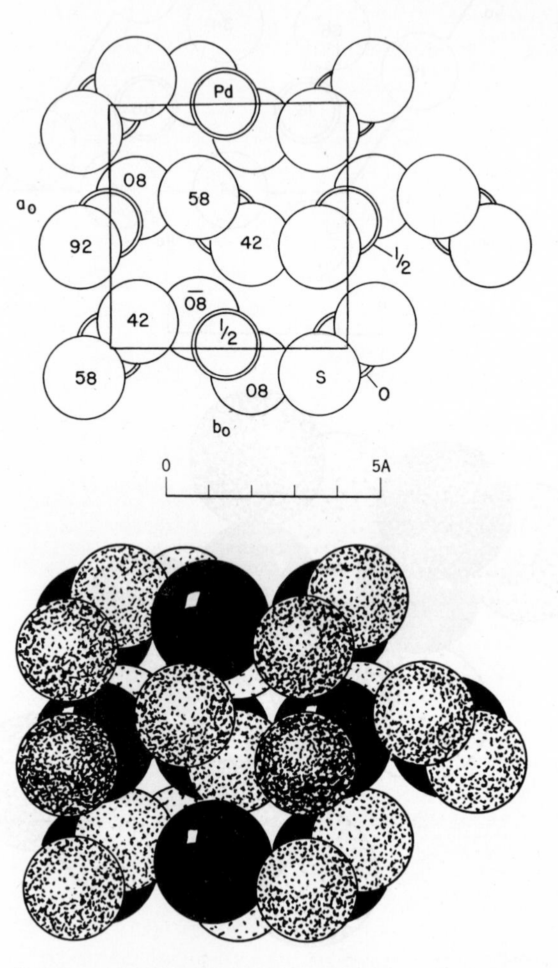

Fig. IV,58a (top). The orthorhombic structure of PdS_2, projected along its c_0 axis. Origin in lower left.

Fig. IV,58b (bottom). A packing drawing of the PdS_2 arrangement viewed along its c_0 axis. Atoms have been given their neutral radii, the palladium atoms being black.

as that which prevails in the monosulfide, PdS (**III,e7**). In PdS_2 the Pd–S distance is 2.30 A.; in the selenide, Pd–Se = 2.44 A. The next nearest Pd–S in PdS_2 is 3.28 A., and the corresponding Pd–Se (involving two Se) is 3.25 A. In PdS_2 the S–S separation in a pair is 2.13 A; in the selenide the close Se–Se = 2.36 A. The interatomic distances in these two crystals, as cited above, are close to the accepted covalent bond distances but, as the figure suggests, there are rather widely separated layers normal to the c_0 axes (corresponding to Pd–S = 3.28 A. and S–S = 3.35 A.). This layered character along c_0 points to bondings that are different in and between these layers.

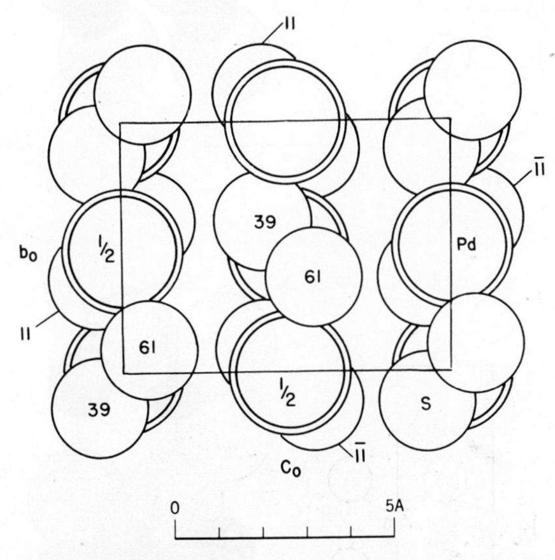

Fig. IV,58c. The PdS_2 structure projected along its a_0 axis rather than its c_0 axis. Comparison with the preceding two drawings brings out the atomic sheets that exist normal to c_0.

IV,f9. Crystals of *palladous chloride*, $PdCl_2$, are orthorhombic, with a bimolecular unit having

$$a_0 = 3.81 \text{ A.};\ b_0 = 11.0 \text{ A.};\ c_0 = 3.34 \text{ A.}$$

In the chosen structure, atoms are in the special positions of V_h^{12} (*Pnnm*):

$$\text{Pd:}\quad (2b)\quad 0\,0\,{}^1/_2;\ {}^1/_2\,{}^1/_2\,0$$

$$\text{Cl:}\quad (4g)\quad \pm(uv0;\ u+{}^1/_2,{}^1/_2-v,{}^1/_2)$$

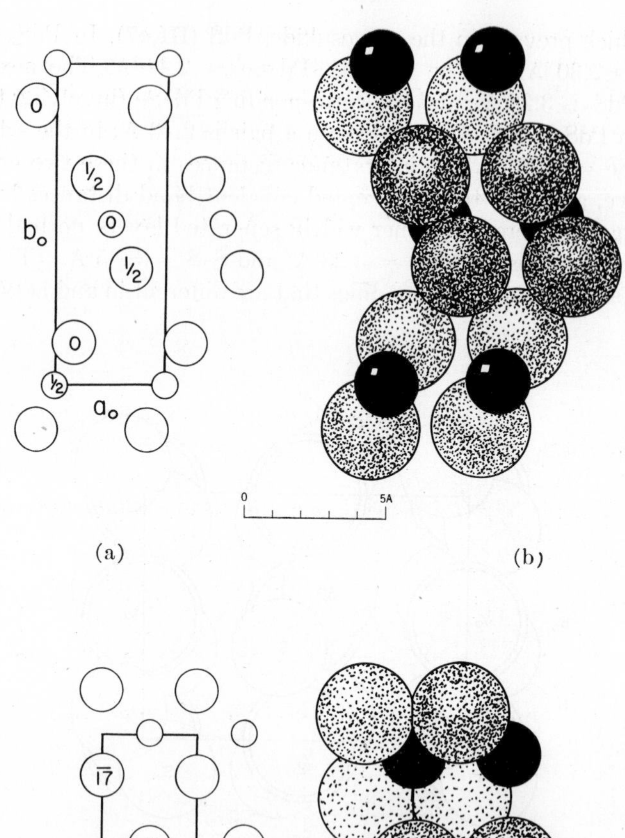

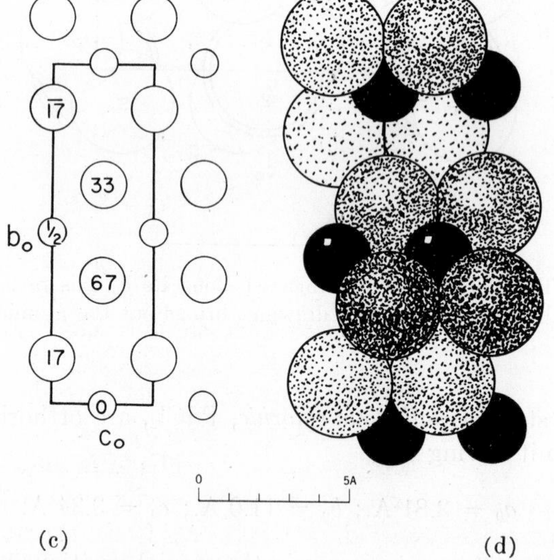

Fig. IV,59. Left: Projections along c_0 (a) and a_0 (c) of the orthorhombic structure of $PdCl_2$. The large circles are chlorine. Origins at lower left. Right: Two packing drawings of the orthorhombic $PdCl_2$ structure viewed in (b) along the c_0 axis and in (d) along the a_0 axis. The chlorine atoms are dotted.

with $u = 0.173$ and $v = 0.132$. Here (Fig. IV,59), each palladium atom has about it four chlorine atoms approximately at the corners of a square (Cl–Pd–Cl = 87°). Two of these chlorine atoms are shared by the square above and two by the square below. These indefinitely extended lines of linked squares have their axes parallel to the c_0 axis of the crystal and are so turned with respect to one another that adjacent lines approach most closely through contacts between chlorine atoms. Within chains, Pd–Cl = 2.31 A. and Cl–Cl = 3.18 and 3.34 A.

Platinous chloride, $PtCl_2$, also has this atomic arrangement, with

$$a_0 = 3.86 \text{ A.}; \; b_0 = 11.05 \text{ A.}; \; c_0 = 3.35 \text{ A.}$$

The parameters for the chlorine atoms have not been established, but presumably they are close to those in $PdCl_2$.

IV,f10. *Cupric bromide*, $CuBr_2$, and the isomorphous *chloride*, $CuCl_2$, form monoclinic crystals which have the bimolecular units:

$$CuBr_2\text{:} \quad a_0 = 7.14 \text{ A.}; \; b_0 = 3.46 \text{ A.}; \; c_0 = 7.18 \text{ A.}; \; \beta = 121°15'$$

$$CuCl_2\text{:} \quad a_0 = 6.70 \text{ A.}; \; b_0 = 3.30 \text{ A.}; \; c_0 = 6.85 \text{ A.}; \; \beta = 121°$$

The structure is described by placing the atoms in the following positions of C_{2h}^3 $(A2/m)$:

$$\text{Cu:} \quad (2b) \quad 0\,{}^1/_2\,0;\; 0\,0\,{}^1/_2$$

$$\text{Br (or Cl):} \quad (4i) \quad \pm(u0v;\; u,{}^1/_2,v+{}^1/_2)$$

For the bromide, $u = 0.240$ and $v = 0.015$. For the chloride, the parameters after an exchange of the a_0 and c_0 axes of the original description and a displacement of the origin through ${}^1/_2\, b_0$ are $u =$ ca. ${}^1/_4$ and $v =$ ca. 0.

This results in a structure (Fig. IV,60) which is somewhat related to that of $PdCl_2$ (**IV,f9**). Each copper atom has four halogen neighbors at the corners of a square, but the arrangement of these squares with respect to one another is different in the copper and in the palladium salts. In $CuBr_2$ each bromine atom occupies a corner of two squares which are in this way linked into chains having their axes along the b_0 axis of the crystal. Each copper atom has four bromine neighbors at a distance of 2.40 A. Within a chain the shortest Br–Br separation is 3.46 A.; between chains it is 3.30 A.

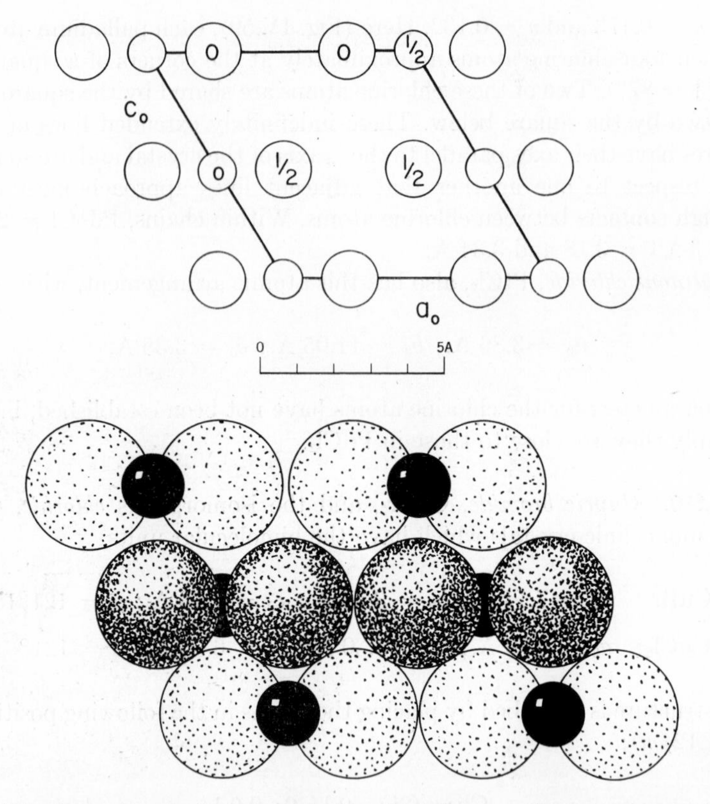

Fig. IV,60a (top). A projection along the b_0 axis of the monoclinic structure of $CuBr_2$. The large circles are bromine. Origin at lower left.

Fig. IV,60b (bottom). A packing drawing of the monoclinic $CuBr_2$ structure viewed along the b_0 axis. The copper atoms are black.

Pyrite-Like and Calcium Carbide-Like Structures

IV,g1. The mineral *pyrite*, FeS_2, has a structure that can be considered as an NaCl-like grouping of iron atoms and S_2 pairs. Its crystals are cubic with four molecules in a cell of the dimensions:

$$a_0 = 5.40667 \text{ A.}$$

Atoms are in the following positions of T_h^6 ($Pa3$):

Fe: (4*a*) $000;\ 0\,{}^1/_2\,{}^1/_2;\ {}^1/_2\,0\,{}^1/_2;\ {}^1/_2\,{}^1/_2\,0$

S: (8*c*) $\pm(uuu;\ u+{}^1/_2,{}^1/_2-u,\bar{u};\ \bar{u},u+{}^1/_2,{}^1/_2-u;\ {}^1/_2-u,\bar{u},u+{}^1/_2)$

with $u = 0.386$. This distributes dumbbell shaped pairs of sulfur atoms around $^1/_2\,^1/_2\,^1/_2$; F.C. (Fig. IV,61). Within a sulfur pair, S–S = 2.14 A.; the shortest Fe–S in the crystal is 2.26 A.

The many compounds with this structure are listed in Table IV,17. Where the parameter has been determined it has had such a value that X–X is about equal to the sum of the neutral atomic radii.

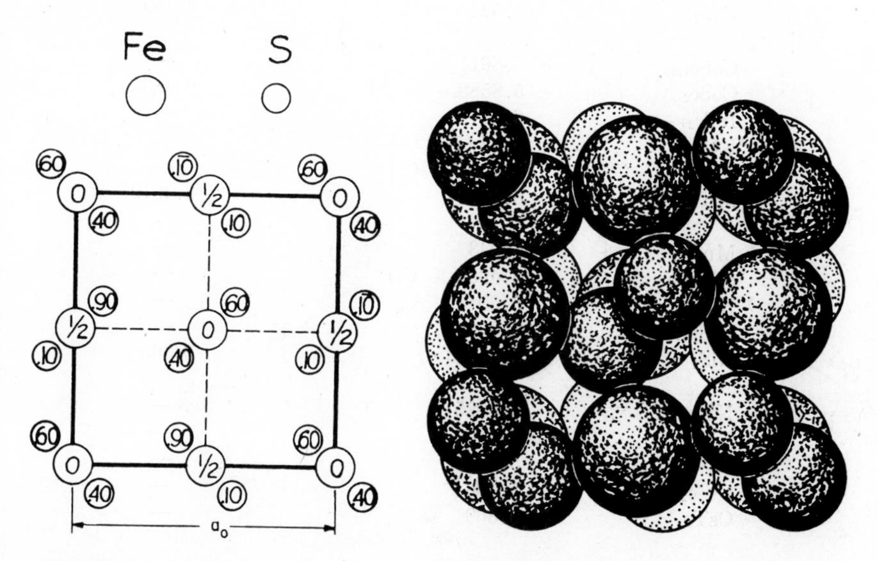

Fig. IV,61a (left). A projection of the unit of the pyrite, FeS_2, structure on a cube face. The metallic atoms are represented by the larger circles.

Fig. IV,61b (right). A packing drawing of the atoms shown in the preceding projection of the pyrite structure. Since the atoms have been given their neutral radii, the iron atoms are the larger.

IV,g2. The structure shown by the mineral *cobaltite*, CoAsS, and by the similar NiAsS and NiSbS (Table IV,17) differs from the pyrite arrangement (**IV,g1**) only to the degree required by the unlikeness of the two more electronegative atoms. The symmetry accordingly is tetartohedral, with all atoms in

$$(4a)\quad uuu;\ u+{}^1/_2,{}^1/_2-u,\bar{u};\ \bar{u},u+{}^1/_2,{}^1/_2-u;\ {}^1/_2-u,\bar{u},u+{}^1/_2$$

of T^4 ($P2_13$).

Accurate parameters have been determined for *ullmanite*, NiSbS. For it, they are: u(Ni) = −0.024, u(Sb) = −0.375, and u(S) = 0.390. The re-

TABLE IV,17
Crystals with the Pyrite and Cobaltite Arrangements (IV,g1, g2)

Crystal[a]	a_0, A.	u
$AuSb_2$	6.659	0.386
CaC_2	5.880 (477°C.)	—
CdO_2	5.313	0.4192
CoAsS	5.65	—
CoS_2	5.524	0.389
$CoSe_{1.90}$	5.8611	—
$CoSe_2$	5.8588	0.380
$IrTe_{2+x}$	6.411	—
$*KO_2$	6.05 (150°C.)	—
	6.12 (300°C.)	—
MnS_2	6.1008	0.4012
$MnTe_2$	6.943	—
$*\beta$-NaO_2	5.490 (20°C.)	—
NiAsS	5.60	—
NiSbS	5.881	—
NiS_2	5.677	0.395
$NiSe_{1.975}$	5.9625	—
$NiSe_2$	5.9604	—
OsS_2	5.6075	—
$OsSe_2$	5.933	—
$OsTe_2$	6.369	—
$PdAs_2$	5.970	—
$PdBi_2$	6.68	—
$PdSb_2$	6.439	—
$PtAs_2$	5.957	0.39
$PtBi_2$	6.683	0.38
PtP_2	5.694	—
$PtSb_2$	6.428	—
RhS_2	5.574	—
$RhSe_2$	6.002 (66.7 at.-% Se)	0.380
	5.985 (71.4 at.-% Se)	—
$RhTe_2$ (low)	6.441	0.362
$RhTe_{2.5}$	6.428	—
RuS_2	5.59	0.39
$RuSe_2$	5.921	—
$RuTe_2$	6.360	—

[a] Asterisk denotes disordered O_2 groups.

sulting atomic separations are Ni–Sb = 2.57 A., Ni–S = 2.34 A., and S–Sb = 2.40 A.—in substantial agreement with the sums of the neutral atomic radii. In connection with this determination the absolute configuration of the structure with respect to its axes was established.

IV,g3. A structure has been assigned to the beta modification of *mercury peroxide*, HgO_2. It is orthorhombic, with four molecules in a cell of the dimensions:

$$a_0 = 6.080 \text{ A.}; \; b_0 = 6.010 \text{ A.}; \; c_0 = 4.800 \text{ A.}$$

The mercury atoms have been placed in positions

$$(4a) \quad 000; \text{ F.C.}$$

of the space group V_h^{15} (*Pbca*). The oxygen positions could not be unequivocally located from the x-ray data, but there is agreement with the data and satisfactory interatomic distances result when the oxygens are placed in

$$(8c) \quad \pm(xyz;\ x+{}^1/_2,{}^1/_2-y,\bar{z};\ \bar{x},y+{}^1/_2,{}^1/_2-z;\ {}^1/_2-x,\bar{y},z+{}^1/_2)$$

with the parameters $x = 0.075$, $y = 0.062$, and $z = 0.405$.

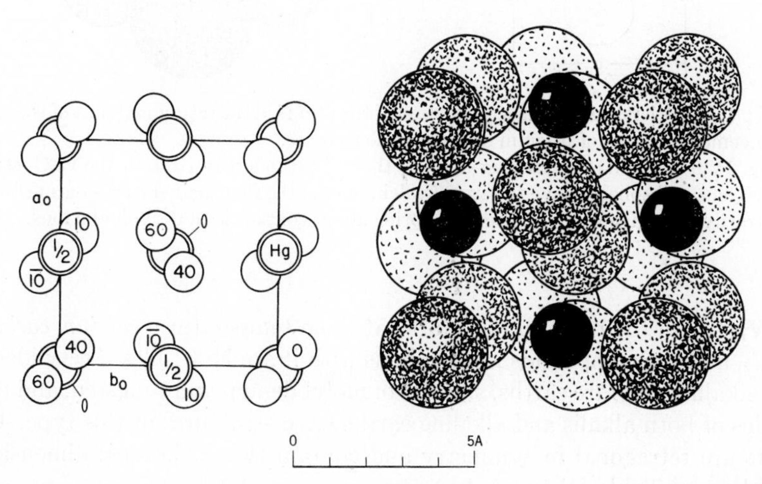

Fig. IV,62a (left). The orthorhombic HgO_2 structure projected along its c_0 axis. The slight measure of its distortion from the pyrite structure is evident by comparing this figure with Figure IV,61. Origin in lower left.

Fig. IV,62b (right). A packing drawing of the HgO_2 structure viewed along its c_0 axis. The atoms have been given their ionic dimensions, the mercury atoms being black.

In this atomic arrangement (Fig. IV,62), the oxygens are in pairs with an O–O separation of 1.5 A. Each mercury atom has two oxygen atoms at a distance of 2.06 A. and four more distant oxygens at 2.66 and 2.68 A. Together these give the mercury atoms a distorted octahedral environment. The structure as a whole is obviously a NaCl-like arrangement of Hg and O_2 pairs, distorted by flattening along the c_0 direction.

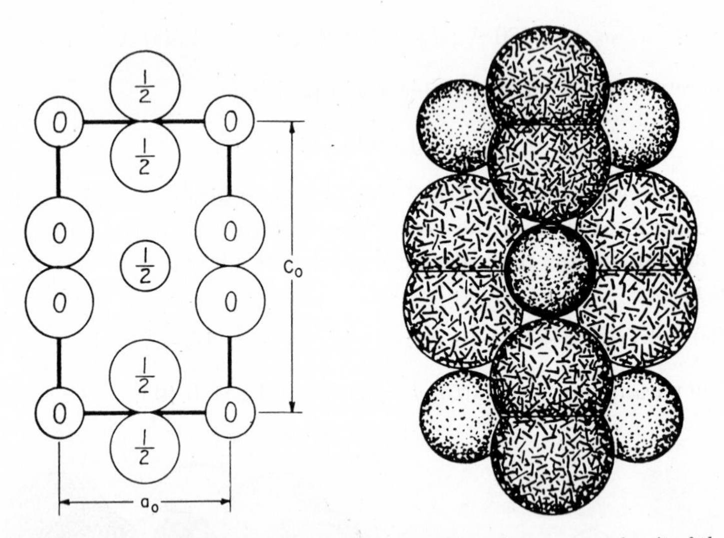

Fig. IV,63a (left). A side-face projection of the atoms in the tetragonal unit of the CaC arrangement. Atoms of calcium are the smaller circles.

Fig. IV,63b (right). A drawing to show the way atoms may pack in the CaC_2 structure. The small spheres have the size of calcium ions; the line-shaded carbon atoms have arbitrarily been given the size needed to bring about contact with the calcium ions.

IV,g4. In the modification stable at room temperature, *calcium carbide*, CaC_2, and the compounds isomorphous with it are NaCl-like. Acetylides of the alkaline and rare earths, silicides of molybdenum and tungsten, and peroxides of both alkalis and alkaline earths have structures of this type. The units are tetragonal in symmetry and contain two molecules; dimensions are those of Table IV,18.

For calcium carbide:

$$a_0 = 3.87 \text{ A.}, \qquad c_0 = 6.37 \text{ A.}$$

TABLE IV,18
Crystals with the Tetragonal CaC_2 Arrangement (**IV,g4**)

Crystal	a_0, A.	c_0, A.
BaC_2	4.39	7.04
CeC_2	3.878	6.488
DyC_2	3.669	6.176
ErC_2	3.620	6.094
GdC_2	3.718	6.275
HoC_2	3.643	6.139
KHC_2	4.28	8.42
LaC_2	3.934	6.572
LuC_2	3.563	5.964
MgC_2	4.86	5.76
$NaHC_2$	3.82	8.17
NdC_2	3.823	6.405
PrC_2	3.855	6.434
SmC_2	3.770	6.331
SrC_2	4.11	6.68
TbC_2	3.690	6.217
TmC_2	3.600	6.047
UC_2	3.517	5.987
YC_2	3.664	6.169
YbC_2	3.637	6.109
BaO_2	3.8154	6.8513 (25°C.)
CaO_2	3.54	5.92
CsO_2	4.44	7.20
KO_2	4.033	6.699 (25°C.)
RbO_2	4.24	7.03
SrO_2	3.5670	6.6161
$MoSi_2$	3.200	7.861
$(Na,Al)Si_2$	4.130	7.400
WSi_2	3.212	7.880

Its atoms are in the following special positions of D_{4h}^{17} ($I4/mmm$):

Ca: (2a) 000; $\frac{1}{2}\frac{1}{2}\frac{1}{2}$

C: (4e) 00u; 00$\bar{u}$; B.C. with $u = 0.406$

In this structure (Fig. IV,63), the significant Ca–C distances are 2.59 and 2.82 A.; the C–C separation within a C_2 group is 1.20 A. The diagonal tetramolecular pseudocell of this grouping is nearly cubic; it shows best the resemblance between this structure and the NaCl arrangement. For it, $a_0' = \sqrt{2}\, a_0 = 5.48$ A. and $c_0' = c_0 = 6.37$ A.

Parameters have been established for four other compounds with this structure. They are:

For BaO_2:	$u = 0.3911$	This gives rise to O–O = 1.49 A.
For KO_2:	$u = 0.4045$	This results in an O–O = 1.28 A.
For UC_2:	$u = 0.388$	From this, C–C = 1.34 A.
For LaC_2:	$u = 0.403$	From this, C–C = 1.28 A., La–2C = 2.65 A., and La–8C = 2.85 A. This dicarbide has the interesting property of being as good a metallic conductor as lanthanum metal itself.

The O–O = 1.49A. found in BaO_2 is the same as that determined in H_2O_2 (**III,g14**) and considerably greater than the 1.28 A. established for KO_2.

Calcium carbide has, in fact, at least four modifications whose occurrence depends on both the temperature of formation and the impurities that are present. The highest temperature form, CaC_2 IV, stable only above 450°C., is cubic, with

$$a_0 = 5.880 \text{ A. } (477°\text{C.})$$

There has been debate as to the structure. It is generally agreed that the calcium atoms are face-centered, in 000; F.C. Centers of the C_2 pairs seem to be in $\frac{1}{2}\frac{1}{2}\frac{1}{2}$; F.C., thus giving the whole an NaCl arrangement. But according to one observer, the C_2 group is "rotating"; according to another its atoms have the positions of the sulfur atoms in pyrite (**IV,g1**).

The lower temperature form designated as CaC_2 III is monoclinic. Its structure has been described in terms of a large, nonprimitive cell having the dimensions:

$$a_0 = 8.36 \text{ A.};\ b_0 = 4.20 \text{ A.};\ c_0 = 11.25 \text{ A.};\ \beta = 96°18'$$

Eight molecules are contained in this cell. The chosen space group is C_{2h}^5 in the orientation $P2_1/c$. All atoms are in general positions:

$$(8e) \quad \pm(xyz;\ \bar{x},y+1/2,1/2-z)$$

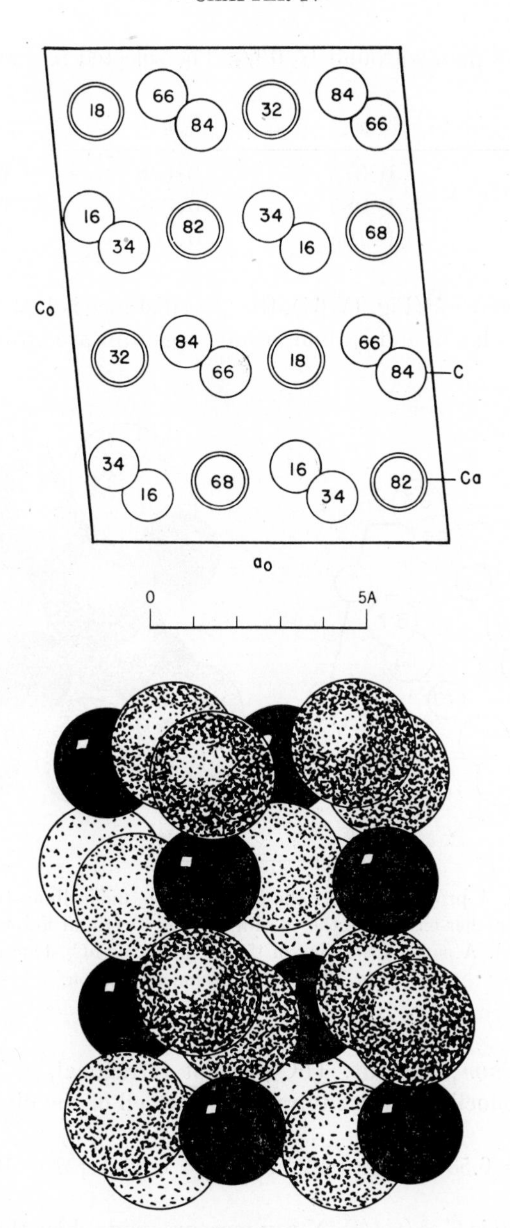

Fig. IV,64a (top). The monoclinic structure of CaC_2 III projected along its b_0 axis. Origin in lower left.

Fig. IV,64b (bottom). A packing drawing of the structure of CaC_2 III viewed along its b_0 axis. The calcium atoms are black.

plus four similar points around $\frac{1}{2}\,0\,\frac{1}{2}$. The selected parameters are:

Atom	x	y	z
Ca	0.375	0.676	0.125
C(1)	0.085	0.338	0.155
C(2)	0.165	0.163	0.095

In this arrangement (Fig. IV,64), the C–C distance is 1.24 A. The separations between calcium and carbon atoms range upwards from 2.46 A.

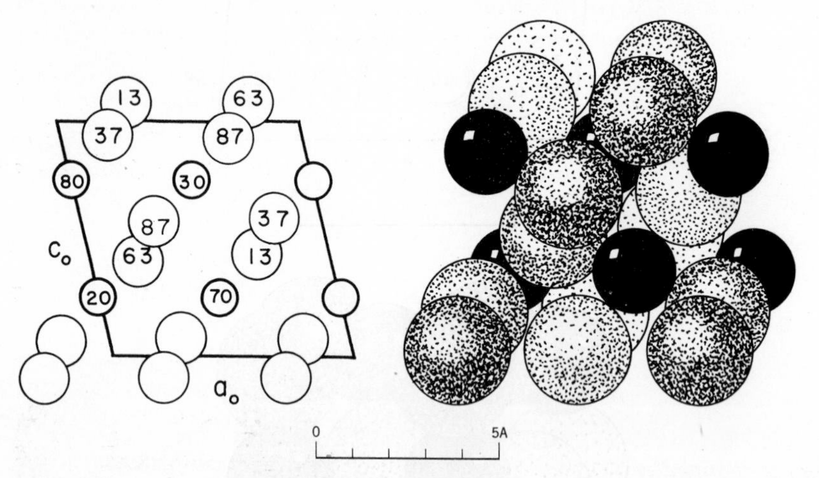

Fig. IV,65a (left). A projection along the b_0 axis of the monoclinic structure of ThC_2. The smaller circles are the thorium atoms. Origin at lower left.

Fig. IV,65b (right). A packing drawing of the monoclinic ThC_2 structure viewed along the b_0 axis. The dotted atoms are carbon.

IV,g5. *Thorium dicarbide*, ThC_2, is not tetragonal, as previously reported, but monoclinic with a tetramolecular cell of the dimensions:

$$a_0 = 6.53\text{ A.};\ b_0 = 4.24\text{ A.};\ c_0 = 6.56\text{ A.};\ \beta = 104°$$

The space group is C_{2h}^6 ($C2/c$). X-ray measurements show that the thorium atoms are in

$$(4e)\quad \pm(0\,u\,\tfrac{1}{4};\ \tfrac{1}{2},u+\tfrac{1}{2},\tfrac{1}{4}) \qquad \text{with } u = 0.202$$

By neutron diffraction it was found that the carbon atoms could satisfactorily be placed in general positions:

$$(8f) \quad \pm(xyz;\ x,\bar{y},z+{}^1/_2;\ x+{}^1/_2,y+{}^1/_2,z;\ x+{}^1/_2,{}^1/_2-y,z+{}^1/_2)$$

with the parameters $x = 0.290$, $y = 0.132$, and $z = 0.082$.

The resulting structure (Fig. IV,65) shows pairs of carbon atoms with a C–C separation of 1.47 A. The shortest Th–C = 2.38 A. Evidently the carbon atoms in ThC_2 are too far apart for their bonding to be like that in CaC_2 and the other acetylides.

IV,g6. There has been much discussion of the structure of the second modification of FeS_2, the orthorhombic mineral *marcasite*. Most of the data point to a bimolecular unit of the dimensions:

$$a_0 = 4.436 \text{ A.};\ b_0 = 5.414 \text{ A.};\ c_0 = 3.381 \text{ A.}$$

Faint reflections have, however, been described which have seemed to call for a tetramolecular cell. The latest study adopts the smaller unit and selects the following arrangement based on V_h^{12} (*Pnnm*):

Fe: (2*a*) $000;\ {}^1/_2\,{}^1/_2\,{}^1/_2$

S: (4*g*) $\pm(uv0;\ {}^1/_2-u,v+{}^1/_2,{}^1/_2)$ with $u = 0.200$, $v = 0.378$

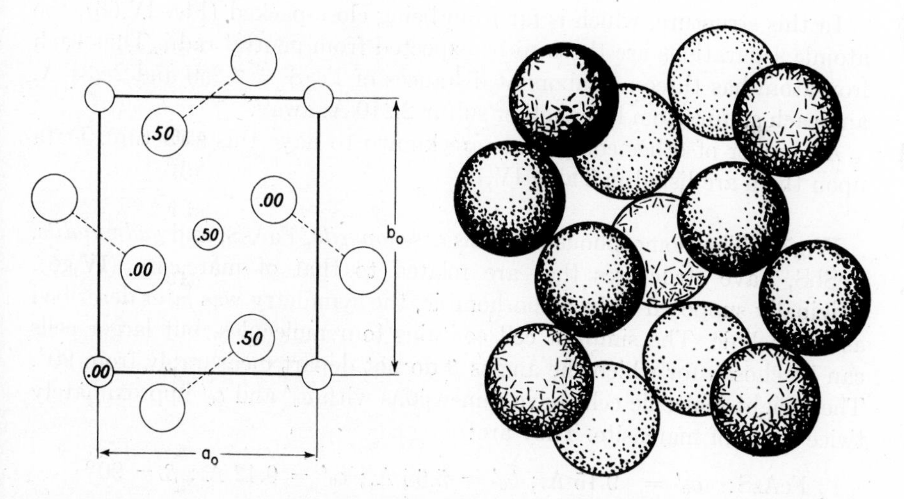

Fig. IV,66a (left). A projection along its c_0 axis of the structure found for marcasite, the orthorhombic form of FeS_2.

Fig. IV,66b (right). A packing drawing of the atoms of Figure IV,66a in which they are given the sizes of their neutral radii. The iron atoms are line-shaded.

TABLE IV,19
Crystals with the Orthorhombic FeS_2 Structure[a] (**IV,6g**)

Crystal	a_0, A.	b_0, A.	c_0, A.	u	v
$CoSe_2$	4.84	5.72	3.60	—	—
$CoTe_2$	5.301	6.298	3.882	0.22	0.36
$CrSb_2$	6.019	6.861	3.271	—	—
$FeAs_2$	5.25	5.92	2.85	0.175	0.361
FeP_2	4.975	5.657	2.725	0.16	0.37
$FeSb_2$	5.819	6.520	3.189	—	—
$FeSe_2$	4.791	5.715	3.575	0.21	0.37
$FeTe_2$	5.340	6.260	3.849	0.22	0.36
III-NaO_2	4.26	5.54	3.44 (−100°C.)	—	—
$NiAs_2$	4.78	5.78	3.53	0.215	0.370
OsP_2	5.098	5.898	2.918	—	—
$OsSb_2$	5.912	6.653	3.196	—	—
RuP_2	5.115	5.888	2.870	—	—
$RuSb_2$	5.930	6.637	3.168	0.18	0.36

[a] Several compounds with this structure show a rather wide homogeneity range with a corresponding change in cell dimensions. As an example, $FeTe_{1.90}$ has been given the dimensions a_0 = 5.261 A., b_0 = 6.264 A., c_0 = 3.872 A., and $FeTe_{2.10}$ the dimensions a_0 = 5.279 A., b_0 = 6.275 A., c_0 = 3.864 A.

In this structure, which is far from being close-packed (Fig. IV,66), the atomic separations are those to be expected from neutral radii. Thus each iron atom has three neighbors at distances of Fe–S = 2.250 and 2.231 A. and each sulfur atom has another sulfur 2.210 A. away.

A number of other compounds are known to have this structure. Data upon them are listed in Table IV,19.

IV,g7. The isomorphous minerals *arsenopyrite*, FeAsS, and *gudmundite*, FeSbS, have structures that are related to that of marcasite (**IV,g6**). Originally supposed to be orthorhombic, the symmetry was later described as monoclinic. The simplest cell contains four molecules, but larger cells can be chosen for which the angles β do not depart measurably from 90°. These eight-molecule cells have dimensions with a_0' and c_0' approximately twice those of marcasite. They are:

$$\text{FeAsS:}\quad a_0' = 9.15\text{ A.};\ b_0' = 5.65\text{ A.};\ c_0' = 6.42\text{ A.};\ \beta = 90°$$

$$\text{FeSbS:}\quad a_0' = 10.04\text{ A.};\ b_0' = 5.93\text{ A.};\ c_0' = 6.68\text{ A.};\ \beta = 90°$$

In the structure that has been developed around this cell (Fig. IV,67), all atoms are in general positions of C_{2h}^5 ($B2_1/d$):

$$(8e)\quad \pm(xyz;\ {}^1/_4-x,y+{}^1/_2,{}^1/_4-z;\ x+{}^1/_2,y,z+{}^1/_2;\ {}^3/_4-x,y+{}^1/_2,{}^3/_4-z)$$

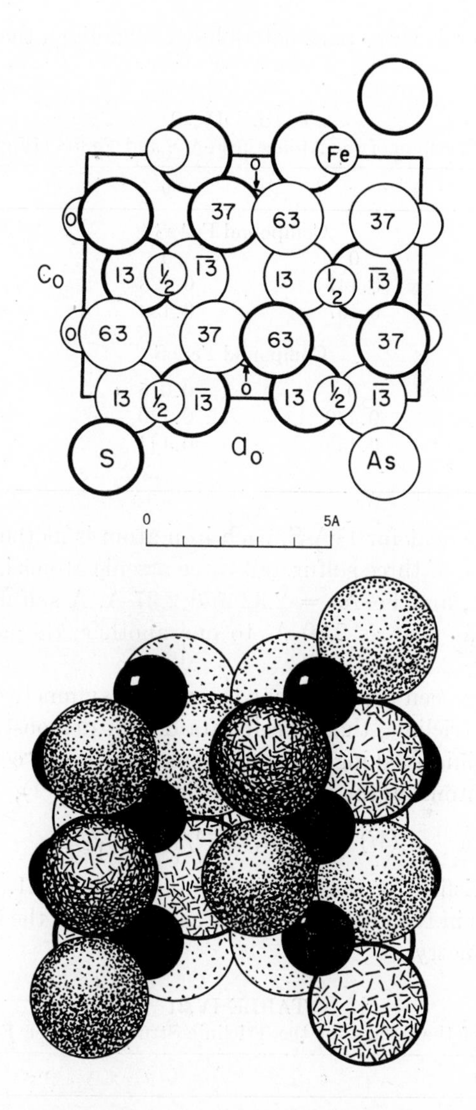

Fig. IV,67a (top). A projection along b_0 of the monoclinic, pseudo-orthorhombic structure assigned arsenopyrite, FeAsS. Iron are the small circles, the arsenic and sulfur the large circles.

Fig. IV,67b (bottom). A packing drawing of the monoclinic structure of arsenopyrite, FeAsS, viewed along the b_0 axis. The iron atoms are black, the larger sulfur atoms are dotted.

For the two minerals these parameters have been given the values listed in Table IV,20.

TABLE IV,20
Parameters of the Atoms in FeAsS and FeSbS (**IV,g7**)

Atom	x	y	z
	Compound FeAsS		
Fe	0	0	0.275
As	0.147	0.128	0
S	0.167	0.132	0.500
	Compound FeSbS		
Fe	0	0.015	0.300
Sb	0.140	0.130	−0.008
S	0.155	0.144	0.511

In this arrangement for FeAsS, each iron atom is at the center of a distorted octahedron of three sulfur and three arsenic atoms in which Fe–S = 2.23 and 2.27 A., and Fe–As = 2.32 and 2.37 A. A sulfur and an arsenic atom approach as near as 2.30 A. to one another; the nearest As–As or S–S distances are 3.15 A.

It has recently been asserted that the correct symmetry of arsenopyrite is no more than triclinic, though cell data do not demonstrate a departure from the monoclinic. Making use of the same B-centered cell defined above, but placing the atoms in the general positions of C_i^1 ($P\bar{1}$),

$$(4i) \quad \pm(xyz,\ x+{}^1/_2,y,z+{}^1/_2)$$

the data are satisfied by the parameters of Table IV,21. These represent only slight shifts in the atomic positions as defined for the structure having monoclinic symmetry.

TABLE IV,21
Parameters of the Atoms for the Triclinic Structure Given FeAsS (**IV,g7**)

Atom	x	y	z
Fe(1)	−0.006	0	0.295
Fe(2)	0.243	$^1/_2$	−0.021
As(1)	0.133	0.128	−0.013
As(2)	0.094	0.628	0.251
S(1)	0.166	0.132	0.499
S(2)	0.584	0.632	0.239

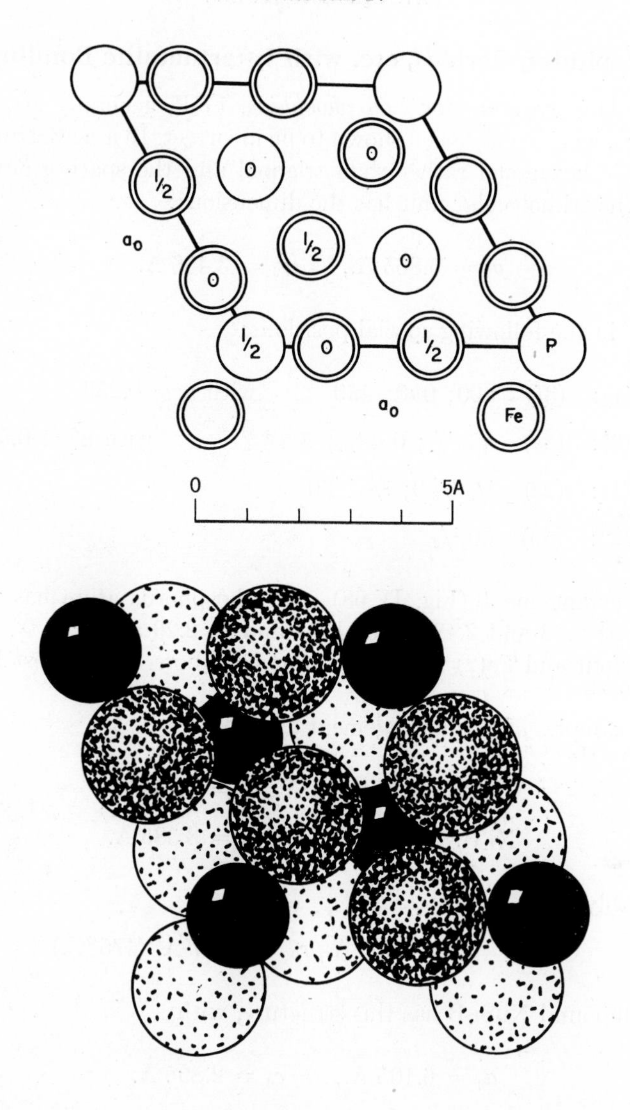

Fig. IV,68a (top). The hexagonal structure of Fe_2P projected along its principal, c_0, axis.

Fig. IV,68b (bottom). A packing drawing of the Fe_2P structure seen along its c_0 axis. Atoms have the relative sizes of the neutral radii, the phosphorus being the black circles.

Phosphides, Borides, etc. with Intermetallic Bondings

IV,h1. The structure for *iron phosphide*, Fe_2P, generally accepted for many years, has recently been shown to be incorrect. In a new structure the symmetry is hexagonal rather than trigonal and the space group is D_{3h}^3 ($P\bar{6}2m$). The trimolecular unit has the dimensions:

$$a_0 = 5.865 \text{ A.}, \qquad c_0 = 3.456 \text{ A.}$$

Atoms are in the following special positions:

Fe(1): (3*f*) $u00;\ 0u0;\ \bar{u}\bar{u}0$ with $u = 0.256$

Fe(2): (3*g*) $u\,0\,{}^1/_2;\ 0\,u\,{}^1/_2;\ \bar{u}\,\bar{u}\,{}^1/_2$ with $u' = 0.594$

P(1): (2*c*) ${}^1/_3\,{}^2/_3\,0;\ {}^2/_3\,{}^1/_3\,0$

P(2): (1*b*) $00\,{}^1/_2$

In this arrangement (Fig. IV,68) each phosphorus atom has nine iron neighbors at 2.22 and 2.48 A. for P(1) and at 2.29 and 2.38 A. for P(2). Fe(1) has four and Fe(2) five phosphorus neighbors. The closest Fe–Fe = 2.60 A.

Other examples of this arrangement are:

Mn_2P: $a_0 = 6.074$ A., $c_0 = 3.454$ A.

Ni_2P: $a_0 = 5.864$ A., $c_0 = 3.385$ A.

and probably

Co_2As: $a_0 = 6.066$ A., $c_0 = 3.557$ A. (475°C.)

The compound Ni_6Si_2B has this structure, with

$$a_0 = 6.105 \text{ A.}, \qquad c_0 = 2.895 \text{ A.}$$

and the parameters u(Ni in 3*f*) = 0.247, u(Ni in 3*g*) = 0.608, Si in (2*c*) and B in (1*b*).

IV,h2. The tetragonal crystals isomorphous with *iron arsenide*, Fe_2As, have bimolecular units. For Fe_2As the cell edges are

$$a_0 = 3.627 \text{ A.}, \qquad c_0 = 5.973 \text{ A.}$$

Its atoms are in the following special positions of D_{4h}^7 ($P4/nmm$):

$$\begin{array}{llll} Fe(1): & (2a) & 000;\ ^1/_2\,^1/_2\,0 & \\ Fe(2): & (2c) & 0\,^1/_2\,u;\ ^1/_2\,0\,\bar{u} & \text{with } u = 0.33 \\ As: & (2c) & \text{with } u' = -0.265 & \end{array}$$

In this arrangement (Fig. IV,69), as in typically intermetallic compounds, chemically like atoms are as closely associated with one another as are the unlike. In Fe_2As each Fe(1) atom has 12 neighbors—four arsenic at 2.40 A., four Fe(2) at 2.39 A., and four Fe(1) at 2.57 A. Each Fe(2) has nine neighbors—four arsenic at 2.60 A., one arsenic at 2.41 A., and four Fe(1) at 2.39 A. Each arsenic atom also has nine neighbors—five iron atoms at 2.40 A. and four at 2.60 A.

Other compounds shown to have this structure are:

Cr_2As: $a_0 = 3.613$ A., $c_0 = 6.333$ A., $u(Cr) = 0.33$, $u(As) = -0.265$

Cu_2Sb: $a_0 = 3.992$ A., $c_0 = 6.091$ A., $u(Cu) = 0.27$, $u(Sb) = -0.30$

$Cu_{4-x}Te_2$: $a_0 = 3.98$ A., $c_0 = 6.12$ A.

Mn_2Sb: $a_0 = 4.078$ A., $c_0 = 6.557$ A., $u(Mn) = 0.295$, $u(Sb) = -0.28$

$ThAs_2$: $a_0 = 4.086$ A., $c_0 = 8.575$ A., $u(As) = 0.36$, $u(Th) = -0.28$

$ThSb_2$: $a_0 = 4.353$ A., $c_0 = 9.172$ A., $u(Sb) = 0.363$, $u(Th) = -0.275$

$ThBi_2$: $a_0 = 4.492$ A., $c_0 = 9.298$ A., $u(Bi) = 0.37$, $u(Th) = -0.28$

UP_2: $a_0 = 3.800$ A., $c_0 = 7.762$ A.

UAs_2: $a_0 = 3.954$ A., $c_0 = 8.116$ A.

USb_2: $a_0 = 4.272$ A., $c_0 = 8.741$ A., $u(Sb) = 0.365$, $u(U) = -0.28$

UBi_2: $a_0 = 4.445$ A., $c_0 = 8.908$ A., $u(Bi) = 0.365$, $u(U) = -0.280$

IV,h3. In the essentially intermetallic arrangement typified by *iron boride*, Fe_2B, there are four molecules in the tetragonal unit. For Fe_2B this unit has the edge lengths:

$$a_0 = 5.099 \text{ A.}, \qquad c_0 = 4.240 \text{ A.}$$

Its atoms have been found to be in the following special positions of D_{4h}^{18} ($I4/mcm$):

B: (4a) $00\,^1/_4;\ 00\,^3/_4$; B.C.

Fe: (8h) $u,u+{}^1/_2,0;\ \bar{u},{}^1/_2-u,0;\ u+{}^1/_2,\bar{u},0;\ {}^1/_2-u,u,0$; B.C.

with $u = 0.167$

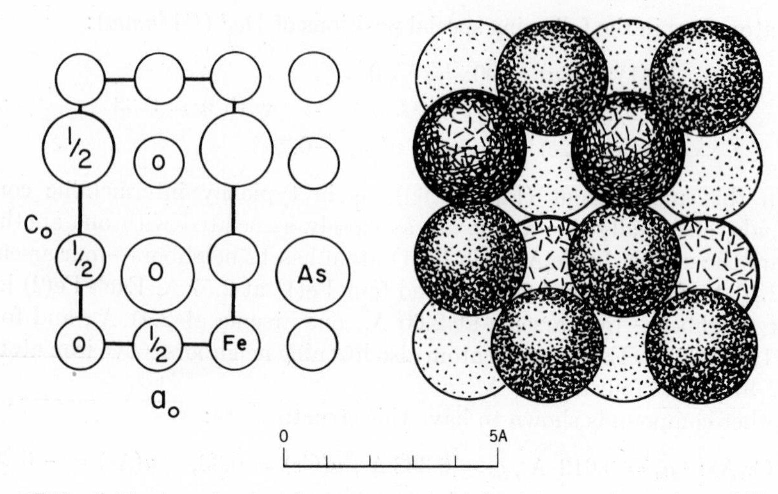

Fig. IV,69a (left). A projection along an a_0 axis of the tetragonal structure of Fe_2As. Origin in the lower left.

Fig. IV,69b (right). A packing drawing of the tetragonal Fe_2As structure viewed along an a_0 axis. The iron atoms are dotted.

In the resulting arrangement (Fig. IV,70) each iron atom has four boron neighbors at a distance of 2.18 A. and an iron atom at 2.40 A. In this structure one does not encounter the closed associations of boron atoms with one another (with B–B = ca. 1.75 A.) which produce the boron chains of FeB (**III,d4**) or the boron layers of the two following structures. Instead, though in Fe_2B each boron atom has two boron neighbors, they lie in a straight line and have a larger B–B separation that depends on the compound: in this case it is 2.12 A.

Other borides and certain intermetallic compounds with this arrangement are listed in Table IV,22.

IV,h4. The compound *aluminum boride*, AlB_2, is hexagonal, with a unimolecular cell having the edges:

$$a_0 = 3.009 \text{ A.}, \qquad c_0 = 3.262 \text{ A.}$$

Its structure is formally related to that of $Cd(OH)_2$ (**IV,c1**), with atoms in the same special positions of D_{3d}^3 ($C\bar{3}m$):

$$\text{Al:} \quad (1a) \quad 000$$

$$\text{B:} \quad (2d) \quad {}^1/_3\,{}^2/_3\,u;\ {}^2/_3\,{}^1/_3\,\bar{u}$$

TABLE IV,22
Crystals with the Fe_2B Structure (**IV,h3**)

Crystal	a_0, A.	c_0, A.	u
Al_2Cu	6.04	4.86	0.158
$AuPb_2$	7.31	5.644	—
Co_2B	5.006	4.212	0.167
Cr_2B	5.18	4.31	—
$FeGe_2$	5.899	4.991	—
$FeSn_2$	6.513	5.295	—
$MnSn_2$	6.659	5.436	—
γ-Mo_2B	5.543	4.735	—
Na_2Au	7.402	5.511	0.16
Ni_2B	4.989	4.246	0.167
$PdPb_2$	6.835	5.821	—
$RhPb_2$	6.651	5.853	—
β-Ta_2B	5.778	4.864	—
VSb_2	6.555	5.635	0.158
γ-W_2B	5.564	4.740	—

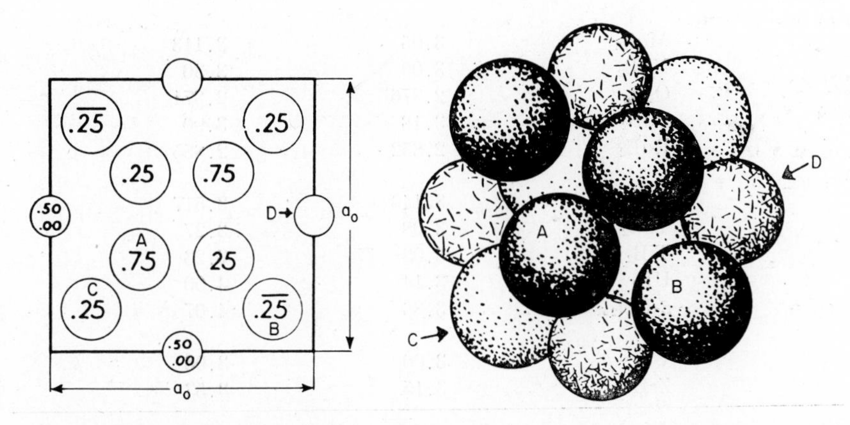

Fig. IV,70a (left). The atomic arrangement in the tetragonal unit of Fe_2B projected on its c face. Large circles represent the iron atoms. The origin of coordinates used in the description of the text lies at the point $^1/_2$, 0, $^3/_4$ if the origin of the drawing is in the lower left-hand corner of the cell projection.

Fig. IV,70b (right). A packing drawing to show the relation of atoms to one another in the tetragonal structure of Fe_2B. Boron atoms are the line-shaded spheres. Corresponding atoms are lettered the same in this and in Figure IV,70a.

In AlB_2, however, u is ca. $^1/_2$ instead of $^1/_4$. This different parameter results in a grouping of alternate layers of aluminum and boron atoms normal to the c_0 axis, each aluminum having 12 equidistant boron neighbors at 2.37 A., six in the plane above and six in the plane below (Fig. IV,71). In the boron planes each atom is surrounded by an equilateral triangle of atoms, with B–B = 1.73 A.

Other borides and several intermetallic compounds with this atomic arrangement are given in Table IV,23.

TABLE IV,23
Crystals with the Hexagonal AlB_2 Structure (**IV,h4**)

Crystal	a_0, A.	c_0, A.
AgB_2	3.00	3.24
AuB_2	3.14	3.51
$CaGa_2$	4.314	4.314
$CeGa_2$	4.303	4.307
CrB_2	2.97	3.07
HfB_2	3.14	3.47
$HfBe_2$	3.787	3.159
$LaGa_2$	4.320	4.396
MgB_2	3.084	3.522
MnB_2	3.007	3.037
MoB_2	3.05	3.113
NbB_2	3.09	3.30
OsB_2	2.876	2.871
PuB_2	3.18	3.90
RuB_2	2.852	2.855
ScB_2	3.146	3.517
TaB_2	3.08	3.27
TiB_2	3.03	3.23
UB_2	3.14	4.00
β-USi_2	3.86	4.07
VB_2	3.00	3.06
ZrB_2	3.15	3.53

IV,h5. *Rhenium diboride*, ReB_2, though hexagonal, has a structure different from the simple AlB_2 arrangement of the preceding paragraph. Its bimolecular unit has the edges:

$$a_0 = 2.900 \text{ A.,} \qquad c_0 = 7.478 \text{ A.}$$

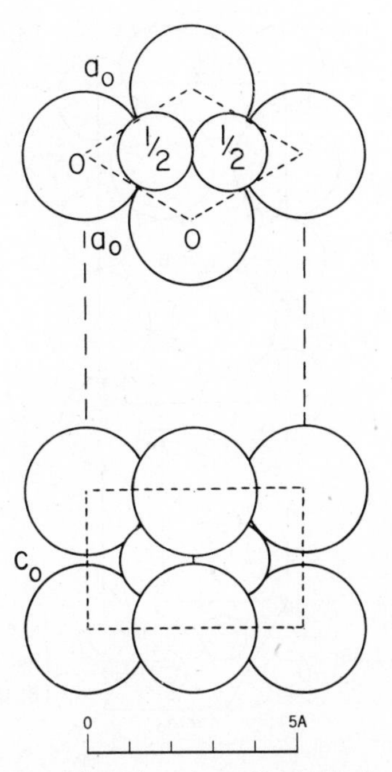

Fig. IV,71. Projections along c_0 (above), and parallel to the base diagonal and c_0 (below) of the hexagonal structure of AlB_2. The aluminum atoms are the larger circles.

The chosen space group is D_{6h}^4 ($P6_3/mmc$), with atoms in the positions:

Re: (2*c*) $\pm(^1/_3\ ^2/_3\ ^1/_4)$

B: (4*f*) $\pm(^1/_3\ ^2/_3\ u;\ ^2/_3,^1/_3,u+^1/_2)$ with $u = 0.548$

In the resulting structure (Fig. IV,72), each rhenium atom has eight boron neighbors at 2.26 or 2.23 A. and each boron atom has four rhenium atoms at these distances. The boron atoms are in puckered layers normal to c_0 with each boron atom 1.82 A. distant from three others.

This substance has the same x-ray pattern as the preparation which had earlier been given the formula of a triboride.

IV,h6. The alloy-type structure found many years ago for Cu_2Mg is possessed by other compounds involving relatively electronegative metals. The symmetry is cubic and the unit contains eight molecules. For Cu_2Mg,

$$a_0 = 7.02 \text{ A.}$$

The homogeneity range of this alloy is only about one per cent with a_0 = 7.06 A. on the magnesium-rich side and 7.00 A. on the copper-rich side.

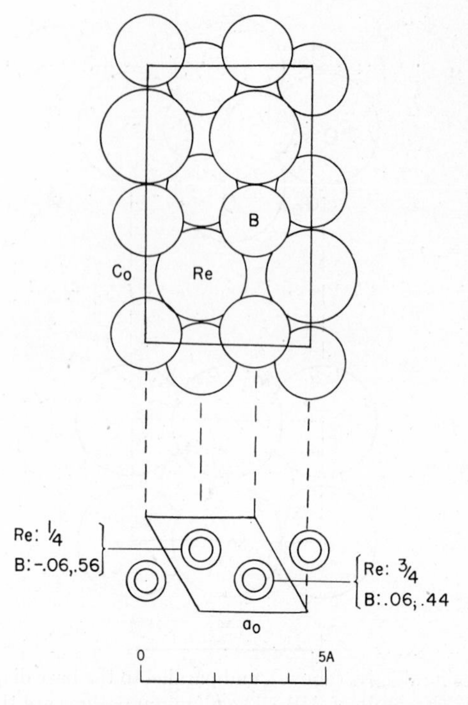

Fig. IV,72. Two projections of the hexagonal structure of ReB_2.

Atoms are in the following special positions of O_h^7 ($Fd3m$):

Mg: (8a) 000; $^1/_4\,^1/_4\,^1/_4$; F.C.

Cu: (16d) $^1/_8\,^3/_8\,^7/_8$; $^7/_8\,^1/_8\,^3/_8$; $^3/_8\,^7/_8\,^1/_8$; $^5/_8\,^5/_8\,^5/_8$; F.C.

These are together the positions (Fig. IV,73) occupied by the metal atoms in spinel, $MgAl_2O_4$ (Chapter VIII). The atomic separations of this structure are those to be expected if metallic radii prevail: Cu–Mg = 3.08 A. and Cu–Cu = 2.49 A.

Other compounds with this structure are

$CsBi_2$: a_0 = 9.760 A.

$RbBi_2$: a_0 = 9.601 A.

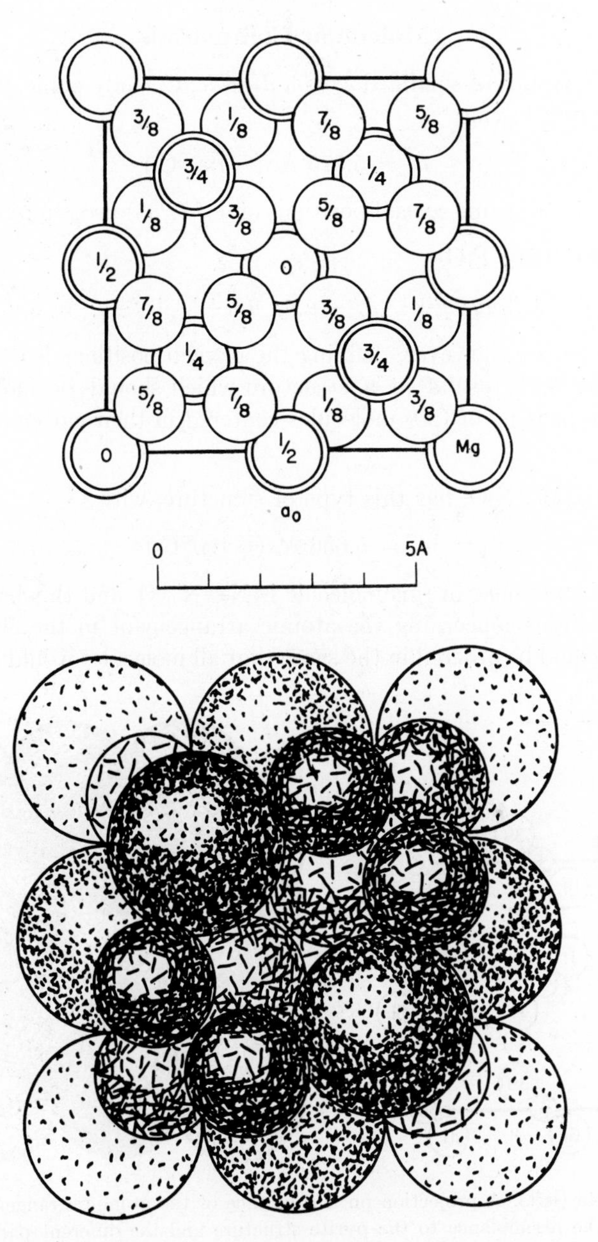

Fig. IV,73a (top). The cubic Cu_2Mg structure projected along a cubic axis.
Fig. IV,73b (bottom). A packing drawing of the Cu_2Mg structure with atoms having their metallic radii. The copper atoms are line-shaded.

Molecular Compounds

IV,i1. Crystals of solidified *carbon dioxide*, CO_2, are cubic, with a tetramolecular unit having

$$a_0 = 5.575 \text{ A. } (-190°\text{C.})$$

Atoms are in the same special positions of T_h^6 ($Pa3$) used in pyrite (**IV,g1**):

C: (4b) 000; F.C.

O: (8c) $\pm(uuu;\ u+{}^1/_2,{}^1/_2-u,\bar{u};\ \bar{u},u+{}^1/_2,{}^1/_2-u;\ {}^1/_2-u,\bar{u},u+{}^1/_2)$

The parameter u, however, defining the oxygen positions is different. The chosen $u = 0.11$ results in a structure which should be looked on as a cubic close-packing of CO_2 molecules centered in their carbon atoms (Fig. IV,74).

Nitrous oxide, N_2O, has this type of structure, with

$$a_0 = 5.656 \text{ A. } (-190°\text{C.})$$

The atomic sequence in this molecule is N≡N═O, and this has presented two possibilities concerning the atomic arrangement in the crystals. The structure could be ordered in the sense that all molecules would be similarly

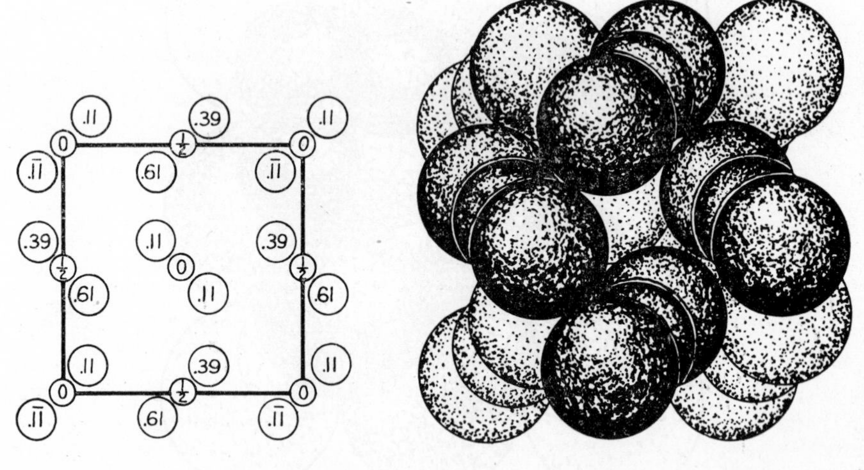

Fig. IV,74a (left). A projection on a cube face of the atomic arrangement in solid CO_2. Both the resemblance to the pyrite structure and the different parameters that make this a molecular crystal are evident.

Fig. IV,74b (right). A packing drawing which shows the way the linear CO_2 molecules pack in the crystalline state. The larger spheres are oxygen atoms.

oriented. In that case, since the N≡N would be expected to be somewhat different from N=O, the space group would be T^4 ($P2_13$), of lower symmetry than T_h^6, and u for the noncentral nitrogen would be somewhat different from $-u$ for O. On the other hand, the structure could be disordered to the degree that the molecules would be haphazardly oriented within the crystal, to yield a mean value of u within the framework of T_h^6. The scattering powers of nitrogen and oxygen for x rays are not sufficiently different to permit a decision between these possibilities, but one has been made using the data of neutron diffraction. In this way it has been shown that the "disordered" structure is to be preferred with a "mean" u parameter of 0.1196.

The following other solidified gases probably have the CO_2 arrangement. Their cubic cells have:

D_2S: $a_0 = 5.712$ A.

D_2Se: $a_0 = 5.973$ A.

H_2S: $a_0 = 5.778$ A. (at -170°C.). Below this temperature electron diffraction has shown the existence of a stable tetragonal form having a large unit cell.

For H_2Se: $a_0 = 6.020$ A. (-170°C.)

IV,i2. Solidified *sulfur dioxide*, SO_2, is orthorhombic, with a tetramolecular unit which at -130°C. has the edge lengths:

$$a_0 = 6.07 \text{ A.};\ b_0 = 5.94 \text{ A.};\ c_0 = 6.14 \text{ A.}$$

Atoms are in the following positions of C_{2v}^{17} (Aba):

S: (4a) $00u;\ \frac{1}{2}\frac{1}{2}u;\ 0,\frac{1}{2},u+\frac{1}{2};\ \frac{1}{2},0,u+\frac{1}{2}$ with $u = 0$

O: (8b) $xyz;\ \frac{1}{2}-x,y+\frac{1}{2},z;\ x+\frac{1}{2},\frac{1}{2}-y,z;\ x,y+\frac{1}{2},z+\frac{1}{2};$

$\bar{x}\bar{y}z;\ \bar{x},\frac{1}{2}-y,z+\frac{1}{2};\ \frac{1}{2}-x,y,z+\frac{1}{2};\ x+\frac{1}{2},\bar{y},z+\frac{1}{2}$

with $x = 0.140$ (0.151), $y = 0.150$ (0.141), and $z = 0.118$ (0.117). The values in parentheses are those stated in 1950: S&K.

The resulting structure is shown in Figure IV,75. In its SO_2 molecules, S–O = 1.43 A. and the angle O–S–O = 119°30′. Between molecules the shortest S–O = 3.10 A. and the shortest O–O = 3.32 A.

IV,i3. Crystals of *hydrazine*, N_2H_4, photographed at -15°C. to -40°C., are monoclinic, with a bimolecular cell of the dimensions:

$$a_0 = 3.56 \text{ A.};\ b_0 = 5.78 \text{ A.};\ c_0 = 4.53 \text{ A.};\ \beta = 109°30'$$

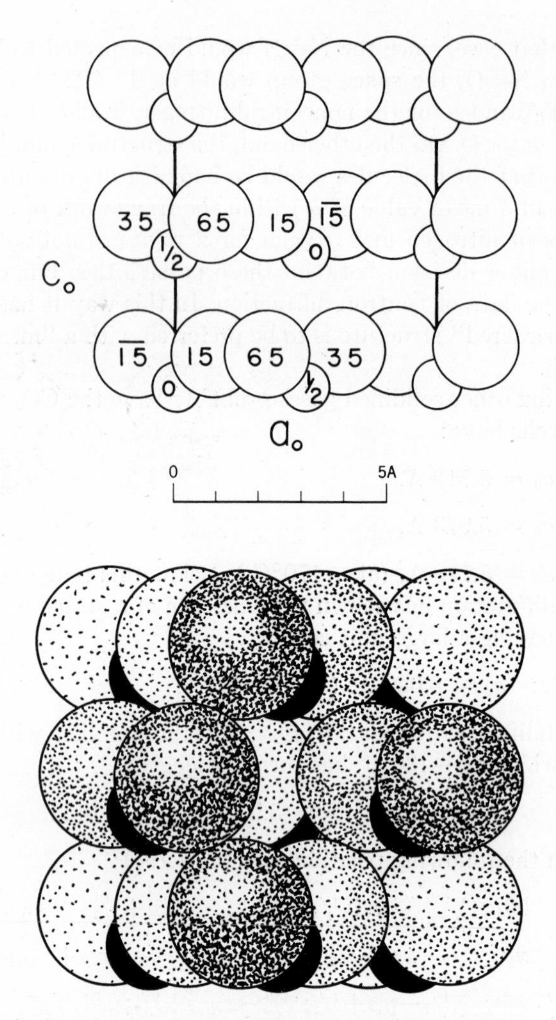

Fig. IV,75a (top). A projection along b_0 of the orthorhombic structure of solid SO_2. The larger circles are oxygen. Origin in lower left.

Fig. IV,75b (bottom). A packing drawing of the orthorhombic structure of solid SO_2 viewed along the b_0 axis. The dotted circles are oxygen.

Nitrogen atoms are in two sets of special positions of C_{2h}^2 ($P2_1/m$):

$$(2e) \quad u0v; \; \bar{u}\,{}^1/_2\,\bar{v}$$

with $u = 0.037$, $v = 0.362$ for N(1) and $u' = 0.736$, $v' = 0.050$ for N(2).

In this arrangement (Fig. IV,76), the N(1)–N(2) within a molecule is 1.46 A. The shortest N–N separations between molecules are 3.19, 3.25, and 3.30 A. Hydrogen positions could not be established from the available x-ray data.

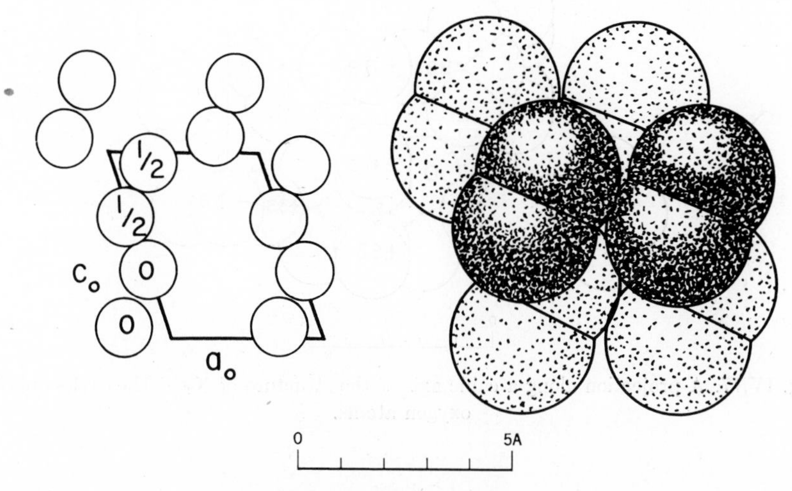

Fig. IV,76a (left). A projection along b_0 of the monoclinic structure of hydrazine, N_2H_4. Origin in lower left.

Fig. IV,76b (right). A packing drawing of the monoclinic N_2H_4 structure viewed along the b_0 axis. The nitrogen atoms only are shown.

IV,i4. Photographs of single crystals of solidified *nitrogen tetroxide*, N_2O_4, at $-40°$C. have led to a structure which differs from that previously proposed on the basis of powder data alone. Six molecules are contained in a unit cube having

$$a_0 = 7.77 \text{ A.}$$

Atoms are placed in the following special positions of T_h^5 ($Im3$):

N: (12d) $\pm(u00;\ 0u0;\ 00u)$; B.C. with $u = 0.394$

O: (24g) $\pm(0vw;\ w0v;\ vw0;\ 0v\bar{w};\ \bar{w}0v;\ v\bar{w}0)$; B.C. with $v = 0.326$ and $w = 0.134$

This is a structure (Fig. IV,77) built up of planar molecules having the dimensions shown in Figure IV,78. The long N–N separation of 1.64 A. is especially noteworthy.

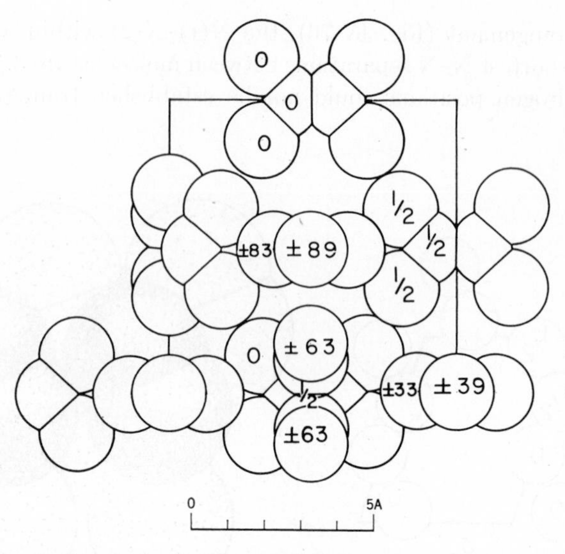

Fig. IV,77. A projection along a cubic axis of the structure of N_2O_4. The circles are the oxygen atoms.

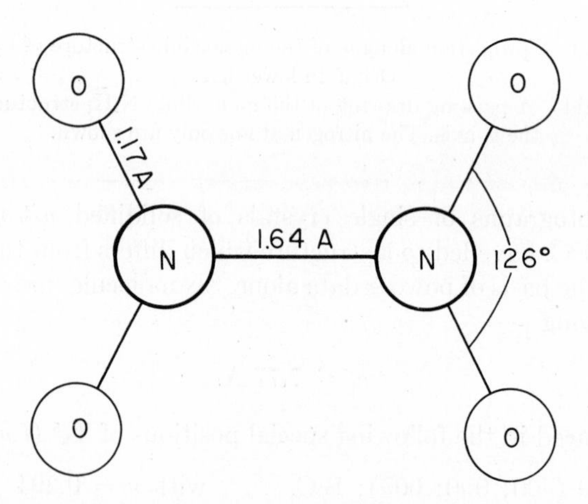

Fig. IV,78. The dimensions of the molecule of N_2O_4.

IV,i5. Earlier x-ray studies of the structure of *mercuric cyanide*, $Hg(CN)_2$, have been confirmed and extended through neutron diffraction. The symmetry is tetragonal, with an eight-molecule cell of the dimensions:

$$a_0 = 9.643 \text{ A.}, \qquad c_0 = 8.880 \text{ A.}$$

Atoms are in the following positions of V_d^{12} ($I\bar{4}2d$):

Mercury is in (8*d*) $u\,{}^1/_4\,{}^1/_8;\ \bar{u}\,{}^3/_4\,{}^1/_8;\ {}^3/_4\,u\,{}^7/_8;\ {}^1/_4\,\bar{u}\,{}^7/_8;$ B.C., with $u = 0.2125$.

The other atoms are in the general positions:

(16*e*) $xyz;\ \bar{x}\bar{y}z;\ \bar{x},y+{}^1/_2,{}^1/_4-z;\ x,{}^1/_2-y,{}^1/_4-z;$

$\bar{y}x\bar{z};\ y\bar{x}\bar{z};\ y,x+{}^1/_2,z+{}^1/_4;\ \bar{y},{}^1/_2-x,z+{}^1/_4;$ B.C.

with $x = 0.197$, $y = 0.047$, $z = 0.159$ for carbon and $x' = 0.212$, $y' = -0.073$, $z' = 0.183$ for nitrogen.

This leads to an arrangement (Fig. IV,79) consisting of $Hg(CN)_2$ molecules in which carbon rather than nitrogen atoms are in contact with the mercury atoms. In the molecule, Hg–C = 1.986 A. and C–N = 1.186 A. The molecule is not strictly linear; instead, the angle N–C–Hg = 173° and C–Hg–C = 171°. Between molecules the shortest distance is Hg–N = 2.70 A.

This arrangement bears no obvious relation to the anti-cuprite structures reported for the corresponding zinc and cadmium cyanides (**IV,f1**).

IV,i6. Crystals of *boron tetrafluoride*, B_2F_4, are monoclinic, with a bimolecular cell of the dimensions:

$$a_0 = 5.49 \text{ A.};\ b_0 = 6.53 \text{ A.};\ c_0 = 4.83 \text{ A.};\ \beta = 102°30'\ (-120°\text{C.})$$

The space group is C_{2h}^5 ($P2_1/c$) with all atoms in the general positions:

(4*e*) $\pm(xyz;\ x,{}^1/_2-y,z+{}^1/_2)$

The parameters are those listed below:

Atom	x	y	z
B	0.631	0.063	0.493
F(1)	0.7290	0.2024	0.6781
F(2)	0.7494	0.0344	0.2824

The structure (Fig. IV,80) is a distorted body-centered packing of the B_2F_4 molecules. Within each molecule, B–B = 1.67 A., B–F = 1.32 A.,

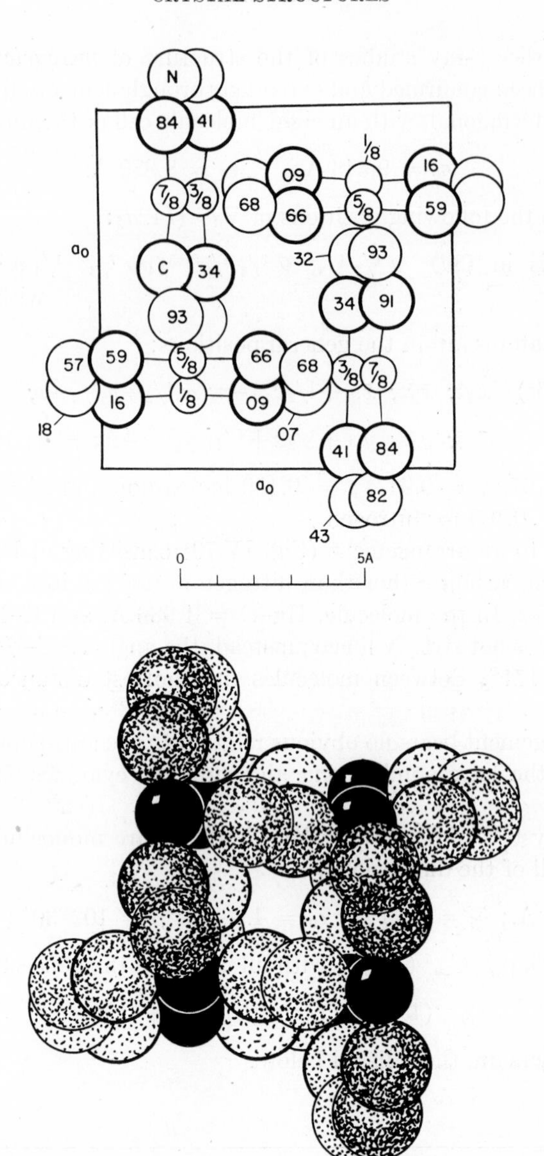

Fig. IV,79a (top). The tetragonal structure of $Hg(CN)_2$ projected along its c_0 axis. The smallest circles are mercury.

Fig. IV,79b (bottom). A packing drawing of the $Hg(CN)_2$ structure viewed along its c_0 axis. The mercury atoms are black. The carbon atoms are represented by the more heavily ringed dotted circles.

and the angle F–B–F = 120°. Between molecules, the shortest F–F distances are 3.27 and 3.32 A.

IV,i7. Crystals of *diboron tetrachloride*, B_2Cl_4, are orthorhombic, unlike the fluoride, with a unit containing four molecules:

$$a_0 = 11.900 \text{ A.};\ b_0 = 6.281 \text{ A.};\ c_0 = 7.690 \text{ A.}\ (-165°\text{C.})$$

The space group is D_{2h}^{15} ($Pbca$). All atoms are in the general positions:

$$(8c)\quad \pm(xyz;\ x+\tfrac{1}{2},\tfrac{1}{2}-y,\bar{z};\ \bar{x},y+\tfrac{1}{2},\tfrac{1}{2}-z;\ \tfrac{1}{2}-x,\bar{y},z+\tfrac{1}{2})$$

with the parameters:

Atom	x	y	z
B	−0.0140	−0.1168	0.0584
Cl(1)	0.0897	−0.2356	0.1815
Cl(2)	−0.1474	−0.2252	0.0502

The resulting structure is clearly built up of molecules of the composition B_2Cl_4 (Fig. IV,81). Within the molecule, which is planar, B–B = 1.752 A., B–Cl(1) = 1.725 A., B–Cl(2) = 1.730 A., and all bond angles are within a degree of 120°. Between molecules, the shortest interatomic distances are Cl–Cl = 3.61 A. and B–Cl = 3.25 and 3.28 A.

IV,i8. Crystals of *phosphorus diiodide*, P_2I_4, are triclinic with the unimolecular cell:

$$a_0 = 4.56 \text{ A.};\ b_0 = 7.06 \text{ A.};\ c_0 = 7.40 \text{ A.}$$

$$\alpha = 80°\ 12';\ \beta = 106°\ 58';\ \gamma = 98°\ 12'$$

All atoms are in the general positions

$$(2i)\quad \pm(xyz)$$

of C_i^1 ($P\bar{1}$), with the parameters:

Atom	x	y	z
P	0.397	0.639	0.463
I(1)	0.557	0.730	0.165
I(2)	0.820	0.803	0.695

The resulting structure is a parallel array of molecules having the dimensions of Figure IV,82.

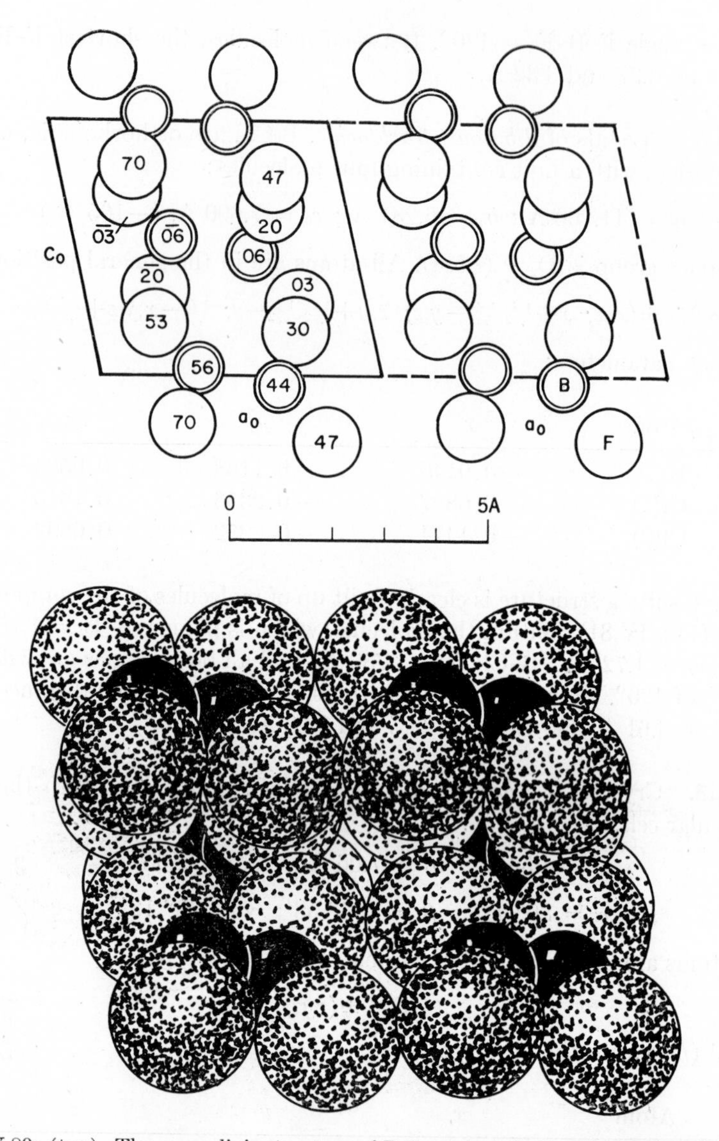

Fig. IV,80a (top). The monoclinic structure of B_2F_4 projected along its b_0 axis. The contents of a second cell are shown to the right of the first. Origin in lower left.

Fig. IV,80b (bottom). A packing drawing of the B_2F_4 structure seen along its b_0 axis. The boron atoms are the black circles.

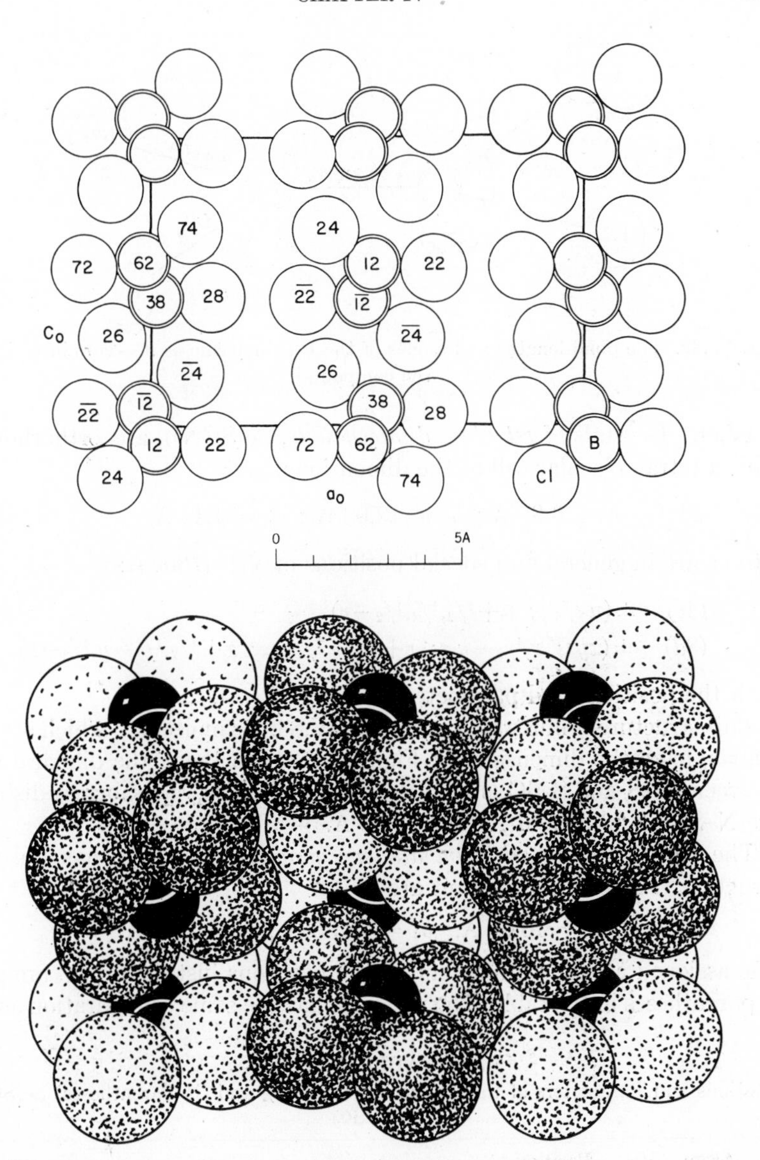

Fig. IV,81a (top). The orthorhombic B_2Cl_4 structure projected along its b_0 axis. Origin in lower left.

Fig. IV,81b (bottom). A packing drawing showing the distribution of the B_2Cl_4 molecules in its crystals, as viewed along the b_0 axis. The boron atoms are the black circles.

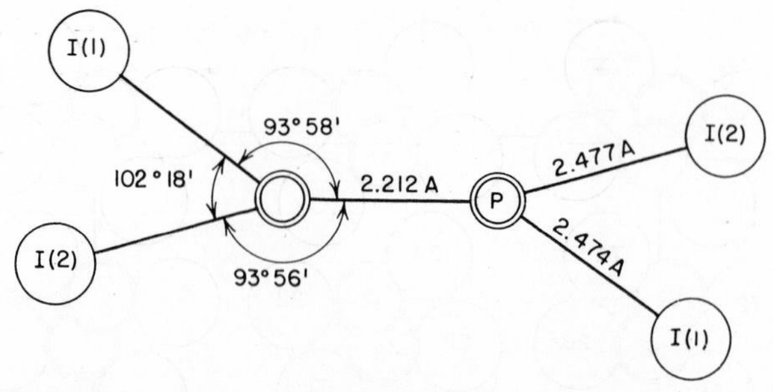

Fig. IV,82. The bond lengths and angles of the P_2I_4 molecule as it occurs in crystals of the compound.

IV,i9. Crystals of *selenium dithiocyanate*, $Se(SCN)_2$, are orthorhombic, with a tetramolecular cell of the dimensions:

$$a_0 = 9.87 \text{ A.};\ b_0 = 13.03 \text{ A.};\ c_0 = 4.44 \text{ A.}$$

Atoms are in general and special positions of V_h^{16} (*Pnma*):

(4c) $\pm(u\ {}^1/_4\ v;\ u+{}^1/_2,{}^1/_4,{}^1/_2-v)$

(8d) $\pm(xyz;\ x,{}^1/_2-y,z;\ x+{}^1/_2,y,{}^1/_2-z;\ x+{}^1/_2,{}^1/_2-y,{}^1/_2-z)$

with the positions and parameters of Table IV,24.

This structure is illustrated in Figure IV,83. Its molecules, which are non-planar, have the dimensions of Figure IV,84. In them, the dihedral angle between SSeS and SeSC is 79°. Between molecules, the shortest distances are N–S = 3.03 A. and N–Se = 2.98 A.

The corresponding *selenium diselenocyanate*, $Se(SeCN)_2$, has the same structure. The edges of its unit are

$$a_0 = 10.07 \text{ A.};\ b_0 = 13.35 \text{ A.};\ c_0 = 4.48 \text{ A.}$$

The assigned atomic parameters, in the positions just described, are given in parentheses in the table. The molecule has the same configuration as that

TABLE IV,24
Positions and Parameters of the Atoms in $Se(SeCN)_2$ and (in Parentheses) $Se(SCN)_2$[a] (**IV,i9**)

Atom	Position	x	y	z
Se(1)	(4c)	0.540 (0.547)	1/4	0.492 (0.500)
Se(2) or S	(8d)	0.442 (0.442)	0.115 (0.119)	0.249 (0.285)
C	(8d)	0.295 (0.294)	0.112 (0.111)	0.488 (0.474)
N	(8d)	0.203 (0.205)	0.095 (0.105)	0.586 (0.600)

[a] Parameters for x given here for $Se(SCN)_2$ are $^1/_2 - x$ those stated in the original.

of the thiocyanate, with the same central bond angle (Se–Se–Se) of 101°. The two C–Se–Se angles equal 95°, and the dihedral angle between Se–Se–Se and Se–Se–C is 94°. The length of the Se–Se bond is 2.33 A.

The *sulfur thiocyanate*, $S(CNS)_2$, has the same structure, with

$$a_0 = 10.12 \text{ A.};\ b_0 = 12.83 \text{ A.};\ c_0 = 4.34 \text{ A.}$$

Atomic positions have not yet been established.

IV,i10. A partial atomic arrangement has been described for *mercuric chlorothiocyanate*, HgCl(CNS). Its orthorhombic unit contains four molecules and has the cell edges:

$$a_0 = 10.16 \text{ A.};\ b_0 = 4.23 \text{ A.};\ c_0 = 10.40 \text{ A.}$$

All atoms are in general positions of V^4 ($P2_12_12_1$):

$$(4a)\quad xyz;\ ^1/_2-x,\bar{y},z+{}^1/_2;\ x+{}^1/_2,{}^1/_2-y,\bar{z};\ \bar{x},y+{}^1/_2,{}^1/_2-z$$

with the parameters:

Atom	x	y	z
Hg	0.155	0.25	0.359
Cl	0.134	0.25	0.579
S	0.199	0.25	0.141
N	0.473	0.25	0.141
C	Not determined		

A partial structure has also been stated for the analogous bromide, HgBr(SCN). It is monoclinic rather than orthorhombic, and its bimolecular unit has the dimensions:

$$a_0 = 6.24 \text{ A.};\ b_0 = 4.28 \text{ A.};\ c_0 = 8.74 \text{ A.};\ \beta = 91°20'$$

Atoms have been put in special positions

$$(2e)\quad \pm(u\ {}^1/_4\ v)$$

of C_{2h}^2 ($P2_1/m$), with the parameters:

Atom	u	v
Hg	−0.049	0.202
Br	0.263	0.027
S	−0.351	0.364

No parameters were established for carbon and nitrogen.

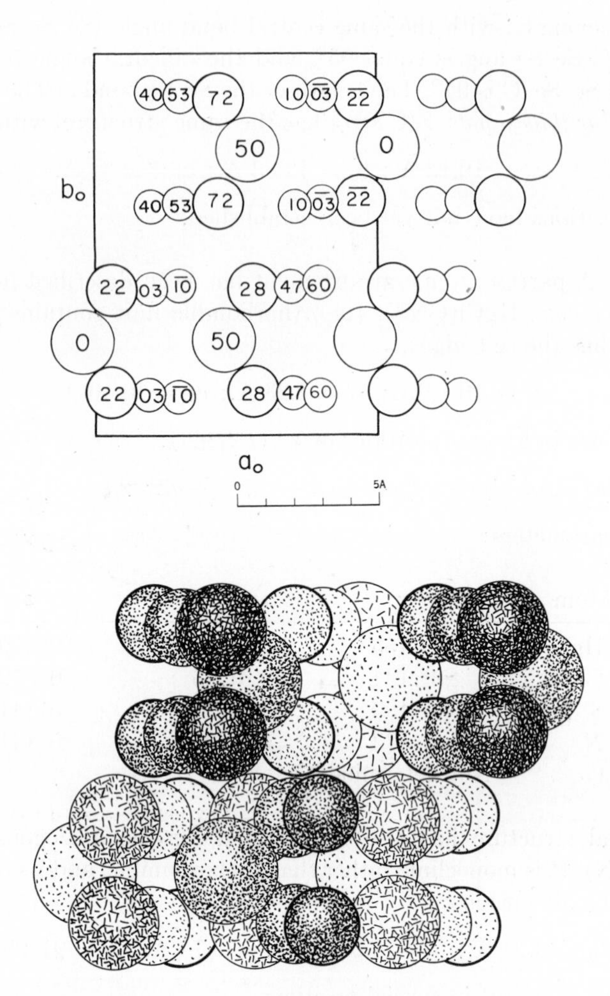

Fig. IV,83a (top). A projection along the c_0 axis of the orthorhombic structure of $Se(SCN)_2$. The largest circles are selenium; of the smallest, the nitrogen atoms are ringed. Origin in lower right.

Fig. IV,83b (bottom). A packing drawing of the orthorhombic $Se(SCN)_2$ structure viewed along the c_0 axis. The sulfur atoms are line-shaded. Of the dotted atoms the largest are selenium.

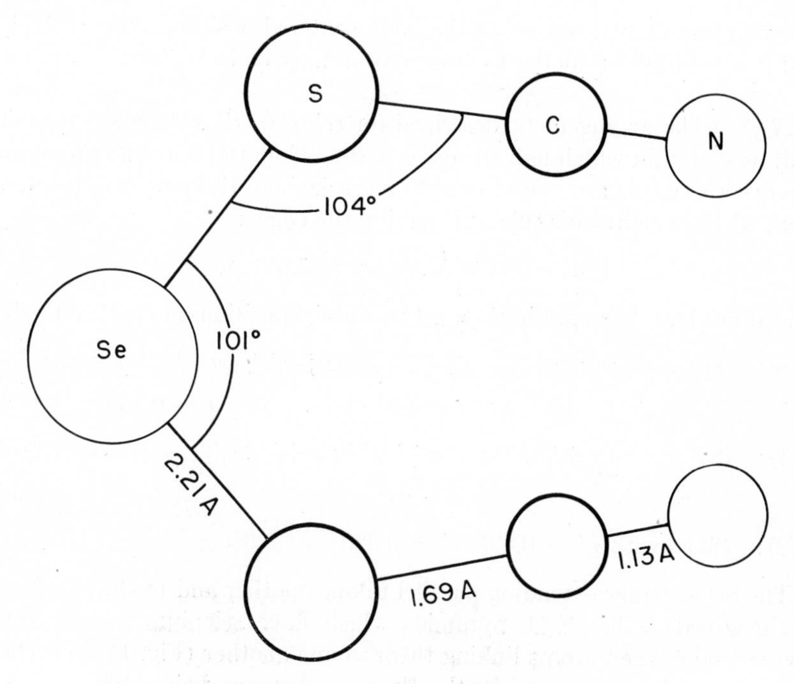

Fig. IV,84. Dimensions of the nonplanar $Se(SCN)_2$ molecule.

Miscellaneous Compounds

IV,j1. A partially disordered structure has been proposed for gamma-Mo_2N based on electron diffraction data. Its cubic pattern corresponds to a bimolecular cell, having

$$a_0 = 4.165 \text{ A.}$$

The four molybdenum atoms are considered to be in a face-centered array: 000; F.C., with some of the face-centering positions empty to correspond to a certain departure from the stoichiometric composition. One nitrogen atom is said to be in $^1/_2\ ^1/_2\ ^1/_2$, the other statistically distributed over the positions $0\ 0\ ^1/_2$; $0\ ^1/_2\ 0$; $^1/_2\ 0\ 0$.

IV,j2. The hydride of tantalum, Ta_2H, has been given a tetragonal unit which is only a slight distortion of the body-centered cubic structure of metallic tantalum (**II,c**). Its unimolecular cell has the edges:

$$a_0 = 3.38 \text{ A.}, \qquad c_0 = 3.41 \text{ A.}$$

The two tantalum atoms are in the body-centered positions 000; $^1/_2\,^1/_2\,^1/_2$, and it is suppysed that the hydrogen atom may be in $^1/_2\,^1/_2\,0$.

IV,j3. The arrangement described for *selenium dioxide*, SeO_2, contains endless strings in which half the oxygen atoms join the selenium atoms and the other oxygens are bound only to one selenium. Its symmetry is tetragonal, with an eight-molecule unit having the edges:

$$a_0 = 8.353 \text{ A.}, \qquad c_0 = 5.051 \text{ A.}$$

The atoms have been put in three sets of special positions of D_{4h}^{13} ($P4/mbc$):

Se: (8*h*) $\pm(uv0;\ v\bar{u}\,^1/_2;\ u+^1/_2,^1/_2-v,0;\ v+^1/_2,u+^1/_2,^1/_2)$ with $u = 0.133,\ v = 0.207$

O(1): (8*g*) $\pm(w,w+^1/_2,\,^1/_4;\ w,w+^1/_2,^3/_4;\ w+^1/_2,\bar{w},^1/_4;\ w+^1/_2,\bar{w},^3/_4)$ with $w = 0.358$

O(2): (8*h*) with $u' = 0.425,\ v' = 0.320$

The SeO_2 strings extending parallel to one another and to the c_0 axis can be imagined as flat SeO_3 pyramids which have selenium atoms at the apexes and oxygen atoms linking them to one another (Fig. IV,85). These pyramids are nearly regular, the Se–O(2) distance being 1.75 A. while Se–O(1), which is the shared oxygen, is 1.79 A. Between neighboring chains the O–O separation is 2.70 A., which is the sum of the ionic radii.

IV,j4. Crystals of *gallium dichloride*, $GaCl_2$, are orthorhombic, with an eight-molecular cell having the edges:

$$a_0 = 7.24 \text{ A.};\ b_0 = 9.72 \text{ A.};\ c_0 = 9.50 \text{ A.}$$

The space group is V_h^6 ($Pnna$), with atoms in the positions:

(4*c*) $\pm(^1/_4\,0\,u;\ ^3/_4,^1/_2,u+^1/_2)$

(4*d*) $\pm(u\,^1/_4\,^1/_4;\ u+^1/_2,^1/_4,^3/_4)$

(8*e*) $\pm(xyz;\ ^1/_2-x,\bar{y},z;\ x,^1/_2-y,^1/_2-z;\ ^1/_2-x,y+^1/_2,^1/_2-z)$

The chosen parameters are listed in Table IV,25.

The resulting structure, as shown in Figure IV,86, is unlike that of any other RX_2 compound. It demonstrates $GaCl_2$ as being $Ga^+(Ga^{3+}Cl_4)$. Each Ga^{3+} is at the center of a tetrahedron of chlorine atoms distant 2.19 A. Each Ga^+ has eight chlorine neighbors, four at 3.18 A. and four more at 3.27 A.

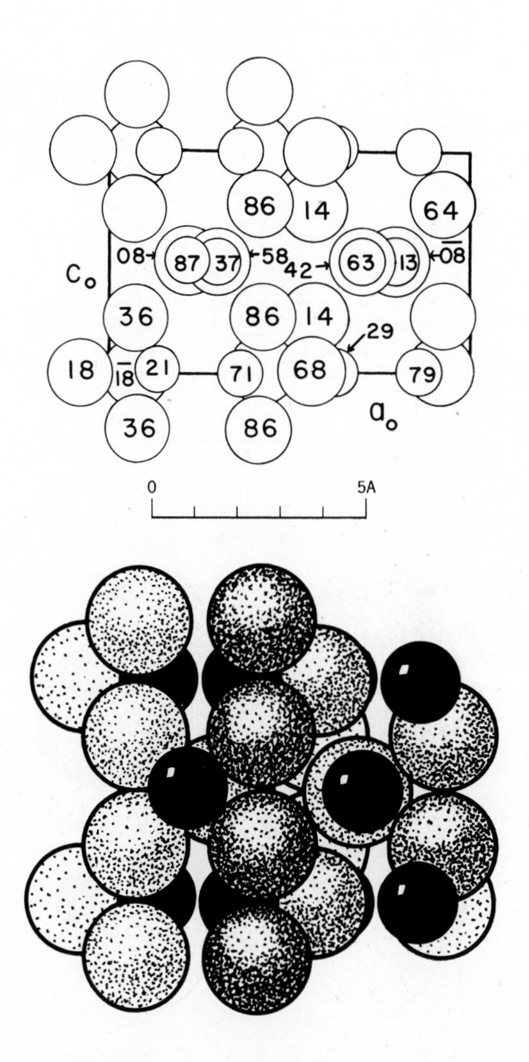

Fig. IV,85a (top). A projection along an a_0 axis of the tetragonal structure of SeO_2 The larger circles are oxygen. Origin in lower left.

Fig. IV,85b (bottom). A packing drawing showing some of the atoms in the tetragonal SeO_2 arrangement viewed along an a_0 axis. The oxygen atoms are dotted.

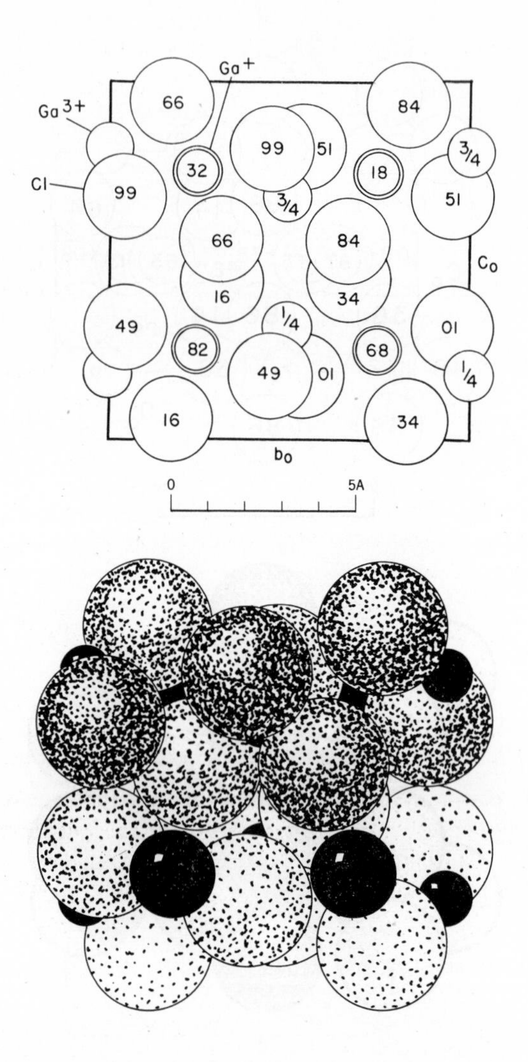

Fig. IV,86a (top). The orthorhombic structure of $GaCl_2$ projected along its a_0 axis. Origin in lower right.

Fig. IV,86b (bottom). A packing drawing of the $GaCl_2$ arrangement seen along its a_0 axis. The Ga^+ ions are the large, the Ga^{3+} the small black circles.

TABLE IV,25
Positions and Parameters of the Atoms in $GaCl_2$ (**IV,j4**)

Atom	Position	x	y	z
Ga^+	(4*d*)	0.681	$^1/_4$	$^1/_4$
Ga^{3+}	(4*c*)	$^1/_4$	0	0.183
Cl(1)	(8*e*)	0.339	0.174	0.054
Cl(2)	(8*e*)	0.010	0.048	0.315

IV,j5. Crystals of *bismuth thiochloride*, BiSCl, are orthorhombic, with a tetramolecular unit of the edge lengths:

$$a_0 = 7.70 \text{ A.}; \; b_0 = 4.00 \text{ A.}; \; c_0 = 9.87 \text{ A.}$$

Atoms are in the following special positions of V_h^{16} (*Pnma*):

$$(4c) \quad \pm(u\,{}^1/_4\,v;\; u+{}^1/_2,{}^1/_4,{}^1/_2-v)$$

with the parameters of Table IV,26. In this arrangement (Fig. IV,87) with its rather complex environment, the shortest Bi–S = 2.77 and 3.01 A., and the shortest Bi–Cl = 2.71 A.

This structure is possessed by a number of chemically related compounds. Those for which cell dimensions and atomic parameters have been determined are listed in Tables IV,26 and IV,27.

Recently, a particularly exact determination has been made of the parameters of one of these compounds, *antimony thiobromide*, SbSBr. These lead to Sb–S distances of 2.49 and 2.67 A. and to a nearest Sb–Br = 2.94 A. The shortest S–Br = S–S = 3.46 A. As the figure suggests, this arrangement can be considered as a group of Sb–S chains running roughly in the c_0 direction; each link is an approximate rectangle with the angles 96° and 84°, and the links are nearly at right angles to one another (96°). From this point of view, the bromine atoms, perhaps as ions, bind the chains together within the crystal.

TABLE IV,26
Parameters of the Atoms in BiSCl and Related Substances (**IV,j5**)

	Bi or Sb		S or Se		Cl, Br, or I	
Substance	u	v	u	v	u	v
BiSCl	0.140	0.138	0.77	0.04	0.50	0.79
BiSBr	0.138	0.135	0.80	0.04	0.48	0.82
BiSI	0.130	0.127	0.825	0.04	0.48	0.82
BiSeBr	0.135	0.130	0.80	0.04	0.47	0.82
SbSBr	0.1205	0.1326	0.8376	0.0456	0.5133	0.8227
SbSI	0.118	0.130	0.835	0.060	0.515	0.825

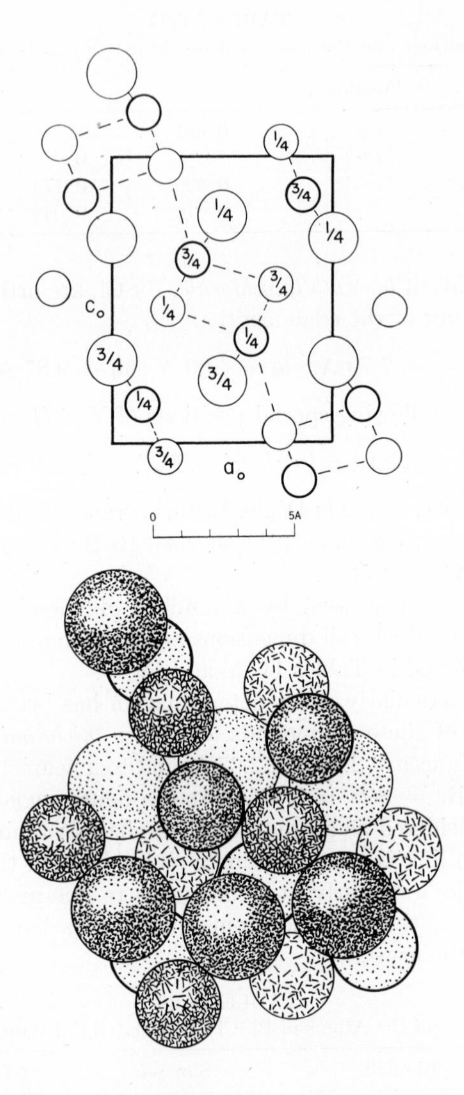

Fig. IV,87a (top). A projection along b_0 of the orthorhombic structure of BiSCl. The largest circles are chlorine, the smallest bismuth. Origin in lower left.

Fig. IV,87b (bottom). A packing drawing of the orthorhombic BiSCl structure viewed along the b_0 axis. The sulfur atoms are line-shaded; the bismuth atoms are the smaller, heavily ringed and dotted circles.

TABLE IV,27
Cell Dimensions of Substances with the BiSCl Structure (**IV,j5**)

Substance	a_0, A.	b_0, A.	c_0, A.
BiSBr	8.02	4.01	9.70
BiSI	8.46	4.14	10.15
BiSeBr	8.18	4.11	10.47
BiSeI	8.71	4.19	10.54
SbSBr	8.26	3.97	9.79
SbSI	8.49	4.16	10.10
SbSeBr	8.30	3.95	10.20
SbSeI	8.65	4.12	10.38

IV,j6. The unit cell assigned the monoclinic *antimony oxychloride*, SbOCl, contains the extraordinarily large number of 12 molecules. Its cell dimensions are

$$a_0 = 9.54 \text{ A.}; \; b_0 = 10.77 \text{ A.}; \; c_0 = 7.94 \text{ A.}; \; \beta = 103°36'$$

All atoms have been placed in general positions of C_{2h}^5 ($P2_1/a$):

$$(4e) \quad \pm(xyz;\; x+{}^1/_2,{}^1/_2-y,z)$$

with the parameters of Table IV,28.

In this arrangement (Fig. IV,88), the environments of the three kinds of antimony atoms are very different. Atom Sb(1) has four oxygen neighbors with Sb–O = 1.99–2.17 A.; the nearest chlorine atoms are 3.20 A. away. Atom Sb(2) has two oxygens at distances of 1.99 and 2.11 A. and a chlorine at 2.47 A. For Sb(3) there are two oxygens at 2.11 and 2.17 A. and a chlorine at 2.29 A. The closest O–Cl is 2.82 A. and the shortest Cl–Cl is 3.25 A.

TABLE IV,28
Parameters of the Atoms in SbOCl (**IV,j6**)

Atom	x	y	z
Sb(1)	0.4278	0.0238	0.2808
Sb(2)	0.0902	0.2418	0.0818
Sb(3)	0.2310	0.1730	0.5208
Cl(1)	0.401	0.404	0.125
Cl(2)	0.360	0.354	0.584
Cl(3)	0.343	0.118	0.887
O(1)	0.450	0.109	0.535
O(2)	0.250	0.144	0.264
O(3)	0.086	0.355	0.279

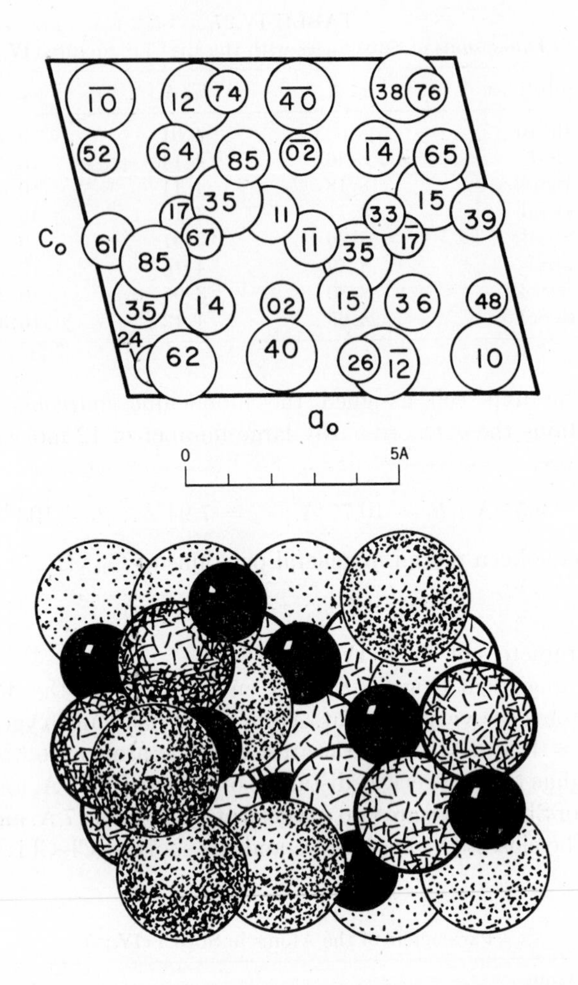

Fig. IV,88a (top). A projection along b_0 of the monoclinic structure of SbOCl. The largest circles are chlorine, the smallest antimony. Origin in lower left.

Fig. IV,88b (bottom). A packing drawing of the monoclinic SbOCl structure viewed along the b_0 axis. The chlorine atoms are dotted and the oxygen atoms line-shaded.

IV,j7. Crystals of *mercuric amidochloride*, $HgNH_2Cl$, are orthorhombic, with a bimolecular cell of the dimensions:

$$a_0 = 5.167 \text{ A.}; \; b_0 = 4.357 \text{ A.}; \; c_0 = 6.690 \text{ A.}$$

Atoms have been given the following positions derived from the space group C_{2v}^4 (Pcm):

Hg: (2a) $0u0$; $0\,u\,{}^1/_2$ with $u = 0$

Cl: (2c) $v\,w\,{}^1/_4$; $\bar{v}\,w\,{}^3/_4$ with $v = 0.32$, $w = {}^1/_2$

N: (2c) with $v' = 0.23$, $w' = 0$

This structure can equally well be deduced from V^2 ($P222_1$). The chlorine and nitrogen parameters stated above are not provided by the x-ray data; they have been based on the assumption that N–Cl should be 3.2 A. and that Cl–Cl should be 3.8 A. The structure that results (Fig. IV, 89) contains endless Hg–N–Hg chains.

The corresponding *bromide*, $HgNH_2Br$, has the same arrangement, with a cell of the dimensions:

$$a_0 = 5.439 \text{ A.}, \; b_0 = 4.487 \text{ A.}, \; c_0 = 6.761 \text{ A.}$$

The bromine parameters were given as $v = 0.32$, $w = {}^1/_2$ and the nitrogen parameters as $v' = 0.22$, $w' = 0$.

IV,j8. A determination of structure has been carried through on one of the two orthorhombic forms of *uranyl hydroxide*, beta-$UO_2(OH)_2$. There are four molecules in a cell of the dimensions:

$$a_0 = 6.295 \text{ A.}, \; b_0 = 5.636 \text{ A.}, \; c_0 = 9.929 \text{ A.}$$

The space group is V_h^{23} ($Fmmm$), with atoms in the following special positions:

U: (4a) 000; F.C.

O(1): (8i) $00u$; $00\bar{u}$; F.C. with $u = 0.20$

O(2): (8e) ${}^1/_4\,{}^1/_4\,0$; ${}^1/_4\,{}^1/_4\,{}^1/_2$; F.C.

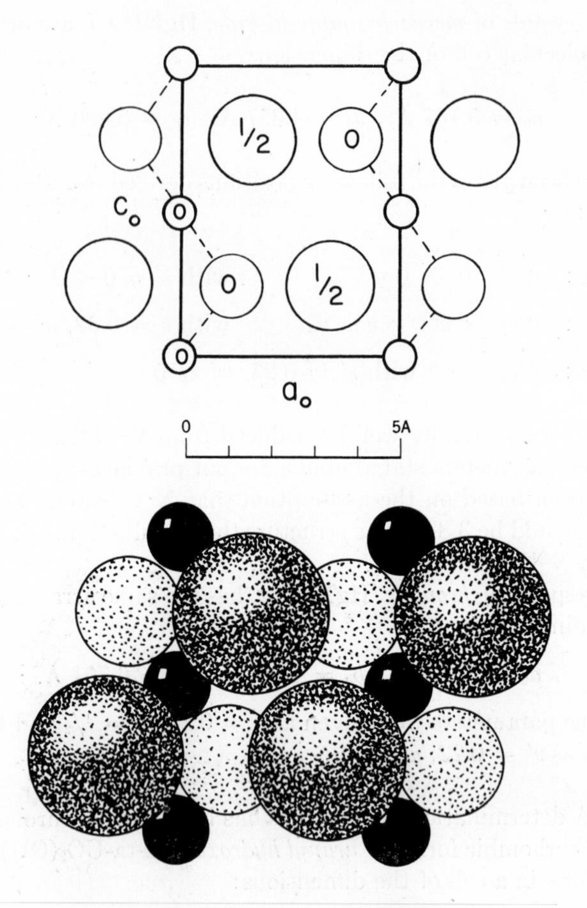

Fig. IV,89a (top). A projection along b_0 of the orthorhombic structure of $HgNH_2Cl$. Mercury are the small and chlorine the larger circles. Origin in lower left.

Fig. IV,89b (bottom). A packing drawing of the orthorhombic structure of $HgNH_2Cl$ viewed along the b_0 axis. The small black atoms are mercury, the larger dotted atoms are chlorine.

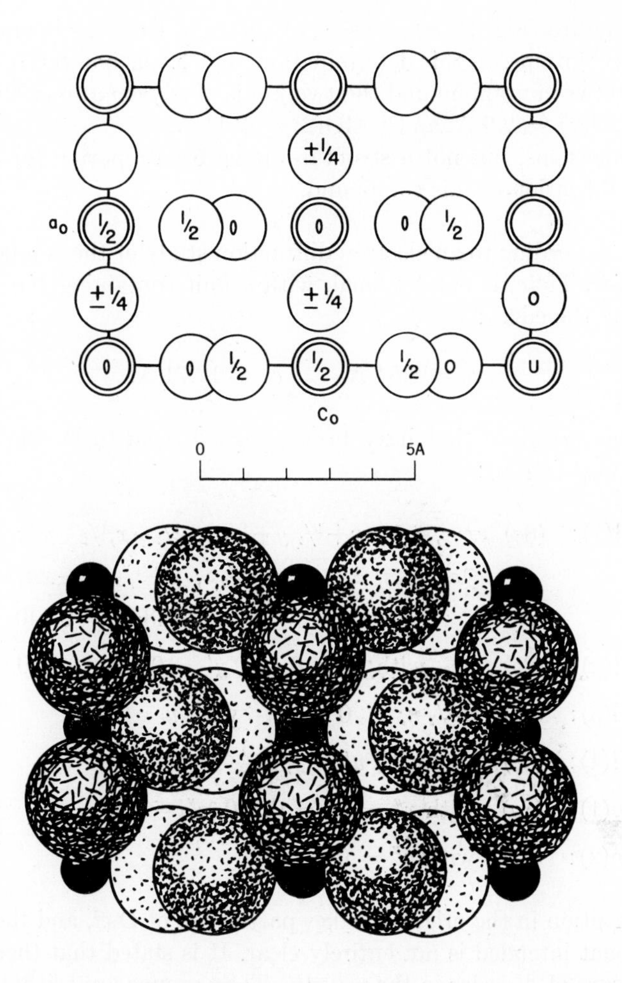

Fig. IV,90a (top). The orthorhombic structure of $UO_2(OH)_2$ projected along its b_0 axis. Origin in the lower left.

Fig. IV,90b (bottom). A packing drawing of the $UO_2(OH)_2$ arrangement viewed along its b_0 axis. The uranium atoms are the small black circles. The U–O distances are all so nearly equal that no clear distinction can be made between the oxygen atoms of the UO_2 ions and the hydroxyl groups.

The structure (Fig. IV,90) consists of sheets of UO_2 ions bound together by hydroxyl groups. In this determination, which was not very precise, the distance between uranium and the two kinds of oxygen atoms is nearly the same: U–2O(1) = 2.0 A. and U–4O(2) = 2.11 A.

Cell dimensions, but not a structure, have been reported for the other, alpha, modification of this compound.

IV,j9. According to an electron diffraction study of the compound *thallous selenide*, Tl_2Se, it is tetragonal, with a unit containing ten molecules and having the edges:

$$a_0 = 8.54 \text{ A.}, \qquad c_0 = 12.71 \text{ A.}$$

The atomic positions that have been assigned seem to be the following, based on C_{4h}^3 ($P4/n$):

Tl(1): (8g) $xyz;\ \bar{x}\bar{y}z;\ x+{}^1/_2,y+{}^1/_2,\bar{z};\ {}^1/_2-x,{}^1/_2-y,\bar{z};$
$\bar{y}x\bar{z};\ y\bar{x}\bar{z};\ {}^1/_2-y,x+{}^1/_2,z;\ y+{}^1/_2,{}^1/_2-x,z$, with
$x = 0.148,\ y = 0.140,\ z = 0.405$

Tl(2): (8g) with $x' = 0.140,\ y' = 0.148,\ z' = 0.081$

Tl(3): (2c) $0\ {}^1/_2\ u;\ {}^1/_2\ 0\ \bar{u}$ with $u = {}^1/_4$

Tl(4): (2c) with $u' = {}^3/_4$

Se(1): (8g) with $x'' = y'' = 0.340,\ z'' = {}^1/_4$

Se(2): (2c) with $u'' = 0$

The description in the original makes partial use of D_{4h}^3, and therefore the arrangement intended is not entirely clear. It is stated that there is a certain measure of disorder in the way the Tl_2Se groups are oriented within a crystal.

IV,j10. A structure has been proposed for *platinum dioxide*, PtO_2, based on a unimolecular hexagonal cell of the dimensions:

$$a_0 = 3.08 \text{ A.}, \qquad c_0 = 4.19 \text{ A.}$$

The platinum atom was placed at the origin, 000, and the two oxygen atoms were described as being in $\pm(00u)$ with $u = 0.35$. This results in the very short O–O = 1.26 A. Obviously, additional work is required.

BIBLIOGRAPHY TABLE, CHAPTER IV

Compound	Paragraph	Literature
AcH_2	a1	1961: F,G,B&M
AcOBr	d3	1949: Z
AcOCl	d3	1949: Z
AcOF	a1	1949: Z; 1951: Z
AgB_2	h4	1961: O
AgCuS	f3	1955: F
AgCuSe	f4	1950: E; 1957: F,C&K
Ag_2F	c1	1928: O&S; T&D
AgMgAs	a1	1941: N&S
Ag_2O	f1	1922: D; N; W; 1924: L&Q; 1926: G; 1960: NBS
Ag_2S	f2	1931: G; P&S; 1939: R; 1955: B,H&T; 1958: F; 1959: NBS
Ag_2Se	f2	1936: R; 1955: B,H&T
Ag_2Te	f2, f7	1932: T; 1936: R; 1939: K; 1951: R&B; 1955: B,H&T; 1959: F
AgZnAs	a1	1951: N&G
AlB_2	h4	1935: H&J; 1936: H&J; 1956: F
AlOCl	d4	1959: R
α-AlO(OH), diaspore	d1	1930: dJ; 1932: D; 1933: T; 1941: H; 1942: H; 1945: G; 1952: E; NBS; 1958: B&L
γ-AlO(OH), boehmite	d2	1936: G; 1946: R&Y; 1952: NBS; 1956: M&MA
$(Al,Na)Si_2$	g4	1947: N&S
AmO_2	a1	1949: Z; 1953: T&D; 1954: S,F,E&Z; 1955: A,E,F&Z
AmOCl	d3	1953: T&D
$AuAl_2$	a1	1931: E&K; 1934: W&P
AuB_2	h4	1961: O
$AuGa_2$	a1	1937: Z,H&H
$AuIn_2$	a1	1937: Z,H&H
$AuPb_2$	h3	1943: W
$AuSb_2$	a1, g1	1928: O; 1931: N,A&W; 1932: B&J; 1933: J&B; 1952: G&K; 1954: S
$AuTe_2$	f5, f6	1935: T&K; 1936: T&K; 1950: T&M
B_2Cl_4	i7	1955: A,L&W; 1957: A,W&L
B_2F_4	i6	1958: T&L

(continued)

BIBLIOGRAPHY TABLE, CHAPTER IV (*continued*)

Compound	Paragraph	Literature
$BaBr_2$	d5	1939: D&K
BaC_2	g4	1930: vS
$BaCl_2$	a1, d5	1939: D&K; 1948: V
BaF_2	a1	1922: D; 1926: G; O; 1927: T; 1929: B,O&P; 1933: S; 1949: V; 1953: NBS
BaH_2	d5	1935: Z&H
BaHBr	d3	1956: E&G
BaHCl	d3	1956: E,A&G; E&G
BaHI	d3	1956: E&K
BaI_2	d5	1939: D&K
BaO_2	g4	1935: B,D,K,R&W; 1954: A&K; 1955: NBS
Be_2B	a1	1955: M,K&K
Be_2C	a1	1931: vS; 1934: vS&Q
$BeCl_2$	e14	1952: R&L
BeF_2	e3, e6, e7	1932: B; 1944: N,L,S&Z; 1952: N,S&Y; 1956: K,N&S; N,B,K&S
$Be(OH)_2$	e11	1950: S,R&S
BiOBr	d3	1935: B&H; 1941: S; 1958: NBS
BiOCl	d3	1935: B&H; 1941: S; 1953: NBS
BiO(OH,Cl)	d3	1935: B&H
BiOF	d3	1948: A
BiOI	d3	1935: B&H; 1941: S; 1958: NBS
BiSBr	j5	1950: D
BiSCl	j5	1950: D
BiSI	j5	1950: D
BiSeBr	j5	1950: D
BiSeI	j5	1950: D
BiTeBr	c1	1951: D
BiTeI	c1	1951: D
CO_2	i1	1924: dS&K; 1925: ML&W; M,B&P; M&P; dS&K; 1926: K; M&P; 1931: V; 1934: K&K
$CaBr_2$	b2	1939: D&K
CaC_2	g1, g4	1926: D&G; 1927: H; 1930: vS; 1959: A&M; 1961: V; B
$CaCl_2$	b2	1935: vB&N

(*continued*)

BIBLIOGRAPHY TABLE, CHAPTER IV (*continued*)

Compound	Paragraph	Literature
CaF_2	a1	1914: B; G; 1915: G; 1921: G; 1922: D; G; 1926: C; G; 1930: R; 1933: S; 1939: Z&U; 1940: C; 1952: A; H&L; 1953: NBS; 1955: S&K
$CaGa_2$	h4	1943: L
CaH_2	d5	1935: Z&H
CaHBr	d3	1956: E&G
CaHCl	d3	1956: E,A&G; E&G
CaHI	d3	1956: E&K
CaI_2	c1	1933: B
CaO_2	g4	1941: K&R
$Ca(OH)_2$	c1	1924: L; 1927: H; R; 1928: N; N&P; R; 1933: T; 1935: B,C&C; 1957: B&L; P; 1961: P
Ca(OH)Cl	c3	1959: F,O&F
$CdBr_2$	c2, c4	1929: F&G; 1933: B&N; 1942: P; H,K&L; H&L; 1959, NBS
CdBrI	c6	1942: H&L; H,K&L
$Cd(CN)_2$	f1	1945: S&Z; 1961: NBS
$CdCl_2$	c2	1925: B&F; 1926: B&F; 1929: P; 1930: P&H; 1941: P&T; 1959: NBS
CdCl(OH)	c3	1934: H&G; 1937: F&G; 1959: F,O&F
CdF_2	a1	1924: K; 1951: H&B
CdI_2	c1, c3	1922: B; 1932: A; 1933: H; 1941: P; S,B&Z; 1942: H,K&L; 1943: H&H; 1956: M
CdO_2	g1	1958: W&J; 1959: H,R&M
$Cd(OH)_2$	c1, c4	1925: N; 1928: N; N&P; 1938: F
CeC_2	g4	1930: vS; 1958: S,G&D
$CeGa_2$	h4	1943: L
CeH_2	a1	1955: H,M,E,K&Z; 1958: A
CeO_2	a1	1923: G&T; 1924: D; 1939: Z&C; 1950: B; 1953: NBS; 1955: B; G; 1958: M&E
CeOCl	d3	1953: T&D
CeOF	a1	1950: F&S
CmO_2	a1	1955: A,E,F&Z
Co_2As	h1	1957: H&C; 1959: R&J
CoAsS	g2	1921: M; 1925: R
Co_2B	h3	1933: B

(*continued*)

BIBLIOGRAPHY TABLE, CHAPTER IV (*continued*)

Compound	Paragraph	Literature
$CoBr_2$	c1	1929: F&G
$CoCl_2$	c2	1927: F; 1929: F,C&G; P; 1934: G&S
CoF_2	b1	1926: F; G; 1953: E; 1954: S&R; 1957: B; 1958: B
CoI_2	c1	1929: F&G
CoMnSb	a1	1952: N&G
$Co(OH)_2$	c1	1926: N&R; 1928: N; N&P; N&S; 1936: L&F; 1938: F
Co(OH)Br	c10	1961: L,L&I
Co(OH)Cl	c10	1959: F,O&F
Co(OH)O	VI	1954: K&F
Co_2P	d5	1947: N; 1959: R&J; 1960: R
CoS_2	g1	1927: dJ&W; 1940: L&E; 1945: K; K,H&K; 1960: E; 1961: D,A&B
$CoSe_2$	g1, g6	1938: T; 1940: L&E; 1955: B,G,H&P; R
$CoSi_2$	a1	1950: S&P
Co_2Si	d5	1955: G&W; 1959: R&J
$CoTe_2$	c1, g6	1938: T
Cr_2As	h2	1938: N&A
CrB_2	h4	1949: K; 1954: P,G&M
Cr_2B	h3	1953: B&B
$CrCl_2$	b2	1951: H&G; 1960: C,W&W; 1961: O; T,G,L,D,M,R,S,Y&W
CrF_2	b9	1957: J&M; 1960: C,W&W
CrO_2	b1	1943: M&B
$(Cr,Mo)O_2$	b1	1958: S,A,M&M
$CrSb_2$	g6	1943: H&R
$CsBi_2$	h6	1957: Z,M&Z; 1961: G,D&K
CsO_2	g4	1938: H; 1939: H&K
Cs_2O	c2	1939: H&K; 1956: T,H&L
$CuAl_2$	h3	1923: O&P; 1924: J,P&W; W&P; 1927: F
$CuBr_2$	f10	1947: H
CuCdSb	a1	1942: N
$CuCl_2$	f10	1947: W
CuF_2	b9	1933: E&W; 1953: NBS; 1957: B&H
Cu_2Mg	h6	1927: F
CuMgBi	a1	1941: N&S

(*continued*)

BIBLIOGRAPHY TABLE, CHAPTER IV *(continued)*

Compound	Paragraph	Literature
CuMgSb	a1	1941: N&S
CuMnSb	a1	1952: N&G
Cu_2O	f1	1922: N; 1924: G; 1925: B&B; 1926: Q; 1931: N; 1932: W&M; 1949: O; 1953: NBS
$Cu(OH)_2$	d2	1960: O&J
Cu(OH)Cl	c11	1952: N&M; 1959: F,O&F; 1961: I,L&O
Cu_2S	f2	1926: B; 1931: A; 1935: K; 1936: R; 1944: B&B; O; 1946: B; B&B; 1958: D
Cu_2Sb	h2	1927: J&E; 1929: H; W,H&E; 1930: H&MJ; 1935: E,H&W
Cu_2Se	f2	1923: D; 1926: H; 1936: R; 1945: B; 1950: E; 1954: M; 1955: B,H&T
Cu_2Te	f2, h2	1946: N; 1949: F&P; 1954: A&S
CuZnAs	f2	1946: N
D_2O	e10	1934: M; 1942: V&H; 1957: P&L
D_2S	i1	1942: H; V&O; 1943: V
D_2Se	i1	1942: H; V&O; 1943: V
DyC_2	g4	1958: S,G&D
DyH_2	a1	1962: P&W
DyOCl	d3	1953: T&D
ErC_2	g4	1958: S,G&D
ErH_2	a1	1962: P&W
ErOCl	d3	1953: T&D
$EuCl_2$	d5	1939: D&K
EuD_2	d5	1956: K&W
EuF_2	a1	1938: B&N; 1939: D&K
EuOCl	d3	1953: T&D
EuOF	a5	1954: B,H,K&P
$FeAs_2$	g6	1926: dJ; 1932: B; 1959: NBS; 1960: H&C
FeAsS	g7	1926: dJ; 1928: dJ; 1936: B; 1961: B&W
Fe_2As	h2	1928: H; 1929: H; 1935: E,H&W; 1957: H&C
Fe_2B	h3	1929: B&A; W; W&M; 1930: H; W&M; 1931: H; 1933: B
$FeBr_2$	c1	1929: F&G
$FeCl_2$	c2	1929: F,C&G; 1930: P&H
FeF_2	b1	1926: F; G; 1953: E; 1954: S&R; 1957, B; 1958: B
$FeGe_2$	h3	1943: W

(continued)

BIBLIOGRAPHY TABLE, CHAPTER IV (*continued*)

Compound	Paragraph	Literature
FeI_2	c1	1929: F&G
FeOCl	d4	1934: G; 1935: G
Fe(OH)Cl	c10	1959: F,O&F
$Fe(OH)_2$	c1	1927: N&C; 1928: N; 1950: F&K; 1959: MK
α-FeO(OH), goethite	d1	1930: dJ; 1931: G; 1932: G; 1935: G; 1941: H
γ-FeO(OH), lepidochrosite	d2	1931: G; 1935: E; G
FeP_2	g6	1934: M
Fe_2P	h1	1928: H; K; 1929: H; O&K; 1930: F; H&K; 1959: R&J
FeS_2, pyrite	g1	1914: B; E; E&F; 1925: R; 1928: O; 1932: P&W; 1942: N; 1945: K; K,H&K; 1951: G; 1954: NBS; 1955: M; 1956: L; 1959: S; 1960: E
FeS_2, marcasite	g6	1926: F; dJ; 1931: B; 1937: B; 1942: N
$FeSb_2$	g6	1927: dJ&W; O; 1928: H; 1929: H; O
FeSbS	g7	1936: B; 1938: B; 1939: B
$FeSe_2$	g6	1938: T; 1955: B&K; 1959: C
$FeSn_2$	h3	1943: W
$FeTe_2$	g6	1938: T; 1946: T; 1954: G,H&V
$GaCl_2$	j4	1957: G&P
GdC_2	g4	1958: S,G&D
GdH_2	a1	1962: P&W
GdOCl	d3	1953: T&D; 1960: NBS
GdOF	a5	1954: B,H,K&P
$GeAs_2$	e13	1962: B
GeI_2	c1	1938: P&B
GeO_2	b1, e3	1928: Z; 1932: G; 1956: B
GeS_2	e12	1934: J&W; 1936: Z; 1940: I
H_2O	e10	1917: R; 1918: SJ; 1919: G; 1921: D; 1922: B; 1929: B; 1930: B; 1933: B&F; K&S; 1934: B; M; 1935: B&O; 1936: MF; 1942: K; V&H; 1944: K; 1947: V; 1949: W,D&S; 1955: T; 1957: H&S; B&L; S&C; 1958: L; 1960: D & R; K & D; S
H_2S	i1	1930: N; V; 1931: N; V; 1942: H; 1943: V; 1961: K,K,H&H
H_2Se	i1	1930: N; V; 1931: N; V; 1942: H; 1943: V
HfB_2	h4	1954: P,G&M

(*continued*)

BIBLIOGRAPHY TABLE, CHAPTER IV *(continued)*

Compound	Paragraph	Literature
$HfBe_2$	**h4**	1961: Z,S,B&K
HfO_2	**a1, a2**	1930: P; 1959: A&R
HfS_2	**c1**	1958: MT&W
$HfSe_2$	**c1**	1958: MT&W
$HgBr_2$	**d7**	1931: V&B; 1932: B
$HgBr(SCN)$	**i10**	1952: Z&Z
$Hg(CN)_2$	**i5**	1926: H; 1928: F&H; 1929: H; 1944: Z&S; 1945: Z&S; 1958: H
$Hg(Br,Cl)_2$	**d8**	1941: S&B
$HgCl_2$	**d6**	1928: B&H; 1932: N&B; 1934: B&S; 1950: G
$HgCl(SCN)$	**i10**	1952: Z&Z
HgF_2	**a1**	1933: E&W; 1961: NBS
HgI_2	**d7, e1**	1925: H; 1926: B,C&K; C; 1927: H&M; 1934: G; 1949: B&M
$HgNH_2Br$	**j7**	1952: N&L; R&B; 1954: B&R
$HgNH_2Cl$	**j7**	1951: L
HgO_2	**g3**	1959: V
HoC_2	**g4**	1958: S,G&D
HoH_2	**a1**	1962: P&W
$HoOCl$	**d3**	1953: T&D
$HoOF$	**a1**	1953: Z&T
$InOBr$	**d4**	1956: F
$InOCl$	**d4**	1956: F
IrO_2	**b1**	1926: G; 1927: L
Ir_2P	**a1**	1940: Z
$IrSe_2$	**b10**	1958: B
$IrSn_2$	**a1**	1947: N
$IrTe_2$	**c1, g1**	1960: H&W
KHC_2	**g4**	1930: vS
KO_2	**g1, g4**	1936: K&K; 1937: K&K; 1938: H; 1939: H&K; 1952: C,M&T; Z&Z; 1955: A&K
K_2O	**a1**	1934: Z,H&D
K_2S	**a1**	1934: W; Z,H&D
K_2Se	**a1**	1934: Z,H&D
K_2Te	**a1**	1934: Z,H&D
LaC_2	**g4**	1930: vS; 1958: A,G,D,R&S; S,G&D; 1959: A&M; 1960: B

(continued)

BIBLIOGRAPHY TABLE, CHAPTER IV (*continued*)

Compound	Paragraph	Literature
$LaGa_2$	h4	1943: L
LaH_2	a1	1955: H,M,E,K&Z; 1961: F,G,B&M
LaOBr	d3	1941: S&N
LaOCl	d3	1941: S&N; 1953: T&D
LaOF	a1, a5, a6	1941: K&K; 1943: C; 1951: Z; 1954: B,H, K&P
LaOI	d3	1941: S&N
LiMgN	a1	1946: J&H
Li_2NH	a1	1951: J&O
Li_2O	a1	1923: B; 1924: B&K; 1926: B,C&K; C; 1934: Z,H&D; 1960: NBS
Li_2S	a1	1925: C; 1934: Z,H&D
Li_2Se	a1	1934: W; Z,H&D
Li_2Te	a1	1934: W; Z,H&D
LiZnN	a1	1946: J&H
LuC_2	g4	1958: S,G&D
LuH_2	a1	1962: P&W
MgB_2	h4	1953: R,H,K&K
$MgBr_2$	c1	1929: F&G
MgC_2	g4	1943: B; R
$MgCl_2$	c2	1925: B&F; 1926: B&F; 1927: F; 1929: P
MgF_2	b1	1925: vA; B&V; F; 1926: G; 1953: NBS; 1956: B; 1958: B; D&S; 1961: B
Mg_2Ge	a1	1933: Z&K
MgI_2	c1	1933: B
MgNiBi	a1	1952: N&G
MgNiSb	a1	1952: N&G
Mg(OH)Cl	c2	1944: F&H; 1959: F,O&F
$Mg(OH)_2$	c1	1919: A; 1921: A; 1924: L&F; 1928: N; N&P; 1953: NBS
Mg_2Pb	a1	1925: S; 1926: F; 1933: Z&K
Mg_2Si	a1	1924: O&P
Mg_2Sn	a1	1923: P; 1925: S; 1933: Z&K
MnB_2	h4	1960: A; B&P
$MnBr_2$	c1	1929: F&G
$MnCl_2$	c2	1926: B&F; 1929: F,C&G; P; 1930: P&H
MnF_2	b1	1926: vA; F; G; 1950: G&S; 1953: E; 1954: S&R; 1957: B
MnI_2	c1	1929: F&G

(*continued*)

BIBLIOGRAPHY TABLE, CHAPTER IV *(continued)*

Compound	Paragraph	Literature
MnO_2	b1, d1	1923: SJ; 1926: F; G; 1947: C,W&W; 1949: B; B&B; B&H; M&K; 1950: B; B&B; 1951: K&Z; 1952: B&T; S; 1959: dW
$Mn(OH)_2$	c1	1919: A; 1928: N; N&P
$Mn(OH)Cl$	c10	1959: F,O&F
$MnO(OH)$	d1	1930: dJ; 1931: F&S; 1935: G; 1936: B; 1947: G; 1949: C&L
Mn_2P	h1	1937: A&N; 1959: R&J
MnS_2	g1	1914: E; E&F; 1930: O; 1932: O; S; 1933: S; 1934: P&H; O; 1936: B&W; 1951: G; 1959: S
Mn_2Sb	h2	1936: H&N; 1955: H&G
$MnSn_2$	h3	1947: N
$MnTe_2$	g1	1928: O
MoB_2	h4	1951: B&B; 1954: P,G&M
Mo_2B	h3	1947: K
Mo_2N	j1	1959: T&P
MoO_2	b1, b6	1926: G; 1944: H&M; 1946: M; 1955: M&A
MoS_2	c9	1923: D&P; 1925: H; 1926: N; 1949: B&M; 1954: NBS; 1957: B&H
$MoSi_2$	g4	1927: Z
$MoTe_2$	c9	1961: K&MD; P&N
N_2H_4	i3	1950: C&L; 1951: C&L
$NH_3(OH)Br$	c12	1947: J; 1948: J
$NH_3(OH)Cl$	c12	1947: J; 1948: J
$(N_2H_6)Cl_2$	a3	1923: W; 1947: D&L
$(N_2H_6)F_2$	c13	1942: K&H
N_2O	i1	1924: dS&K; 1931: V; 1961: H&P
N_2O_4	i4	1930: V; 1931: H; V; 1949: B&R
Na_2Au	h3	1937: H
$NaHC_2$	g4	1930: vS
NaO_2	g1, g6	1950: T&D; 1952: Z&Z; 1953: C&T
Na_2O	a1	1934: Z,H&D
Na_2S	a1	1925: C; 1934: Z,H&D
Na_2Se	a1	1934: Z,H&D
$(Na,Al)Si_2$	g4	1947: N&S
Na_2Te	a1	1934: Z,H&D
NaZnAs	a1	1951: N&G
NbB_2	h4	1954: P,G&M; 1959: V&K

(continued)

BIBLIOGRAPHY TABLE, CHAPTER IV (*continued*)

Compound	Paragraph	Literature
NbH_2	**a1**	1961: B&M
NbO_2	**b1, b8**	1926: G; 1961: M
NbS_2	**c2, c9**	1960: J,B&M
NdC_2	**g4**	1930: vS; 1931: vS; 1958: S,G&D
NdH_2	**a1**	1955: H,M,E,K&Z; 1962: P&W
NdOBr	**d3**	1949: Z
NdOCl	**d3**	1949: Z; 1953: T&D; 1958: NBS
NdOF	**a1, a5**	1950: M&I; 1954: B,H,K&P
$NiAs_2$	**g6**	1939: P; 1946: K; 1959: NBS; 1960: H&C
NiAsS	**g2**	1925: O; R; 1954: B&T; 1960: NBS
Ni_2B	**h3**	1933: B
$NiBr_2$	**c2, c4**	1934: K
$NiCl_2$	**c2**	1927: F; 1929: P
NiF_2	**b1**	1926: F; G; 1952: H,P&B; 1953: E; 1954: S&R; 1957: B; 1958: B
NiI_2	**c2**	1934: K
$Ni(OH)_2$	**c1**	1925: N; 1926: N&R; 1928: N; N&P; 1933: C&O; 1936: L&F
Ni(OH)Cl	**c2**	1936: F&C; 1959: F,O&F
Ni_2P	**h1**	1938: N&H; 1959: R&J
NiS_2	**g1**	1927: dJ&W; 1945: K; K,H&K; 1953: G, S&A; 1960: E
NiSbS	**g2**	1925: R; 1948: P&H; 1954: B&T
$NiSe_2$	**g1**	1938: T; 1950: E
$NiSi_2$	**a1**	1950: S&P
Ni_2Si	**d5**	1952: T; 1959: R&J
$NiTe_2$	**c1**	1938: T; 1946: P&T; 1957: S&I
NpO_2	**a1**	1949: Z; 1954: S,F,E&Z; 1955: A,E,F&Z; 1958: C&P
$(NpO_2)F_2$	**c14**	1949: Z
NpOS	**d3**	1949: Z
OsB_2	**h4**	1961: K&F
OsO_2	**b1**	1926: G
OsP_2	**g6**	1960: R
OsS_2	**g1**	1928: O; 1934: M
$OsSb_2$	**g6**	1960: K,Z&L
$OsSe_2$	**g1**	1929: T
$OsTe_2$	**g1**	1929: T

(*continued*)

BIBLIOGRAPHY TABLE, CHAPTER IV *(continued)*

Compound	Paragraph	Literature
P_2I_4	**i8**	1956: L&W
PaO_2	**a1**	1949: Z; 1950: E,F,S&Z; 1954: S,F,E&Z
PaOS	**d3**	1954: S,F,E&Z
$PbBr_2$	**d5**	1928: B&H; 1932: N&B; 1939: D&K; 1952: NBS
$PbCl_2$	**d5**	1928: B&H; 1931: M; 1932: B; 1939: D&K; 1942: S&S; 1961: S&Z
PbF_2	**a1, d5**	1924: K; 1932: K; 1933: S; 1947: B; 1951: S; 1954: NBS; 1961: Y,N&B
PbFBr	**d3**	1932: N&B
PbFCl	**d3**	1932: N&B; 1933: N; 1934: B; 1953: NBS
PbI_2	**c1, c2, c5**	1926: vA; T&W; 1929: F&G; 1943: P,T&N; 1953: NBS; 1959: M
PbO_2	**b1, b7**	1925: vA; F; 1926: G; 1932: D; 1945: B; 1950: Z,K&T; 1952: Z&T; 1957: W&F; 1958: T
Pb_2O	**f1**	1926: F
Pb(OH)Cl	**d5**	1937: G; 1939: G; 1940: B; 1950: P
Pb(OH)I	**d5**	1960: M&Z
$PdAs_2$	**g1**	1929: T
$PdBi_2$	**g1**	1957: F; 1958: H&B
$PdCl_2$	**f9**	1938: W
PdF_2	**b1**	1931: E; 1956: B&H
$PdPb_2$	**h3**	1943: W
PdS_2	**f8**	1956: G&R; 1957: G&R
$PdSb_2$	**g1**	1928: T
$PdSe_2$	**f8**	1956: G&R; 1957: G&R; S,B,B,G,H,L,V, W&W
$PdTe_2$	**c1**	1929: T; 1956: G&R
PoO_2	**a1**	1954: B&DE; M
PrC_2	**g4**	1930: vS; 1931: vS; 1958: S,G&D
PrH_2	**a1**	1955: H,M,E,K&Z; 1962: P&W
PrO_2	**a1**	1926: G; 1928: S&P; 1957: S&E
PrOCl	**d3**	1949: Z; 1953: T&D
PrOF	**a1, a5**	1950: M&I; 1954: B,H,K&P
$PtAl_2$	**a1**	1937: Z,H&H
$PtAs_2$	**g1**	1925: dJ; R; 1928: A&P; 1929: T; 1932: B
$PtBi_2$	**g1**	1943: W
$PtCl_2$	**f9**	1958: F&R

(continued)

BIBLIOGRAPHY TABLE, CHAPTER IV (*continued*)

Compound	Paragraph	Literature
$PtGa_2$	a1	1937: Z,H&H
$PtIn_2$	a1	1937: Z,H&H
PtO_2	j10	1951: G; 1952: B,G,R&C
PtP_2	g1	1929: T
PtS_2	c1	1929: T
$PtSb_2$	g1	1929: T
$PtSe_2$	c1	1929: T
$PtSn_2$	a1	1943: W; 1947: N
$PtTe_2$	c1	1929: T
PuB_2	h4	1960: MD&S
PuO_2	a1	1949: Z; 1955: A,E,F&Z; 1958: B,G,M&R; M&E
PuOBr	d3	1949: Z
PuOCl	d3	1949: Z
PuOF	a1, a6	1949: Z; 1951: Z
$(PuO_2)F_2$	c14	1958: A,Z,L,N,F&C
PuOI	d3	1949: Z
PuOSe	d3	1957: G
RaF_2	a1	1936: S
$RbBi_2$	h6	1957: Z,M&Z; 1961: G,D&K
RbO_2	g4	1938: H; 1939: H&K
Rb_2O	a1	1939: H&K
Rb_2S	a1	1936: M
ReB_2	h5	1962: LP&P
ReO_2	b6, b7	1955: M&A; 1956: M; 1957: M
Re_2P	d5	1961: R
Rh_2Ge	d5	1955: G; 1959: R&J
Rh_2P	a1	1940: Z
$RhPb_2$	h3	1943: W
RhS_2	g1	1929: T
$RhSe_2$	g1	1955: G&C
Rh_2Si	d5	1959: A,A&S
$RhTe_2$	c1, g1	1955: G
RuB_2	h4	1961: K&F
RuO_2	b1	1926: G; 1927: L
RuP_2	g6	1960: R
Ru_2P	d5	1960: R

(*continued*)

BIBLIOGRAPHY TABLE, CHAPTER IV (*continued*)

Compound	Paragraph	Literature
RuS_2	g1	1927: dJ&H; 1928: O; 1932: B
$RuSb_2$	g6	1960: K,Z&L
$RuSe_2$	g1	1929: T
$RuTe_2$	g1	1929: T
SO_2	i2	1950: S&K; 1952: P,S&F; 1954: S
$S(SCN)_2$	i9	1956: F
SbOCl	j6	1953: E
SbSBr	j5	1950: D; 1959: C&MC
SbSI	j5	1950: D
SbSeBr	j5	1950: D
SbSeI	j5	1950: D
SbTeI	d5	1951: D
ScB_2	h4	1958: Z&S
ScH_2	a1	1960: MG&K
γ-ScO(OH)	d2	1956: M&MA
SeO_2	j3	1937: MC
$Se(SCN)_2$	i9	1954: O&V; 1956: F
$Se(SeCN)_2$	i9	1954: A&F
SiO_2, quartz	e3, e4	1914: B; 1915: R; 1916: H&J; 1920: B; 1922: S&D; 1923: MK; 1925: B; B&G; G; W; 1926: G; W; 1927: G; 1930: B; 1933: B&J; J; K&K; 1934: J; 1935: B,B&G; N; W; 1937: B&H; N; 1939: B,H&P; 1944: J; 1948: B; B&F; 1950: K; B&H; 1953: B; 1955: F&H; 1958: dV; 1961: FK&K
SiO_2, cristobalite	e6, e7	1924: S; S&K; 1925: W; 1932: B; 1935: B, B&G; N; 1937: N; 1944: J; 1958: T; 1959: NBS
SiO_2, tridymite	e5	1926: G
SiO_2, coesite	e8	1955: R; 1959: S,K&V; Z&B
SiO_2, keatite	e9	1959: S,K&V
SiO_2, fibrous	e14	1954: W&W
SiS_2	e14	1935: Z&L; B,F&G
$SiSe_2$	e14	1952: W&W
$SiTe_2$	c1	1953: W&W
SmC_2	g4	1931: vS; 1958: S,G&D
$SmCl_2$	d5	1939: D&K
SmH_2	a1	1955: H,M,E,K&Z; 1962: P&W

(*continued*)

BIBLIOGRAPHY TABLE, CHAPTER IV (*continued*)

Compound	Paragraph	Literature
Sm_2O	**III,a1**	1956, E,B&E
SmOCl	**d3**	1953: T&D; 1960: NBS
SmOF	**a1, a5**	1950: M&I; 1954: B,H,K&P
SmOI	**d3**	1961: K,A&M
SnO_2	**b1**	1916: H&J; V; 1917: W; 1922: H; 1923: Y; 1924: D; 1926: G; V; 1952: N; 1953: NBS; 1956: B; 1959: S
SnS_2	**c1**	1926: O; 1928: O
SnSSe	**c1**	1961: B,F,H&S
$SnSe_2$	**c1**	1961: B,F,H&S
$SrBr_2$	**d5**	1939: K
SrC_2	**g4**	1930: vS
$SrCl_2$	**a1**	1925: M&T; 1926: O; 1946: C&B; 1953: NBS
SrF_2	**a1**	1924: vA; 1926: G; 1927: T; 1930: R; 1933: S; 1937: K&W; 1939: Z&U; 1954: NBS
SrH_2	**d5**	1935: Z&H
SrHBr	**d3**	1956: E&G
SrHCl	**d3**	1956: E&G; E,A&G
SrHI	**d3**	1956: E&K
SrO_2	**g4**	1935: B,D,K,R&W; 1955: NBS
TaB_2	**h4**	1954: P,G&M
Ta_2B	**h3**	1949: K
Ta_2H	**j2**	1956: W,W&C
TaO_2	**b1**	1954: S
TaS_2	**c1, c2, c3, c7**	1938: B&K; 1954: H&S
TbC_2	**g4**	1958: S,G&D
TbH_2	**a1**	1962: P&W
TbO_2	**a1**	1961: B,E,S&E
TbOCl	**d3**	1953: T&D
TbOF	**a5**	1954: T&D
TcO_2	**b6**	1955: M&A
TeO_2	**b1, b4, b5**	1926: G; 1939: I&S; 1948: S&B; 1949: S&B; 1955: M&A; 1959: NBS
$ThAs_2$	**h2**	1955: F
$ThBi_2$	**h2**	1957: F
ThC_2	**g5**	1930: vS; 1951: H&R
ThI_2	**c1**	1949: A&DE
ThO_2	**a1**	1923: G&T; 1924: vA; D; 1941: B; 1952: S&E; 1953: L,M&G; NBS; 1954: S,F,E&Z; 1958: K; M&E; 1959: V&K

(*continued*)

BIBLIOGRAPHY TABLE, CHAPTER IV (*continued*)

Compound	Paragraph	Literature
ThOS	d3	1949: Z
ThOSe	d3	1952: DE,S&M
ThOTe	d3	1954: DE&S; 1955: F
ThS_2	d5	1949: Z; 1960: G&MT
$ThSb_2$	h2	1956: F
$ThSe_2$	d5	1952: DE,S&M; DE; 1953: DE; 1960: G&MT
TiB_2	h4	1949: Z; 1954: P,G&M
$TiBr_2$	c1	1961: E,G&S
$TiCl_2$	c1	1948: B&R
TiI_2	c1	1941: K&G
TiO_2	b1, b3, b4	1916: V; 1917: W; 1923: P; 1924: D; G; 1926: G; H; V; 1927: T; 1928: P&S; S; 1929: S&P; 1942: S; 1953: L&D; NBS; 1954: M&B; 1955: B; C&H; 1956: B; 1959: W; 1961: B; S,E&J
TiOCl	d4	1958: S,W&W
TiS_2	c1	1928: O; 1954: H&S; 1958: MT&W; 1959: J&B
$TiSe_2$	c1	1928: O; 1949: E; 1958: MT&W
$TiTe_2$	c1	1928: O; 1949: E; 1958: MT&W
Tl_2S	c8	1939: K&G
Tl_2Se	j9	1958: S&V
TmC_2	g4	1958: S,G&D
TmH_2	a1	1962: P&W
TmI_2	c1	1960: A&K
TmOI	d3	1961: K,A&M
UAs_2	h2	1952: I; 1953: F
UB_2	h4	1954: P,G&M; 1960: MD&S
UBi_2	h2	1952: F; 1953: F
UC_2	g4	1948: L,G&C; R,B,W&MD; 1959: A&M; 1960: B; W; 1961: H
UN_2	a1	1948: R,B,W&MD
UO_2	a1	1923: G&T; 1924: vA; 1929: H&vA; 1935: S&B; 1948: R,B,W&MD; 1951: B; T; 1952: H&P; S&E; 1953: S; NBS; 1954: S,F,E&Z; 1958: R,B&W; 1959: L; 1960: A,S,W,W&B
UOCl	d3	1957: S&E
$(UO_2)F_2$	c14	1948: Z

(*continued*)

BIBLIOGRAPHY TABLE, CHAPTER IV (*continued*)

Compound	Paragraph	Literature
$UO_2(OH)_2$	j8	1956: B&L
UOS	d3	1949: Z
UOSe	d3	1954: F; 1957: K
UP_2	h2	1952: I; 1953: F
US_2	d5	1953: P&F
USb_2	h2	1952: F; 1953: F
USe_2	d5	1957: K
USi_2	h4	1949: Z
UTe_2	d3	1954: F
VB_2	h4	1954: P,G&M
VBr_2	c1	1941: K&G
VCl_2	c1	1959: E&S
VI_2	c1	1941: K&G
VO_2, paramontroseite	b6, d1	1926: G; 1953: A; 1954: A; 1955: E&M, M&A; 1956: A
VO(OH), montroseite	d1	1953: E&B; 1955: E&M
VSb_2	h3	1951: N,F&P
W_2B	h3	1947: K
W_2C	c1	1926: W&P; 1928: B; 1960: B&P
WO_2	b1, b6	1926: G; 1944: H&M; 1946: M; 1955: M&A
WS_2	c9	1926: vA; 1948: G,S&K; 1958: NBS
WSe_2	c9	1948: G,S&K
WSi_2	g4	1927: Z
YC_2	g4	1931: vS
YH_2	a1	1960: D&F; 1962: P&W
YOCl	d3	1949: Z; 1953: T&D; 1960: NBS
YOF	a1, a5, a6	1951: H; Z
YbC_2	g4	1958: S,G&D
YbD_2	d5	1956: K&W
YbI_2	c1	1939: D&K; 1960: A&K
YbOI	d3	1961: K,A&M
$ZnBr_2$	c2, e1	1942: Y; 1960: O; 1961: B
$Zn(CN)_2$	f1	1941: Z; 1954: NBS
$ZnCl_2$	c2, e1, e2	1926: B&F; 1929: P; 1959: B; 1960: O&J
ZnF_2	b1	1926: F; G; 1952: H,P&B; 1954: S&R; 1957: B; 1958: B

(*continued*)

BIBLIOGRAPHY TABLE, CHAPTER IV (*continued*)

Compound	Paragraph	Literature
ZnI_2	c1, c2	1942: Y; 1946: P,T&N; 1948: B; 1959: NBS; 1960: O
$Zn(OH)_2$	e11	1927: G&M; 1928: N; 1932: F; 1933: C&W; 1935: M; 1936: L&F; 1938: F
$Zn(OH)Cl$	c10	1959: F&N; F,O&F
$ZrAs_2$	d5	1958: T,W&L
ZrB_2	h4	1936: MK; 1953: G&P; 1954: P,G&M; 1959: V&K
ZrO_2	a1, a2	1924: vA; 1925: B; 1926: B; D; G; Y; 1929: R&E; R,E&S; 1930: C&T; P; 1936: NS; 1957: F&H; 1959: A&R; MC&T
ZrOS	a4	1948: MC,B&B
ZrS_2	c1	1924: vA; 1954: H&S; 1958: MT&W
$ZrSe_2$	c1	1924: vA; 1958: MT&W; 1959: H&N
$ZrTe_2$	c1	1958: MT&W; 1959: H&N

BIBLIOGRAPHY

1914

Bragg, W. H., "X-Ray Spectra Given by Sulfur and Quartz," *Proc. Roy. Soc.* (*London*), **89A,** 575.

Bragg, W. L., "The Analysis of Crystals by the X-Ray Spectrometer," *Proc. Roy. Soc.* (*London*), **89A,** 468.

Ewald, P. P., "Crystal Structure from Interference Photographs by X-Rays," *Physik. Z.*, **15,** 399.

Ewald, P. P., and Friedrich, W., "X-Ray Spectra of Cubic Crystals, Especially of Iron Pyrite," *Ann. Physik,* **44,** 1183.

Glocker, R., "The Interference of X-Rays," *Physik. Z.*, **15,** 401.

1915

Glocker, R., "Crystal Structure and Interference of X-Rays," *Ann. Physik,* **47,** 377.

Rinne, F., "X-Ray Photographs of Crystals," *Ber. Sachs. Gesell. Wiss. Leipzig* (*Math.-Phys. Kl.*), **67,** 303.

1916

Haga, H., and Jaeger, F. M., "The Symmetry of X-Ray Patterns of Tetragonal Crystals," *Proc. Acad. Sci. Amsterdam,* **18,** 1350.

Haga, H., and Jaeger, F. M., "The Symmetry of X-Ray Patterns of Triclinic and some Rhombic Crystals; Remarks upon the Diffraction Images of Quartz," *Proc. Acad. Sci. Amsterdam,* **18,** 1552.

Vegard, L., "Results of Crystal Analysis," *Phil. Mag.*, **32**, 65.

Vegard, L., "Results of Crystal Analysis III," *Phil. Mag.*, **32**, 505.

1917

Rinne, F., "The Crystallography of Ice," *Ber. Sachs. Gesell. Wiss. Leipzig* (*Math.-Phys. Kl.*), **69**, 57.

Williams, C. M., "X-Ray Analysis of the Crystal Structure of Rutile and Cassiterite," *Proc. Roy. Soc.* (*London*), **93A**, 418.

1918

St. John, A., "The Crystal Structure of Ice," *Proc. Natl. Acad. Sci.*, **4**, 193.

1919

Aminoff, G., "The Crystal Structure of Pyrochroite," *Geol. Foren. Stockholm Forh.*, **41**, 407.

Gross, R., "Laue Diagram of Ice," *Central. Mineral. Geol.*, **1919**, 201.

1920

Beckenkamp, J., "Atomic Grouping and the Optical Rotation of Quartz and Sodium Chlorate," *Z. Anorg. Chem.*, **110**, 290.

1921

Aminoff, G., "On the Crystal Structure of Magnesium Hydroxide," *Z. Krist.*, **56**, 506.

Dennison, D. M., "The Crystal Structure of Ice from its X-Ray Pattern," *Phys. Rev.*, **17**, 20; *Chem. News*, **122**, 54.

Gerlach, W., "Crystal Lattice Structure Investigations with X-Rays and a Simple X-Ray Tube," *Physik. Z.*, **22**, 557.

Mechling, M., "The Crystal Structure of Cobaltite," *Abhandl. Math.-Phys. Kl. Sachs. Akad. Wiss. Leipzig*, **38**, 37.

1922

Bozorth, R. M., "The Crystal Structure of Cadmium Iodide," *J. Am. Chem. Soc.*, **44**, 2232.

Bragg, W. H., "The Crystal Structure of Ice," *Proc. Phys. Soc.* (*London*), **34**, 98.

Davey, W. P., "The Absolute Sizes of Certain Monovalent and Bivalent ions," *Phys. Rev.*, **19**, 248.

Gerlach, W., "The *K*-Alpha Doublet, Including a New Determination of the Lattice Constants of Some Crystals," *Physik. Z.*, **23**, 114.

Hedvall, J. A., "Changes in Properties Produced by Different Methods of Preparation of Some Ignited Oxides, as Studied by X-Ray Interference," *Arkiv Kemi Mineral. Geol.*, **8**, No. 11; *Z. Anorg. Chem.*, **120**, 327.

Niggli, P., "The Crystal Structures of Several Oxides," *Z. Krist.*, **57**, 253.

Siegbahn, M., and Dolejsek, V., "Increase of Accuracy of Measurement in X-Ray Spectra II," *Z. Physik*, **10**, 159.

Wyckoff, R. W. G., "The Crystal Structure of Silver Oxide," *Am. J. Sci.*, **3**, 184.

1923

Bijvoet, J. M., "X-Ray Investigation of the Crystal Structure of Lithium and Lithium Hydride," *Rec. Trav. Chim.*, **42**, 859.

Davey, W. P., "Crystal Structure and Densities of Cu_2Se and ZnSe," *Phys. Rev.*, **21**, 380.

Dickinson, R. G., and Pauling, L., "The Crystal Structure of Molybdenite," *J. Am. Chem. Soc.*, **45**, 1466.

Goldschmidt, V. M., and Thomassen, L., "The Crystal Structure of Natural and Synthetic Oxides of Uranium, Thorium and Cerium," *Skrifter Norske Videnskaps-Akad. Oslo I. Mat.-Naturv. Kl.*, **1923**, No. 2.

McKeehan, L. W., "The Crystal Structure of Quartz," *Phys. Rev.*, **21**, 503.

Owen, E. A., and Preston, G. D., "X-Ray Analysis of Solid Solutions," *Proc. Phys. Soc. (London)*, **36**, 14.

Parker, R. L., "The Crystallography of Anatase and Rutile II. Structure of Anatase," *Z. Krist.*, **59**, 1.

Pauling, L., "The Crystal Structure of Magnesium Stannide," *J. Am. Chem. Soc.*, **45**, 2777.

St. John, A., "The Crystal Structure of Manganese Dioxide," *Phys. Rev.*, **21**, 389.

Wyckoff, R. W. G., "The Crystal Structure of Hydrazine Dihydrochloride," *Am J. Sci.*, **5**, 15.

Yamada, M., "Study of Metal Oxides and Their Hydrates by X-Rays. I. X-Ray Analysis of Stannic Oxide," *J. Chem. Soc. Japan*, **44**, 210.

1924

Arkel, A. E. van, "Crystal Structure and Physical Properties," *Physica*, **4**, 286.

Bijvoet, J. M., and Karssen, A., "X-Ray Investigation of the Crystal Structure of Lithium Oxide," *Rec. Trav. Chim.*, **43**, 680.

Davey, W. P., "Crystal Structures and Densities of Oxides of the Fourth Group," *Phys. Rev.*, **23**, 763.

Greenwood, G., "The Crystal Structure of Cuprite and Rutile," *Phil. Mag.*, **48**, 654.

Jette, E. R., Phragmén, G., and Westgren, A., "X-Ray Studies on the Copper–Aluminium Alloys," *J. Inst. Metals*, **31**, 193.

Kolderup, N.-H., "The Crystal Structure of Lead, Galena, Lead Fluoride and Cadmium Fluoride," *Bergens Museums Aarbok, Naturvidensk Raek.*, No. 2.

Levi, G. R., "The Crystal Structure of Calcium Hydrate," *G. Chim. Ind. Appl.*, **6**, 333.

Levi, G. R., and Ferrari, A., "Crystalline Lattices of Magnesium Hydroxide and Carbonate," *Rend. Accad. Lincei*, **33**, 397.

Levi, G. R., and Quilico, A., "On the Nonexistence of the Suboxide of Silver," *Gazz. Chim. Ital.*, **54**, 598.

Owen, E. A., and Preston, G. D., "The Atomic Structure of Two Intermetallic Compounds," *Proc. Phys. Soc. (London)*, **36**, 341.

Seljakov, N., Strutinsky, L., and Krasnikov, A., "On the Structure of Glass," *Mitt. Wiss.-Tech. Arbeit. Republik (Russ.)*, **13**, 18; *Z. Physik*, **33**, 53 (1925); *Rept. Phys.-Tech. Roentgen Inst. Leningrad Phys.-Tech. Lab.*, **1918–1926**, p. 101.

Smedt, J. de, and Keesom, W. H., "The Structure of Solid Nitrous Oxide and Carbon Dioxide," *Proc. Acad. Sci. Amsterdam*, **27**, 839; *Verslag Akad. Wetenschap. Amsterdam*, **33**, 571, 888; *Rept. Commun. First Intern. Comm. Intern. Inst. Refrigeration*, p. 117.

Westgren, A., and Phragmén, G., "On the Structure of Solid Solutions," *Nature*, **113**, 122.

1925

Arkel, A. E. van, "The Crystal Structure of Magnesium Fluoride and Analogous Substances," *Physica*, **5**, 162.

Böhm, J., "The Glowing of the Oxides of Certain Metals," *Z. Anorg. Chem.*, **149**, 217.

Bragg, W. H., "The Structure of Quartz," *J. Soc. Glass Tech.*, **9**, 272.

Bragg, W. H., and Bragg, W. L., *X-Rays and Crystal Structure*, 5th Ed., G. Bell & Sons, London.

Bragg, W. H., and Gibbs, R. E., "The Structure of α and β Quartz," *Proc. Roy. Soc.* (*London*), **109A**, 405.

Bruni, G., and Ferrari, A., "Solid Solutions between Compounds of Elements with Different Valences—LiCl and $MgCl_2$," *Rend. Accad. Lincei*, **2**, 457.

Buckley, H. E., and Vernon, W. S., "The Crystal Structure of Magnesium Fluoride," *Phil. Mag.*, **49**, 945.

Claassen, A., "The Crystal Structure of the Anhydrous Alkali Monosulfides," *Rec. Trav. Chim.*, **44**, 790.

Ferrari, A., "The Crystalline Lattices of Lithium and Magnesium Fluorides and Their Isomorphism," *Rend. Accad. Lincei*, **1**, 664.

Ferrari, A., "The Crystal Structure of Lead Dioxide Determined by X-Rays," *Rend. Accad. Lincei*, **2**, 186.

Gibbs, R. E., "The Variation with Temperature of the Intensity of Reflection of X-Rays from Quartz and its Bearing on the Crystal Structure," *Proc. Roy. Soc.* (*London*), **107A**, 561.

Hassel, O., "The Crystal Structure of Molybdenite, MoS_2," *Z. Krist.*, **61**, 92.

Havighurst, R. J., "X-Ray Reflections from Mercuric Iodide," *Am. J. Sci.*, **10**, 556.

Jong, W. F. de, "The Structure of Sperrylite," *Physica*, **5**, 292.

McLennan, J. C., and Wilhelm, J. O., "The Crystal Structure of Carbon Dioxide," *Trans. Roy. Soc. Canada, Sect. III*, **19**, 51.

Mark, H., Basche, W., and Pohland, E., "Determination of the Structure of Some Simple Inorganic Substances," *Z. Elektrochem.*, **31**, 523.

Mark, H., and Pohland, E., "The Structure of Solid Carbon Dioxide," *Z. Krist.*, **61**, 293.

Mark, H., and Tolksdorf, S., "Scattering Power of Atoms for X-Rays," *Z. Physik*, **33**, 681.

Natta, G., "The Crystalline Structure of the Hydrates of Cadmium and Nickel," *Rend. Accad. Lincei*, **2**, 495.

Olshausen, S., "Crystal Structure Studies using the Debye-Scherrer Method," *Z. Krist.*, **61**, 463.

Ramsdell, L. S., "The Crystal Structure of Some Metallic Sulfides," *Am. Mineralogist*, **10**, 281.

Sacklowski, A., "An X-Ray Examination of Some Alloys," *Ann. Physik*, **77**, 241.

Smedt, J. de, and Keesom, W. H., "The Structure of Solid Carbon Dioxide," *Z. Krist.*, **62**, 312.

Wyckoff, R. W. G., "The Crystal Structure of the High Temperature Form of Cristobalite (SiO_2)," *Am. J. Sci.*, **9**, 448; *Z. Krist.*, **62**, 189.

Wyckoff, R. W. G., "The Crystal Structure of the High Temperature (β-) Modification of Quartz," *Science*, **62**, 496; *Am. J. Sci.*, **11**, 101 (1926).

1926

Arkel, A. E. van, "The Crystal Structure of Manganese Fluoride, Lead Iodide and Tungsten Sulfide," *Rec. Trav. Chim.*, **45**, 437.

Barth, T., "The Regular Crystal Type of Cu_2S," *Central. Mineral. Geol.*, **1926A**, 284.

Becker, K., "X-Ray Method of Determining Coefficient of Expansion at High Temperatures," *Z. Physik*, **40**, 37.

Bijvoet, J. M., Claassen, A., and Karssen, A., "Crystal Structure of Red Mercuric Iodide," *Proc. Acad. Sci. Amsterdam*, **29**, 529.

Bijvoet, J. M., Claassen, A., and Karssen, A., "Determination of the Scattering Power for X-Rays of Lithium and Oxygen from the Diffraction Intensities of Powdered Lithium Oxide," *Verslag Akad. Wetenschap. Amsterdam*, **35**, 891.

Bruni, G., and Ferrari, A., "The Crystal Structures of Some Divalent Chlorides," *Rend. Accad. Lincei*, **4**, 10.

Claassen, A., "The Crystal Structure of Red Mercuric Iodide," thesis, Amsterdam.

Clausen, H., "Laue Diagram of Fluorite," *Medd. Dansk. Geol. Foren.*, **7**, 40; *Neues Jahrb. Mineral.*, **1927A**, II, 114.

Davey, W. P., "The Crystal Structure of Zirconium Oxide," *Phys. Rev.*, **27**, 798.

Dehlinger, U., and Glocker, R., "The Crystal Structure of Calcium Carbide," *Z. Krist.*, **64**, 296.

Ferrari, A., "The Crystal Structures of Manganese Fluoride and Manganese Dioxide," *Rend. Accad. Lincei*, **3**, 224.

Ferrari, A., "The Crystal Structures of Some Alkali Fluorides of FeF_2, CoF_2, NiF_2 and ZnF_2," *Rend. Accad. Lincei*, **3**, 324.

Ferrari, A., "The Suboxide of Lead," *Gazz. Chim. Ital.*, **56**, 630.

Friauf, J. B., "The Crystal Structure of Magnesium Plumbide," *J. Am. Chem. Soc.*, **48**, 1906.

Frielinghaus, W., thesis, Greifswald.

Gibbs, R. E., "Structure of α-Quartz," *Proc. Roy. Soc. (London)*, **110A**, 443.

Gibbs, R. E., "The Polymorphism of Silicon Dioxide and the Structure of Tridymite," *Proc. Roy. Soc. (London)*, **113A**, 351.

Goldschmidt, V. M., "Crystal Structures of the Rutile Type with Remarks on the Geochemistry of the Bivalent and Quadrivalent Elements," *Skrifter Norske Videnskaps-Akad. Oslo I. Mat.-Naturv. Kl.*, **1926**, No. 1.

Goldschmidt, V. M., "The Laws of Crystal Chemistry," *Skrifter Norske Videnskaps-Akad. Oslo I. Mat.-Naturv. Kl.*, **1926**, No. 2; *Naturwiss.*, **14**, 477.

Hartwig, W., "The Crystal Structure of Berzelianite (Cu_2Se)," *Z. Krist.*, **64**, 503; *Mineral. Petrogr. Mitt.*, **37**, 248; *Fortschr. Mineral. Kryst. Petrog.*, **11**, 307 (1927).

Hassel, O., "X-Ray Investigation of Tetragonal Mercury Cyanide," *Z. Krist.*, **64**, 217.

Huggins, M. L., "The Crystal Structures of Anatase and Rutile, the Tetragonal Forms of TiO_2," *Phys. Rev.*, **27**, 638.

Jong, W. F. de, "Determination of the Absolute Axial Lengths of Marcasite and Isomorphous Minerals," *Physica*, **6**, 325.

Krüner, H., "The Crystal Structure of Solid Carbon Dioxide," *Z. Krist.*, **63**, 275.

Mark, H., and Pohland, E., "The Crystal Structure of Solid Carbon Dioxide," *Z. Krist.*, **64**, 113.

Natta, G., "Applications of X-Rays in Analytical Chemistry. I. Analysis of the Molybdenite of Zovon," *Gazz. Chim. Ital.*, **56,** 651.

Natta, G., and Reina, A., "The Crystal Structures of Cobaltous Oxide and Hydroxide," *Rend. Accad. Lincei,* **4,** 48.

Oftedal, I., "The Crystal Structure of SnS_2," *Norsk Geol. Tidsskr.*, **9,** 225; *Neues Jahrb. Mineral.*, **1927A,** II, 115.

Ott, H., "The Structures of MnO, MnS, AgF, NiS, SnI_4, $SrCl_2$, BaF_2; Precision Measurements upon Various Alkali Halides," *Z. Krist.*, **63,** 222.

Quilico, A., "X-Ray Examination of Metallic Hydrides; Hydrides of Copper," *Rend. Accad. Lincei,* **4,** 57.

Smedt, J. de, "The Crystal Structure of Solid Carbon Disulfide," *Natuurw. Tijdschr.*, **8,** 13.

Terpstra, P., and Westenbrink, H. G. K., "The Crystal Structure of Lead Iodide," *Proc. Acad. Sci. Amsterdam,* **29,** 431; *Verslag Akad. Wetenschap. Amsterdam,* **35,** 75.

Vegard, L., "Results of Crystal Analysis," *Phil. Mag.*, **1,** 1151.

Westgren, A., and Phragmén, G., "X-Ray Analysis of the Systems Tungsten–Carbon and Molybdenum–Carbon," *Z. Anorg. Chem.*, **156,** 27.

Wyckoff, R. W. G., "Criteria for Hexagonal Space Groups and the Crystal Structure of β-Quartz," *Z. Krist.*, **63,** 507.

Yardley, K., "The Structure of Baddeleyite and of Prepared Zirconia," *Mineral. Mag.*, **21,** 169.

1927

Ferrari, A., "On the Crystalline Structure of Bivalent Chlorides: Anhydrous Chlorides of Cobalt and of Nickel," *Rend. Accad. Lincei,* **6,** 56.

Friauf, J. B., "The Crystal Structures of Two Intermetallic Compounds," *J. Am. Chem. Soc.*, **49,** 3107.

Gossner, B., "The Structure of Quartz," *Central. Mineral. Geol.*, **1927A,** 329.

Gottfried, C., and Mark, H., "The Structure of Zinc Hydroxide," *Z. Krist.*, **65,** 416.

Harrington, E. A., "X-Ray Diffraction Measurements on Some of the Pure Compounds Concerned in the Study of Portland Cement," *Am. J. Sci.*, **13,** 467.

Hermann, C., "The Space Lattice of Calcium Carbide," *Z. Krist.*, **66,** 314.

Huggins, M. L., and Magill, P. L., "The Crystal Structures of Mercuric and Mercurous Iodides," *J. Am. Chem. Soc.*, **49,** 2357.

Jones, W. M., and Evans, E. J., "The Crystal Structure of Cu_3Sn and Cu_3Sb," *Phil. Mag.*, **4,** 1302.

Jong, W. F. de, and Hoog, A., "The Compound RuS_2 and its Structure," *Rec. Trav. Chim.*, **46,** 173.

Jong, W. F. de, and Willems, H. W. V., "Compounds of the Lattice Type of Pyrrhotite (FeS)," *Physica,* **7,** 74.

Jong, W. F. de, and Willems, H. W. V., "The Existence and Structure of the Disulfides NiS_2 and CoS_2," *Z. Anorg. Chem.*, **160,** 185.

Lunde, G., "The Existence and Preparation of Certain Oxides of the Platinum Metals (with a Supplement Regarding Amorphous Oxides)," *Z. Anorg. Chem.*, **163,** 345.

Natta, G., and Casazza, E., "The Crystalline and Atomic Structure of Ferrous Hydroxide," *Rend. Accad. Lincei,* **5,** 803.

Oftedal, I., "Some Crystal Structures of the Type NiAs," *Z. Physik. Chem. (Leipzig),* **128,** 135.

Reina, A., "The Crystalline Structure of Calcium Hydroxide," *Rend. Accad. Lincei*, **5**, 1008.
Thilo, F., "X-Ray Investigation of the Alkaline Earth Fluorides," *Z. Krist.*, **65**, 720.
Tokody, L., "The Structure of Rutile," *Mathematik. Természettudományi Értesito*, **44**, 247.
Zachariasen, W. H., "The Crystal Structure of $MoSi_2$ and WSi_2," *Z. Physik. Chem. (Leipzig)*, **128**, 39.

1928

Aminoff, G., and Parsons, A. L., "The Crystal Structure of Sperrylite," *Univ. Toronto Studies, Geol. Ser.*, No. 26.
Becker, K., "Crystal Structure and Linear Coefficient of Thermal Expansion of Tungsten Carbides," *Z. Physik*, **51**, 481.
Becker, K., "The Constitution of Tungsten Carbide," *Z. Elektrochem.*, **34**, 640.
Bräkken, H., and Harang, L., "The Crystal Structure of Some Orthorhombic Compounds of the Type MX_2," *Z. Krist.*, **68**, 123.
Fricke, R., and Havestadt, L., "The Crystal Structure of Mercuric Cyanide," *Z. Anorg. Chem.*, **171**, 344.
Hägg, G., "X-Ray Studies of the Binary Systems of Iron with Phosphorus, Arsenic, Antimony and Bismuth," *Z. Krist.*, **68**, 470.
Jong, W. de, "The Crystal Structures of Arsenopyrite, Bornite and Tetrahedrite," thesis, Delft.
Kreutzer, C., "X-Ray Diffraction Measurements in the Systems Iron–Silicon, Iron–Chromium and Iron–Phosphorus," *Z. Physik*, **48**, 556.
Natta, G., "Constitution of Hydroxides and of Hydrates I," *Gazz. Chim. Ital.*, **58**, 344.
Natta, G., and Passerini, L., "Solid Solutions by Precipitation," *Gazz. Chim. Ital.*, **58**, 597.
Natta, G., and Strada, M., "Oxides and Hydroxides of Cobalt," *Gazz. Chim. Ital.*, **58**, 419.
Oftedal, I., "X-Ray Investigations of SnS_2, TiS_2, $TiSe_2$, $TiTe_2$," *Z. Physik. Chem. (Leipzig)*, **134**, 301.
Oftedal, I., "Crystal Structures of the Compounds RuS_2, OsS_2, $MnTe_2$, $AuSb_2$," *Z. Physik. Chem. (Leipzig)*, **135**, 291.
Ott, H., and Seyfarth, H., "The Structure of Silver Subfluoride, Ag_2F," *Z. Krist.*, **68**, 239.
Pauling, L., and Sturdivant, J. H., "The Crystal Structure of Brookite," *Z. Krist.*, **68**, 239.
Rumpf, E., "On the Lattice Constants of Calcium Oxide and Calcium Hydroxide," *Ann. Physik*, **87**, 595.
Scherrer, P., and Palacios, J., "Crystalline Structure of Praeseodymium Dioxide," *Anal. Soc. Espan. Fis. Quim.*, **26**, 309.
Schröder, A., "X-Ray Investigation of the Structure of Brookite and the Physical Properties of the Three Titanium Dioxides," *Z. Krist.*, **66**, 493.
Terrey, H., and Diamond, H., "The Crystal Structure of Silver Subfluoride," *J. Chem. Soc.*, **1928**, 2820.
Thomassen, L., "The Preparation and Crystal Structure of the Mono- and Diantimonides of Palladium," *Z. Physik. Chem. (Leipzig)*, **135**, 383.
Zachariasen, W. H., "The Crystal Structure of the Water-Soluble Modification of Germanium Dioxide," *Z. Krist.*, **67**, 226.

1929

Barnes, W. H., "The Crystal Structure of Ice between 0°C and −183°C," *Proc. Roy. Soc. (London)*, **125A**, 670.

Bjurström, T., and Arnfelt, H., "X-Ray Analysis of the System Iron–Boron," *Z. Physik. Chem.*, **4B**, 469.

Broch, E., Oftedal, I., and Pabst, A., "New Determinations of the Lattice Constants of Potassium Fluoride, Cesium Chloride and Barium Fluoride," *Z. Physik. Chem.*, **3B**, 209.

Ferrari, A., Celeri, A., and Giorgi, F., "The Importance of Crystalline Form in the Formation of Solid Solutions V. Thermal and X-Ray Analysis of the Systems $CoCl_2$–$FeCl_2$ and $MnCl_2$–$FeCl_2$," *Rend. Accad. Lincei*, **9**, 782.

Ferrari, A., and Giorgi, F., "The Crystal Structure of Bromides of Bivalent Metals," *Rend. Accad. Lincei*, **9**, 1134.

Ferrari, A., and Giorgi, F., "The Crystal Structure of the Anhydrous Iodides of Bivalent Metals I. Iodides of Cobalt, Iron and Manganese," *Rend. Accad. Lincei*, **10**, 522.

Hadding, A., and Aubel, R. van, "The Structure of Crystalline Uraninite from Katanga (Belgian Congo)," *Compt. Rend.*, **188**, 716.

Hägg, G., "An X-Ray Study of the System Iron–Arsenic," *Z. Krist.*, **71**, 134.

Hägg, G., "X-Ray Studies on the Binary Systems of Iron with Nitrogen, Phosphorus, Arsenic, Antimony and Bismuth," *Nova Acta Reg. Soc. Sci. Upsaliensis IV*, **7**, No. 1.

Hassel, O., "Is the Lattice of Tetragonal Mercuric Cyanide a Molecular or Radical Lattice?" *Z. Anorg. Chem.*, **180**, 370; cf. Fricke, R., *ibid.*, **180**, 374.

Oberhoffer, P., and Kreutzer, C., "The Systems Iron–Silicon, Iron–Chromium and Iron–Phosphorus," *Arch. Eisenhüttenw.*, **2**, 449; *Stahl Eisen*, **49**, 189.

Oftedal, I., "Observations on the Lattice Dimensions of the System Fe_x–Sb_y," *Z. Physik. Chem.*, **4B**, 67.

Pauling, L., "The Crystal Structure of the Chlorides of Certain Bivalent Elements," *Proc. Natl. Acad. Sci.*, **15**, 709.

Ruff, O., and Ebert, F., "The Forms of Zirconium Dioxide," *Z. Anorg. Chem.*, **180**, 19.

Ruff, O., Ebert, F., and Stephan, E., "The System ZrO_2–CaO," *Z. Anorg. Chem.*, **180**, 215.

Smedt, J. de, "X-Ray Analysis of Solid Carbon Disulfide," *Physica*, **9**, 5.

Sturdivant, J. H., and Pauling, L., "Note on the Paper of A. Schröder: Beiträge zur Kenntnis des Feinbaues des Brookits, usw.," *Z. Krist.*, **69**, 557.

Thomassen, L., "The Crystal Structure of Some Binary Compounds of the Platinum Metals," *Z. Physik. Chem.*, **2B**, 349.

Thomassen, L., "The Crystal Structure of Some Binary Compounds of the Platinum Metals II," *Z. Physik. Chem.*, **4B**, 277.

Westgren, A., Hägg, G., and Eriksson, S., "X-Ray Analysis of the Copper–Antimony and Silver–Antimony Systems," *Z. Physik. Chem.*, **4B**, 453.

Wever, F., "Iron–Beryllium and Iron–Boron Alloys; The Structure of Iron Boride," *Z. Tech. Physik*, **10**, 137.

Wever, F., and Müller, A., "The Binary Systems Iron–Boron, Iron–Beryllium and Iron–Aluminum," *Mitt. Kaiser Wilhelm-Inst. Eisenforsch. Düsseldorf*, **11**, 193.

1930

Bergquist, O., "The Grating Constant of Quartz," *Z. Physik*, **66**, 494.

Brandenberger, E., "On the Crystal Structure of Ice," *Z. Krist.*, **73**, 429.

Cohn, W. M., and Tolksdorf, S., "The Forms of ZrO_2 and Their Previous Treatment," *Z. Physik. Chem.*, **8B**, 331.

Friauf, J. B., "The Crystal Structure of an Iron Phosphide (Fe_2P)," *Trans. Am. Soc. Steel Treating*, **17**, 499.

Hägg, G., "Crystal Structure of the Compound Fe_2B," *Z. Physik. Chem.*, **11B**, 152.

Hendricks, S. B., and Kosting, P. R., "The Crystal Structure of Fe_2P, Fe_2N, Fe_3N and FeB," *Z. Krist.*, **74**, 511.

Howells, E. V., and Morris-Jones, W., "An X-Ray Investigation of the Copper–Antimony System of Alloys," *Phil. Mag.*, **9**, 993.

Jong, W. F. de, "Goethite, Stainierite, Diaspore and Heterogenite," *Natuurw. Tijdschr.*, **12**, 69.

Natta, G., "The Crystalline Structure of Hydrogen Sulfide and Hydrogen Selenide I and II," *Rend. Accad. Lincei*, **11**, 679, 749.

Onorato, E., "The Symmetry and Structure of Hauerite," *Periodico Mineral.* (*Rome*), **1**, No. 2.

Passerini, L., "Isomorphism among Oxides of Quadrivalent Metals. The Systems CeO_2–ThO_2, CeO_2–ZrO_2 and CeO_2–HfO_2," *Gazz. Chim. Ital.*, **60**, 762.

Pauling, L., and Hoard, J. L., "The Crystal Structure of Cadmium Chloride," *Z. Krist.*, **74**, 546.

Rumpf, E., "The Mixed Crystal Series CaF_2–SrF_2," *Z. Physik. Chem.*, **7B**, 148.

Stackelberg, M. v., "Investigations on Carbides I. The Crystal Structure of Carbides MC_2," *Z. Physik. Chem.*, **9B**, 437; *Naturwiss.*, **18**, 305.

Vegard, L., "Structure of Hydrogen Sulfide, Hydrogen Selenide and Nitrogen Peroxide at Liquid Air Temperature," *Nature*, **126**, 916; *Naturwiss.*, **18**, 1098.

Wever, F., and Müller, A., "The Binary System Iron–Boron and the Structure of the Iron Boride Fe_4B_2," *Z. Anorg. Chem.*, **192**, 317.

1931

Alsén, N., "Crystal Structure of Covellite (CuS) and Copper Glance (Cu_2S)," *Geol. For. Forh.*, **53**, 111.

Buerger, M. J., "The Crystal Structure of Marcasite," *Am. Mineralogist*, **16**, 361.

Ebert, F., "The Crystal Structure of Some Fluorides of the Eighth Group of the Periodic System," *Z. Anorg. Chem.*, **196**, 395.

Eisenhut, O., and Kaupp, E., "On the Existence of Three Compounds $AuAl_2$, AuAl and Au_2Al," *Z. Elektrochem.*, **37**, 466.

Ferrari, A., and Scherillo, A., "The Crystal Structure of Manganite," *Z. Krist.*, **78**, 496.

Garrido, J., "Structural Relations between Argentite and Acanthite," *Anales Soc. Espan. Fis. Quim.*, **29**, 505.

Goldsztaub, S., "Dehydration of Natural Ferric Hydrates," *Compt. Rend.*, **193**, 533.

Hägg, G., "Correction of the Article 'Crystal Structure of the Compound Fe_2B,'" *Z. Physik. Chem.*, **12B**, 413.

Hendricks, S. B., "The Crystal Structure of N_2O_4," *Z. Physik*, **70**, 699.

Miles, F. D., "The Apparent Hemihedrism of Crystals of Lead Chloride and Some Other Salts," *Proc. Roy. Soc.* (*London*), **132A**, 266.

Natta, G., "Structure of Hydrogen Sulfide and Hydrogen Selenide," *Nature*, **127**, 129.

Neuburger, M. C., "Precision Measurements of the Lattice Constant of Cuprous Oxide," *Z. Physik*, **67**, 845.

Neuburger, M. C., "The Lattice Constant of Cuprous Oxide," *Z Krist.*, **77**, 169.

Nial, O., Almin, A., and Westgren, A., "X-Ray Analysis of the Systems: Gold–Antimony and Silver–Tin," *Z. Physik. Chem.*, **14B**, 81.

Palacios, J., and Salvia, R., "Crystalline Structure of Argentite and Acanthite," *Anales Soc. Espan. Fis. Quim.*, **29**, 269, 514.

Stackelberg, M. v., "The Crystal Structures of Several Carbides and Borides," *Z. Elektrochem.*, **37**, 542.

Vegard, L., "The Structure of Solid N_2O_4 at the Temperature of Liquid Air," *Z. Physik*, **68**, 184.

Vegard, L., "The Crystal Structure of N_2O_4," *Z. Physik*, **71**, 299.

Vegard, L., "The Structure of Solid Hydrogen Sulfide and Hydrogen Selenide at the Temperature of Liquid Air," *Z. Krist.*, **77**, 23.

Vegard, L., "The Structure of Solid COS at the Temperature of Liquid Air," *Z. Krist.*, **77**, 411.

Vegard, L., "Mixed-Crystal Formation in Molecular Lattices by Irregular Exchange of Molecules," *Naturwiss.*, **19**, 443.

Vegard, L., "Mixed-Crystal Formation in Molecular Lattices by Means of Exchange of Molecules," *Z. Physik*, **71**, 465.

Verweel, H. J., and Bijvoet, J. M., "The Crystal Structure of Mercury Bromide," *Z. Krist.*, **77**, 122.

1932

Arnfelt, H., "On the Formation of Layer-Lattices," *Arkiv Mat. Astron. Fysik*, **23B**, No. 2.

Bannister, F. A., "Determination of Minerals in Platinum Concentrates from the Transvaal by X-Ray Methods," *Mineral. Mag.*, **23**, 188.

Barth, T. F. W., "The Cristobalite Structures I. High-Cristobalite," *Am. J. Sci.*, **23**, 350.

Barth, T. F. W., "The Cristobalite Structures II. Low-Cristobalite," *Am. J. Sci.*, **24**, 97.

Bottema, J. A., and Jaeger, F. M., "On the Law of Additive Atomic Heats in Intermetallic Compounds IX. The Compounds of Tin and Gold, and of Gold and Antimony," *Proc. Acad. Sci. Amsterdam*, **35**, 916.

Braekken, H., "The Crystal Structure of Mercuric Bromide," *Z. Krist.*, **81**, 152.

Braekken, H., "The Crystal Structure of Lead Chloride," *Z. Krist.*, **83**, 222.

Brandenberger, E., "The Crystal Structure of Beryllium Fluoride," *Schweiz. Mineral.-Petrog. Mitt.*, **12**, 243.

Buerger, M. J., "The Crystal Structure of Löllingite," *Z. Krist.*, **82**, 165.

Darbyshire, J. A., "An X-Ray Examination of the Oxides of Lead," *J. Chem. Soc.*, **1932**, 211.

Deflandre, M., "Crystal Structure of Diaspore," *Bull. Soc. Franc. Min.*, **55**, 140.

Feitknecht, W., "The Structure of α-Zinc Hydroxide," *Z. Krist.*, **84**, 173.

Goldschmidt, V. M., "The Rutile Modification of Germanium Dioxide," *Z. Physik. Chem.*, **17B**, 172.

Goldsztaub, S., "Crystalline Structure of Goethite," *Compt. Rend.*, **195**, 964.

Ketelaar, J. A. A., "The Crystal Structure of PbF_2," *Z. Krist.*, **84**, 62.

Linde, J. O., "The Lattice Constants of Copper–Palladium Mixed Crystals," *Ann. Physik*, **15**, 249.

Nieuwenkamp, W., and Bijvoet, J. M., "The Crystal Structure of Lead Fluochloride," *Z. Krist.*, **81**, 469.

Nieuwenkamp, W., and Bijvoet, J. M., "The Crystal Structure of Lead Fluobromide," *Z. Krist.*, **82**, 157.

Nieuwenkamp, W., and Bijvoet, J. M., "The Crystal Structure of Lead Bromide," *Z. Krist.*, **84**, 49.

Onorato, E., "Symmetry and Structure of Hauerite," *Periodico Mineral.*, **1**, 109 (1930); *Neues Jahrb. Mineral. Geol. Referate I*, **1932**, 26.

Parker, H. M., and Whitehouse, W. J., "X-Ray Analysis of Iron Pyrites by the Method of Fourier Series," *Phil. Mag.*, **14**, 939.

Schnaase, H., "The Crystal Structure of Red Manganese Sulfide," *Naturwiss.*, **20**, 640.

Tokody, L., "The Structure of Hessite," *Z. Krist.*, **82**, 154.

Tutiya, H., "X-Ray Observation of Molybdenum Carbides Formed at Low Temperatures," *Bull. Inst. Phys. Chem. Res. (Tokyo)*, **11**, 1150.

Wrigge, F. W., and Meisel, K., "Molecular and Atomic Volumes XXXVI. The Density of Cuprous Oxide," *Z. Anorg. Chem.*, **203**, 312.

1933

Bernal, J. D., and Fowler, R. H., "The Theory of Water," *J. Chem. Phys.*, **1**, 515.

Bijvoet, J. M., and Nieuwenkamp, W., "The 'Variable Structure' of Cadmium Bromide," *Z. Krist.*, **86**, 466.

Bjurström, T., "X-Ray Analysis of the Iron–Boron, Cobalt–Boron and Nickel–Boron Systems," *Arkiv Kemi Mineral. Geol.*, **11A**, No. 5.

Blum, H., "Crystal Structure of Water-Free Magnesium Iodide and Calcium Iodide," *Z. Physik. Chem.*, **22B**, 298.

Bradley, A. J., and Jay, A. H., "Quartz as a Standard for Accurate Lattice Spacing Measurements," *Proc. Phys. Soc. (London)*, **45**, 507.

Cairns, R. W., and Ott, E., "X-Ray Studies of the System Nickel–Oxygen–Water I. Nickelous Oxide and Hydroxide," *J. Am. Chem. Soc.*, **55**, 527.

Corey, R. B., and Wyckoff, R. W. G., "The Crystal Structure of Zinc Hydroxide," *Z. Krist.*, **86**, 8.

Ebert, F., and Woitinek, H., "Crystalline Structure of Fluorides II. HgF, HgF_2, CuF and CuF_2," *Z. Anorg. Chem.*, **210**, 269.

Hassel, O., "The Crystal Structure of Cadmium Iodide," *Z. Physik. Chem.*, **22B**, 333.

Jaeger, F. M., and Bottema, J. A., "The Exact Determination of Specific Heats at Elevated Temperatures IV. Law of Neumann-Joule-Kopp-Regnault Concerning the Additive Property of Atomic Heats of Elements in Their Chemical Combinations," *Rec. Trav. Chim.*, **52**, 89.

Jay, A. H., "The Thermal Expansion of Quartz by X-Ray Measurements," *Proc. Roy. Soc. (London)*, **142A**, 237.

Kinsey, E. L., and Sponsler, O. L., "The Molecular Structure of Ice and Liquid Water," *Proc. Phys. Soc. (London)*, **45**, 768.

Kunzl, V., and Koppel, J., "The Constant of the Crystalline Grating of the Rhombohedral Face of Quartz," *Compt. Rend.*, **196**, 787.

Nieuwenkamp, W., "The Chemical Composition of Matlockite," *Z. Krist.*, **86**, 470.

Schnaase, H., "Crystal Structure of Manganous Sulfide and its Mixed Crystals with Zinc Sulfide and Cadmium Sulfide," *Z. Physik. Chem.*, **20B**, 89.

Schumann, H., "The Dimorphism of Lead Fluoride," *Centr. Mineral. Geol.*, **1933A**, 122.

Takané, K., "Crystal Structure of Diaspore," *Proc. Imp. Akad. Tokyo*, **9**, 113.

Tilley, C. E., "Portlandite, $Ca(OH)_2$," *Mineral. Mag.*, **23**, 419.

Zintl, E., and Kaiser, H., "Metals and Alloys VI. Ability of Elements to Form Negative Ions," *Z. Anorg. Chem.*, **211**, 113.

1934

Bannister, F. A., "The Crystal Structure and Optical Properties of Matlockite," *Mineral. Mag.*, **23**, 587.

Bernal, J. D., "Discussion on Heavy Hydrogen," *Proc. Roy. Soc.* (*London*), **144A**, 24.

Braekken, H., and Scholten, W., "The Crystal Structure of Mercuric Chloride," *Z. Krist.*, **89**, 448.

Burgers, W. G., and Basart, J. C. M., "Formation of High-Melting Metallic Carbides by Igniting a Carbon Filament in the Vapor of a Volatile Halogen Compound of the Metal," *Z. Anorg. Chem.*, **216**, 209.

Goldsztaub, S., "Crystal Structure of Ferric Oxychloride," *Compt. Rend.*, **198**, 667.

Gorsky, W. S., "The Crystal Structure of Yellow HgI_2," *Physik. Z. Sowjetunion*, **5**, 367.

Grime, H., and Santos, J. A., "The Structure and Color of Anhydrous Cobalt Chloride at Room and Very Low Temperatures," *Z. Krist.*, **88**, 136.

Hoard, J. L., and Grenko, J. D., "The Crystal Structure of Cadmium Hydroxychloride," *Z. Krist.*, **87**, 110.

Jay, A. H., "The Thermal Expansion of Silver, Quartz and Bismuth by X-Ray Measurements," *Z. Krist.*, **89**, 282.

Johnson, W. C., and Wheatley, A. C., "The Sulfides of Germanium," *Z. Anorg. Chem.* **216**, 273.

Keesom, W. H., and Köhler, J. W. L., "New Determination of the Lattice Constant of Carbon Dioxide," *Physica*, **1**, 167.

Keesom, W. H., and Köhler, J. W. L., "The Lattice Constant and Expansion Coefficient of Solid Carbon Dioxide," *Physica*, **1**, 655.

Ketelaar, J. A. A., "The Crystal Structure of Nickel Bromide and Iodide," *Z. Krist.*, **88**, 26.

Megaw, H. D., "On Ordinary and Heavy Ice," *Nature*, **134**, 900.

Meisel, K., "The Crystal Structure of FeP_2," *Z. Anorg. Chem.*, **218**, 360.

Meisel, K., "The Lattice Constant of OsS_2," *Z. Anorg. Chem.*, **219**, 141.

Offner, F., "A Redetermination of the Parameter for Hauerite," *Z. Krist.*, **89**, 182.

Pauling, L., and Huggins, M. L., "Covalent Radii of Atoms and Interatomic Distances in Crystals Containing Electron-Pair Bonds," *Z. Krist.*, **87**, 205.

Stackelberg, M. v., and Quatram, F., "The Structure of Beryllium Carbide," *Z. Physik. Chem.*, **27B**, 50.

West, C. D., "The Crystal Structures of Some Alkali Hydrosulfides and Monosulfides," *Z. Krist.*, **88**, 97.

West, C. D., and Peterson, A. W., "The Crystal Structure of $AuAl_2$," *Z. Krist.*, **88A**, 93.

Zintl, E., Harder, A., and Dauth, B., "Lattice Structure of the Oxides, Sulfides, Selenides and Tellurides of Lithium, Sodium and Potassium," *Z. Elektrochem.*, **40**, 588.

1935

Bannister, F. A., and Hey, M. H., "The Crystal Structure of BiOCl, BiOBr and BiOI," *Nature*, **134**, 855 (1934); *Mineral. Mag.*, **24**, 49.

Bernal, J. D., Diatlowa, E., Kasarnowsky, J., Reichstein, S., and Ward, A. G., "The Crystal Structure of BaO_2 and SrO_2," *Z. Krist.*, **92A**, 344.

Bever, A. K. van, and Nieuwenkamp, W., "The Crystal Structure of $CaCl_2$," *Z. Krist.*, **90A**, 374.

Bunn, C. W., Clark, L. M., and Clifford, J. L., "The Constitution of Bleaching Powder," *Proc. Roy. Soc. (London)*, **151A**, 141.

Burton, E. F., and Oliver, W. F., "The Crystal Structure of Ice at Low Temperatures," *Nature*, **135**, 505; *Proc. Roy. Soc. (London)*, **153A**, 166.

Büssem, W., Bluth, M., and Grochtmann, G., "The Structure of Cristobalite at Various Temperatures," *Ber. Deut. Keram. Ges.*, **16**, 381.

Büssem, W., Fischer, H., and Gruner, E., "The Structure of SiS_2," *Naturwiss.*, **23**, 740.

Elander, M., Hägg, G., and Westgren, A., "The Crystal Structure of Cu_2Sb and Fe_2As," *Arkiv Kemi Mineral. Geol.*, **12B**, No. 1.

Ewing, F. J., "The Crystal Structure of Lepidochrosite, γ-FeO(OH)," *J. Chem. Phys.*, **3**, 420.

Feitknecht, W., and Fischer, G., "On Basic Cobalt Chlorides," *Helv. Chim. Acta*, **18**, 555.

Feitknecht, W., and Lotmar, W., "The Structure of Green Basic Cobalt Bromide," *Z. Krist.*, **91A**, 136.

Garrido, J., "The Structure of Manganite," *Bull. Soc. Franc. Mineral.*, **58**, 224.

Goldsztaub, S., "The Structure of Goethite," *Bull. Soc. Franc. Mineral.*, **58**, 6.

Hofmann, W., and Jänicke, W., "The Structure of AlB_2," *Naturwiss.*, **23**, 851.

Jurriaanse, T., "The Crystal Structure of Au_2Bi," *Z. Krist.*, **90A**, 322.

Tunell, G., and Ksanda, C. J., "The Crystal Structure of Calaverite, $AuTe_2$," *J. Wash. Acad. Sci.*, **25**, 32.

Kurz, W., "X-Ray Study of Blue Copper Sulfide," *Z. Krist.*, **92A**, 408.

Megaw, H. D., "The Crystal Structure of Zinc Hydroxide," *Z. Krist.*, **90A**, 283.

Nieuwenkamp, W., "The Crystal Structure of Low Cristobalite," *Z. Krist.*, **92A**, 82.

Schoep, A., and Billiet, V., "The Structure of Uraninite," *Bull. Soc. Geol. Belg.*, **58**, 198.

Stackelberg, M. v., and Paulus, R., "Phosphides and Arsenides of Zinc and Cadmium," *Z. Physik. Chem.*, **28B**, 427.

Wei, P. H., "The Structure of Alpha-Quartz," *Z. Krist.*, **92A**, 355.

Weiser, H. B., and Milligan, W. O., "X-Ray Studies on the Hydrous Oxides. V. Beta-Ferric Oxide Monohydrate," *J. Am. Chem. Soc.*, **57**, 238.

Zintl, E., and Harder, A., "The Constitution of Alkaline Earth Hydrides," *Z. Elektrochem.*, **41**, 34.

Zintl, E., and Loosen, K., "The Structure of SiS_2," *Z. Physik. Chem.*, **174A**, 301.

1936

Aiken, J. K., Haley, J. B., and Terrey, H., "On the Structure of $InCl_2$ and $SnCl_2$," *Trans. Faraday Soc.*, **32**, 1617.

Biltz, W., and Wiechmann, F., "The System Mn–S; Synthesis and Decomposition of Hauerite," *Z. Anorg. Allgem. Chem.*, **228**, 268.

Buerger, M. J., "The Crystal Structure of Arsenopyrite," *Z. Krist.*, **95A**, 83.

Buerger, M. J., "The Crystal Structure of the Arsenopyrite Group," *Am. Mineralogist*, **21**, 203.

Buerger, M. J., "The Crystal Structure of Manganite, MnO(OH)," *Z. Krist.*, **95A**, 163.

Feitknecht, W., "Basic Cobalt Halides," *Helv. Chim. Acta*, **19**, 467; *Z. Angew. Chem.*, **49**, 24.

Feitknecht, W., and Collet, A., "Basic Nickel Halides," *Helv. Chim. Acta*, **19**, 831.

Ferrari, A., and Curti, R., "Oxyhalides of Nickel Obtained by the Synthesis of Senarmontite," *Gazz. Chim. Ital.*, **66**, 104.

Goldsztaub, S., "Some Observations on Boehmite," *Bull. Soc. Franc. Mineral.*, **59**, 348.

Halla, F., and Nowotny, H., "X-Ray Study of the System Mn–Sb," *Z. Physik. Chem.*, **34B**, 141.

Hartmann, H., and Orban, J., "Electrolysis of Phosphate Melts, II. A New Tungsten Phosphide," *Z. Anorg. Allgem. Chem.*, **226**, 257.

Hofmann, W., and Jänicke, W., "The Structure of AlB_2," *Z. Physik. Chem.*, **31B**, 214.

Kassatoschkin, W., and Kotow, W., "The Structure of K_2O_4," *J. Chem. Phys.*, **4**, 458.

Lotmar, W., and Feitknecht, W., "Hydroxide Layer Lattices," *Z. Krist.*, **93A**, 368.

McFarlan, R. L., "The Structure of Ice III," *J. Chem. Phys.*, **4**, 253; *Phys. Rev.*, **49**, 644.

McFarlan R. L., "The Structure of Ice II," *J. Chem. Phys.*, **4**, 60; *Phys. Rev.*, **49**, 199.

McKenna, P. M., "Tantalum Carbide and its Relation to Other Hard Refractory Compounds," *J. Ind. Eng. Chem.*, **28**, 767.

May, K., "The Crystal Structure of Rb_2S," *Z. Krist.*, **94A**, 412.

Naray-Szabo, S. v., "The Structure of Baddeleyite, ZrO_2," *Z. Krist.*, **94A**, 414.

Rahlfs, P., "The Cubic High Temperature Forms of Copper and Silver Sulfides," *Z. Physik. Chem.*, **31B**, 157.

Schulze, G. E. R., "The Crystal Structure of RaF_2," *Z. Physik. Chem.*, **32**, 430.

Tunell, G., and Ksanda, C. J., "The Crystal Structure of Krennerite, $AuTe_2$," *J. Wash. Acad. Sci.*, **26**, 507.

Zachariasen, W. H., "The Crystal Structure of GeS_2," *Phys. Rev.*, **49**, 884; *J. Chem. Phys.*, **4**, 618.

1937

Årstad, O., and Nowotny, H., "X-Ray Investigation of the System Mn–P," *Z. Physik. Chem.*, **38B**, 356.

Blattman, S., and Hägele, G., "The Lattice Constant of Quartz and Chalcedony," *Zentr. Mineral. Geol.*, **1937A**, 313.

Buerger, M. J., "Interatomic Distances in Marcasite and Notes on the Bonding in Crystals of Löllingite, Arsenopyrite and Marcasite," *Z. Krist.*, **97A**, 504.

Feitknecht, W., "On the Constitution of Solid Basic Salts of Divalent Metals, III. Basic Cobalt Nitrates," *Helv. Chim. Acta*, **20**, 177.

Feitknecht, W., and Gerber, W., "The Structure of Basic Cadmium Chloride," *Z. Krist.*, **98A**, 168.

Goldsztaub, S., "The Crystal Structure of Laurionite, Pb(OH)Cl," *Compt. Rend.*, **204**, 702.

Haucke, W., "The Constitution of Sodium–Gold Alloys," *Z. Elektrochem.*, **43**, 712.

Kasatochkin, V., and Kotov, V., "X-Ray Investigation of the Structure of KO_2," *J. Tech. Phys. USSR*, **7**, 1468.

Ketelaar, J. A. A., and Willems, P. J. H., "Anomalous Mixed Crystals in the System SrF_2–LaF_3," *Rec. Trav. Chim.*, **56**, 29.

McCullough, J. D., "The Crystal Structure of SeO_2," *J. Am. Chem. Soc.*, **59**, 789.

Nieuwenkamp, W., "On the Structure of High Cristobalite," *Z. Krist.*, **96A**, 454.

Zintl, E., Harder, A., and Haucke, W., "Alloy Phases with the Fluorite Structure," *Z. Physik. Chem.*, **35B**, 354.

1938

Beck, G., and Nowacki, W., "Preparation and Crystal Structure of EuS and EuF_2," *Naturwiss.*, **26,** 495.

Biltz, W., and Köcher, A., "The System Tantalum-Sulfur," *Z. Anorg. Allgem. Chem.*, **238,** 81.

Buerger, M. J., "The Crystal of Gudmundite, FeSbS," *Am. Mineralogist*, **23,** No. 12, Pt. 2, p. 4.

Feitknecht, W., "The Alpha-Form of Bivalent Metal Hydroxides," *Helv. Chim. Acta*, **21,** 766.

Helms, A., "The Structure of Alkali Tetroxides," *Z. Angew. Chem.*, **51,** 498.

Kratky, O., and Nowotny, H., "The Crystal Structure of Beta-FeO(OH)," *Z. Krist.*, **100A,** 356.

Lundqvist, D., and Westgren, A., "X-Ray Study of the System Cobalt–Sulfur," *Z. Anorg. Allgem. Chem.*, **239,** 85.

Nowotny, H., and Årstad, O., "X-Ray Investigation of the System Cr–CrAs," *Z. Physik. Chem.*, **38B,** 461.

Nowotny, H., and Henglein, E., "X-Ray Investigation of the System Ni–P," *Z. Physik. Chem.*, **40B,** 281.

Powell, H. M., and Brewer, F. M., "The Structure of GeI_2," *J. Chem. Soc.*, **1938,** 197.

Tengner, S., "Diselenides and Ditellurides of Iron, Cobalt and Nickel," *Z. Anorg. Allgem. Chem.*, **239,** 126.

Wells, A. F., "The Crystal Structure of $PdCl_2$," *Z. Krist.*, **100A,** 189.

1939

Brill, R., Hermann, C., and Peters, C., "Studies of Chemical Bonds by Means of Fourier Analysis, III. The Bond in Quartz," *Naturwiss.*, **27,** 676.

Buerger, M. J., "The Crystal Structure of Gudmundite, FeSbS, and its Bearing on the Existence Field of the Arsenopyrite Structural Type," *Z. Krist.*, **101A,** 290.

Döll, W., and Klemm, W., "Measurements on Di- and Quadrivalent Compounds of the Rare Earths, VII. The Structures of Several Dihalides," *Z. Anorg. Allgem. Chem.*, **241,** 239.

Feitknecht, W., "On the Chemistry and Morphology of Basic Salts of Divalent Metals, VII. Basic Nickel Chlorides," *Helv. Chim. Acta*, **22,** 1428.

Feitknecht, W., "On the Chemistry and Morphology of Basic Salts of Divalent Metals, VIII. Basic Nickel Bromides," *Helv. Chim. Acta*, **22,** 1444.

Goldsztaub, S., "The Positions of the Atoms in Laurionite, Pb(OH)Cl," *Compt. Rend.*, **208,** 1234.

Helms, A., and Klemm, W., "The Crystal Structure of Rb_2O and Cs_2O," *Z. Anorg. Allgem. Chem.*, **242,** 33.

Helms, A., and Klemm, W., "The Structure of the So-Called Alkali Tetroxides," *Z. Anorg. Allgem. Chem.*, **241,** 97.

Ito, T., and Sawada, H., "The Crystal Structure of Tellurite, TeO_2," *Z. Krist.*, **102A,** 13.

Kamermans, M. A., "The Crystal Structure of $SrBr_2$," *Z. Krist.*, **101A,** 406.

Ketelaar, J. A. A., and Gorter, E. W., "The Crystal Structure of Tl_2S," *Z. Krist.*, **101A,** 367.

Koern, V., "The Binary Alloy System Silver–Tellurium," *Naturwiss.*, **27**, 432.
Peacock, M. A., "Rammelsbergite and Para-Rammelsbergite, Distinct Orthorhombic Forms of $NiAs_2$," *Am. Mineralogist*, **24**, No. 12, Pt 2, p. 10.
Ramsdell, L. S., "The Crystal System and Unit Cell of Acanthite, Ag_2S," *Am. Mineralogist*, **24**, No. 12, Pt. 2, 11.
Weil, R., and Hocart, R., "On Lautite, CuAsS," *Compt. Rend.*, **209**, 444.
Zintl, E., and Croatto, U., "The Fluorite Lattice with Empty Places," *Z. Anorg. Allgem. Chem.*, **242**, 79.
Zintl, E., and Udgard, A., "Mixed Crystal Formation between Fluorides of Different Formulas," *Z. Anorg. Allgem. Chem.*, **240**, 150.

1940

Brasseur, H., "X-Ray Study of Laurionite PbOHCl," *Bull. Soc. Sci. Liege*, **9**, 166.
Brauer, G., "The Problem of the Lower Oxide of Columbium, A Columbium Subnitride," *Z. Elektrochem.*, **46**, 397.
Chatterjee, N., "Structural Investigations of Natural and Synthetic Yttrofluorite," *Z. Krist.*, **102A**, 245.
Feitknecht, W., "Laminary Disperse Hydroxides and Basic Salts of Bivalent Metals," *Kolloid Z.*, **92**, 257; **93**, 66.
Ivanov-Emin, B. N., "On Selenides of Germanium," *J. Gen. Chem., USSR*, **10**, 1813.
Lewis, B., and Elliott, N., "Interatomic Distances in $CoSe_2$," *J. Am. Chem. Soc.*, **62**, 3180.
Mazza, L., Iandelli, A., and Botti, E., "Oxyhalides of Rare Earths and their Diffraction Spectra, Oxyhalides of the Cerium Metals," *Gazz. Chim. Ital.*, **70**, 57.
Zumbusch, M., "The Structure of Uranium Subsulfide and of the Subphosphides of Iridium and Rhodium," *Z. Anorg. Allgem. Chem.*, **243**, 322.

1941

Bespalov, M. M., "Discovery of a New Mineral of the Thorianite Group," *Soviet Geol.*, **1941**, No. 6, 105.
Hoppe, W., "The Structure of Alpha AlO(OH) and of Alpha FeO(OH)," *Z. Krist.*, **103A**, 73.
Klemm, W., and Grimm, L., "Dihalides of Titanium and Vanadium," *Z. Anorg. Allgem. Chem.*, **249**, 198.
Klemm, W., and Klein, H. A., "Lanthanum Oxyfluoride," *Z. Anorg. Allgem. Chem.*, **248**, 167.
Kotov, V., and Raikhshtein, S., "The Structure of CaO_2," *J. Phys. Chem. USSR*, **15**, 1057.
Nowotny, H., and Sibert, W., "Ternary Valence Compounds in Magnesium Systems," *Z. Metallk.*, **33**, 391.
Pinsker, Z. G., "Electron Diffraction Investigation of the Structure of CdI_2," *Acta Physicochim. URSS*, **14**, 503.
Pinsker, Z. G., "Electron Diffraction Investigation of the Structure of CdI_2," *J. Phys. Chem. USSR*, **15**, 559.
Pinsker, Z. G., and Tatarinova, I., "Electron Diffraction Investigation of $CdCl_2$," *Acta Physicochim. URSS*, **14**, 737; *J. Phys. Chem. USSR*, **15**, 1005.
Scholten, W., and Bijvoet, J. M., "The Crystal Structure of $Hg(Cl,Br)_2$," *Z. Krist.*, **103A**, 415.
Sillen, L. G., "X-Ray Studies on BiOCl, BiOBr and BiOI," *Svensk Kem. Tidskr.*, **53**, 39.

Sillen, L. G., and Nylander, A. L., "The Crystal Structure of LaOCl, LaOBr and LaOI," *Svensk Kem. Tidskr.*, **53**, 367.

Smirnova, L., Brager, A., and Zhdanov, H., "X-Ray Examination of CdI_2," *Acta Physicochim. URSS*, **15**, 255.

Zhdanov, G. S., "The Crystal Structure of $Zn(CN)_2$," *Compt. Rend. Acad. Sci. URSS*, **31**, 352.

1942

Feitknecht, W., "Formation of Double Hydroxides between Bivalent and Trivalent Metals," *Helv. Chim. Acta*, **25**, 555.

Feitknecht, W., and Gerber, M., "Double Hydroxides and Basic Double Salts, III. Magnesium–Aluminum Double Hydroxides," *Helv. Chim. Acta*, **25**, 131.

Hägg, G., "MX_2 Layer Structures with Close-Packed X Atoms," *Arkiv Kemi Mineral. Geol.*, **16B**, No. 3.

Hägg, G., Kiessling, R., and Linden, E., "The Crystal Structure of $CdBr_2$ and CdI_2," *Arkiv Kemi Mineral. Geol.*, **B16**, No. 4, 1.

Hägg, G., and Linden, E., "X-Ray Investigation of the System $CdBr_2$–CdI_2," *Arkiv Kemi Mineral. Geol.*, **B16**, No. 5, 1.

Hoppe, W., "The Crystal Structure of Alpha-AlO(OH), II. Fourier Analysis," *Z. Krist.*, **104A**, 11.

Konig, H., "Electron Interference in Ice," *Nachr. Akad. Wiss., Goettingen, Math.-Physik. Kl.*, **1942**, No. 1, 6 pp.

Kronberg, M. L., and Harker, D., "The Crystal Structure of Hydrazinium Difluoride," *J. Chem. Phys.*, **10**, 309.

Neuhaus, A., "The Behavior of As in Compact Iron Pyrites (Melnikovitepyrite, Gelpyrite) from Wiesloch, Baden and Deutsch-Bleischarley (Upper Silesia)," *Metall u. Erz*, **39**, 157, 187.

Nowotny, H., "Intermetallic Compounds of the CaF_2 Type," *Z. Metallk.*, **34**, 237.

Pinsker, Z. G., "Electron Diffraction Determination of the Structure of $CdBr_2$," *Acta Physicochim. URSS*, **16**, 148; *J. Phys. Chem. USSR*, **16**, 1.

Schossberger, F., "On the Inversion of Titanium Dioxide," *Z. Krist.*, **104**, 358.

Straumanis, M., and Sauka, J., "Precise Determination of the Lattice Constants of Rhombic Crystals—$PbCl_2$ as Example," *Z. Physik. Chem.*, **51B**, 219.

Vegard, L., and Hillesund, S., "Structure of a Few D Compounds and Comparison with that of the Corresponding H Compounds," *Avhandl. Norske Videnskaps-Akad. Oslo. I. Mat.-Naturv. Kl.*, **1942**, No. 8, 24 pp.

Vegard, L., and Oserød, L. S., "Isotope Effect for the Exchange of H and D in Solid H_2S and H_2Se," *Avhandl. Norske Videnskaps. Akad. Oslo, I. Mat.-Naturv. Kl.*, **1942**, No. 7.

Yamaguchi, S., "Determining the Crystal Structure of Hygroscopic Substances by Electron Diffraction. $ZnBr_2$," *Sci. Papers Inst. Phys. Chem. Res. (Tokyo)*, **39**, 291.

Yamaguchi, S., "Determining the Crystal Structure of Hygroscopic Substances by Electron Diffraction. ZnI_2," *Sci. Papers Inst. Phys. Chem. Res. (Tokyo)*, **39**, 357.

1943

Bredig, M. A., "The Crystal Structure of Magnesium Carbide," *J. Am. Chem. Soc.*, **65**, 1482.

Croatto, U., "Crystal Structures with Lattice Disturbances. Lanthanum Oxyfluoride," *Gazz. Chim. Ital.*, **73**, 257.

Feitknecht, W., and Bucher, H., "The Chemistry and Morphology of the Basic Salts of Bivalent Metals, XII. The Hydroxyfluorides of Cadmium," *Helv. Chim. Acta*, **26**, 2177; "The Hydroxyfluorides of Zinc," *ibid.*, 2196.

Feitknecht, W., and Weidmann, H., "The Chemistry and Morphology of the Basic Salts of Bivalent Metals, X. The Highly Basic Zinc Hydroxychloride, III," *Helv. Chim. Acta*, **26**, 1560.

Hägg, G., and Hermansson, E., "On the Crystal Structure of Cadmium Iodide," *Arkiv Kemi Mineral. Geol.*, **17B**, No. 10.

Haraldson, H., and Rosenqvist, T., "Magnetic Relations and Binding Relation in Cr Antimonides," *Tidsskr. Kjemi, Bergvesen Met.*, **3**, 81.

Hassel, O., "The Lattice Constants of the Solid Hydrides and Deuterides of S and Se. Remarks on a Paper by L. Vegard and L. Sinding Oserød," *Avhandl. Norske Videnskaps-Akad. Oslo I Mat.-Naturv. Kl.*, **1942**, No. 10, 4 pp.

Klemm, W., and Fartini, N., "The System NiTe–$NiTe_2$," *Z. Anorg. Allgem. Chem.*, **251**, 222.

Laves, F., "The Crystal Structures of $CaGa_2$, $LaGa_2$ and $CeGa_2$," *Naturwiss.*, **31**, 145.

Michel, A., and Bénard, J., "The Preparation, Properties and Formula of the Magnetic Oxide of Chromium," *Bull. Soc. Chim. France*, **10**, 315.

Pinsker, Z. G., Tatarinova, L. I., and Novikova, V. A., "Electronographic Investigation of the Structure of PbI_2," *Acta Physicochim. URSS*, **18**, 378.

Rueggeberg, W. H. C., "The Carbides of Magnesium," *J. Am. Chem. Soc.*, **65**, 602.

Schwarz, R., and Haschke, E., "The Chemistry of Germanium, XIX. The Polymorphism of GeO_2," *Z. Anorg. Allgem. Chem.*, **252**, 170.

Vegard, L., "The Isotope Effect in the Replacement of H by D in Solid H_2S and H_2Se," *Avhandl. Norske Videnskaps-Akad. Oslo I. Mat.-Naturv. Kl.*, **1943**, 3.

Wallbaum, H. J., "A_2B Compounds of the Al_2Cu Type," *Z. Metallk.*, **35**, 218.

Wallbaum, H. J., "The Crystal Structures of Bi_2Pt and Sn_2Pt," *Z. Metallk.*, **35**, 200.

1944

Buerger, M. J., and Buerger, N. W., "Low-Chalcocite and High-Chalcocite," *Am Mineralogist*, **29**, 55.

Feitknecht, W., and Held, F., "Hydroxy Salts of Bivalent Metals. XXV. Hydroxy Chlorides of Magnesium," *Helv. Chim. Acta*, **27**, 1480.

Hägg, G., and Magneli, A., "X-Ray Studies on Molybdenum and Tungsten Oxides," *Arkiv Kemi Mineral. Geol.*, **19A**, 14 pp.

Jay, A. H., "Roentgen-Ray Pattern of Low-Temperature Cristobalite," *Mineral. Mag.*, **27**, 54.

König, H., "A Cubic Ice Modification," *Z. Krist.*, **105**, 279.

Novoselova, A. V., Levina, M. E., Simanov, Y. P., and Zhasmin, A. G., "Thermal Analysis of the System NaF–BeF_2," *J. Gen. Chem. (USSR)*, **14**, 385.

Oftedal, I., "Unit Cell and Space Group of Chalcocite," *Norsk Geol. Tidsskr.*, **24**, 114.

Zhdanov, G. S., and Shugam, E. A., "The Crystal Structure of Mercuric Cyanide," *Dokl. Akad. Nauk USSR*, **45**, 312; *Compt. Rend. Acad. Sci. URSS*, **45**, 295.

1945

Borchert, W., "Lattice Transformations in the System $Cu_{2-x}Se$," *Z. Krist.*, **106**, 5.

Byström, A., "The Decomposition Products of Lead Peroxide and Oxidation Products of Lead Oxide," *Arkiv Kemi Mineral. Geol.*, **A20**, No. 11, 31 pp.

Feitknecht, W., "Structure of Cadmium Hydroxy Halides $CdCl_{0.67}(OH)_{1.33}$, $CdBr_{0.6}$-$(OH)_{1.4}$, $CdI_{0.5}(OH)_{1.5}$," *Experientia*, **1**, 230.

Garrido, J., "The Existence of Forbidden Spectra on the X-Ray Diagrams of Diaspore," *Compt. Rend.*, **221**, 148.

Juza, R., and Sachsze, W., "Metal Amides and Metal Nitrides. XIII. The System Cobalt–Nitrogen," *Z. Anorg. Chem.*, **253**, 95.

Kerr, P. F., "Cattierite and Vaesite: New Co–Ni Minerals from the Belgian Congo," *Am. Mineralogist*, **30**, 483.

Kerr, P. F., Holmes, R. J., and Knox, M. S., "Lattice Constants in the Pyrite Group," *Am. Mineralogist*, **30**, 498.

Shugam, E. A., and Zhdanov, G. S., "The Crystal Structure of Cadmium Cyanide," *Acta Physicochim. URSS*, **20**, 247.

Zhdanov, G. S., and Shugam, E. A., "Crystal Structure of Cyanides. Determination of the Position of the Cyanide Group in Mercuric Cyanide," *J. Phys. Chem., USSR*, **19**, 433.

1946

Belov, N. V., and Butuzov, V. P., "The Structure of High Chalcocite, Cu_2S," *Compt. Rend. Acad. Sci. URSS*, **54**, 717.

Butuzov, V. P., "The Structure of High Chalcocite, Cu_2S," *Compt. Rend. Acad. Sci. URSS*, **54**, 717.

Croatto, U., and Bruno, M., "Crystalline Structures with Lattice Disturbances. III. Strontium Chloride," *Gazz. Chim. Ital.*, **76**, 246.

Epprecht, W., "The Manganese Minerals of Gonzen and Their Paragenesis," *Schweiz. Mineral. Petrog. Mitt.*, **26**, 19.

Juza, R., and Hund, F., "The Crystal Structures of LiMgN, LiZnN, Li_3AlN_2, and Li_3GaN_2," *Naturwiss.*, **33**, 121.

Kaiman, S., "Crystal Structure of Rammelsbergite, $NiAs_2$," *Univ. Toronto Studies, Geol. Ser.*, **51**, 49.

Magnéli, A., "The Crystal Structure of the Dioxides of Molybdenum and Tungsten," *Arkiv Kemi Mineral. Geol.*, **24A**, No. 2, 11 pp.

Nowotny, H., "The Crystal Structure of CuZnAs," *Metallforsch.*, **1**, 38.

Nowotny, H., "The Crystal Structure of Cu_2Te," *Metallforsch.*, **1**, 40.

Peacock, M. A., and Thompson, R. M., "Melonite from Quebec and the Crystal Structure of $NiTe_2$," *Univ. Toronto Studies, Geol. Ser.*, **50**, 63.

Pinsker, Z. G., Tatarinova, L. I., and Novikova, V. A., "Electronographic Determination of the Structure of Zinc Iodide," *J. Phys. Chem.* (*USSR*), **20**, 1401.

Reichertz, P. P., and Yost, W. J., "The Crystal Structure of Synthetic Boehmite," *J. Chem. Phys.*, **14**, 495.

Thompson, R. M., "Frohbergite, $FeTe_2$—New Member of the Marcasite Group," *Univ. Toronto Studies, Geol. Ser.*, **51**, 35.

1947

Byström, A., "The Structure of the Fluorides and Oxyfluorides of Bivalent Lead," *Arkiv Kemi Mineral. Geol.*, **24A**, No. 33, 18 pp.

Cole, W. F., Wadsley, A. D., and Walkley, A., "An X-Ray Diffraction Study of Manganese Dioxide," *Trans. Elektrochem. Soc.*, **92**, 22 pp.

Donohue, J., and Lipscomb, W. N., "The Crystal Structure of Hydrazinium Dichloride, $N_2H_6Cl_2$," *J. Chem. Phys.*, **15**, 115.

Gruner, J. W., "Groutite, $HMnO_2$, a New Mineral of the Diaspore–Goethite Group," *Am. Mineralogist*, **32**, 654.

Helmholz, L., "The Crystal Structure of Anhydrous Cupric Bromide," *J. Am. Chem. Soc.*, **69**, 886.

Jerslev, B., "Crystal Structure of Hydroxylammonium Chloride and Bromide," *Nature*, **160**, 641.

Kiessling, R., "Crystal Structures of Molybdenum and Tungsten Borides," *Acta Chem. Scand.*, **1**, 893.

Lundqvist, D., "X-Ray Studies on the Binary System Ni–S," *Arkiv Kemi Mineral. Geol.*, **24A**, No. 21, 12 pp.

Nial, O., "X-Ray Studies on Binary Alloys of Tin with Transition Metals," *Svensk Kem. Tidskr.*, **59**, 165, 172, 177.

Nowotny, H., "The Crystal Structure of Co_2P," *Z. Anorg. Allgem. Chem.*, **254**, 31.

Nowotny, H., and Kieffer, R., "X-Ray Investigation of Carbide Systems," *Metallforschung*, **2**, 257.

Nowotny, H., and Scheil, E., "A Ternary Compound in the System Aluminum–Silicon–Sodium," *Metallforschung*, **2**, 76.

Vegard, L., "Investigation into the Structure and Properties of Solid Matter with the Help of X-Rays," *Skrifter Norske Videnskaps-Akad. Oslo. I., Mat.-Naturv. Kl.*, **1947**, No. 2, 83 pp.

Wells, A. F., "The Crystal Structure of Anhydrous Cupric Chloride and the Stereochemistry of the Cupric Atom," *J. Chem. Soc.*, **1947**, 1670.

1948

Aurivillius, B., "X-Ray Investigation of Bismuth Oxyfluorides," *Arkiv Kemi Mineral. Geol.*, **26B**, No. 2.

Baenziger, N. C., and Rundle, R. E., "The Structure of Titanium Chloride," *Acta Cryst.*, **1**, 274.

Balconi, M., "Structure of Zinc Iodide," *Rend. Soc. Mineral. Ital.*, **5**, 49.

Brogren, G., "The Lattice Constants of Some Reflecting Planes in Quartz," *Arkiv Mat. Astron. Fysik*, **36B**, No. 3, 3 pp.

Brogren, G., and Friskopp, K. G., "Determination of the Lattice-Constant of Quartz $10\bar{1}1$," *Arkiv Mat. Astron. Fysik*, **36B**, No. 4, 5 pp.

Glemser, O., Sauer, H., and Konig, P., "The Sulfides and Selenides of Tungsten," *Z. Anorg. Allgem. Chem.*, **257**, 241.

Jerslev, B., "The Structure of Hydroxyl Ammonium Chloride, NH_3OHCl, and of Hydroxyl Ammonium Bromide, NH_3OHBr," *Acta Cryst.*, **1**, 21.

Litz, L. M., Garrett, A. B., and Croxton, F. C., "Preparation and Structure of the Carbides of Uranium," *J. Am. Chem. Soc.*, **70**, 1718.

McCullough, J. D., Brewer, L., and Bromley, L. A., "The Crystal Structure of Zirconium Oxysulfide," *Acta Cryst.*, **1**, 287.

Peacock, M. A., and Henry, W. G., "The Crystal Structures of Cobaltite, Gersdorffite and Ullmannite (NiSbS)," *Univ. Toronto Studies, Geol. Ser.*, **52**: *Contrib. Can. Mineral.*, **1947**, 71.

Rundle, R. E., Baenziger, N. C., Wilson, A. S., and McDonald, R. A., "The Structures of Carbides, Nitrides and Oxides of Uranium," *J. Am. Chem. Soc.*, **70**, 99.

Stehlik, B., and Balak, L., "Crystal Structure of Tellurium Dioxide," *Chem. Zvesti*, **2**, 6, 33, 69.

Vainshtein, B. K., "X-Ray Determination of the Structure of Barium Chloride," *Dokl. Akad. Nauk USSR*, **60**, 1169.

Zachariasen, W. H., "Crystal Chemical Studies of the 5*f*-Series of the Elements. III. A Study of Disorder in the Crystal Structure of the Anhydrous Uranyl Fluoride," *Acta Cryst.*, **1**, 277.

1949

Anderson, J. S., and D'Eye, R. W. M., "The Lower Valency States of Thorium," *J. Chem. Soc.*, **1949**, S 244.

Belov, N. V., and Mokeeva, V. I., "Application of the Methods of Harmonic Analysis for Determination of Crystal Structure Parameters from the Standard Powder Röntgenograms," *Tr. Inst. Kristallogr., Akad. Nauk SSSR*, No. 5, 13.

Brenet, J., and Héraud, A., "A Structure Peculiar to Certain Dioxides of Manganese," *Compt. Rend.*, **228**, 1487.

Broadley, J. S., and Robertson, J. M., "Structure of Dinitrogen, Tetroxide," *Nature*, **164**, 915.

Byström, A. M., "Crystal Structure of Ramsdellite, an Orthorhombic Modification of Manganese Dioxide," *Acta Chem. Scand.*, **3**, 163.

Byström, A., and Byström, A. M., "Crystal Structure of Hollandite," *Nature*, **164**, 1128.

Collin, R. L., and Lipscomb, W. N., "The Crystal Structure of Groutite," *Acta Cryst.*, **2**, 104.

Ehrlich, P., "Titanium Selenides and Tellurides," *Z. Anorg. Allgem. Chem.*, **260**, 1.

Forman, S. A., and Peacock, M. A., "Crystal Structure of Rickardite, $Cu_{4-x}Te_2$," *Am. Mineralogist*, **34**, 441.

Fricke, R., and Dürrwächter, W., "Further Investigations on the Crystalline Hydroxides of the Rare Earths," *Z. Anorg. Allgem. Chem.*, **259**, 305.

Kiessling, R., "The Binary System Chromium–Boron. I. Phase Analysis and Structure of the ζ and θ Phases," *Acta Chem. Scand.*, **3**, 595.

Kiessling, R., "The Borides of Tantalum," *Acta Chem. Scand.*, **3**, 603.

Morozov, I. S., and Kuznetsov, V. G., "The γ Modification of Manganese Dioxide," *Izvest. Akad. Nauk USSR, Otdel. Khim. Nauk*, **1949**, 343.

Okada, T., "The Crystal Structure of Cuprous Oxide," *Proc. Phys. Soc. Japan*, **4**, 140.

Rundle, R. E., and Wilson, A. S., "The Structures of Some Metal Compounds of Uranium," *Acta Cryst.*, **2**, 148.

Stehlik, B., and Balak, L., "The Crystal Structure of Tellurium Dioxide," *Collection Czech. Chim. Commun.*, **14**, 595.

Vainshtein, B. K., "Electron Diffraction Determination of the Structure of Barium Fluoride," *Tr. Inst. Kristallogr., Akad. Nauk SSSR*, **5**, 113.

Wollan, E. O., Davidson, W. L., and Schull, C. G., "Neutron-Diffraction Study of the Structure of Ice," *Phys. Rev.*, **75**, 1348.

Zachariasen, W. H., "Crystal Chemical Studies of the 5*f*-Series of Elements. VIII. Crystal Structure Studies of Uranium Silicides and of $CeSi_2$, $NpSi_2$, $PuSi_2$," *Acta Cryst.*, **2**, 94.

Zachariasen, W. H., "Crystal Chemical Studies of the 5*f*-Series of Elements. X. Sulfides and Oxysulfides," *Acta Cryst.*, **2**, 291.

Zachariasen, W. H., "Crystal Chemical Studies of the 5*f*-Series of Elements. XII. New Compounds Representing Known Structure Types," *Acta Cryst.*, **2**, 388.

1950

Brenet, J., "Structure of Manganese Dioxides Utilized in Electrochemical Cells," *Bull. Soc. Franc. Mineral.*, **73**, 409.

Brogen, G., and Haeggblom, L. E., "The Lattice Constant of Quartz," *Arkiv Fysik*, **2**, No. 1.

Bruno, M., "Anomalous Solid Solutions of the Sesquioxide in the Dioxide of Cerium," *Ric. Sci.*, **20**, 645.

Byström, A., and Byström, A. M., "The Crystal Structure of Hollandite, the Related Manganese Oxide Minerals and α-MnO_2," *Acta Cryst.*, **3**, 146.

Collin, R. L., and Lipscomb, W. N., "Configuration of the Hydrazine Molecule," *J. Chem. Phys.*, **18**, 566.

Dönges, E., "On the Thiohalogenides of Trivalent Antimony and Bismuth," *Z. Anorg. Allgem. Chem.*, **263**, 112.

Dönges, E., "The Selenohalides of Trivalent Antimony and Bismuth and Antimony Selenide," *Z. Anorg. Allgem. Chem.*, **263**, 280.

Early, J. W., "Description and Synthesis of Selenide Minerals," *Am. Mineralogist*, **35**, 337.

Elson, R., Fried, S., Sellers, P., and Zachariasen, W. H., "Quadrivalent and Quinquevalent States of Protoactinium," *J. Am. Chem. Soc.*, **72**, 5791.

Feitknecht, W., and Keller, G., "The Dark Green Hydroxy Compounds of Iron," *Z. Anorg. Allgem. Chem.*, **262**, 61.

Finkelnburg, W., and Stein, A., "Cerium Oxyfluoride and its Lattice Structure," *J. Chem. Phys.*, **18**, 1296.

Grdenić, D., "The Covalent Bond Length of Mercury Chloride," *Arkiv Kemi*, **22**, 14.

Griffel, M., and Stout, J. W., "Single Crystals of Manganous Fluoride. Crystal Structure," *J. Am. Chem. Soc.*, **72**, 4351.

Keith, H. D., "Precision Lattice-Parameter Measurements," *Proc. Phys. Soc. (London)*, **63B**, 1034.

Mazza, L., and Iandelli, A., "On the Crystal Structure and Reflection Spectra of the Oxyfluorides of Praseodymium, Neodymium and Samarium," *Atti Accad. Ligure Sci. Lettere*, **7**, 44.

Palache, C., "Paralaurionite," *Mineral. Mag.*, **29**, 341.

Schubert, K., and Pfisterer, H., "Crystal Chemistry of Alloys of Elements of the Iron Group with Those of the IVB Group," *Z. Metallk.*, **41**, 433.

Seitz, A., Rösler, U., and Schubert, K., "The Crystal Structure of β-$Be(OH)_2$," *Z. Anorg. Allgem. Chem.*, **261**, 94.

Sugawara, T., and Kanda, E., "The Crystal Structure of Sulfur Dioxide," *Sci. Rept. Res. Inst. Tohoku Univ., Ser. A*, **2**, 216.

Templeton, D. H., and Dauben, C. H., "The Crystal Structure of Sodium Superoxide," *J. Am. Chem. Soc.*, **72**, 2251.

Tunell, G., and Murata, K. J., "The Atomic Arrangement and Chemical Composition of Krennerite," *Am. Mineralogist*, **35**, 959.

Zaslavskii, A. I., Kondrashov, Y. D., and Tolkachev, S. S., "New Modification of Lead Dioxide," *Dokl. Akad. Nauk SSSR*, **75**, 559.

1951

Bertaut, F., and Blum, P., "Structure of a New Phase in the System Mo–B," *Acta Cryst.*, **4**, 72.

Bueno, V., A. M., "Comparative Study of the Oxides of Uranium by X-Ray Diffraction and Selection of Characteristic Lines for the Determination of Their Presence in Ores," *Rev. Cienc.* (*Lima*), **53,** No. 475–6, 13.

Clarke, J., and Jack, K. H., "The Preparation and the Crystal Structures of Cobalt Nitride, Co_2N, of Cobalt Carbonitrides, $Co_2(C,N)$, and of Cobalt Carbide," *Chem. Ind.* (*London*), **1951,** 1004.

Collin, R. L., and Lipscomb, W. N., "The Crystal Structure of Hydrazine," *Acta Cryst.*, **4,** 10.

Dönges, E., "The Tellurohalides of Trivalent Antimony and Bismuth," *Z. Anorg. Allgem. Chem.*, **265,** 56.

Goche, O., "Crystal Structure of Platinum Dioxide," *Bull. Classe Sci. Acad. Roy. Belg.*, **37,** 393.

Gordon, R. B., "Some Measurements on Minerals of the Pyrite Group," *Am. Mineralogist*, **36,** 918.

Haendler, H. M., and Bernard, W. J., "The Reaction of Fluorine with Cadmium and Some of its Binary Compounds. The Crystal Structure, Density and Melting Point of Cadmium Fluoride," *J. Am. Chem. Soc.*, **73,** 5218.

Handy, L. L., and Gregory, N. W., "Crystal Structure of Chromium(II) Chloride," *J. Chem. Phys.*, **19,** 1314.

Hund, F., "Yttrium Oxyfluoride," *Z. Anorg. Allgem. Chem.*, **265,** 62.

Hunt, E. B., and Rundle, R. E., "The Structure of Thorium Dicarbide by X Ray and Neutron Diffraction," *J. Am. Chem. Soc.*, **73,** 4777.

Juza, R., and Opp, K., "Metallic Amides and Metallic Nitrides, XXV. Lithium Imide," *Z. Anorg. Allgem. Chem.*, **266,** 325.

Juza, R., and Puff, H., "Crystal Structure of Cobalt Carbide," *Naturwiss.*, **38,** 331.

Kondrashev, Y. D., and Zaslavskii, A. I., "Structure of the Modifications of Manganese Dioxide," *Izvest. Akad. Nauk SSSR, Ser. Fiz.*, **15,** 179.

Lipscomb, W. N., "The Structure of Mercuric Amido Chloride, $HgNH_2Cl$," *Acta Cryst.*, **4,** 266.

Nowotny, H., Funk, R., and Pesl, J., "Crystal Chemical Studies in the Systems Mn–As, V–Sb, Ti–Sb," *Monatsh.*, **82,** 513.

Nowotny, H., and Glatzl, B., "Crystal Structures of AgZnAs and NaZnAs," *Monatsh.*, **82,** 720.

Rowland, J. F., and Berry, L. G., "The Structural Lattice of Hessite," *Am. Mineralogist*, **36,** 471.

Sauka, Y., "Precision Lattice Constants of Lead Fluoride," *Zh. Fiz. Khim.*, **25,** 41.

Tavora, E., "Structural Data of Uraninite of Acari, R. G. N.," *Anais Acad. Brasil. Cienc.*, **23,** 155.

Zachariasen, W. H., "Crystal Chemical Studies of the 5*f*-Series of Elements. XIV. Oxyfluorides, XOF," *Acta Cryst.*, **4,** 231.

1952

Allen, R. D., "Variations in Chemical and Physical Properties of Fluorite," *Am. Mineralogist*, **37,** 910.

Busch, R. H., Galloni, E. E., Raskovan, J., and Cairo, A. E., "Crystal Structure of Platinum Dioxide," *Anais Acad. Brasil Cienc.*, **24,** 185.

Butler, G., and Thirsk, H. R., "Electron Diffraction Evidence for the Existence and Fine Structure of a Cryptomelane Modification of Manganese Dioxide Prepared in the Absence of Potassium," *Acta Cryst.*, **5,** 288.

Carter, G. F., Margrave, J. L., and Templeton, D. H., "A High Temperature Crystal Modification of KO_2," *Acta Cryst.*, **5**, 851.

D'Eye, R. W. M., Sellman, P. G., and Murray, J. R., "The Thorium–Selenium System," *J. Chem. Soc.*, **1952**, 2555.

Ervin, G., Jr., "Structural Interpretation of the Diaspore–Corundum and Boehmite–γ Alumina Transitions," *Acta Cryst.*, **5**, 103.

Ferro, R., "Compounds of Uranium with Antimony," *Atti Accad. Nazl. Lincei, Rend. Classe Sci. Fis. Mat. Nat.*, **13**, 53.

Ferro, R., "On the Alloy of Uranium with Antimony," *Atti Accad. Nazl. Lincei, Rend. Classe Sci. Fis. Mat. Nat.*, **13**, 151.

Graham, A. R., and Kaiman, S., "Aurostibite, $AuSb_2$; A New Mineral in the Pyrite Group," *Am. Mineralogist*, **37**, 461.

Haendler, H. M., Patterson, W. L., Jr., and Bernard, W. J., "The Reaction of Fluorine with Zinc, Nickel and Some of Their Binary Compounds. Some Properties of Zinc and Nickel Fluorides," *J. Am. Chem. Soc.*, **74**, 3167.

Hering, H., and Pério, P., "The Equilibrium of Uranium Oxides between UO_2 and U_3O_8," *Bull. Soc. Chim. France*, **19**, 35.

Hocart, R., and Molé, R., "Synthesis of Two Copper Tellurides by the Compression of Powdered Copper and Tellurium," *Compt. Rend.*, **234**, 111.

Hund, F., and Lieck, K., "The Quinary Fluoride $NaCaCdYF_8$," *Z. Anorg. Allgem. Chem.*, **271**, 17.

Hund, F., and Peetz, U., "The System Uranium Oxide–Erbium Oxide," *Z. Anorg. Allgem. Chem.*, **267**, 189.

Iandelli, A., "Uranium Arsenides I," *Atti Accad. Nazl. Lincei, Rend. Classe Sci. Fis. Mat. Nat.*, **13**, 138.

Iandelli, A., "On the Arsenides of Uranium II. The Crystal Structures of UAs_2 and UP_2," *Atti Accad. Nazl. Lincei, Rend. Classe Sci. Fis. Mat. Nat.*, **13**, 144.

Niggli, E., "Varlamoffite," *Leidsche Geol. Mededeel*, **17B**, 207.

Nijssen, L., and Lipscomb, W. N., "The Structure of Mercuric Amidobromide," *Acta Cryst.*, **5**, 604.

Nishiyama, Z., and Iwanaga, R., "Crystal Structure of Manganese Nitrides," *Mem. Inst. Sci. Ind. Res., Osaka Univ.*, **9**, 74.

Novoselova, A. V., Simanov, Y. P., and Yarembash, E. I., "Thermal and X-Ray Analysis of the Lithium–Beryllium Fluoride System," *Zh. Fiz. Chim.*, **26**, 1244.

Nowacki, W., and Maget, K., "The Crystallography of Cu(OH)Cl," *Experientia*, **8**, 55.

Nowotny, H., and Glatzl, B., "New Representatives of Ternary Compounds with C 1 Structure," *Monatsh.*, **83**, 237.

Post, B., and Glaser, G., "Borides of Some Transition Metals," *J. Chem. Phys.*, **20**, 1050.

Post, B., Schwartz, R. S., and Fankuchen, I., "The Crystal Structure of Sulfur Dioxide," *Acta Cryst.*, **5**, 372.

Ramdohr, P., "Maldonite," *Sitzber. Deut. Akad. Wiss. Berlin Kl. Math. Phys. Tech.*, **1952**, No. 5, p. 3.

Rüdorff, W., and Brodersen, K., "The Structure of Mercuric Amidobromide and the Formation of Mixed Crystals between Diaminomercuric Bromide, Mercuric Amidobromide and Ammonium Bromide," *Z. Anorg. Allgem. Chem.*, **270**, 145.

Rundle, R. E., and Lewis, P. H., "Electron-Deficient Compounds. VI. The Structure of Beryllium Chloride," *J. Chem. Phys.*, **20**, 132.

Schröder, A., "The Unit Cell and Density of Ramsdellite, MnO_2," *Fortschr. Mineral.*, **31**, 11.

Slowinski, E., and Elliott, N., "Lattice Constants and Magnetic Susceptibilities of Solid Solutions of Uranium and Thorium Dioxides," *Acta Cryst.*, **5**, 768.

Stosick, A. J., "Space Group and Unit Cell of Beryllium Borohydride," *Acta Cryst.*, **5**, 151.

Toman, K., "The Structure of Ni_2Si," *Acta Cryst.*, **5**, 329.

Weiss, A., and Weiss, A., "The Crystal Structure of Silicon Diselenide," *Z. Naturforsch.*, **7B**, 483.

Zaslavskii, A. I., and Tolkachev, S. S., "The Structure of the α-Modification of Lead Dioxide," *Zh. Fiz. Khim.*, **26**, 743.

Zhdanov, G. S., and Zvonkova, Z. V., "Crystal Structures of Higher Oxides of Metals of the First Group of the Periodic System," *Dokl. Akad. Nauk SSSR*, **82**, 743.

Zvonkova, Z. V., and Zhdanov, G. S., "The Crystal Structure of Mercury Halothiocyanates," *Zh. Fiz. Khim.*, **26**, 586.

1953

Andersson, G., "X-Ray Studies on Vanadium Oxides," *Research (London)*, **6**, 45S.

Ariya, S. M., Shchukarev, S. A., and Glushkova, V. B., "Chromium Dioxide, Its Preparation and Properties," *Zh. Obshch. Khim.*, **23**, 1241.

Bertaut, F., and Blum, P., "Borides of Chromium," *Compt. Rend.*, **236**, 1055.

Brogren, G., "A Determination of the Avogadro Number," *Arkiv Fysik*, **7**, 47.

Carter, G. F., and Templeton, D. H., "Polymorphism of Sodium Superoxide," *J. Am. Chem. Soc.*, **75**, 5247.

D'Eye, R. W. M., "The Crystal Structures of $ThSe_2$ and Th_7Se_{12}," *J. Chem. Soc.*, **1953**, 1670.

Edstrand, M., "The Structures of Antimony(III) Oxide Halides. II. The Crystal Structure of SbOCl," *Arkiv Kemi*, **6**, 89.

Erickson, R. A., "Neutron-Diffraction Studies of Antiferromagnetism in Manganous Fluoride and Some Isomorphous Compounds," *Phys. Rev.*, **90**, 779.

Evans, H. T., Jr., and Block, S., "The Crystal Structure of Montroseite, a Vanadium Member of the Diaspore Group," *Am. Mineralogist*, **38**, 1242.

Ferro, R., "On the Bonding of Uranium with Bismuth," *Atti Accad. Lincei, Rend. Classe Sci. Fis. Mat. Nat.*, [8] **14**, 89.

Furberg, S., "The System Manganese–Tellurium," *Acta Chem. Scand.*, **7**, 693.

Glagoleva, V. P., and Zhdanov, G. S., "Structure of Superconductors. III. X-Ray Investigation of the Structure and Solubility of Components in BiRh," *Zh. Eksperim. Teoret. Fiz.*, **25**, 248.

Glaser, F. W., and Post, B., "The System: Zirconium–Boron," *J. Metals*, **5**, *AIME Trans.*, **197**, 1117.

Glemser, O., Hauschild, U., and Trüpel, F., "Chromium Oxides between Cr_2O_3 and CrO_3," *Naturwiss.*, **40**, 317.

Gritsaenko, G. S., Sludskaya, N. N., and Aidinyan, N. K., "Synthesis of Vaesite and Polydymite," *Zap. Vses. Mineralog. Obshchestva*, **82**, 42.

Lambertson, W. A., Mueller, M. H., and Günzel, F. H., Jr., "Uranium Oxide Phase Equilibrium Systems," *J. Am. Chem. Soc.*, **36**, 397.

Legrand, C., and Delville, J., "Crystal Parameters of Rutile and Anatase," *Compt. Rend.*, **236**, 944.

Picon, M., and Flahaut, J., "The Properties of Uranium Sulfides, α- and β-US_2," *Compt. Rend.*, **237**, 1160.

Russell, V., Hirst, R., Kanda, F. A., and King, A. J., "An X-Ray Study of the Magnesium Borides," *Acta Cryst.*, **6**, 870.

Susić, M., "Dependence of the Ratio of Oxygen to Uranium in Uranium Dioxide on the Reduction Temperature of Uranium Trioxide," *Rec. Trav. Inst. Recherches Structure Matière (Belgrade)*, **2**, 89.

Templeton, D. H., and Dauben, C. H., "Crystal Structures of Rare Earth Oxychlorides," *J. Am. Chem. Soc.*, **75**, 6069.

Weiss, A., and Weiss, A., "Silicon Chalcogenides, II. Silicon Ditelluride," *Z. Naturforsch.*, **8b**, 104.

Weiss, A., and Weiss, A., "Silicon Chalcogenides, IV. Silicon Ditelluride," *Z. Anorg. Allgem. Chem.*, **273**, 124.

Zalkin, A., and Templeton, D. H., "The Crystal Structures of Yttrium Fluoride and Related Compounds," *J. Am. Chem. Soc.*, **75**, 2453.

Zevin, L. S., Zhdanov, G. S., and Zhuravlev, N. N., "X-Ray Investigation of the Structure of Low (α) Bi_2Pd," *Zh. Eksperim. Teoret. Fiz.*, **25**, 751.

Zhuravlev, N. N., and Zhdanov, G. S., "X-Ray Studies on the System Bi–Pd," *Zh. Eksperim. Teoret. Fiz.*, **25**, 485.

1954

Abrahams, S. C., and Kalnajs, J., "The Crystal Structure of Barium Peroxide," *Acta Cryst.*, **7**, 838.

Aksnes, O., and Foss, O., "The Structure of Selenium Diseleno Cyanate," *Acta Chem. Scand.*, **8**, 702.

Anderko, K., and Schubert, K., "The System Copper–Tellurium," *Z. Metallk.*, **45**, 371.

Andersson, G., "Vanadium Oxides. I," *Acta Chem. Scand.*, **8**, 1599.

Baenziger, N. C., Holden, J. R., Knudson, G. E., and Popov, A. I., "Unit Cell Dimensions of Some Rare Earth Oxyfluorides," *J. Am. Chem. Soc.*, **76**, 4734.

Bagnall, K. W., and D'Eye, R. W. M., "The Preparation of Polonium Metal and Polonium Dioxide," *J. Chem. Soc.*, **1954**, 4295.

Boku, G. B., and Tsinober, L. I., "X-Ray Examination of Cobaltite, Gersdorffite and Ullmannite," *Tr. Inst. Kristallogr., Akad. Nauk SSSR*, **9**, 239.

Brauer, G., Renner, H., and Wernet, J., "Carbides of Columbium," *Z. Anorg. Allgem. Chem.*, **277**, 249.

Brauer, G., and Zapp, K. H., "The Nitrides of Tantalum," *Z. Anorg. Allgem. Chem.*, **277**, 129.

Brodersen, K., and Rüdorff, W., "The Structure of Hg_2NHBr_2," *Z. Naturforsch.*, **9b**, 164.

D'Eye, R. W. M., and Sellman, P. G., "Thorium–Tellurium System," *J. Chem. Soc.*, **1954**, 3760.

Ferro, R., "Several Selenium and Tellurium Compounds of Uranium," *Z. Anorg. Allgem. Chem.*, **275**, 320.

Glemser, O., Hauschild, U., and Trüpel, F., "Oxides of Chromium between Cr_2O_3 and CrO_3," *Z. Anorg. Allgem. Chem.*, **277**, 113.

Grønvold, F., Haraldsen, H., and Vihovde, J., "Phase and Structural Relations in the System Iron–Tellurium," *Acta Chem. Scand.*, **8**, 1927.

Hägg, G., and Schönberg, N., "X-Ray Studies of Sulfides of Titanium, Zirconium, Columbium and Tantalum," *Arkiv Kemi*, **7**, 371.

Kondrashev, Y. D., and Fedorova, N. N., "Crystal Structure of CoO(OH)," *Dokl. Akad. Nauk SSSR*, **94**, 229.

Martin, A. W., "Determination of the Formula of an Oxide of Polonium," *J. Phys. Chem.*, **58,** 911.

Mergault, P., and Branche, G., "Crystallization of Anhydrous Titania Dissolved in Cryolite in the Form of Rutile," *Compt. Rend.*, **238,** 914.

Molé, R., "The Formation of the Nonstoichiometric Sulfides, Selenides and Tellurides of Copper from One or Two Solid Phases," *Ann. Chim. (Paris)*, **9,** 145.

Mooney, R. W., and Welch, A. J. E., "The Crystal Structure of Rh_2B," *Acta Cryst.*, **7,** 49.

Ohlberg, S. M., and Vaughan, P. A., "Crystal Structure of Selenium Dithiocyanate," *J. Am. Chem. Soc.*, **76,** 2649.

Post, B., Glaser, F. W., and Moskowitz, D., "Transition Metal Diborides," *Acta Met.*, **2,** 20.

Schönberg, N., "An X-Ray Investigation of the Tantalum–Oxygen System," *Acta Chem. Scand.*, **8,** 240.

Schönberg, N., "Ternary Metallic Phases in the Ta–C–N, Ta–C–O and Ta–N–O Systems," *Acta Chem. Scand.*, **8,** 620.

Schönberg, N., "Composition of the Phases in the Vanadium–Carbon System," *Acta Chem. Scand.*, **8,** 624.

Schubert, K., Dörre, E., and Günzel, E., "Crystal Chemical Phenomena in Phases of B Elements," *Naturwiss.*, **41,** 448.

Sellers, P. A., Fried, S., Elson, R. E., and Zachariasen, W. H., "Preparation of Some Protoactinium Compounds and the Metal," *J. Am. Chem. Soc.*, **76,** 5935.

Sobotka, J., "Aurostibite, $AuSb_2$, Its First Macroscopic Occurrence," *Rozpravy Cesk. Akad. Ved.*, **64,** No. 7, 43.

Steyn, J. G. D., "Spectrographic and X-Ray Data on Some Fluorites from the Transvaal, South Africa," *Mineral. Mag.*, **30,** 327.

Stout, J. W., and Reed, S. A., "The Crystal Structure of MnF_2, FeF_2, CoF_2, NiF_2 and ZnF_2," *J. Am. Chem. Soc.*, **76,** 5279.

Sugawara, T., "Crystal Structure of Sulfur Dioxide and Acetylene at Low Temperature," *X-Sen (X-Rays)*, **8,** 21.

Templeton, D. H., and Dauben, C. H., "Lattice Parameters of Some Rare Earth Compounds and a Set of Crystal Radii," *J. Am. Chem. Soc.*, **76,** 5237.

Weiss, A., and Weiss, A., "A New Silicon Dioxide Modification with Fiber Structure and Chain Molecules," *Naturwiss.*, **41,** 12.

Weiss, A., and Weiss, A., "Silicon Chalcogenides. VI. A Fibrous Silicon Dioxide Modification," *Z. Anorg. Allgem. Chem.*, **276,** 95.

1955

Abrahams, S. C., and Kalnajs, J., "The Crystal Structure of α-Potassium Superoxide," *Acta Cryst.*, **8,** 503.

Asprey, L. B., Ellinger, F. H., Fried, S., and Zachariasen, W. H., "Evidence for Quadrivalent Curium; X-Ray Data on Curium Oxides," *J. Am. Chem. Soc.*, **77,** 1707.

Atoji, M., Lipscomb, W. N., and Wheatley, P. J., "Molecular Structure of Diborane Tetrachloride, B_2Cl_4," *J. Chem. Phys.*, **23,** 1176.

Baur, W., "Atom Distances and Bond Angle in Rutile," *Naturwiss.*, **42,** 295.

Bevan, D. J. M., "Ordered Intermediate Phases in the System CeO_2–Ce_2O_3," *J. Inorg. Nuclear Chem.*, **1,** 49.

Boettcher, A., Haase, G., and Treupel, H., "The Structures and Structural Changes in the Sulfides and Selenides of Silver and Copper," *Z. Angew. Phys.*, **7,** 478.

Bøhm, F., Grønvold, F., Haraldsen, H., and Prydz, H., "X-Ray and Magnetic Study of the System Cobalt Selenium," *Acta Chem. Scand.*, **9**, 1510.
Bur'yanova, E. Z., and Komkov, A. I., "Ferroselite, a New Mineral," *Dokl. Akad. Nauk SSSR*, **105**, 812.
Cromer, D. T., and Herrington, K., "The Structures of Anatase and Rutile," *J. Am. Chem. Soc.*, **77**, 4708.
Evans, H. T., Jr., and Mrose, M. E., "A Crystal Chemical Study of Montroseite and Paramontroseite," *Am. Mineralogist*, **40**, 861.
Ferro, R., "The Crystal Structure of Thorium Arsenides," *Acta Cryst.*, **8**, 360.
Ferro, R., "Thorium Monotelluride and Thorium Oxytelluride," *Atti Accad. Lincei Rend. Classe Sci. Fis. Mat. Nat.*, **18**, 641.
Frondel, C., and Hurlbut, C. S., Jr., "Determination of the Atomic Weight of Silicon by Physical Measurements on Quartz," *J. Chem. Phys.*, **23**, 1215.
Frueh, A. J., Jr., "The Crystal Structure of Stromeyerite, AgCuS: A Possible Defect Structure," *Z. Krist.*, **106**, 299.
Geller, S., "The Crystal Structure of RhTe and $RhTe_2$," *J. Am. Chem. Soc.*, **77**, 2641.
Geller, S., "The Rhodium–Germanium System. I. The Crystal Structures of Rh_2Ge, Rh_5Ge_3 and RhGe," *Acta Cryst.*, **8**, 15.
Geller, S., and Cetlin, B. B., "The Crystal Structure of $RhSe_2$," *Acta Cryst.*, **8**, 272.
Geller, S., and Wolontis, V. M., "The Crystal Structure of Co_2Si," *Acta Cryst.*, **8**, 83.
Graham, A. R., "Cerianite, CeO_2: A New Rare-Earth Oxide Mineral," *Am. Mineralogist*, **40**, 560.
Groenefeld Meijer, W. O. J., "Synthesis, Structures and Properties of Platinum Metal Tellurides," *Am. Mineralogist*, **40**, 646.
Heaton, L., and Gingrich, N. S., "The Crystal Structure of Mn_2Sb," *Acta Cryst.*, **8**, 207.
Holley, C. E., Jr., Mulford, R. N. R., Ellinger, F. H., Koehler, W. C., and Zachariasen, W. H., "The Crystal Structure of Some Rare Earth Hydrides," *J. Phys. Chem.*, **59**, 1226.
Magnéli, A., and Andersson, G., "On the MoO_2 Structure Type," *Acta Chem. Scand.*, **9**, 1378.
Markovskii, L. Y., Kondrashev, Y. D., and Kaputovskaya, G. V., "The Composition and Properties of the Beryllium Borides," *Zh. Obshchei Khim.*, **25**, 1045.
Menary, J. W., "Lattice Constants," *Acta Cryst.*, **8**, 840.
Ramdohr, P., "Four New Natural Cobalt Selenides from the Trogtal Quarry near Lautental (Harz Mountains)," *Neues Jahrb. Mineral., Monatsh.*, **6**, 133.
Ramsdell, L. S., "The Crystallography of Coesite," *Am. Mineralogist*, **40**, 975.
Smakula, A., and Kalnajs, J., "Precision Determination of Lattice Constants with a Geiger-Counter X-Ray Diffractometer," *Phys. Rev.*, **99**, 1737.
Truby, F. K., "Lattice Constants of Pure and Fluoride-Contaminated Ice," *Science*, **121**, 404.

1956

Andersson, G., "Studies on Vanadium Oxides. II. The Crystal Structure of Vanadium Dioxide," *Acta Chem. Scand.*, **10**, 623.
Bartlett, N., and Hepworth, M. A., "Pure Palladium Difluoride," *Chem. Ind. (London)*, **1956**, 1425.
Baur, W. H., "The Refinement of the Crystal-Structure Determination of Some Representatives of the Rutile Type: TiO_2, SnO_2, GeO_2 and MgF_2," *Acta Cryst.*, **9**, 515.

Bergström, G., and Lundgren, G., "X-Ray Investigations on Uranyl Hydroxides. I. Crystal Structure of β-$UO_2(OH)_2$," *Acta Chem. Scand.*, **10**, 673.

Ehrlich, P., Alt., B., and Gentsch, L., "Alkaline Earth Metal Hydride Chlorides," *Z. Anorg. Allgem. Chem.*, **283**, 58.

Ehrlich, P., and Görtz, H., "Alkaline Earth Metal Hydride Bromides," *Z. Anorg. Allgem. Chem.*, **288**, 148.

Ehrlich, P., and Kulka, H., "Alkaline Earth Metal Hydride Iodides," *Z. Anorg. Allgem. Chem.*, **288**, 156.

Eick, H. A., Baenziger, N. C., and Eyring, L., "Lower Oxides of Samarium and Europium. The Preparation and Crystal Structure of $SmO_{0.4-0.6}$, SmO and EuO," *J. Am. Chem. Soc.*, **78**, 5147.

Felten, E. J., "The Preparation of Aluminum Diboride, AlB_2," *J. Am. Chem. Soc.*, **78**, 5977.

Ferro, R., "The Crystal Structures of Thorium Antimonides," *Acta Cryst.*, **9**, 817.

Forsberg, H. E., "Crystal Structure of Indium Oxychloride and Indium Oxybromide," *Acta Chem. Scand.*, **10**, 1287.

Foss, O., "Space Groups and Molecular Symmetries of Two Thiocyanates," *Acta Chem. Scand.*, **10**, 136.

Grønvold, F., and Jacobsen, E., "X-Ray and Magnetic Study of Nickel Selenides in the Range NiSe and $NiSe_2$," *Acta Chem. Scand.*, **10**, 1440.

Grønvold, F., and Røst, E., "On the Sulfides, Selenides and Tellurides of Palladium," *Acta Chem. Scand.*, **10**, 1620.

Kirkina, D. F., Novoselova, A. V., and Simanov, Y. P., "Polymorphism of Beryllium Fluoride," *Dokl. Akad. Nauk SSSR*, **107**, 837.

Korst, W. L., and Warf, J. C., "The Crystal Structure of the Deuterides of Ytterbium and Europium," *Acta Cryst.*, **9**, 452.

Lepp, H., "Precision Measurements of the Cell Edge of Synthetic Pyrite," *Am. Mineralogist*, **41**, 347.

Leung, Y. C., and Waser, J., "The Crystal Structure of Phosphorus Diiodide, P_2I_4," *J. Phys. Chem.*, **60**, 539.

Magnéli, A., "Orthorhombic Rhenium Dioxide: A Representative of a Hypothetic Structure Type Predicted by Pauling and Sturdivant," *Acta Cryst.*, **9**, 1038.

Milligan, W. O., and McAtee, J. L., "Crystal Structure of γ-AlOOH and γ-ScOOH," *J. Phys. Chem.*, **60**, 273.

Mitchell, R. S., "Polytypism of Cadmium Iodide and its Relationship to Screw Dislocations. I. Cadmium Iodide Polytypes," *Z. Krist.*, **108**, 296.

Novoselova, A. V., Breusov, O. N., Kirkina, D. F., and Simanov, Y. P., "Quartzlike Beryllium Fluoride," *Zh. Neorg. Khim.*, **1**, 2670.

Sturdy, G. E., and Mulford, R. N. R., "The Gadolinium–Hydrogen System," *J. Am. Chem. Soc.*, **78**, 1083.

Tsai, K.-R., Harris, P. M., and Lassettre, E. N., "The Crystal Structure of Cesium Monoxide," *J. Phys. Chem.*, **60**, 338.

Waite, T. R., Wallace, W. E., and Craig, R. S., "Structures and Phase Relationships in the Tantalum–Hydrogen System between −145° and 70°C.," *J. Chem. Phys.*, **24**, 634.

1957

Atoji, M., Wheatley, P. J., and Lipscomb, W. N., "Crystal and Molecular Structure of Diboron Tetrachloride, B_2Cl_4," *J. Chem. Phys.*, **27**, 196.

Baur, W. H., "Structure of Fluorides of the Rutile Type: MnF_2, FeF_2, CoF_2, NiF_2, ZnF_2," *Naturwiss.*, **44**, 349.

Bell, R. E., and Herfert, R. E., "Preparation and Characterization of a New Crystalline Form of Molybdenum Disulfide," *J. Am. Chem. Soc.*, **79**, 3351.

Billy, C., and Haendler, H. M., "Crystal Structure of Copper(II) Fluoride," *J. Am. Chem. Soc.*, **79**, 1049.

Blackman, M., and Lisgarten, N. D., "The Cubic and Other Structural Forms of Ice at Low Temperature and Pressure," *Proc. Roy. Soc. (London)*, **A239**, 93.

Busing, W. R., and Levy, H. A., "Neutron Diffraction Study of Calcium Hydroxide," *J. Chem. Phys.*, **26**, 563.

Ferro, R., "The Crystal Structures of Thorium Bismuthides," *Acta Cryst.*, **10**, 476.

Fischer, W. A., and Hoffmann, A., "Equilibrium Investigations in the System Iron(II) Oxide–Zirconium Oxide," *Arch. Eisenhüttenw.*, **28**, 739.

Frueh, A. J., Jr., Czamanske, G. K., and Knight, C., "The Crystallography of Eucairite, CuAgSe," *Z. Krist.*, **108**, 389.

Garton, G., and Powell, H. M., "The Crystal Structure of Gallium Dichloride," *J. Inorg. Nuclear Chem.*, **4**, 84.

Gorum, A. E., "Crystal Structures of PuAs, PuTe, PuP and PuOSe," *Acta Cryst.*, **10**, 144.

Grønvold, F., and Røst, E., "The Crystal Structure of $PdSe_2$ and PdS_2," *Acta Cryst.*, **10**, 329.

Hahn, H., and Ness, P., "The Structure of the Phases Appearing in the System Zirconium–Tellurium," *Naturwiss.*, **44**, 534.

Heyding, R. D., and Calvert, L. D., "Arsenides of Transition Metals: The Arsenides of Iron and Cobalt," *Can. J. Chem.*, **35**, 449.

Honjo, G., and Shimaoka, K., "Determination of Hydrogen Position in Cubic Ice by Electron Diffraction," *Acta Cryst.*, **10**, 710.

Jack, K. H., and Maitland, R., "The Crystal Structures and Interatomic Bonding of Chromous and Chromic Fluorides," *Proc. Chem. Soc.*, **1957**, 232.

Khodadad, P., "Selenides of Quadrivalent Uranium," *Compt. Rend.*, **245**, 934.

Khodadad, P., "Uranium Oxyselenide, OSeU," *Compt. Rend.*, **245**, 2286.

Magnéli, A., "Studies on Rhenium Oxides," *Acta Chem. Scand.*, **11**, 28.

Petch, H. E., "A Determination of the Hydrogen Positions in $Ca(OH)_2$ by X-Ray Diffraction," *Can. J. Phys.*, **35**, 983.

Peterson, S. W., and Levy, H. A., "A Single-Crystal Neutron Diffraction Study of Ice," *Acta Cryst.*, **10**, 70.

Pinsker, Z. G., Kaverin, S. V., and Troitskaya, N. V., "Electron Diffraction Study of Molybdenum Nitrides," *Kristallografiya*, **2**, 179.

Schneider, A., and Imhagen, K. H., "Temperature Dependence of the Lattice Constants in the Mixed Crystal Series NiTe–$NiTe_2$," *Naturwiss.*, **44**, 324.

Schubert, K., Breiner, H., Burkhardt, W., Günzel, E., Haufler, R., Lukas, H. L., Vetter, H., Wegst, J., and Wilkens, M., "Structural Data on Metallic Phases, II," *Naturwiss.*, **44**, 229.

Shallcross, F. V., and Carpenter, G. B., "X-Ray Diffraction Study of the Cubic Phase of Ice," *J. Chem. Phys.*, **26**, 782.

Shchukarev, S. A., and Efimov, A. I., "Oxychloride of Trivalent Uranium," *Zh. Neorg. Khim.*, **2**, 2304.

Sieglaff, C. L., and Eyring, L., "Praseodymium Oxides. IV. A Study of the Region $PrO_{1.83}$–$PrO_{2.00}$," *J. Am. Chem. Soc.*, **79**, 3024.

Takeuchi, Y., "The Absolute Structure of Ullmanite, NiSbS," *Mineral. J. (Sapporo)*, **2**, 90.

Weiss, R., and Faivre, R., "A New Form of Lead Dioxide and the Products of its Thermal Decomposition," *Compt. Rend.*, **245**, 1629; *Congr. Intern. Chim. Pure Appl. 16e, Paris 1957, Mem. Sect. Chim. Minerale*, 807 (pub. 1958).

Zhuravlev, N. N., Mingazin, T. A., and Zhdanov, G. S., "Superconducting Alloys of Bismuth with Rubidium and Bismuth with Cesium," *Redkie Metally i Splavy, Tr. Pervogo Vses. Soveshch. po Splavam Redkikh Metal. Akad. Nauk SSSR, Inst. Met., Moscow, 1957*, 372 (pub. 1960).

1958

Alenchikova, I. F., Zaitseva, L. L., Lipis, L. V., Nikolaev, N. S., Fomin, V. V., and Chebotarev, N. T., "Physicochemical Properties of Plutonyl Fluoride," *Zh. Neorg. Khim.*, **3**, 951.

Atoji, M., Gschneider, K., Jr., Daane, A. H., Rundle, R. E., and Spedding, F. H., "The Structures of Lanthanum Dicarbide and Sesquicarbide by X-Ray and Neutron Diffraction," *J. Am. Chem. Soc.*, **80**, 1804.

Ayphassorho, C., "Radio Crystallographic Study of the Cerium Hydrogen System," *Compt. Rend.*, **247**, 1597.

Ball, J. G., Greenfield, P., Mardon, P. G., and Robertson, J. A. L., "Crystal Structure of Plutonium, δ and ϵ Phases," At. Energy Research Estab. (Gt. Brit.), M/R 2416, 11 pp.

Barricelli, L. B., "The Crystal Structure of Iridium Diselenide," *Acta Cryst.*, **11**, 75.

Baur, W. H., "Refinement of the Crystal Structure Determination of Some Compounds of the Rutile Type. II. The Difluorides of Manganese, Iron, Cobalt, Nickel and Zinc," *Acta Cryst.*, **11**, 488.

Busing, W. R., and Levy, H. A., "A Single-Crystal Neutron Diffraction Study of Diaspore, AlO(OH)," *Acta Cryst.*, **11**, 798.

Collins, D. A., and Phillips, G. M., "Observations on the Preparation of Neptunium Oxides," *J. Inorg. Nuclear Chem.*, **6**, 67.

Djurle, S., "An X-Ray Study on the System Cu–S," *Acta Chem. Scand.*, **12**, 1415.

Djurle, S., "An X-Ray Study on the System Ag–Cu–S," *Acta Chem. Scand.*, **12**, 1427.

Duncanson, A., and Stevenson, R. W. H., "Some Properties of Magnesium Fluoride Crystallized from the Melt," *Proc. Phys. Soc. (London)*, **72**, 1001.

Falqui, M. T., and Rollier, M. A., "Crystal Structure of Some Halides of Elements of the Eight Group. I. The Crystal Structure of $PtCl_2$," *Ann. Chim. (Rome)*, **48**, 1154.

Frueh, A. J., Jr., "The Crystallography of Silver Sulfide, Ag_2S," *Z. Krist.*, **110**, 136.

Hawley, J. E., and Berry, L. G., "Michenerite and Froodite, Palladium Bismuthide Minerals," *Can. Mineralogist*, **6**, 200.

Hvoslef, J., "A Combined X-Ray and Neutron Diffraction Investigation of Mercuric Cyanide," *Acta Chem. Scand.*, **12**, 1568.

Karabash, A. G., "Several Chemical Properties of Thorium and Uranium," *Zh. Neorg. Khim.*, **3**, 986.

Lonsdale, K., "Structure of Ice," *Proc. Roy. Soc. (London)*, **A247**, 424.

McTaggart, F. K., and Wadsley, A. D., "The Sulfides, Selenides and Tellurides of Titanium, Zirconium, Hafnium and Thorium," *Australian J. Chem.*, **11**, 445.

Mulford, R. N. R., and Ellinger, F. H., "ThO_2–PuO_2 and CeO_2–PuO_2 Solid Solutions," *J. Phys. Chem.*, **62**, 1466.

Rundle, R. E., Baenziger, N. C., and Wilson, A. S., "X-Ray Study of the Uranium–Oxygen System," U.S. At. Energy Comm. TID-5290, Bk. 1, p. 131.

Rundle, R. E., Wilson, A. S., Baenziger, N. C., and Tevebaugh, A. D., "X-Ray Analysis of the Uranium–Carbon System," U.S. At. Energy Comm. TID-5290, Bk. 1, p. 67.

Schäfer, H., Wartenpfuhl, F., and Weise, E., "Titanium Chlorides. V. Titanium(III) Oxychloride," *Z. Anorg. Allgem. Chem.*, **295**, 268.

Spedding, F. H., Gschneider, K., Jr., and Daane, A. H., "The Crystal Structures of Some of the Rare Earth Carbides," *J. Am. Chem. Soc.*, **80**, 4499.

Stasova, M. M., and Vainshtein, B. K., "Determination of the Structure of Thallium Selenide by Electron Diffraction," *Kristallografiya*, **3**, 141.

Sundholm, A., Andersson, S., Magnéli, A., and Marinder, B. O., "Metal–Metal Bonding in a Mixed Chromium Molybdenum Oxide Phase of Rutile Structure," *Acta Chem. Scand.*, **12**, 1343.

Tokuda, T., "Lattice Spacings in Synthetic Cristobalite—Especially in the Back-Reflection Region," *Nippon Kagaku Zasshi*, **79**, 1063.

Tolkachev, S. S., "Structure of β-PbO," *Vestn. Leningr. Univ.*, **13**, No. 4, *Ser. Fiz. Khim.*, No. 1, p. 152.

Trefonas, L., and Lipscomb, W. N., "Crystal and Molecular Structure of Diboron Tetrafluoride, B_2F_4," *J. Chem. Phys.*, **28**, 54.

Trzebiatowski, W., Weglowski, S., and Lukaszewicz, K., "The Crystal Structure of Zirconium Arsenides, ZrAs and $ZrAs_2$," *Roczniki Chem.*, **32**, 189.

Vries, A. de, "Determination of the Absolute Configuration of α-Quartz," *Nature*, **181**, 1193.

Wilhelmi, K. A., and Jonsson, O., "Crystal Structure of Ferromagnetic Chromium Dioxide," *Acta Chem. Scand.*, **12**, 1532.

Zhuravlev, N. N., and Stepanova, A. A., "Structure of ScB_2," *Kristallografiya*, **3**, 83.

1959

Adam, J., and Rogers, M. D., "The Crystal Structure of ZrO_2 and HfO_2," *Acta Cryst.*, **12**, 951.

Aronsson, B., Aselius, J., and Stenberg, E., "Borides and Silicides of the Platinum Metals," *Nature*, **183**, 1318.

Atoji, M., and Medrud, R. C., "Structures of Calcium Dicarbide and Uranium Dicarbide by Neutron Diffraction," *J. Chem. Phys.*, **31**, 332.

Austin, A. E., "Carbon Positions in Uranium Carbides," *Acta Cryst.*, **12**, 159.

Brehler, B., "On α- and β-$ZnCl_2$," *Naturwiss.*, **46**, 554.

Chentsov, I. G., "Selenium in Paleogene Deposits of Central Asia (Ferroselite)," *Tr. Inst. Geol. Rudn. Mestorozhd., Petrogr. Mineralog. i Geokhim.*, **1959**, No. 28, 83.

Christofferson, G. D., and McCullough, J. D., "The Crystal Structure of Antimony(III) Sulfobromide, SbSBr," *Acta Cryst.*, **12**, 14.

Ehrlich, P., and Seifert, H. J., "Vanadium Trichloride," *Z. Anorg. Allgem. Chem.*, **301**, 282.

Feitknecht, W., Oswald, H. R., and Forsberg, H. E., "The Structure of Metallic Hydroxychlorides," *Chimia (Aarau)*, **13**, 113.

Forsberg, H. E., and Nowacki, W., "Crystal Structure of β-ZnOHCl," *Acta Chem. Scand.*, **13**, 1049.

Frueh, A. J., Jr., "The Structure of Hessite, Ag_2Te-III," *Z. Krist.*, **112**, 44.

Hahn, H., and Ness, P., "On the System Zirconium–Tellurium," *Z. Anorg. Allgem. Chem.*, **302**, 136.

NAME INDEX*

* In addition to those elements and compounds reported in this volume, this index contains citations of future volumes of this work and of the first edition. These types of citation are enclosed in brackets. Citations of the first edition (which was published in loose-leaf form) give only the chapter number in roman letters (e.g., [XII]). Citations of future volumes give both the volume number and the chapter number (e.g., [**3**, Ch. IX]).

K

L

M

N

Q

R

S

T

U

FORMULA INDEX*

* In addition to those elements and compounds reported in this volume, this index contains citations of future volumes of this work and of the first edition. These types of citation are enclosed in brackets. Citations of the first edition (which was published in loose-leaf form) give only the chapter number in roman letters (e.g., [XII]). Citations of future volumes give both the volume number and the chapter number (e.g., [**3**, Ch, IX]).

N

U

V

W

X

Y